WELDING

The fundamentals of welding, cutting, brazing, soldering, and surfacing of metals

FUNDAMENTALS OF SERVICE

PUBLISHER
DEERE & COMPANY
JOHN DEERE PUBLISHING
One John Deere Place
Moline, IL 61265
http://www.johndeere.com/publications
1–800–522–7448

Fundamentals of Service is a series of manuals created by Deere & Company. Each book in the series is conceived, researched, outlined, edited, and published by Deere & Company, John Deere Publishing. Authors are selected to provide a basic technical manuscript that could be edited and rewritten by staff editors.

ACKNOWLEDGEMENTS: John Deere gratefully acknowledges help from the following groups or their publications from which information was obtained: Machine Design; Fasteners Handbook, Fasteners Standards; Service Tools, Inc.; Cummins Engine Co.; Caterpillar Tractor Co.; Adhesive Materials—Their Properties and Usage; Introduction to Mechanical Design; Minnesota Mining and Manufacturing Co.; Snap-On Tools Corp. Thanks also to a host of John Deere people for their valuable suggestions and comments.

HOW TO USE THE MANUAL: This manual can be used by anyone — experienced mechanics, shop trainees, vocational students, and lay readers.

Persons not familiar with the topics discussed in this book should begin with Chapter 1 and then study the chapters in sequence. The experienced person can find what is needed on the "Contents" page.

This current EDITION was produced by the technical writers, illustrators, and editors of Almon, Inc. — a full-service technical publications company headquartered in Waukesha, Wisconsin (www.almoninc.com).

CO-AUTHORS OF THIS EDITION:

Richard Alexander
Edward R. Bohnart
Roger W. Witcraft

FOR MORE INFORMATION: This book is one of many books published on agricultural and related subjects. For more information or to request a FREE CATALOG, call 1–800–522–7448 or send your request to address above or:

Visit Us on the Internet
http://www.johndeere.com/publications

JOHN DEERE

We have a
long-range interest in
Your Farming Success

ISBN 0-86691-347-5

Contents

1 INTRODUCTION TO JOINING AND CUTTING METALS

2 OXYFUEL WELDING (OFW) PROCESSES

3 SHIELDED METAL ARC WELDING (SMAW)

4 GAS METAL ARC WELDING AND FLUX-CORED ARC WELDING (GMAW AND FCAW)

5 GAS TUNGSTEN ARC WELDING (GTAW)

6 CUTTING AND GOUGING METALS

10 METAL SURFACING

11 PROPERTIES OF METALS

12 DESIGN OF WELDED UNITS

13 WELDING SAFETY

14 APPENDIX

Introduction to Joining and Cutting Metals

1

Introduction

Joining and cutting of metals plays an important role in the occupation of those employed in the metal working industry. The pictures that follow illustrate the common joining and cutting processes explained in this text.

Chapter 1 of this book provides a general understanding of joining and cutting processes and their relationship to the metal working industry. Professionals who work with metals must understand the features as well as know the limitations of the processes they intend to use to make or repair a metal product. These processes are actually the tools used by everyone in the metal working industry.

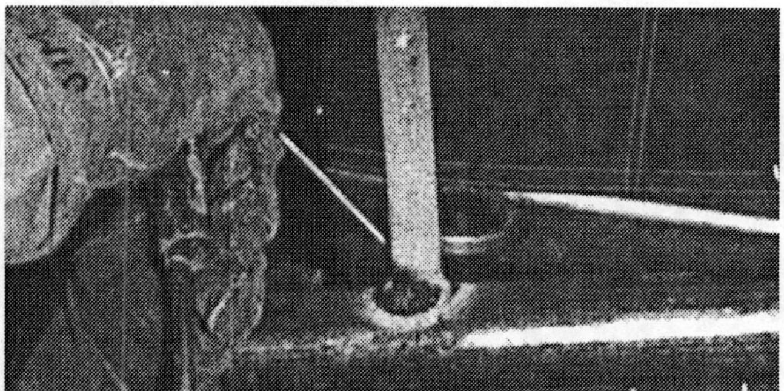

JDPX7252

Fig. 1 — Shielded Metal Arc Welding (SMAW)

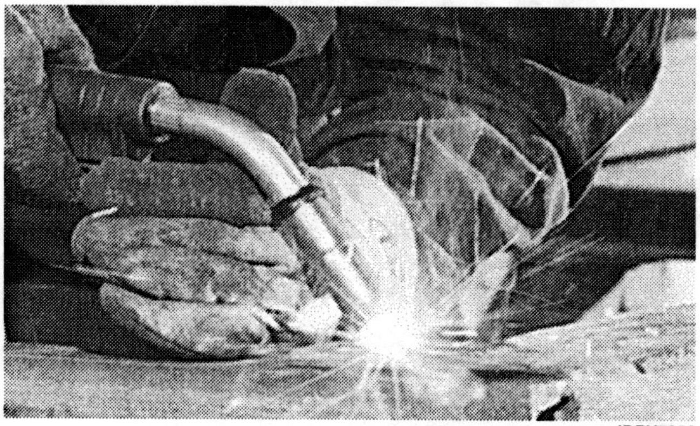

JDPX7253

Fig. 2 — Gas Metal Arc Welding (GMAW)

JDPX7254

Fig. 3 — Oxyacetylene Welding (OAW)

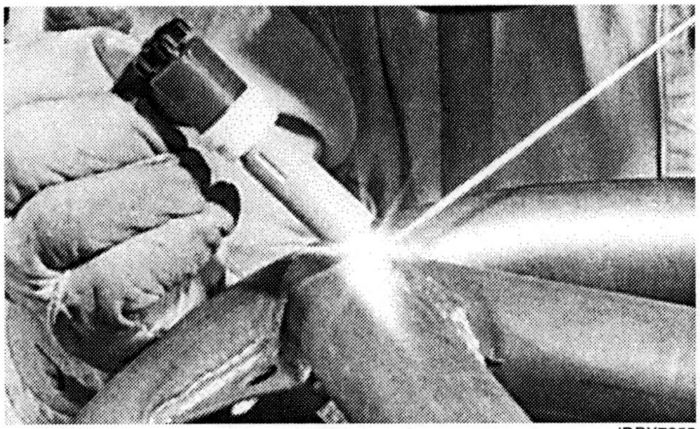

JDPX7255

Fig. 4 — Gas Tungsten Arc Welding (GTAW)

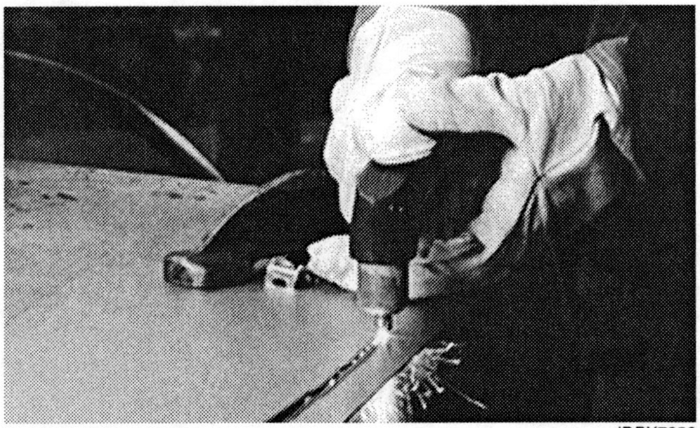

JDPX7256

Fig. 5 — Plasma Arc Cutting (PAC)

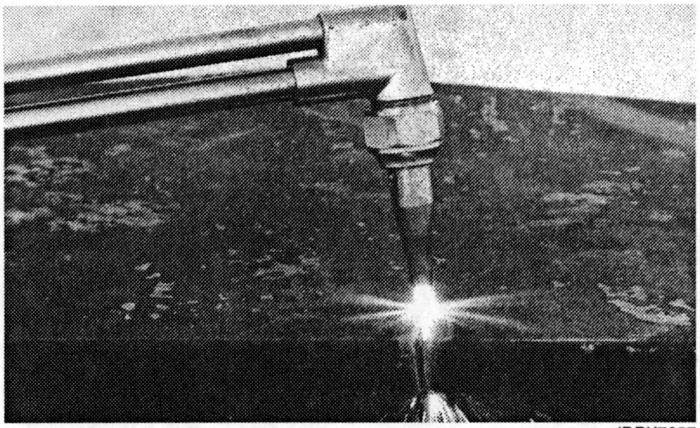

JDPX7257

Fig. 6 — Oxyfuel Cutting (OFC)

The remaining chapters of this book provide more in-depth information about each process. Each part includes a description of basic equipment, working methods, terminology, and special safety precautions.

Process Applications

Applications for the processes in the chart in Table 1 are only limited by the user's understanding of the processes and the ability to apply them to new situations. Earnings from the manufacture of new products or savings from the repair of existing products is proportional to the knowledge and skill of the people using the processes. In other words, the more a person understands these processes, and the more proficient they become in using the processes, the more opportunities there will be to use them in new situations.

Manufacturing New Metal Products

The GMAW (gas metal arc welding) process (Fig. 7) is one of the most widely used welding processes in industry today. In the past 30 years, manufacturers of metal products have replaced many applications that had previously been done with SMAW (shielded metal arc welding) or OF (oxyfuel) with processes in the continuous wire group.

JDPX7258

Fig. 7 — Gas Metal Arc Welding

The plasma arc cutting (PAC) process (Fig. 8) is used for cutting all types of metals. The versatility and speed of this process is responsible for its wide-spread use in today's metal working industry. Unlike the oxyfuel cutting process which is limited to cutting carbon steel, the plasma process can cut almost any metal that will conduct electricity.

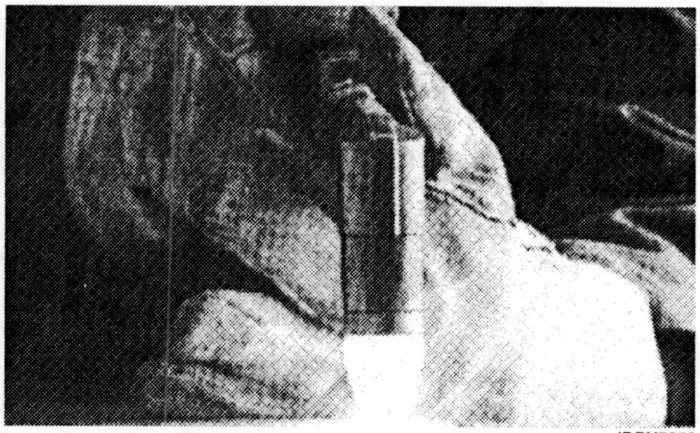

JDPX7259

Fig. 8 — Plasma Arc Cutting

Everyday our lives are touched by products that have been manufactured using one or more of these joining and cutting processes. Almost everything we depend on, from automobiles, trucks, farm equipment, and household appliances like refrigerators, have been fabricated from metal or manufactured using machinery fabricated from metal.

Repairing Worn and Broken Parts

The life of this same equipment can be extended by using joining and cutting processes to repair worn and broken parts. A perfectly good piece of equipment can be rendered useless when a component is damaged by the failure of another component or from some type of accident.

Aluminum air conditioner components and oil coolers on automobiles, trucks, and farm equipment can be repaired to "as good as new" by using a welding, soldering, or brazing process. The problem may be a small puncture hole or a broken line (Fig. 9), but a person with the proper training and equipment can make a repair for a small fraction of the cost for a replacement part.

JDPX7260

Fig. 9 — Aluminum Air Conditioning Fitting Repair

The failure of a bearing is a common example of one component damaging another component. When a bearing seizes on a piece of equipment, it can damage the metal shaft it is supporting in a few minutes if the problem is not identified and the equipment stopped. Welding can be used to rebuild the damaged area to the size needed for a machinist to machine the shaft to fit a new bearing.

Methods of Joining and Cutting Metals					
	Welding	**Cutting**	**Brazing**	**Soldering**	**Surfacing**
Basic Processes					
Oxyfuel (OF)	L	X	X	X	X
Shielded Metal Arc Welding (SMAW)	X	L	----	----	X
Continuous Wire Gas Metal Arc Welding (GMAW)	X	----	----	----	X
Flux-Cored Arc Welding (FCAW)	X	----	-----	----	X
Submerged Arc Welding (SAW)	X	----	----	----	X
Gas Tungsten Arc Welding (GTAW)	X	----	----	----	X
Plasma Arc (PA)	X	X	----	----	L
Resistance (R)	X	-----	----	----	----
Exotic Processes					
Laser Beam (LB)	X	X	----	----	----
Electron Beam	X	X	----	----	----
Others	X	X	X	X	X
X = Currently in Use					
L = Very Limited Use					
---- = Not in Use					

Table 1 — Process Chart

Important Concepts for Joining Metals

The descriptions for the processes which follow focus on the concepts of using heat to either join or cut metals. Joining metals can be achieved by either welding, soldering, or brazing. Cutting metals can be accomplished by either an oxidation process (burning the metal) or by a thermal process (melting the metal). To become proficient in selecting and/or using the best process for a project, a person must have a basic understanding of these concepts and of the differences (Fig. 10).

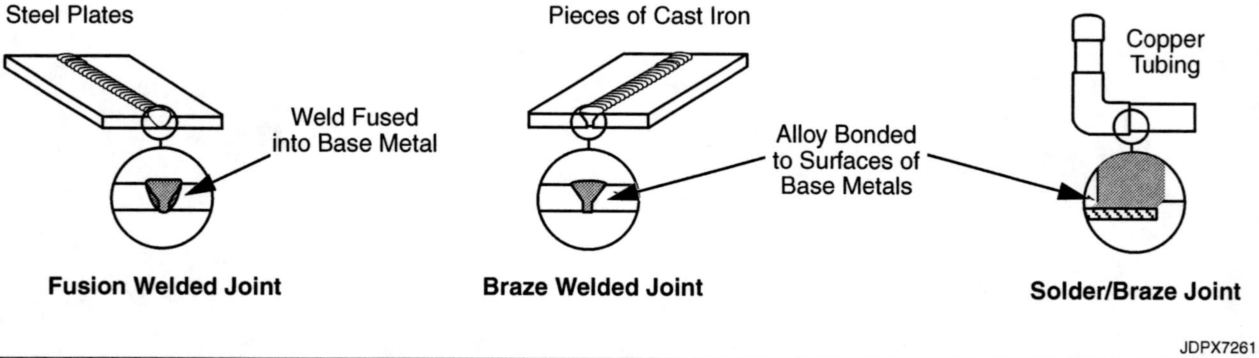

Fig. 10 — A Comparison of Fusion Welding, Braze Welding, Soldering, and Brazing

The Fusion Welding Process

Fusion welding is basically the melting, flowing together, and solidifying together of two substances in a controlled situation. The heat source for this melting can be supplied from something high-tech like the arc from the gas metal arc welding gun (Fig. 7) or from something more basic like the flame from an oxyacetylene torch. Most welding processes are limited to the joining of similar metals or metals which have relatively close melting temperatures.

With any type of fusion welding, the source of heat (an electric arc or a flame) must be capable of producing temperatures which can melt the edges of metal almost instantly — or in a few seconds. As the liquid area flows together, filler metal is usually added to ensure the weld is of the proper size. As the heat is moved along the area to be joined, the liquid puddle of metal immediately begins cooling to room temperature and the pieces of metal solidify into one piece, creating a fusion weld. During the entire process the molten metal must be protected from being contaminated by the air, foreign material on the metal, or by impurities in the metal. This shielding and cleaning action is done somewhat differently with each process.

The Soldering and Brazing Processes

Soldering and brazing can be used to join similar metals or dissimilar metals because the base metals are not melted.

When soldering or brazing, a source of heat is used to raise the temperature of the base metal(s) to the melting point of the filler metal, which is called the soldering or brazing alloy. This alloy flows onto both of the pieces, fills the space between the pieces, and adheres to the surfaces. When the heat source is removed, everything cools to room temperature, the alloy solidifies, and the metal(s) are bonded together. A flux (a cleaning agent) must be applied to the base metal at the beginning of the operation to remove anything on the surface (oil, grease, paint, rust, and/or dirt) that could prevent the alloy from adhering to the metal and to protect the liquid soldering or brazing alloy from the air.

Soldering and brazing rely on the principle of capillary action for the soldering or brazing alloy to flow into the space between two or more tight-fitting parts. There must be enough space for the alloy to flow into the joint, but not so much space that the alloy cannot fill the gap. Some very free-flowing alloys require as little as 0.001 to 0.003 inch (0.025 mm to 0.076 mm) clearance when the alloy is at its melting temperature.

Soldered and brazed joints typically provide much more strength than glued joints on metal. Also, the joints can withstand much more heat than a glued joint on metal.

One of the main distinctions between the soldering and the brazing process is the temperature at which the alloy changes from a solid to a liquid. Soldering alloys melt at a relatively low temperature (840°F/450°C or less), while brazing alloys melt at relatively high temperatures (above 840°F/450°C).

Braze Welding

Braze welding is done by bonding a base metal to a filler metal. It does not rely on capillary action to make the filler metal flow between the pieces to be joined. As with soldering and brazing processes, the base metal is not melted when braze welding.

NOTE: Soldering and brazing refer to the process of joining metals, and not necessarily the composition of the filler metal. For example you can braze or solder aluminum by using the proper alloy. Other common metals that can be joined with these processes include steel, stainless steel, and copper. The main differences between soldering, brazing, and fusion welding is that the base metal is not melted during the soldering and brazing processes, while the base metal must melt during the fusion welding process.

Important Concepts for Cutting Metals

Cutting metals with heat can be done by either flame cutting or by a thermal cutting process (Fig. 11). Most metals that can be readily flame cut are classified as carbon steel. Stainless steel and aluminum are two common metals that require a thermal cutting process. Each of these metals form a tough surface oxide which protects the metal underneath from further oxidation. The thermal cutting processes can also be used on carbon steel and sometimes provide much higher cutting speeds on thicknesses under 3/4 inch (19 mm).

Flame Cutting

To flame cut a piece of carbon steel, the metal must burn (oxidize) when it reaches its kindling temperature in the presence of almost pure oxygen (99% will give acceptable results, but 99.5% gives the best results). The flame used as a heat source must be capable of bringing the metal up to its burning temperature quickly — within a few seconds. The edges of metal that have been flame cut properly are smooth, almost perfectly square, and usually do not require any preweld cleaning. Flame cutting is used to cut very thin metal (1/8 inch/3.2 mm) to very thick metal (40 inches/1.00 m).

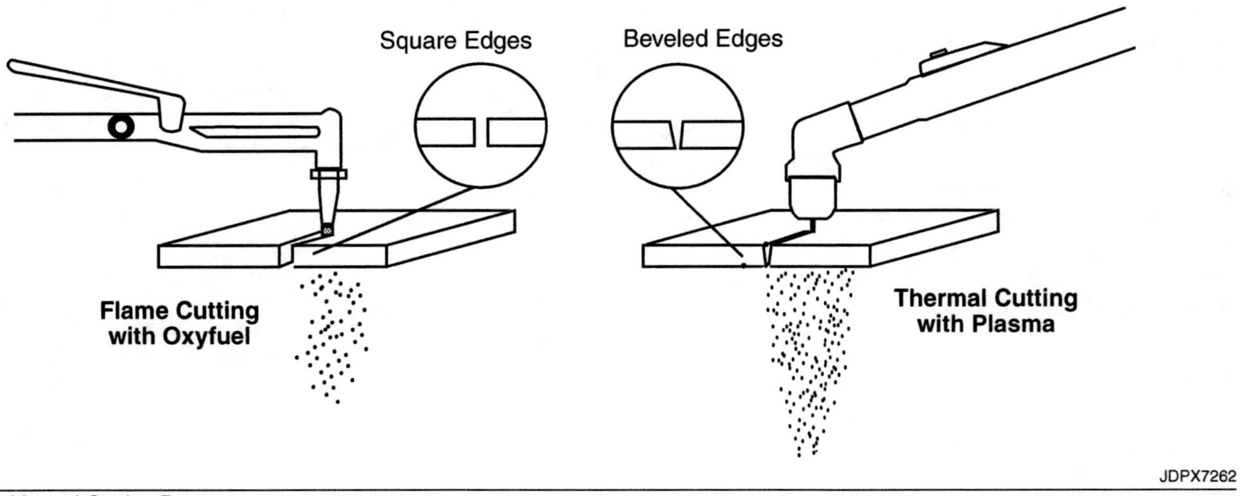

Fig. 11 — Manual Cutting Process

Thermal Cutting

The two most common thermal cutting processes are plasma arc and laser beam. These processes are capable of producing temperatures (50,000°F/27 800°C) many times the melting temperature of the metals used in industry. The heat and force from the source is so intense that it melts the metal instantly and blows it away. A kerf (hole) is formed as metal that is melted and vaporized is removed from the area where the heat is concentrated. The edges of the metal that have been plasma cut properly are smooth. They sometimes require preweld cleaning to remove the dross (slag-like material) and contain a 3° to 7° bevel (one side contains more bevel than the other). Laser cut edges usually do not require preweld cleaning and contain a 1° to 3° bevel. Both processes typically do not warp or distort the metal as much as oxyfuel cutting.

Becoming a Craftsman

Joining and cutting metals is both an art and a science. The processes used in industry today are high tech and require people who have been properly trained to use the equipment. A very high level of knowledge and skill is necessary for these processes to be applied in a craftsman-like manner so that a completed project is recognized as good work and so that it provides the quality expected for its use.

Technology from many of the sciences must be used in selecting the proper combination of manufacturing equipment, base metals, and filler metals for producing products that meet today's demanding service requirements. To be successful today, a craftsman must be capable of using the proper combination of these variables to obtain quality while maintaining productivity at competitive costs.

Learning the Basics

Welding and cutting require good eye-mind-hand coordination and the ability to concentrate on the work at hand. Students must see and then understand the activity in the molten metal as the joining or cutting process is occurring. They must also have a thorough understanding of the operation of the equipment so that it is set and/or adjusted for optimum performance.

Good depth perception is necessary to maintain the proper torch or electrode to work distance. A student must be able to recognize the time and place to apply heat and/or filler metal. This means maintaining a good body position so it is possible to look left and right, and ahead and behind the molten metal — all at the same instant. Fig. 12 illustrates the formation of a GMAW weld and the actions that are under the welder's control.

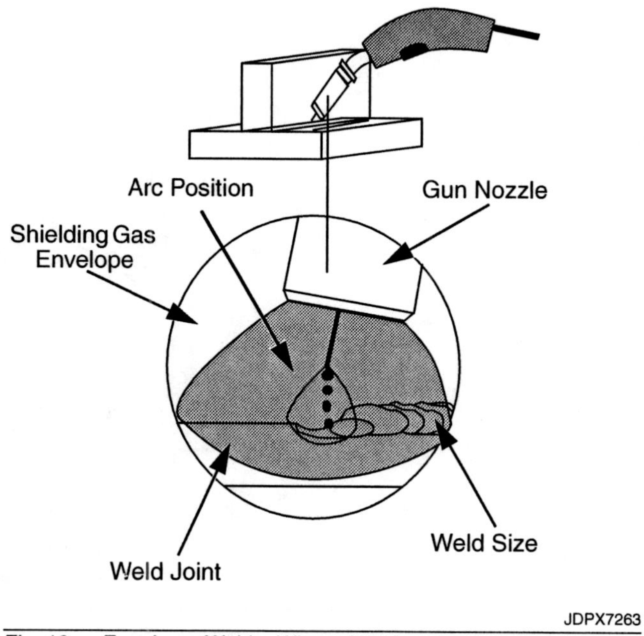

Fig. 12 — Forming a Weld – What a Welder Needs to See

Most of all, the student must know the quality standards for each operation and understand the proper procedures for inspecting the work. The person performing the work sees more of each operation in the process than anyone else, and when properly trained can identify a problem immediately. It is just good business to have someone capable of identifying a problem at its source, rather than by someone else at another operation or after the project has been completed.

Students must teach themselves the process, quality standards, and inspection procedures. Teachers, textbooks, and visual aids only identify the things that must be learned. A student must first identify and learn these basic concepts, along with the necessary related safe work habits. Then it will take a lifetime of experience applying the basics with different processes to various kinds of metal, in all different kinds of situations, to become a true craftsman.

Key Variables to Control for Learning Any Process

There are five variables which must be controlled when joining or cutting metals. Although they are listed separately, each one can affect the other when it comes to the end results — a finished project that has an acceptable appearance and meets quality standards.

1. Heat
2. Distance
3. Angle
4. Speed
5. Manipulation

Learn to observe the ways an experienced craftsman recognizes when each variable is to their satisfaction and how they change each variable to keep the process under control. At first, it may be easier to learn to recognize the upper and lower limit for each variable and the effect each has on controlling the process. Then learn to stay within each limit rather than trying to keep each variable exactly the same each time. Being flexible and being able to adapt to changing conditions is a very important characteristic of a good craftsman.

Heat — The heat necessary for any joining or cutting process is controlled primarily by the equipment size and settings. However, the distance the actual heat source is from the work, the angle it is pointed toward the work, the travel speed across the work, and manipulation of the heat source can also have an effect on the success of the operation. Setting the equipment in the middle of the manufacturer's recommended range is a good starting point.

Distance — The distance to position the heat source from the work is measured from different locations and depends on the process. For example, with the oxyfuel processes the important measurement is from the tip of the inner cone of the flame to the work. The description of each process explains where to measure the distance and what it should be. Proper joining and cutting will not occur if the distance is varied beyond the limits.

Angle — The angle is actually two angles. One angle is in relation to the work piece (or pieces) and the other is in relation to the direction of travel (Fig. 13). For example, more heat can be concentrated on the work by pointing the arc more directly into the work or less heat can be concentrated on the work by pointing the arc at more of an angle away from the work.

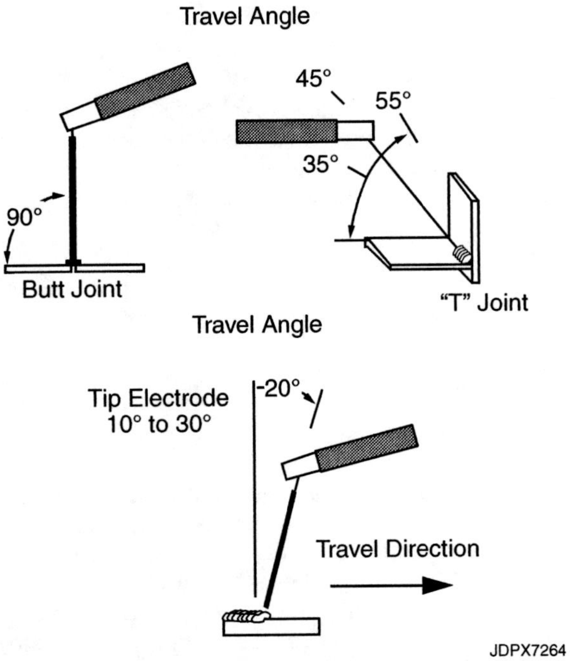

Fig. 13 — Work and Travel Angles — Examples for Shielded Metal Arc Welding (SMAW)

Speed — The travel speed is determined by watching the progress of the operation at hand. For example the person making a weld must recognize the proper size of weld as it is in progress and travel at a speed to keep the weld uniform. To produce a weld of a given size, a very experienced person can use more heat and travel faster than an inexperienced person who typically uses less heat and travels at a relatively slow speed.

Manipulation — The manipulation or movement of the electrode, gun, or torch during a joining or cutting operation is sometimes necessary to control the process when working out-of-position (horizontal, vertical, or overhead), or to get the weld or cut started and/or stopped. Some processes require a movement to control the formation and appearance of a weld or braze joint. Conditions such as a poor-fitting joint that has a large gap to fill require a motion to help make the liquid metal bridge the space between the parts.

Points to Remember about Manipulation

1. Make sure there is a sound reason for using a manipulation with a joining or cutting process. Unnecessary manipulations can also cause a lot of unnecessary wear and tear on the joints in a person's hand and wrist. Do not make unnecessary work for yourself.
2. Match the motion with the proper process and/or filler metal. Weld defects can be caused by using unnecessary motions or by using the wrong motion with a particular process or type of filler metal. Common weld defects caused by unnecessary manipulations include a lack of fusion, undercut, porosity, and slag inclusions.

Types of Welds

Most welds can be classified as either a groove weld or a fillet weld. A groove weld is formed when the sides or ends of two pieces are butted together. Names for the parts of a typical groove weld are illustrated by the single "V" groove in Fig. 14.

Groove and Fillet Welds

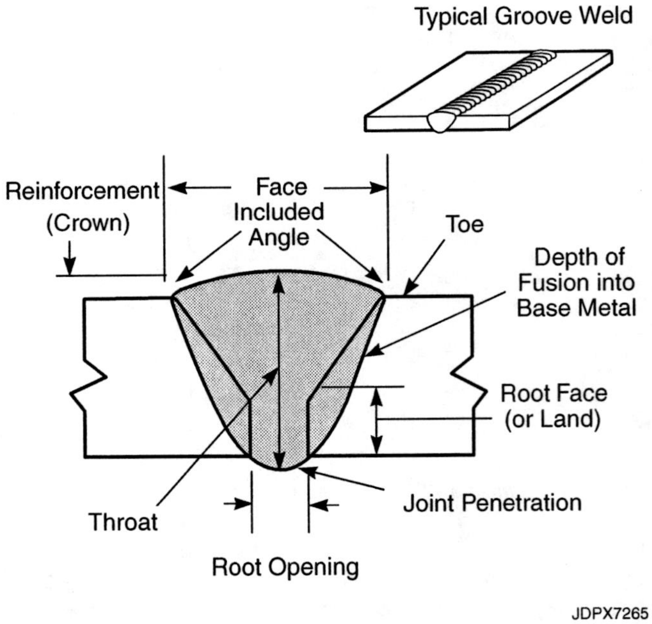

Fig. 14 — "V" Groove Weld Nomenclature

The name for a groove weld is determined by the shape of the edges joined by welding. Common types of grooves are illustrated in Fig. 15.

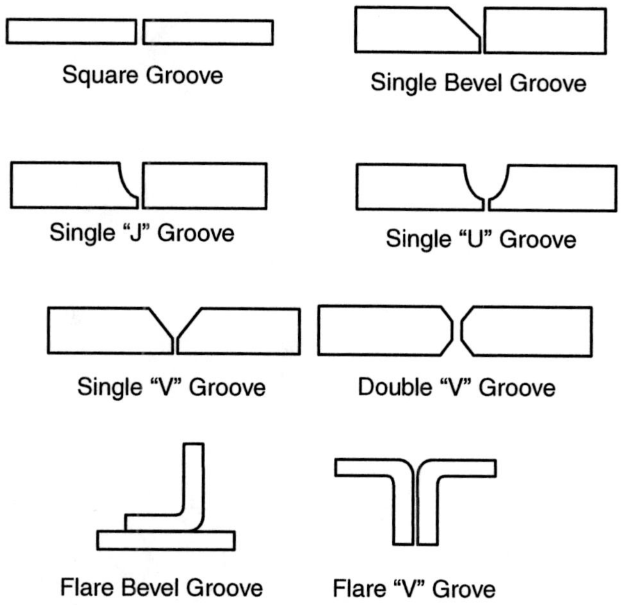

Fig. 15 — Common Types of Grooves

A flare bevel is formed by joining a curved surface to a flat surface. Examples would be a round bar, a piece of tubing, or a formed angle that is joined to a flat surface. A flare "V" is formed when two curved surfaces are joined.

A fillet weld is the name given to the triangle of metal formed by a weld at the intersection of the surfaces of two pieces of metal that come together at a right angle. The names for the main parts of a fillet weld are illustrated in Fig. 16. Fig. 17 illustrates how to use a fillet gauge to measure the legs and the throat of a fillet weld.

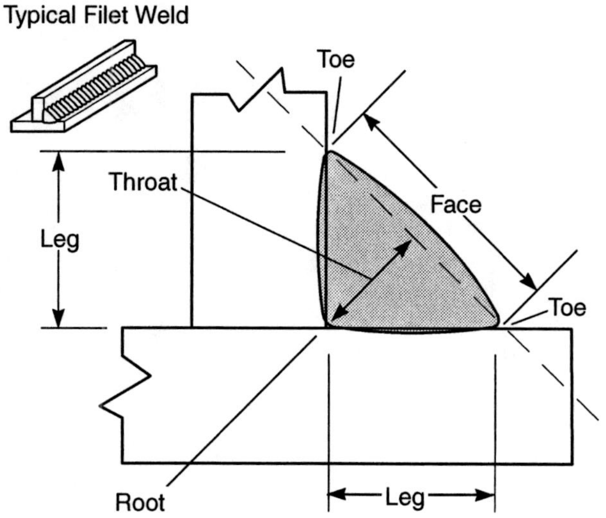

Fig. 16 — Fillet Weld Nomenclature

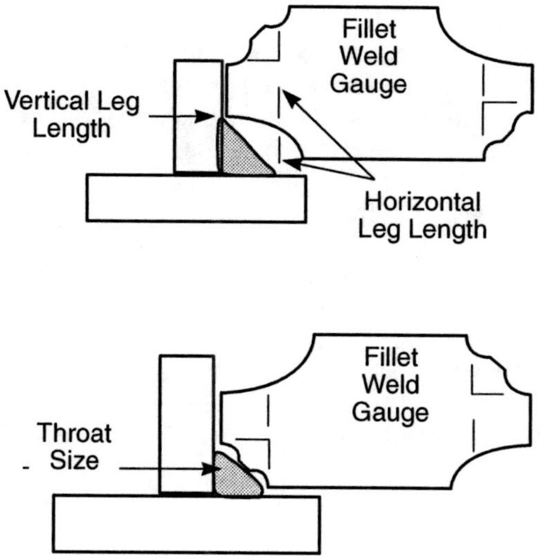

Fig. 17 — Fillet Weld Measurement Using a Fillet Gauge

Weld Joints and Weld Size

Fig. 18 illustrates five common weld joints. Note that a fillet weld joins the two pieces of metal that form a corner joint, Tee (or "T") joint and a lap joint. A groove weld joins the pieces that form a butt joint and an edge joint.

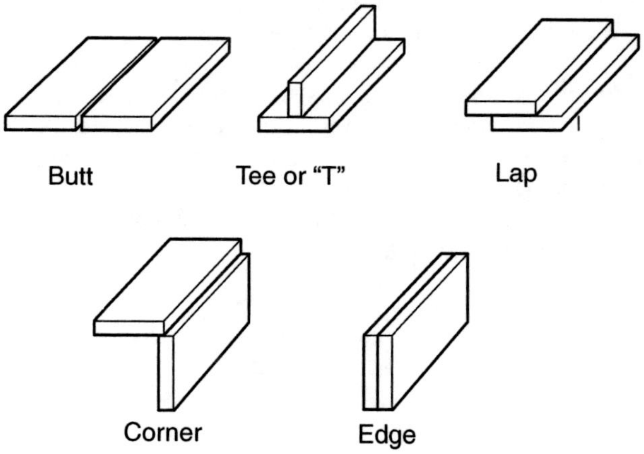

Fig. 18 — Five Common Weld Joints

The relative size for either a groove weld or fillet weld is based on having as much weld metal as the thinnest piece being joined. For example, the throat of a groove weld should equal the thickness of the thinnest piece to be joined. The legs of a fillet weld (when welded on one side) should have leg lengths equal to the thickness of the thinnest piece to be joined and the face should be flat or slightly convex.

Welding Positions

Names for the positions (or locations) for making groove and fillet welds on plate and for making groove welds on pipe are illustrated in Fig. 19. The "G" after each position indicates that it is for a groove weld and the "F" is for a fillet weld.

Plate positions are numbered according to the position of the face of the weld and the direction of welding on the plates. Plate positions are numbered 1 (flat), 2 (horizontal), 3 (vertical — either up or down), and 4 (overhead). Pipe groove welds are numbered 1 (flat welding, pipe rotated), 2 (horizontal welding with pipes in a vertical position), 3 (pipe fixed horizontal and welding vertical—either up or down), and 4 (pipe fixed at a 45° angle and welding vertical — either up or down).

Weld Discontinuities and Defects

A weld discontinuity is something in a weld that keeps it from being perfect but doesn't prevent it from serving the purpose for which it was intended. A weld defect is something in a weld that can cause the weld to not serve the purpose for which it was intended. Usually, when a weld discontinuity reaches a certain size it is considered a defect and the weld is unacceptable. Some things in a weld cannot be tolerated, no matter how minute they are, and immediately cause a weld to be unacceptable. Depending on the circumstances, some weld defects can be repaired, but some require that the entire weld be removed and replaced with another weld.

Common weld defects are illustrated in Fig. 20 and Fig. 21. Measures for preventing these defects are covered in the troubleshooting sections for each of the common welding processes. The extent for which a discontinuity becomes a weld defect is covered briefly in the next section, "Standards for Measuring Weld Quality."

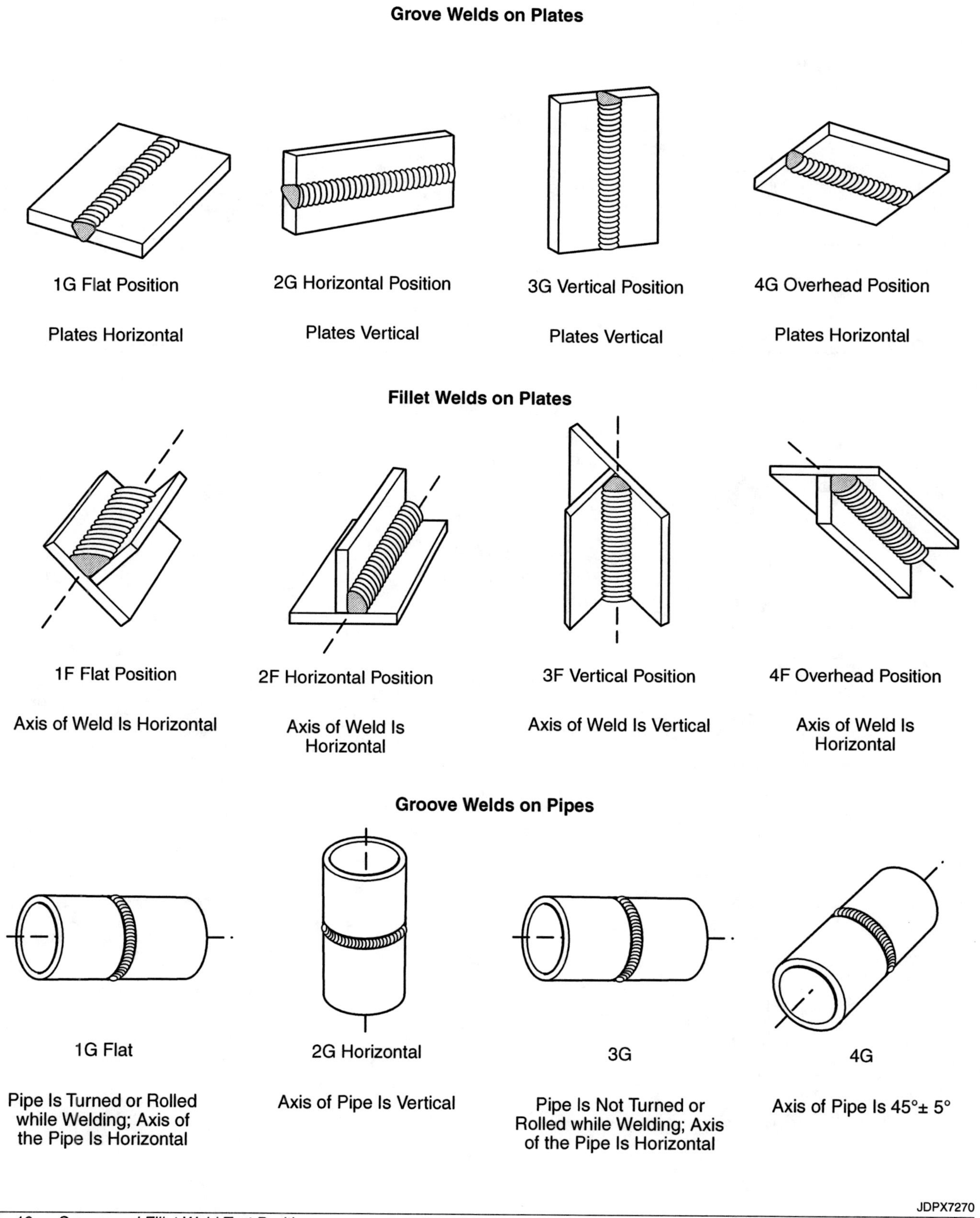

Fig. 19 — Groove and Fillet Weld Test Positions

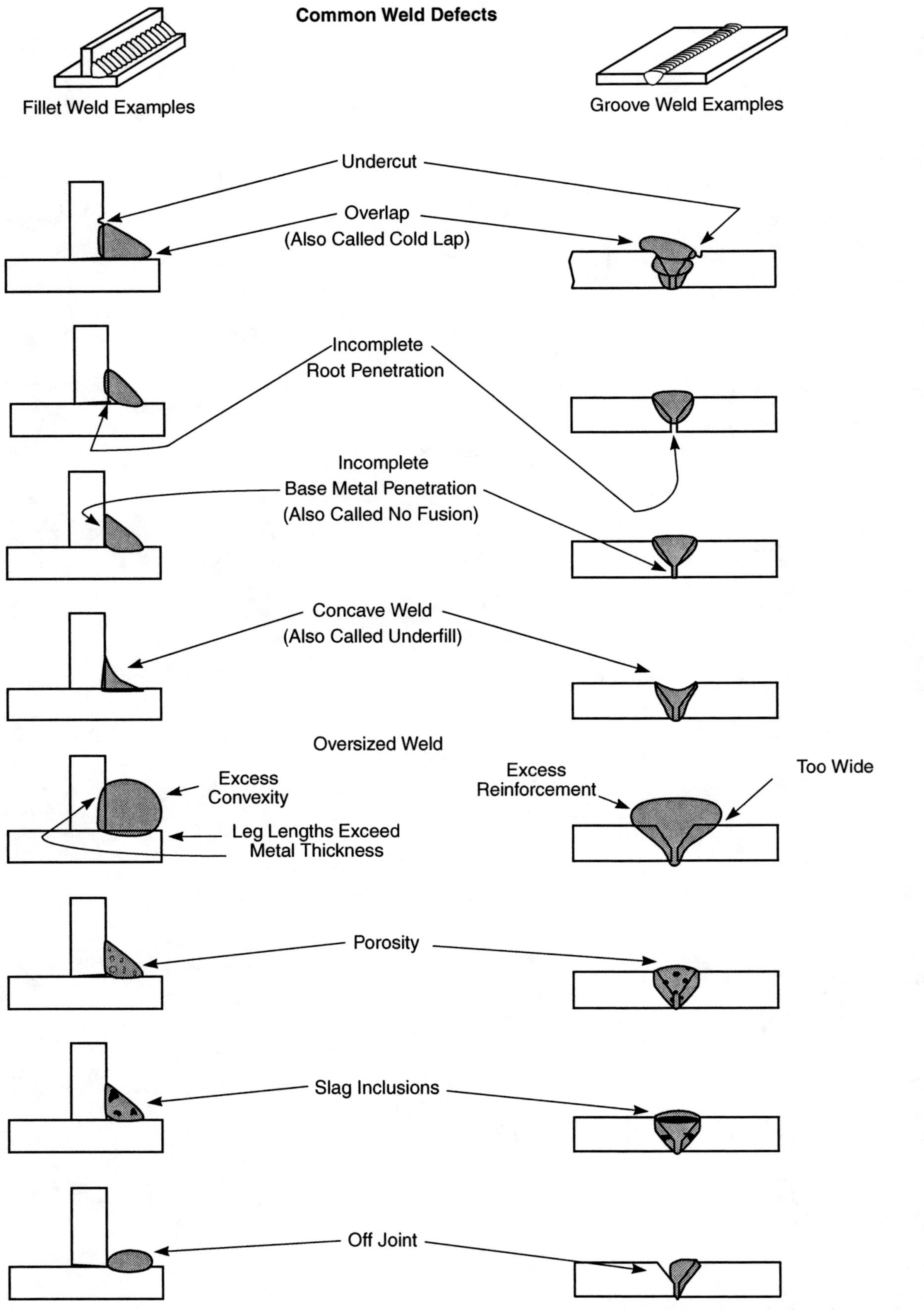

Fig. 20 — Common Weld Defects

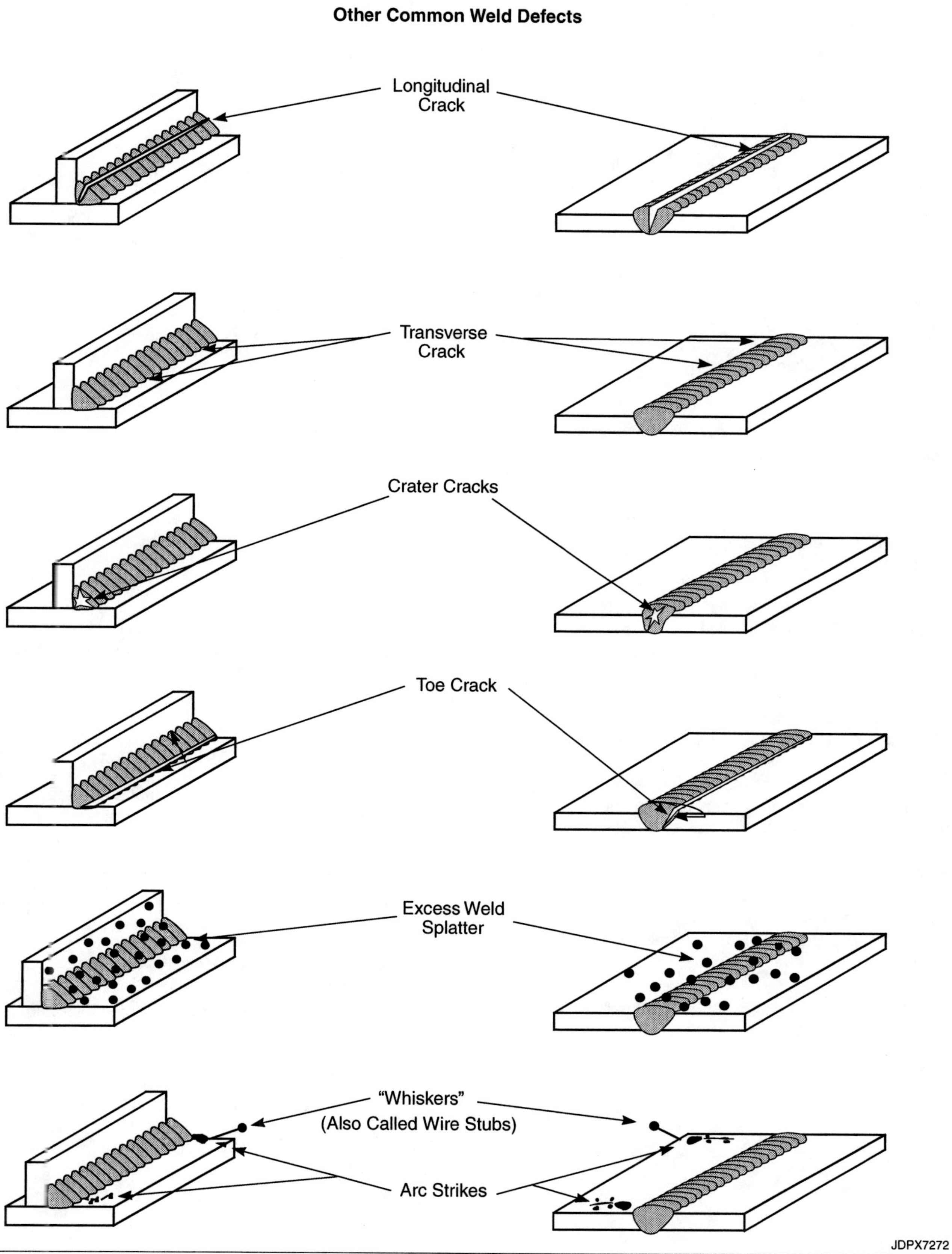

Fig. 21 — Common Weld Defects, Continued

Standards for Measuring Weld Quality

Acceptable weld quality is determined by the appearance of a weld and by the capacity of the weld to hold two or more pieces of metal together under service conditions for the life of the part. A document called a welding code or standard, establishes the rules for measuring weld quality. This code defines a procedure for verifying that the variables used to join two pieces of metal produce a weld that meets the service requirements for the part. The same code defines a process for verifying that a welder (a person) has the ability to follow each established procedure and for determining the acceptability of his/her welds as defined by the code's inspection standards.

The first step in the process is called qualifying the welding procedure. It actually is a two-part process which involves writing the instructions for making the weld and verifying the results by making and testing the weld. This first part is called writing a welding procedure specification (WPS). The second part is called completing a procedure qualification record (PQR).

The second step is called qualifying the welder. A welder must demonstrate his/her ability to follow the welding procedure (WPS) and possess the skill to make a weld that meets the inspection standards of the specific welding code. The results of the welder's test is then recorded on a report form which references the welding procedure that was followed.

The entire process relates to what is called welder certification. The term certification is used interchangeably to mean the same as qualification. There is no such thing as a universal welder qualification that allows a person to weld with any process on any metal in any welding position. Each welder qualification has limits as to each of these items as stated in the welding code that is being followed.

Taking into account the different types of welding codes, welding processes, and base metals, there is almost an unlimited number of different welder qualifications.

Manufacturers of products that use joining and cutting processes apply these rules to their products. Some of the more common welding codes, the organizations which have written the codes, and the general scope of each code follow.

Examples of forms used by the American Welding Society to record this information are included in the Appendix.

Types of Welding Codes

ASME (American Society of Mechanical Engineers)

Provides guidelines for making welds on boilers, pressure vessels, and pressure piping. There are 11 sections to the Pressure Vessel Code. Section IX concerns the qualification of welding procedures and welders.

AWS (American Welding Society)

Provides guidelines for welding steel structures such as bridges, buildings, and industrial equipment. There are presently 30 different codes and standards in the D classification. Two widely used codes are D1.1 and D1.3. D1.1 concerns the welding of structural steels having a thickness of 1/8 inch (3.2 mm) or more. D1.3 concerns the welding of structural steels having a thickness of 1/8 inch (3.2 mm) or less.

API (American Petroleum Institute)

Provides guidelines for welding of cross-country pipelines. The most widely used code is 1104 which concerns the welding of pipe.

Military Codes

Provide guidelines for the construction of military equipment. The two most widely used codes are 1595 (aerospace welding) and 9858 (structural steel).

Methods of Inspecting and Testing Welds

Welds can be inspected with nondestructive tests (NDT) and/or destructive tests. As the names imply, a part that passes a nondestructive test can be put into service. A part that has undergone a destructive test must be either repaired or replaced with another part. A combination of the two types can minimize the overall cost of weld inspection.

Nondestructive Weld Tests

Common nondestructive tests include the following:

1. Visual inspection
2. Liquid penetrant inspection
3. Magnetic particle inspection
4. Ultrasonic inspection
5. Radiographic inspection
6. Leak test

Visual inspection means examining the exposed surfaces of a weld for discontinuities and measuring its size. Visual inspection has to be acceptable before there is any further testing.

Liquid penetrants (usually a dye) are used for locating discontinuities on the surface of the weld area. The penetrant is applied to the weld area (after thorough cleaning to remove any foreign material such as dirt or oil). The penetrant is left on the weld area until it has had adequate time to be drawn into surface discontinuities by capillary action. The surface is wiped clean with a cloth containing a solvent. Next, the entire area is sprayed with a developer. Reverse capillary action draws the penetrant back to the surface where there are discontinuities. A liquid penetrant can be used on any type of metal.

Magnetic particle inspection is also used for locating surface discontinuities. The weld area is covered with a fine iron powder, and a probe is used to magnetize the area to be inspected. The magnetic lines of force concentrate at points where there are surface discontinuities. Magnetic particle inspection is limited to use on metals that can be magnetized.

Ultrasonic inspection transmits sound waves through a transducer to the weld area and can locate internal discontinuities. When the sound wave encounters a discontinuity, it is reflected back to the transducer. The reflections of the sound waves are shown as vertical lines on a screen similar to a television monitor. The relative height of the lines can be used to find the location and the size of a discontinuity.

Radiographic inspection can be used to locate internal discontinuities in welds. It involves transmitting either X-rays from a special electronic device or rays from a radioactive isotope through the object to be examined. The rays expose a special film placed inside hollow objects or behind solid objects. When developed, the film shows light and dark areas that represent the size and location of internal discontinuities.

Leak tests are performed on weld joints subjected to both low and high pressures. To locate a leak, the part in question is usually pressurized to a predetermined pressure (3 – 2500 psi/21 – 17,237 kPa). The pressure must be maintained for the test period that ranges from a few minutes to several hours after the valve from the source is closed and the pressure stabilized.

On relatively small components, such as a 500-gallon (1893-liter) fuel storage tank, the seams are checked with soapy water at low pressures (usually under 5 psi/34 kPa). Large containers, such as a cross-country pipe line, will be pressurized well above service pressure (500 – 600 psi/3447 – 4137 kPa) for several hours. If there is pressure loss, each seam is checked with a leak detection fluid until the leak is located.

On open containers that cannot be pressurized, a liquid penetrant is sprayed on one side of the weld and the developer on the other side. Penetrant on the other side means there is a leak.

Destructive Weld Tests

Common destructive weld tests include the following:

1. Bend test
2. Tensile test
3. Nick break test

Bend tests are given to welds to determine if the weld joint will bend without breaking when it is under pressure. When the top of a weld is made to stretch, it is called a face bend, and when the bottom of the weld is made to stretch, it is called a root bend. Pieces having a thickness greater than 3/8 inch are given side bends. Figure 22 illustrates the size and location of bend test coupons. Figure 23 illustrates the proper preparation after it has been removed from the test plates. Figure 24 illustrates the proper positioning of each piece for bending.

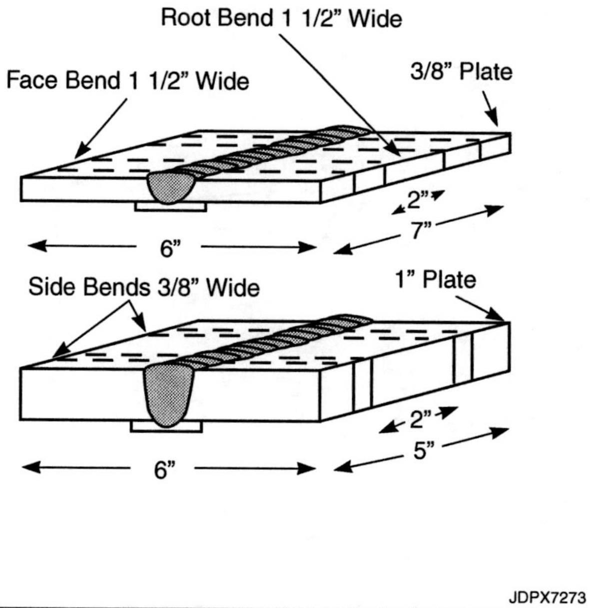

Fig. 22 — Bend Test Coupon Size and Locations

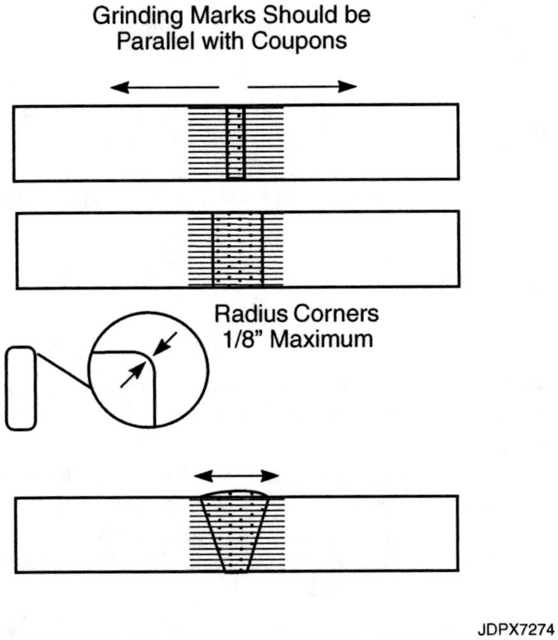

Fig. 23 — Bend Test Coupon Finishing

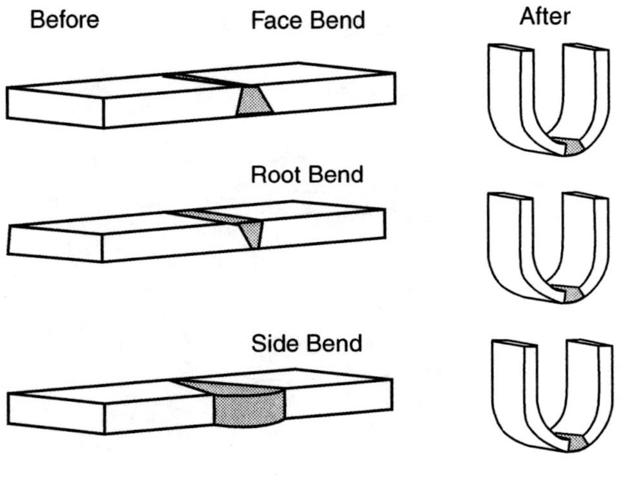

Fig. 24 — Bend Test Procedures

The amount each test piece is expected to bend has been predetermined by welding codes based on specifications provided by the manufacturers of the weld filler metal and the steel. AWS guidelines for making bend tests are given in the Appendix.

Sections from the weld taken for bend testing are of a designated width and are cut from specific areas of a butt joint. The surfaces of each piece are prepared for bending by grinding and/or machining. The sections are then placed on a special "U" fixture and bent approximately to a "U" shape or about 180°. Fig. 24 illustrates the preparation of pieces for bend testing and the proper positioning of the weld during each type of test.

The test is considered unacceptable if the piece breaks or if there are openings in the weld that exceed the limits set by the welding code that is being followed.

NOTE: Each welding code has different limits of acceptability. However, most codes do not allow a single defect in a bend test to be greater than 1/8 inch (3.2 mm) in size.

One type of tensile test is used to measure the strength of the weld metal and the base metal. A section of metal is taken from the weld area and machined to a specific size. Fig. 25 illustrates the way a piece is prepared for a standard tensile test.

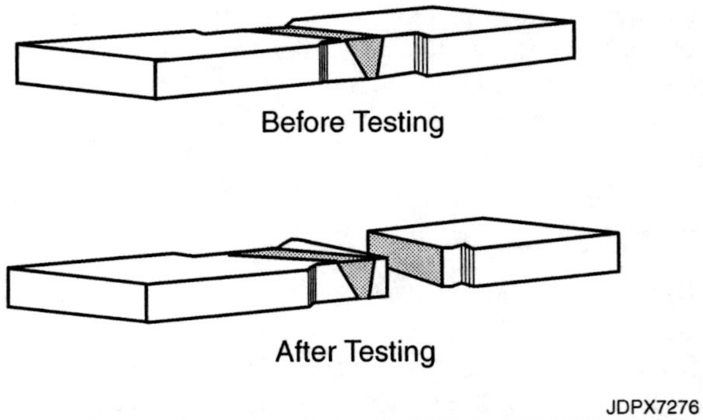

Fig. 25 — Typical Tensile Coupon Preparation and Test

The piece is secured in the jaws of a tensile testing machine and pressure is used to pull on the piece until either the metal or the weld fails. This pressure is recorded and compared to the minimum tensile strength specified by the steel manufacturer or the filler metal manufacturer. It is acceptable for the metal or weld to fail at a value equal to or above the manufacturer's specification. A failure at a lower value indicates there is a problem(s) with the welding procedure that must to be corrected.

The API has a slightly different method of doing a tensile test for a welder qualification test. The weld is not machined from the section taken from the weld joint. The marks made from the cutting process are smoothed along the edges. The piece is pulled until it breaks. The weld is considered to be acceptable if the break is in the base metal and not in the weld. This is illustrated in Fig. 26.

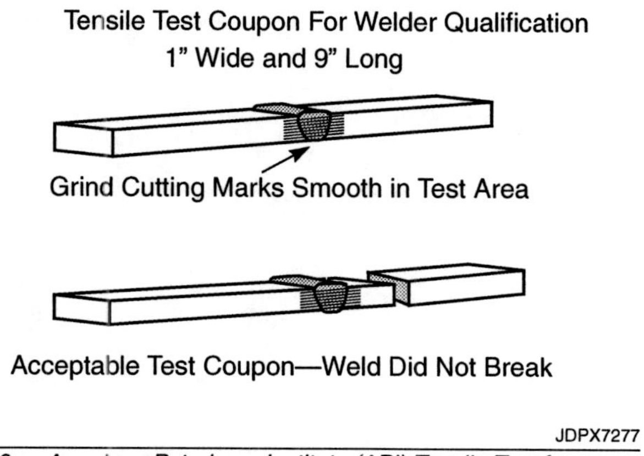

Fig. 26 — American Petroleum Institute (API) Tensile Test for Welder Qualification

Nick break tests are used by the API to determine if there are internal discontinuities in a weld. A piece for testing is taken from the completed weld. As illustrated in (Fig. 27) a "nick" is made in each side of the piece in the center of the weld by sawing a 1/8 inch (3.2 mm) deep groove. The piece is either pulled apart or bent back and forth in a vise until the weld fails. The fractured surface is examined for discontinuities. For a test to be acceptable, there cannot be a defect larger than 1/8 inch (3.2 mm). To be considered a defect, the discontinuity has to be 1/32 inch (2.4 mm) or larger. If there are two defects in the same piece, they cannot be within a 1/2 inch (12.5 mm) of each other. Fig. 27 shows a test piece that has been prepared properly for a nick break test.

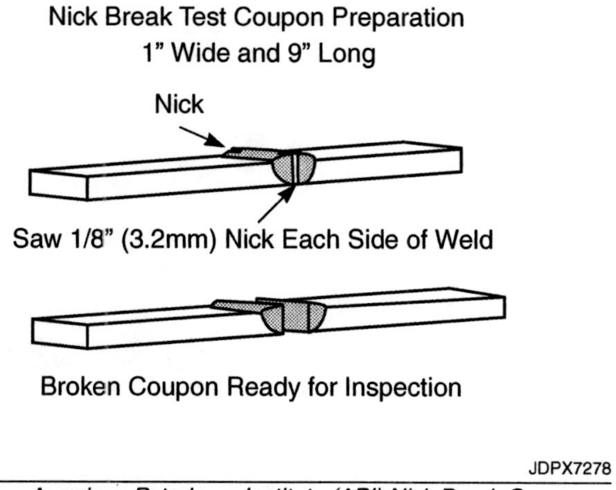

Fig. 27 — American Petroleum Institute (API) Nick Break Coupon Preparation and Testing

Test Yourself

Questions

1. Name two metal cutting processes used in metal fabrication.
2. Name an arc welding process that uses a non-consumable electrode.
3. What arc welding process stands for SMAW?
4. What type of welding process is gas metal arc welding (GMAW)?
   a. Manual
   b. Semi-Automatic
   c. Automatic
5. What gases are is used for oxyacetylene welding (OAW) fusion welding?
6. What type of metal can be cut using the oxyfuel process?
7. Name the type of welding that combines the base metal and filler metal into one solid piece of metal.
8. At what temperature do soldering alloys and brazing alloys melt?
9. Soldering and brazing rely on the principle of __________ __________ for the soldering or brazing alloys to flow into the space between two or more tight-fitting parts.
10. What cutting process is used to cut stainless steel and aluminum?
11. List five key variables for learning a welding process.
12. Name the five parts of a fillet weld.
13. The amount of build-up above the base metal on a "V" groove weld is called __________.
14. The space between two plates on a "V" groove weld is called __________.
15. List the five basic joints in welding metal fabrication.
16. The size of a fillet weld is measured by ____________.
17. The amount of melt-through at the root of a "V" groove is called __________.
18. A fillet weld gauge measures________and __________.
19. What weld test position is identified by 4F?
20. When welding pipe in the 2G position, in what direction is the welding performed?
21. Name the weld defects that can be visually inspected on fillet and groove welds.
22. Name two types of testing done on weld joints.
23. Name three common types of destructive testing performed on welds.
24. List four organizations that provide weld codes for the metal fabrication industry.
25. Name three types of weld-guided bend tests.

Oxyfuel Welding (OFW) Processes

2

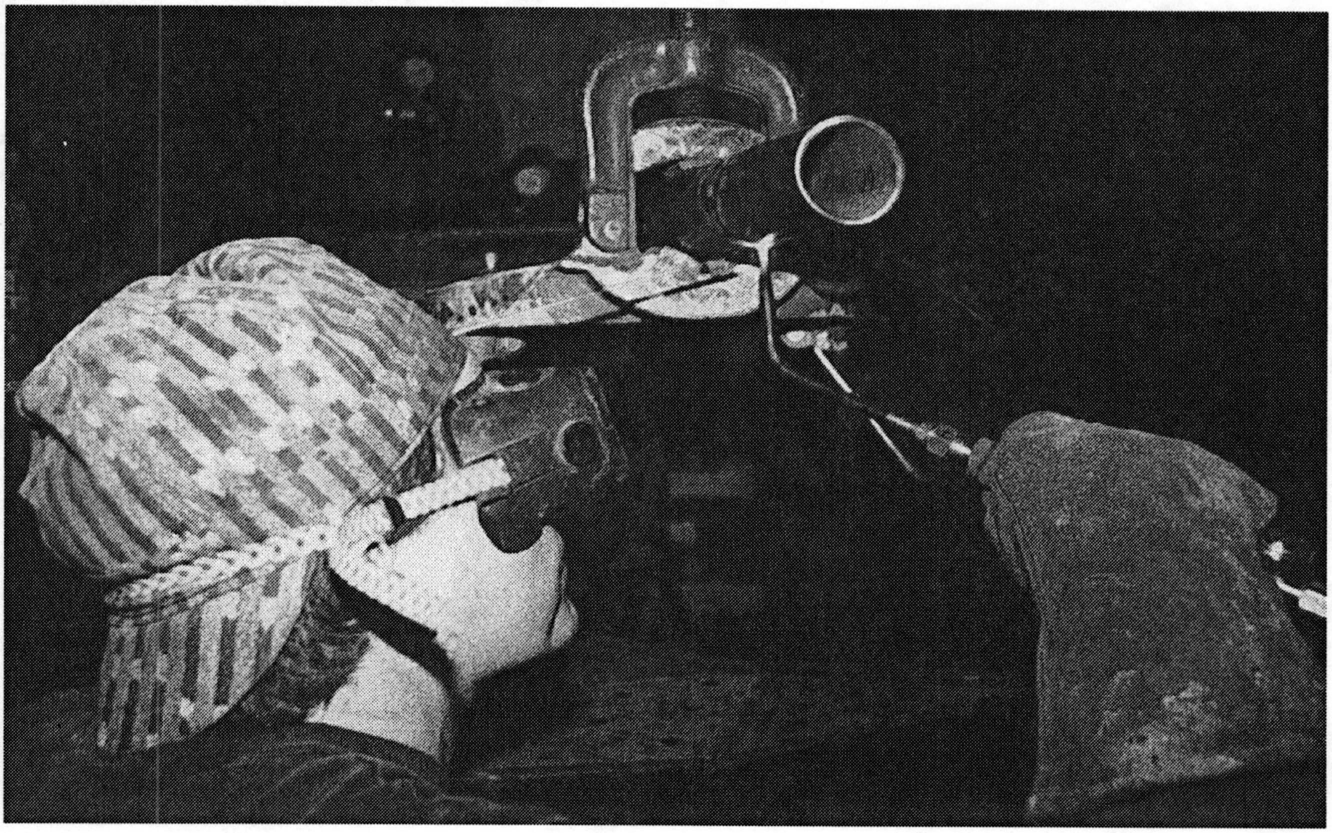

JDPX7209

Fig. 1 — Oxyacetylene Welding (OAW)

Introduction

Oxyfuel flames are used for brazing, flame hardening, metalizing, soldering, stress relieving, cutting, heating, and bending. Common fuel gases used in industry today include acetylene, MAPP® (this mixture of stabilized methylacetylene and propadiene is a registered trademark of Air Reduction Company Inc.), propylene and propylene based gases, natural gas (methane), and propane.

There is very little oxyfuel welding done in the metal working industry today because the GMAW and GTAW processes are so much more efficient. Acetylene is the fuel gas used for the limited applications of oxyfuel fusion welding.

The local welding gas supplier should be contacted for more detailed information regarding the best fuel gas for a specific application and the corresponding equipment that provides the best combination of economy and safety in the work environment. (SeeTable 1.)

Oxyacetylene Flame

The oxyacetylene flame is best suited for oxyfuel welding because of its high flame temperature when it is adjusted to what is known as a neutral flame (see later in Chapter 2). The flame is generated by burning a mixture of acetylene and oxygen. A ratio of 2-1/2 parts oxygen to 1 part acetylene produces a very hot flame which is about 6000°F (3315°C). Combustion occurs in two stages (Fig. 2):

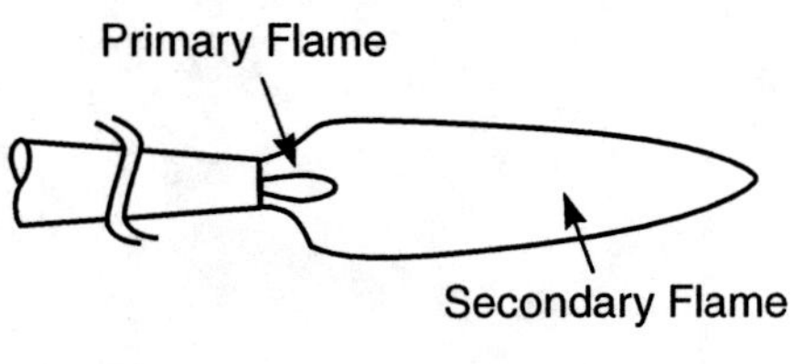

Fig. 2 — The Neutral Flame

Primary combustion takes place at the end of the tip. This area forms the inner cone of the flame and is the hottest part of the flame.

Secondary combustion takes place in the outer portion of the flame and is called the outer envelope. This outer envelope serves as a protective shield to prevent the atmospheric air from contaminating the molten puddle and preheats the metal for welding.

Fuel	Maximum Flame Temperature in Oxygen		Maximum Burning Velocity in Oxygen	
	(°F)	(°C)	Ft./Sec.	m/Sec.
Acetylene	5589	3087	9.8	3
MAPP® Gas	5301	2927	7.9	2.4
Propane	4579	2526	5.5	1.7
Propylene	5250	2900		
Natural Gas	4600	2528		

Table 1 — Flame Temperature Comparison of Fuel Gasses

The primary flame results when oxygen and acetylene are mixed in the welding torch in a one-to-one ratio. This mixture flowing from the torch tip burns to form the inner cone when one volume of acetylene combines with one volume of oxygen to produce two volumes of carbon monoxide and one volume of hydrogen.

The carbon monoxide and hydrogen produced by the first reaction combine with oxygen in the atmosphere to form the secondary flame. Thus, a complete combustion of acetylene results from one volume of oxygen provided by the torch and 1-1/2 volumes provided from the surrounding atmosphere.

Fig. 3 shows a cutaway of the oxyfuel torch and flame.

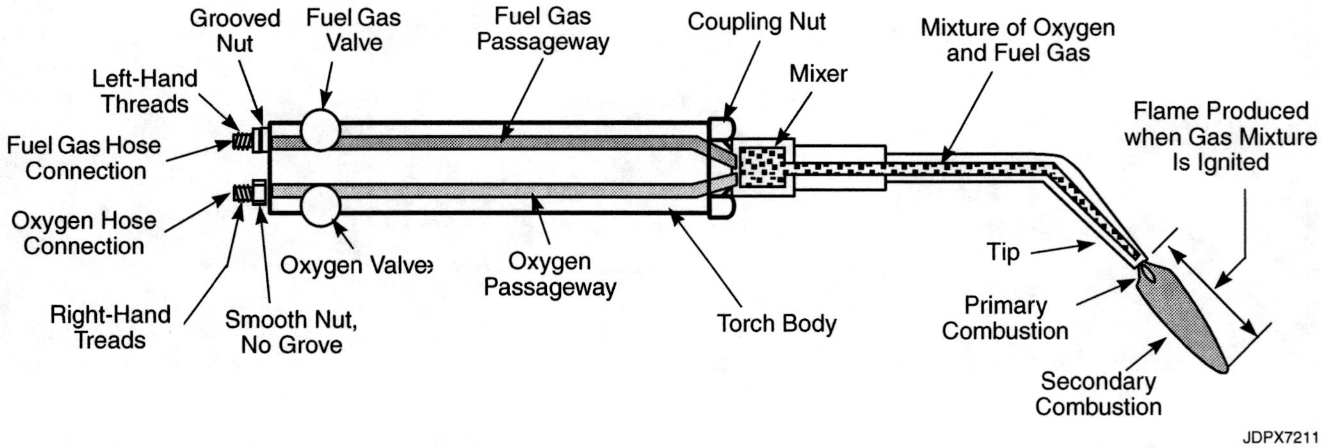

Fig. 3 — Basic Principle of an Oxyfuel Torch and Flame

Oxyfuel Equipment

Following is a list of basic oxyfuel welding and safety equipment:

- Cylinders and cart
- Pressure regulators
- Torch and tips
- Accessory check valves and flash arrestor
- Hoses
- Welder's clothing

Fig. 4 illustrates the basic oxyfuel equipment.

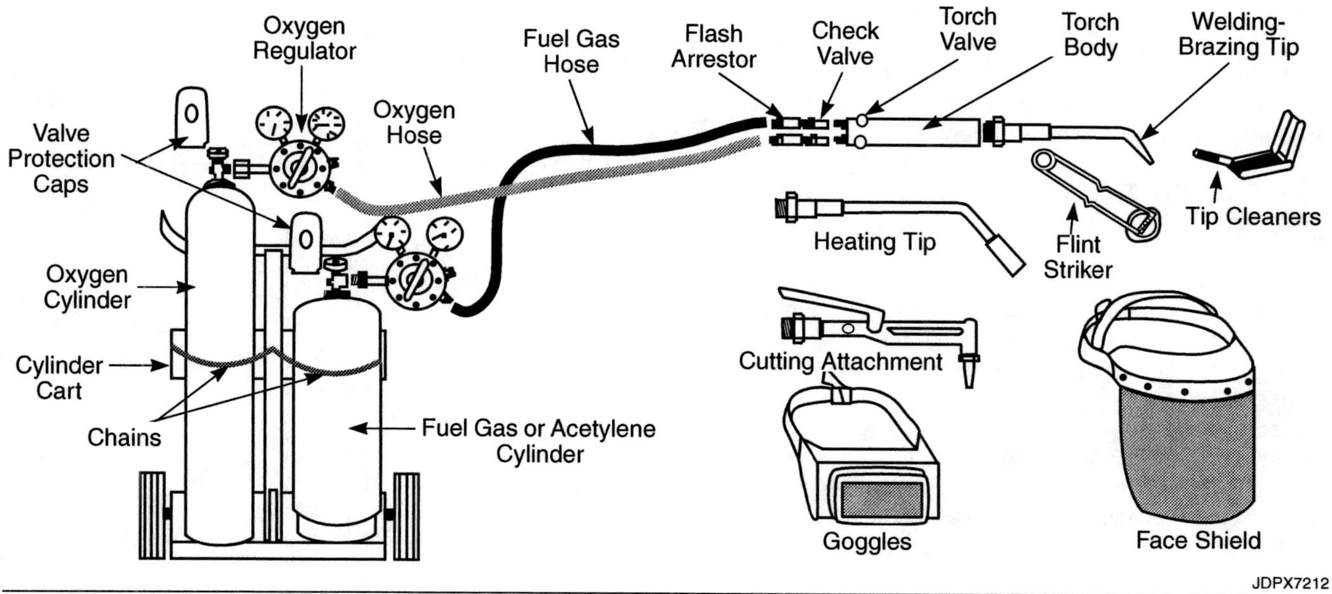

Fig. 4 — Basic Oxyfuel Equipment

Cylinders and Cart

Strict standards must be followed when handling the various sizes and shapes of high-strength steel cylinders. All cylinders used for the transportation of liquefied or compressed gases are regulated by the Hazardous Materials Regulations of the Department of Transportation (DOT). This includes regulations requiring periodical inspection and testing of each cylinder by the gas supplier throughout the life of the cylinder. Each test date is stamped permanently on the cylinder after it has successfully passed inspection requirements. Every cylinder contains the dates it was tested since being put into service.

CAUTION: Cylinders can explode if damaged. Welding cylinders contain gas under high pressure. If damaged, a cylinder can explode. Since gas cylinder are normally part of the welding process, be sure to treat them carefully.

The Compressed Gas Association (CGA) has specifications and standards for welding gas fittings and connections. Cylinder inlet connections, where the regulators attach, conform to CGA regulations and are marked with an identifying number. All fittings are made of brass to prevent any sparks while working with the connections. All oxygen cylinder, regulator, and hose connections have right-hand threads.

All acetylene connections are identified by grooves in the connecting nuts and all acetylene hose connecting nuts have left-hand threads.

CAUTION: Acetylene cylinder and regulator connections can have right-hand or left-hand threads, male or female fittings. It is important that you have the proper acetylene cylinder and regulator connections to match.

There are special safety precautions for cylinders. Always fasten them securely in a vertical position on a cart or in a cylinder rack with a chain or other holder. Always attach the metal valve protection cap (on cylinders with a threaded neck ring) when a cylinder is being stored or transported. Never try to use high-pressure gases directly from a cylinder without a suitable pressure-reducing regulator.

High-Pressure Oxygen Cylinders20

Cylinders of oxygen (Fig. 5) contain very high pressures (over 2200 psi at 70°F/15,200 kPa at 20°C) when filled to capacity. The cylinders are made of a special alloy steel formed into a seamless steel shell. An oxygen cylinder is specially built and will not rupture under normal circumstances. However, just in case a cylinder accidentally gets close to high temperatures, as from a fire, the cylinders are protected from rupturing by a safety disk located on the valve assembly in Fig. 6. This safety disk will rupture at about 3000 psi/20,700 kPa and release the entire contents of the cylinder out several small openings through an assembly that looks like a cap screw

CAUTION: This safety device should never be tampered with.

The flow of oxygen from the cylinder is controlled by a special double-seating valve illustrated in Fig. 6. The double-seating valve prevents high-pressure gas from leaking around the stem when opened all the way (when the valve has been turned as far as it will turn open, the valve is seated in the open position). The hand wheel is turned in a counterclockwise direction to open the valve and in a clockwise direction to close the valve. Always open high-pressure oxygen cylinder valves slowly to allow a gradual build-up of pressure inside the regulator.

The three common capacities of high-pressure oxygen cylinders for a shop are 244, 150, and 80 cubic feet (6.9, 4.25, and 2.25 cubic liters). Factories and other places that are large quantity users, obtain oxygen in low-pressure (125 psi/862 kPa), specially insulated containers of liquid oxygen. These liquid storage cylinders can release a much as 4500 cubic feet (127 cubic meters) of oxygen when the liquid is allowed to vaporize as it is withdrawn during use.

Acetylene Cylinders

Unlike most gases, acetylene in its free state (as it comes out of a torch) is unstable at pressures above 15 psi (100 kPa). This characteristic makes storage in cylinders with 250 psi at 70°F (1724 kPa at 21°C) impossible without using special processing procedures that are explained next.

Acetylene can only be stored safely above 15 psi (100 kPa) pressure by dissolving it in liquid acetone in special cylinders containing a porous filler material. The acetone absorbs large quantities of acetylene under pressure, just as soda pop absorbs carbon dioxide, and the porous material acts like a sponge to keep the liquid from splashing around when the cylinder is moved.

Acetylene is also shock-sensitive and could explode if the gas mixture were free to splash around in liquid form like soda pop in a bottle or can. Even when it is dissolved in acetone and contained by this porous material, these special cylinders should be handled carefully and not dropped or given hard bumps.

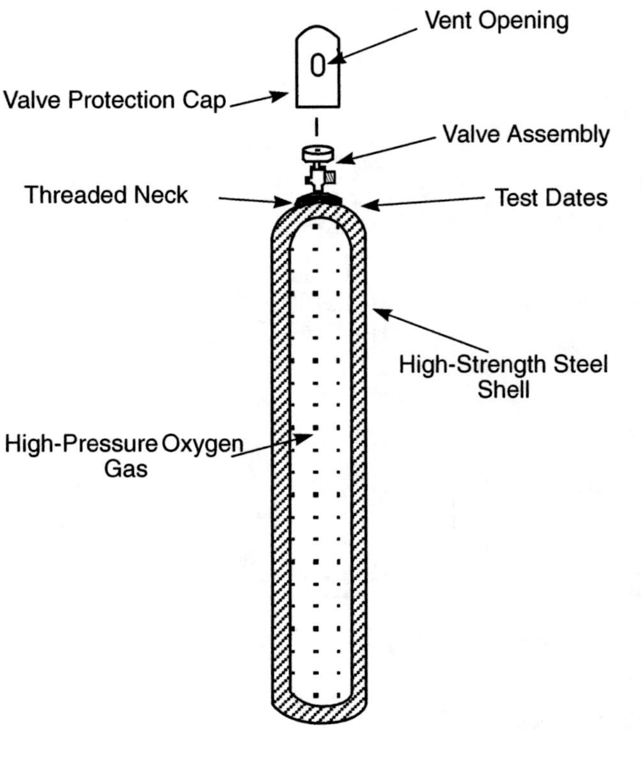

Fig. 5 — Typical High-Pressure Oxygen Cylinder

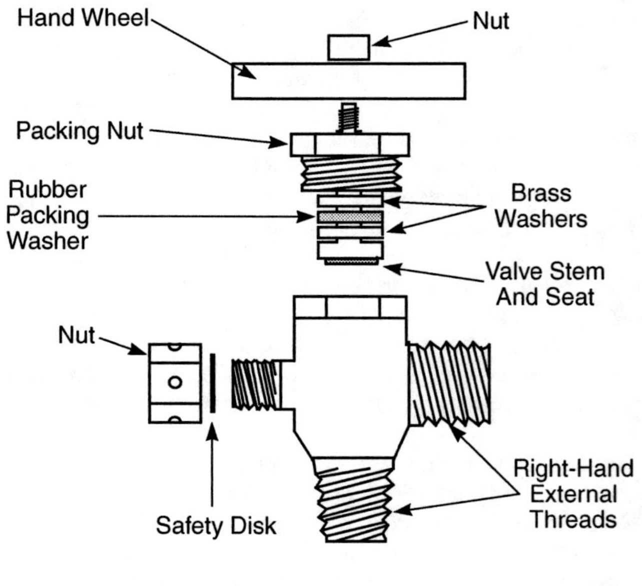

Fig. 6 — Double-Seating Oxygen Safety Valve Assembly

Acetylene cylinders are protected from dangerous internal pressures by plugs filled with a special fusible metal that melts about 212°F (100°C). These "fuse plugs" are threaded into the top and bottom of the cylinders. If the cylinder of acetylene is exposed to temperatures above 212°F (100°C) the metal melts, allowing the acetylene gas to escape slowly.

CAUTION: Do not use boiling water in the winter to melt ice that may have frozen in the valve opening or top of the acetylene cylinder. It is okay to use warm water, but boiling water could accidentally melt the metal inside these fuse plugs and cause the contents of the acetylene cylinder to escape.

Acetylene cylinders similar to the one in Fig. 7 are equipped with valves that must be opened and closed with a special wrench. Other types of acetylene cylinders look similar to fuel gas cylinders and have hand wheel valves.

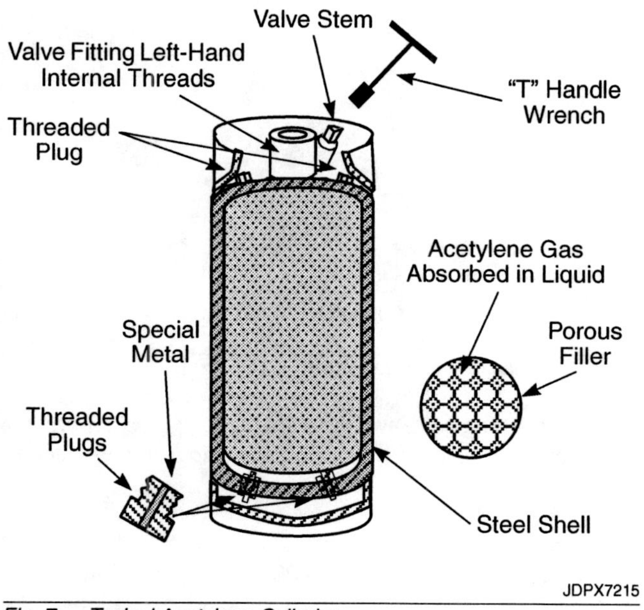

Fig. 7 — Typical Acetylene Cylinder

The acetylene cylinder valve should be opened 1/2 to 3/4 of a turn. This is a safety practice and permits closing the valve quickly in case of an emergency. The wrench should always be left on the valve when the cylinder is in use.

The contents of a cylinder of acetylene is indirectly determined by weight. The weight of a full cylinder is subtracted from the combined weight of the cylinder and the liquid acetone. This value is converted to cubic feet and/or cubic liters and varies somewhat between cylinders of equal size. Three common capacities of acetylene cylinders are 300, 120, and 75 cubic feet (8.5, 3.4, and 2.12 cubic liters).

Acetylene should never be withdrawn from a cylinder at a rate per hour greater than 1/7 the cylinder's capacity. For example, no more than 19 cubic feet (0.54 cubic liters) per hour should be withdrawn from a cylinder with a capacity of 120 cubic feet (3.4 cubic liters). These amounts were calculated by dividing the cylinder's capacity by 7.

Acetylene usage and corresponding cylinder capacity for a project should be determined before beginning a project. If this 1/7 withdrawal rate is exceeded, the hose and regulator can be damaged when liquid acetone is drawn into the system. This acetone will also burn with the acetylene and can change the chemistry of the flame. Always refer to the manufacturer's recommendations for the tip size needed and corresponding pressure settings.

Projects which need rosebud heating tips do use large volumes of acetylene that can exceed the 1/7 rule. Also, some medium and large cutting tips use a lot of acetylene. If the acetylene usage exceeds the 1/7 rule for the cylinder at hand, obtain a larger size cylinder with adequate capacity before starting the job. If a single cylinder is not capable of delivering an adequate flow, contact a local gas supplier for information about a gas manifold system for attaching two or more cylinders together.

Remember three very important things about acetylene:

1. Acetylene should never be used at pressures above 15 psi (100 kPa).
2. The cylinders should always be used in an upright position to prevent liquid acetone from getting into the equipment.
3. Filling cylinders is a dangerous operation and should be done only by persons who have the proper training and equipment. Don't try to fill an empty cylinder with a full cylinder.

Fuel Gas Cylinders

Fuel gas cylinders are relatively simple hollow steel containers which contain both vapors and liquid. Fig. 8 illustrates the relative location of the two substances inside the metal container and the location of spring-loaded relief valve in the cylinder valve body.

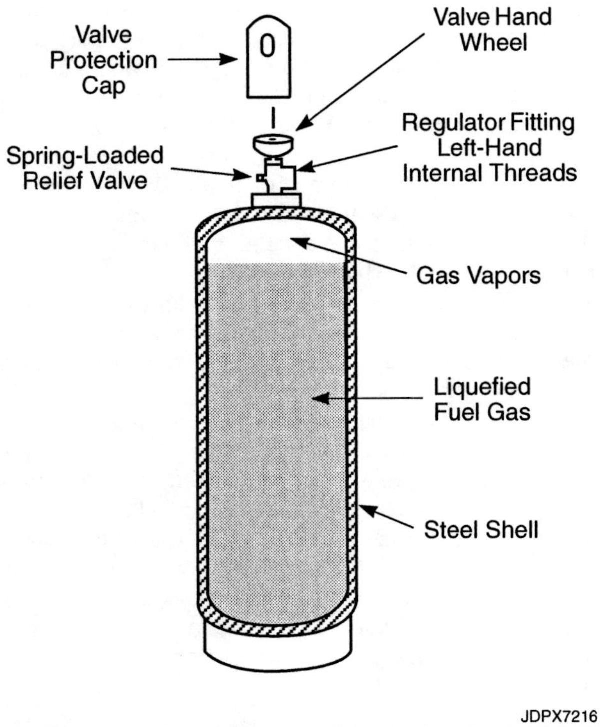

Fig. 8 — Liquefied Fuel Gas Cylinder and Contents

Like oxygen and acetylene cylinders exposed to above normal temperatures, the contents of a fuel gas cylinder expand to create unsafe pressures when in high-temperature places (like a fire). If this happens, the cylinder will eventually rupture unless there is some type of safety device to keep the pressure at a safe level.

Fuel gas cylinder valves are equipped with spring-loaded relief valves preset to release pressure before it gets high enough to rupture the cylinder. When the internal pressure drops below the safety limit, the valve closes.

The contents of a fuel gas cylinder are measured by weight. Three common cylinder capacities are 30, 70, and 100 lb (13.6, 31.7, and 45.4 kg).

Fuel gases do not have the working pressure or withdrawal limits of acetylene and they are not shock sensitive. Like acetylene, fuel gas cylinders must be used in the upright position so that liquid does not get into the regulator. The regulator will not operate properly and there could be a serious accident if the gas gets out of control.

IMPORTANT: Always secure cylinders in a vertical position on a cart or a cylinder rack. When transporting, always secure cylinders to a suitable support with a chain or other holder.

Regulators

Gas regulators are needed to reduce high cylinder pressures to safe working pressures. They are attached to the cylinder valve with special fittings and thread patterns that prevent attaching fuel regulators to oxygen cylinders and vice versa. Gas flow from a cylinder is regulated by an adjusting screw. The tension on this screw must always be released before a cylinder valve is opened, otherwise the regulating mechanism and pressure gauges can be damaged by the sudden rush of high-pressure gas into the regulator. The external parts of a typical oxygen regulator are illustrated in Fig. 9.

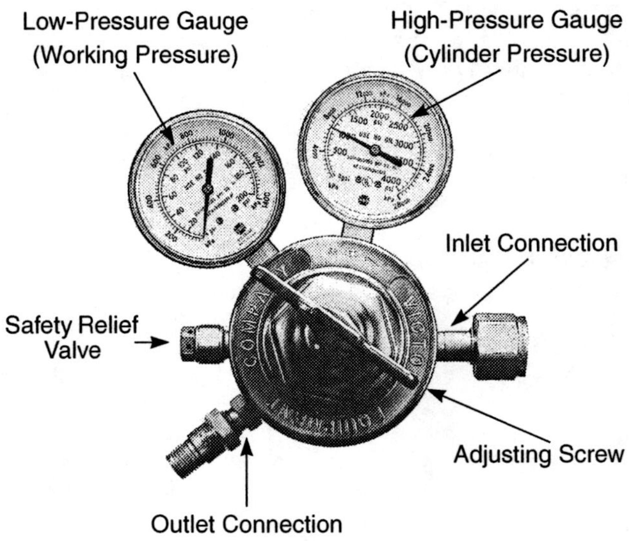

Fig. 9 — Typical Oxygen Regulator

Most regulators have two gauges. One gauge shows the actual cylinder pressure and the other shows the working or line pressure.

For example, an oxygen regulator must reduce a full cylinder's pressure of about 2200 psi (15,420 kPa) to less than 10 psi (70 kPa) for most welding operations. The regulator must also maintain an even working pressure as the cylinder's contents are used.

Regulators are classified into two groups according to the number of steps or stages used to reduce high cylinder pressures to safe working pressures.

The single-stage regulator illustrated in Fig. 10 reduces the gas pressure in one stage. The two-stage regulator illustrated in Fig. 11 reduces the gas pressure in two steps or stages.

Single-Stage Regulators

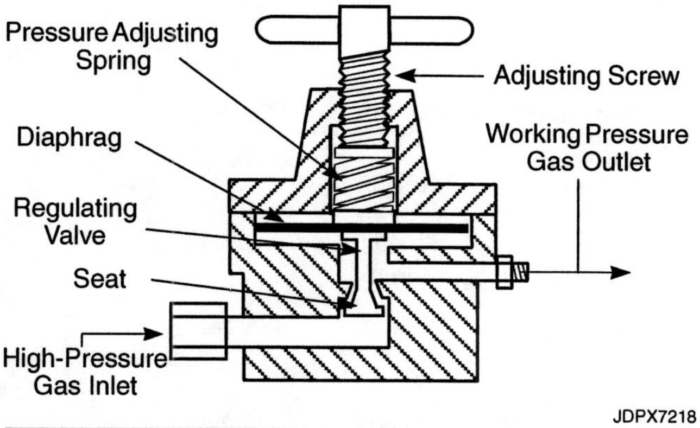

Fig. 10 — Simple Single-Stage Regulator

Single-stage regulators are the most common type of regulator found in general purpose shops. These regulators have a minimum number of working parts and restrictions. This means they are less expensive to own and maintain. It also means they can deliver a relatively high gas flow.

The only disadvantage of a single-stage regulator is the fact that a significant drop in cylinder pressure will cause a slight change in working pressure. This change in pressure will cause the flame to go out of adjustment.

For example, a cylinder pressure drop from 3000 psi (2070 kPa) to 400 psi (2760 kPa) will cause there to be a 3 psi/ 21 kPa to 5 psi (43 kPa) change in working pressure.

Single-stage regulators do an excellent job for most hand oxyfuel operations like heating, brazing, gas welding and cutting. In most situations the torch is not on continuously for a period of time for the drop in cylinder pressure to affect the flame. If the flame gets slightly out of adjustment, the operator will notice a change in the flame and make the necessary readjustment.

Two-Stage Regulators

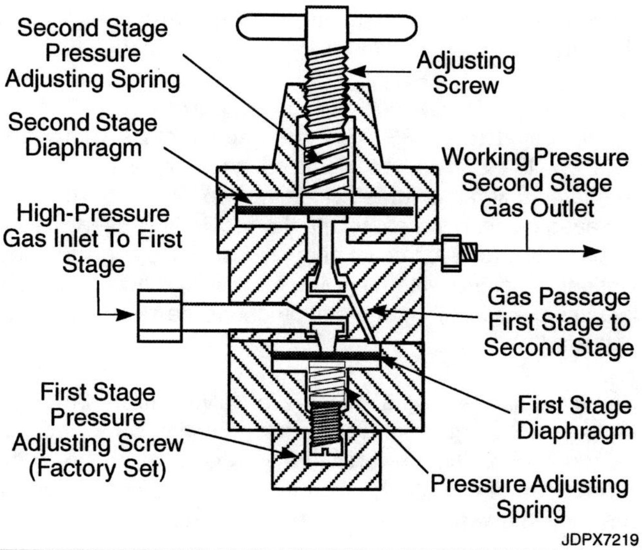

Fig. 11 — Simple Two-Stage Regulator

Two-stage regulators have a greater number of working parts and restrictions. This means they are more expensive to own and maintain. It also means they cannot deliver as high a gas flow as the same size single-stage regulator.

The main advantage of a two-stage regulator is its ability to provide a very constant working pressure as the cylinder's contents are used and the cylinder pressure drops. This is because the first stage is preset by the manufacturer to reduce the cylinder pressure to about 200 psi (1380 kPa) when it enters the second stage. The pressure delivered to the second stage is then reduced to the desired working pressure. The pressure entering the second stage remains fairly constant until the cylinder pressure drops below 200 psi (1380 kPa). The end result is a very consistent working pressure until almost all of the cylinder's contents are used.

Two-stage regulators are the best insurance for maintaining uniform results in mechanized oxyfuel operations. For a mechanized operation, the parts being produced could be ruined if the flame gets out of adjustment and the machine operator does not notice the change and make an immediate readjustment.

Two-stage regulators are typically found on multiple-station gas systems called manifolds and on silver brazing and cutting operations that are mechanized.

Regulator Creep

Regulator creep can be identified a few minutes after the torch valves are closed and the flame is extinguished. The reading on the working pressure gauge gradually rises or "creeps" to a pressure higher than the original setting. Regulator creep is a common problem with gas regulators that have not been handled or operated properly.

When a regulator first begins to creep, the pressure increase is only slight (5 psi/35 kPa) and usually stops within five minutes. If creep is not identified as a problem and the regulator taken out of service, the working pressure will increase a little more each time the regulator is used. Within a few weeks (or even days) of operation in this condition, the working pressure, with the torch off, will increase until there is a complete failure in the equipment or the safety device is activated.

Some regulators have a safety relief valve similar to the one in Fig. 9. On equipment not equipped with some type of safety device, the low-pressure gauge can explode, the diaphragm can rupture, and/or the gas hose can burst, resulting in a serious accident from regulator creep.

The regulator malfunctions because there is something preventing the regulator valve from making a tight seal on the seat as indicated in Fig. 12.

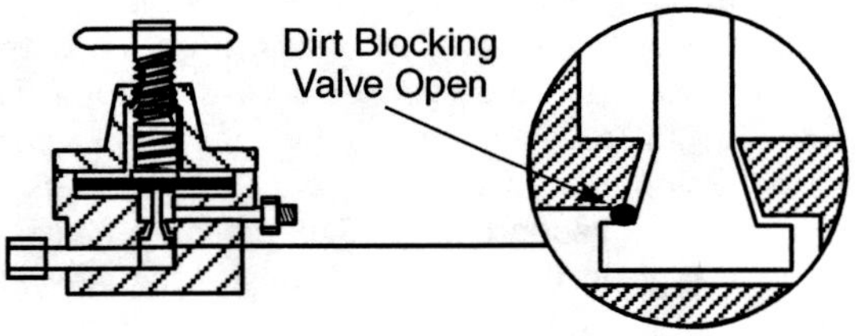

Fig. 12 — Regulator Valve Blocked Open

The two most common causes for the valve not making a tight seal are foreign material (dirt) getting inside the regulator, or the hard impacting damage of the valve against the seat when the cylinder valve is opened without releasing the tension on the regulator spring and diaphragm.

Regulator creep (Fig. 13) should not be confused with the slight change (1–3 psi/7–21 kPa) in the working pressure reading (Fig. 14) when the torch is turned on and off.

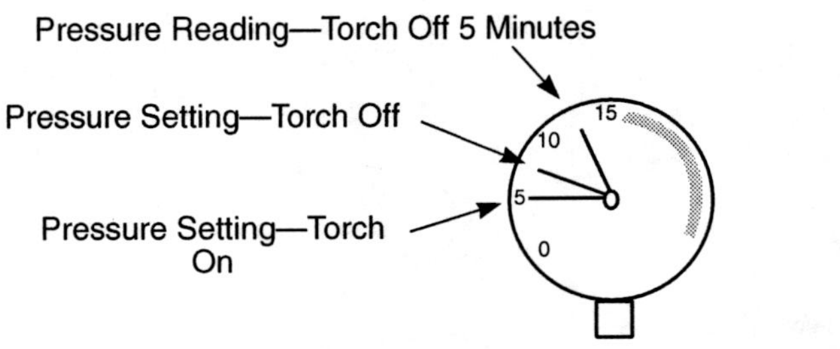

Fig. 13 — Fuel Gauge Showing Pressure Creep

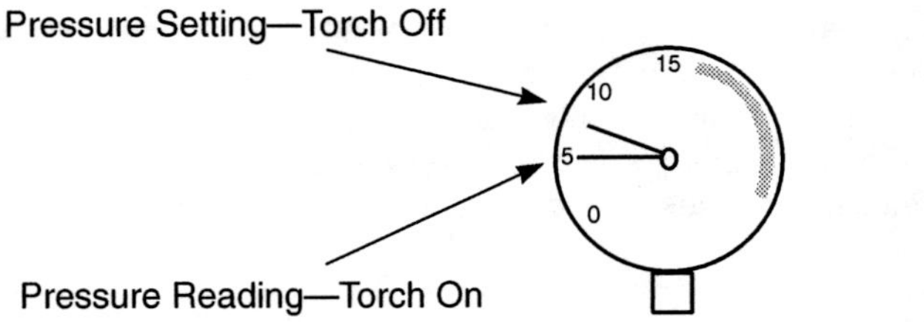

Fig. 14 — Fuel Gauge Showing Normal Pressure Fluctuation

Oxygen and fuel gas pressures should be set according to the equipment manufacturer's recommendations for the type of torch, tips, hose size, and hose length for each application like brazing, cutting, heating, or welding. Using the incorrect pressures could result in damage to the equipment and cause a serious accident like a flashback, which is explained later in this section.

Torches

The essential elements of an oxyfuel torch are a mixing chamber, where the two gases are brought together and mixed; two control valves, which control the flow of oxygen and fuel gas; and hose connections. The torch in Fig. 15 is also factory-equipped with internal reverse-flow check valves. (A complete explanation for these valves is provided later under "Reverse-Flow Check Valves.")

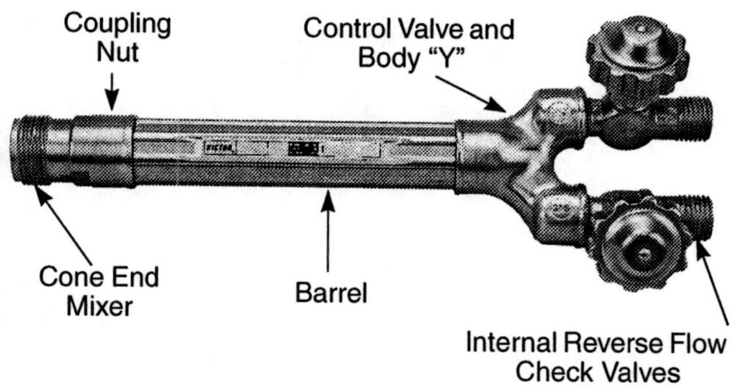

Fig. 15 — Parts of Typical Oxyfuel Torch Body and Welding Nozzle

NOTE: Not all oxyfuel equipment manufacturers build these check valves into the torches.

Tips Used with Torches

The last piece attached to the group of parts commonly called "the torch" is a copper part called a tip. A passageway inside the tip carries gases from the mixer to the outside opening at the end where the flame burns. The flame formed by the burning gases has a variety of applications and includes welding, brazing, soldering, heating, and hardfacing.

Tips are made with different-sized openings or "orifices" for a range of job applications. Manufacturers of welding equipment provide charts for matching tip size to metal thickness and for setting the proper pressures.

Other than the tips used for cutting, which will be covered in another section, tips can be divided into a group that has a single orifice and a group that has multiple orifices. Tips in the first group are commonly used for welding, brazing, soldering, hardfacing, and for heating small pieces of metal. These tips are called "welding" tips even though they are used for other applications.

Most of these tips can also be used interchangeably between acetylene and a group of other fuel gases that include MAPP® gas and propylene-based gases. Because these other fuel gases are more dense than acetylene, it typically takes a tip one or two sizes larger to obtain the same size flame.

Tips made especially for heating large pieces of metal are commonly called "rosebuds" and are illustrated in Fig. 16. They are classified into two groups: one-piece tips for acetylene and two-piece tips for fuel gases.

NOTE: Fuel gas heating tips should never be used with acetylene.

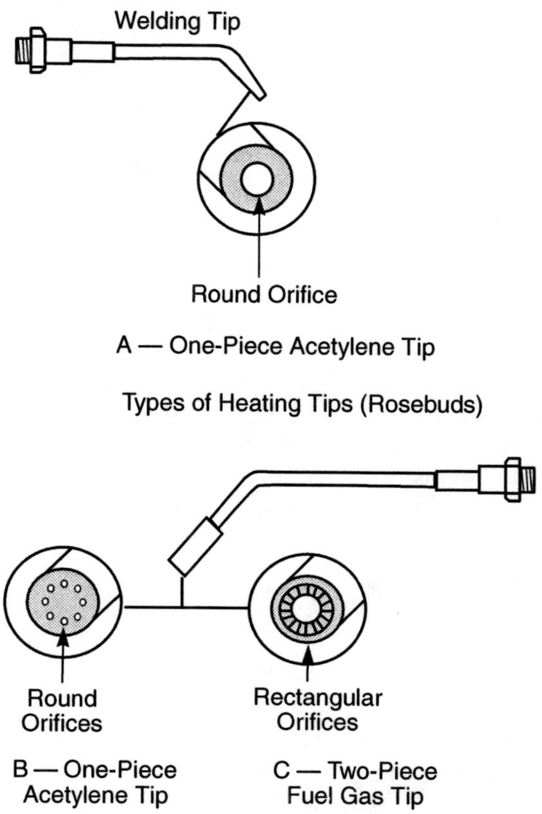

Fig. 16 — Types of Welding and Heating Tips

Tip Cleaning Procedures

Even when used properly, the end of a tip needs periodic cleaning because the orifice gets partially blocked or dirty from small bits of hot metal coming from the work. When this happens, the flame does not concentrate the heat in one area properly and can cause the job at hand to be ruined. Therefore, it is important for a person using the torch to immediately recognize the flame produced from a dirty tip.

A flame coming from a dirty tip does not have a uniform shape. Fig. 17 illustrates an evenly-shaped flame (A) from a clean tip and an irregularly-shaped flame (B) from a dirty or damaged tip.

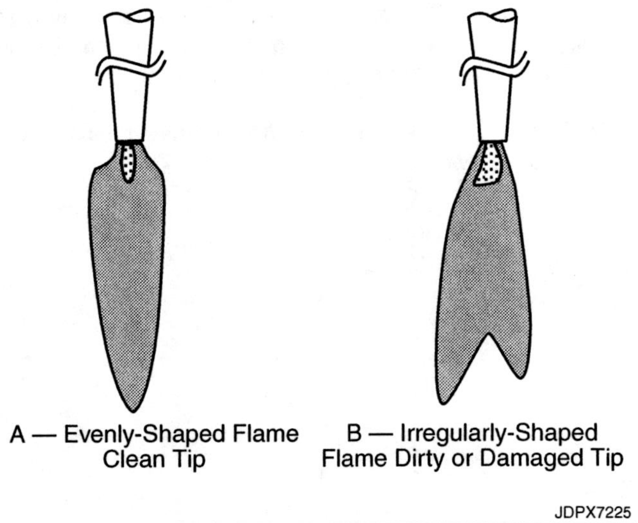

Fig. 17 — Flames from Clean and Dirty or Damaged Tips

A special set of tools called tip cleaners (Fig. 18) should be used to clean a dirty tip. The flat file portion of the set should be used to remove the accumulation of material on the outside of the tip as illustrated in Fig. 19. The round orifice cleaners should be used to clean the inside of the tip as illustrated in Fig. 20.

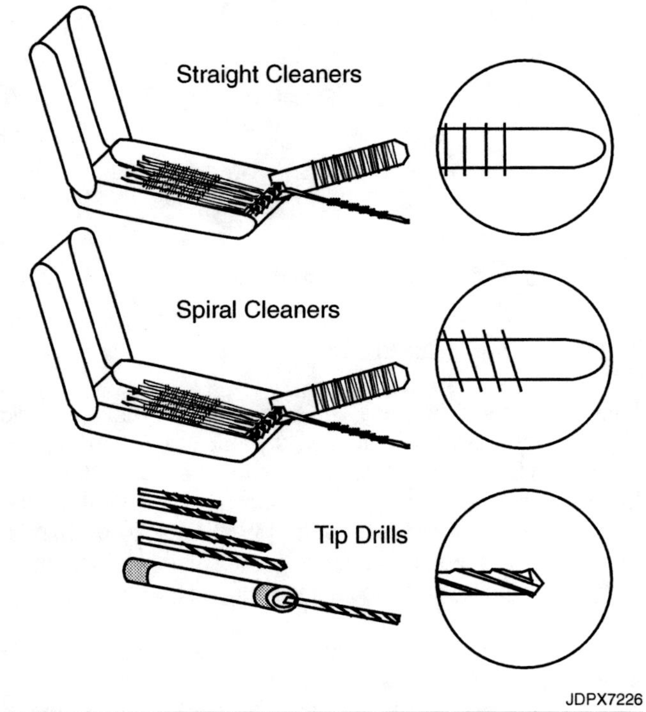

Fig. 18 — Tip Cleaning Tools

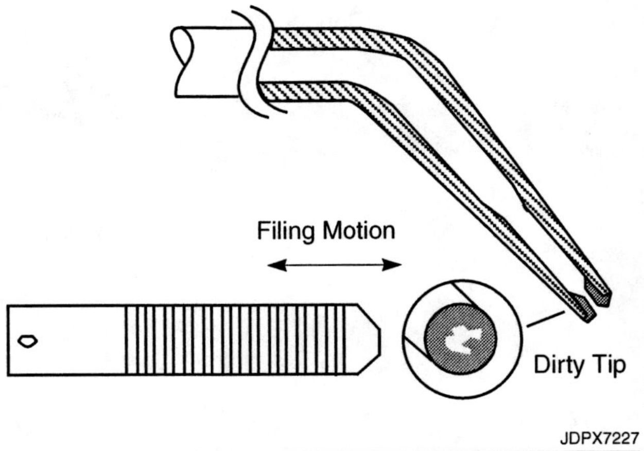

Fig. 19 — Using the Tip File to Clean a Tip

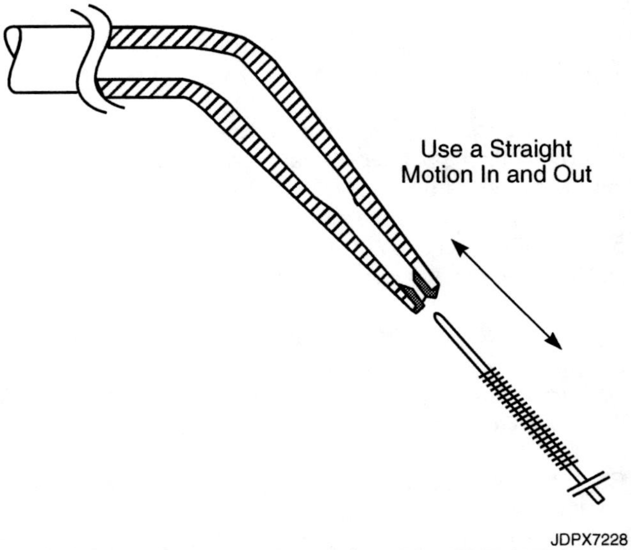

Fig. 20 — Using a Round Orifice Cleaner

Start with the smallest size that will easily go into the opening. Gently use a straight motion to move the cleaner in and out of the orifice. Work up one size at a time until the orifice cleaner size matches the orifice opening as in Fig. 21.

NOTE: *Be careful when cleaning the tip, the orifice cleaners can lodge in the tip and/or break off inside the tip very easily.*

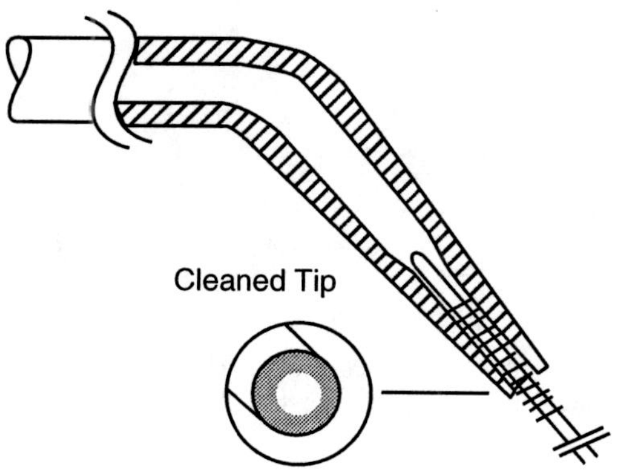

Fig. 21 — Round Orifice Cleaner Matches Tip Size

Using side pressure instead of the straight in-and-out motion will damage the tip by making the outside of the orifice larger than the inside. This condition is called a bell mouthed tip (Fig. 22) and makes it very difficult to adjust the flame so it doesn't keep popping during use.

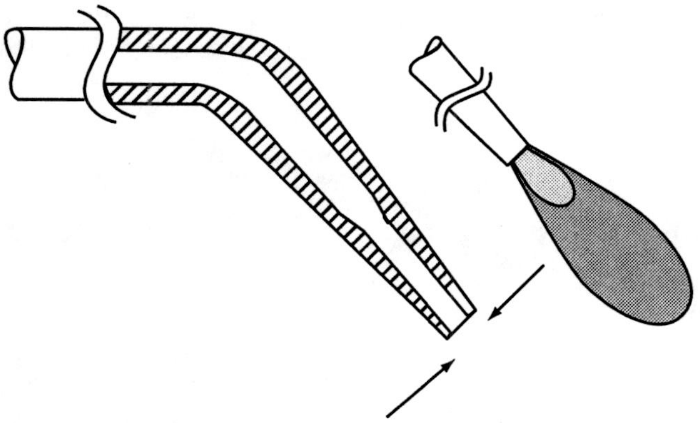

Fig. 22 — Damaged Bell Mouthed Tip and Flame

Care and Cleaning of Tips

Tips are subjected to mechanical wear, flame erosion, and plugging. Any of the tools in Fig. 18 can be used to restore a dirty or damaged tip to a usable condition.

IMPORTANT: Use the cleaning tools very carefully; they can lodge in the tip or break off inside very easily.

Recommendations for the proper care of torch tips:

1. Inspect threads on threaded tips periodically. (Never use pliers; use only box-end wrenches.)
2. Never try to remove a hot tip from its torch tube. Allow it to cool first. Also, do not install a cold tip onto a hot torch tube.
3. Use a sized tip cleaner to unplug tip orifices (Fig. 21). When welding, molten metal droplets may be thrown into the tip orifices. Also, carbon soot may deposit in the tip. A good sign that the tip is clogged is difficulty in getting a neutral flame.
4. Check tips for flame erosion periodically. Always use tips made for a particular torch.
5. On tips with gasket seals, tighten only hand tight. A wrench is not recommended.
6. Avoid dropping a tip as the seat which seals the joint may be damaged.
7. Try to avoid contacting the torch tip with the welding work, the bench, or firebricks. This may damage the end of the tip and cause the flame to burn with a "fishtail."

Check Valves and Arrestors

Backfire and Flashback

A backfire is the name given to loud snap or popping sound when the torch goes out. When this happens, close the torch valves, check the connections and operation, and then relight using a friction striker.

CAUTION: Do not try to relight the flame from hot metal — it is very dangerous.

Always try to find the cause if the torch repeatedly backfires. A backfire may be caused by touching the tip against the work, by overheating the tip, by operating the torch at improper pressures, by a loose tip or head, or by dirt on the seat.

If the flame burns back inside the tip and/or torch body it is called a flashback and the oxygen torch valve should be closed as quickly as possible. Then close the acetylene valve. A flashback can occur when the flame burns back inside the tip and/or torch body, usually with a shrill hissing, squealing sound. Closing the torch oxygen valve stops the flashback at once.

With the oxygen and acetylene regulators closed, let the torch cool before relighting the flame. Also, blow oxygen through the torch tip for a few seconds to clear out soot that may have accumulated in the passages.

Flashbacks may extend back into the hose or regulators. When a flashback occurs, it means that something is radically wrong with the torch or with the way it is being operated. In any case, it should be investigated to determine the cause before relighting the torch. A clogged orifice or incorrect oxygen or acetylene pressures are often responsible. Only use gas pressures within the range recommend by the equipment manufacturer.

Reverse-Flow Check Valves

When installed and maintained properly, reverse-flow check valves (commonly called check valves) allow forward flow of gas and close if conditions are created that try to make the gas flow in the wrong direction.

Conditions that can cause reverse flow in units without check valves:

1. Blocked tip.
2. Excess fuel gas or oxygen pressure.
3. A lack of pressure when one cylinder becomes empty.
4. Unsafe start-up and/or shut-down procedures.

A reverse gas flow can cause the mixing of the fuel gas and oxygen in the torch body and/or other parts of the equipment besides the mixer and the tip. If the mixed gases inside any part of the torch (including the tip and mixer) are ignited because of poor operating procedures and/or poorly maintained equipment, it will cause a flashback.

The operation of a simple reverse-flow check valve is illustrated in Fig. 23. When gas pressure on the inlet side of the disk overcomes the spring pressure on the outlet side of the disk, gas flows through the valve. The reverse-flow check valve is designed to close when the gas tries to flow in the wrong or opposite direction. Reverse-flow check valves only prevent the reverse flow of gas and do not stop a flashback.

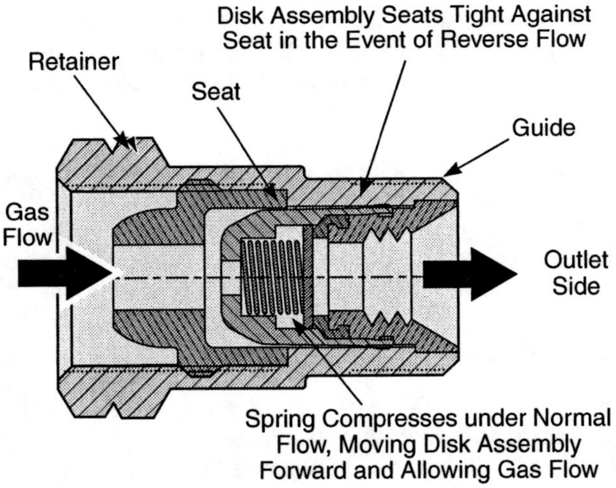

JDPX7231

Fig. 23 — Simple Reverse-Flow Check Valve

Flashback Arrestor

A flashback arrestor is a device that can be installed on oxyfuel equipment to stop a flashback from traveling past the point of installation. The simple flashback arrestor illustrated in Fig. 24 is constructed with a sintered filter that stops and extinguishes a flame. The sintered filter contains very small openings that allow the flow of gas but stop and extinguish a flame.

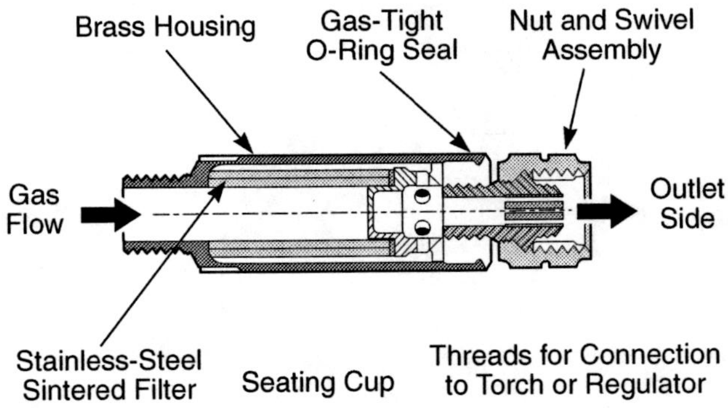

JDPX7232

Fig. 24 — Simple Sintered Flashback Arrestor

Installation Procedures and Precautions

Accessory reverse-flow check valves and flashback arrestors can be installed at the gas hose-to-torch connections and/or at the regulator outlet hose connections of both gases (Fig. 25). The flow capacity rating of these devices should match the equipment manufacturer's recommendations for operation. Installing units of insufficient capacity can also create a safety hazard. Flash arrestors only stop a flashback. They do not prevent the reverse flow of gas.

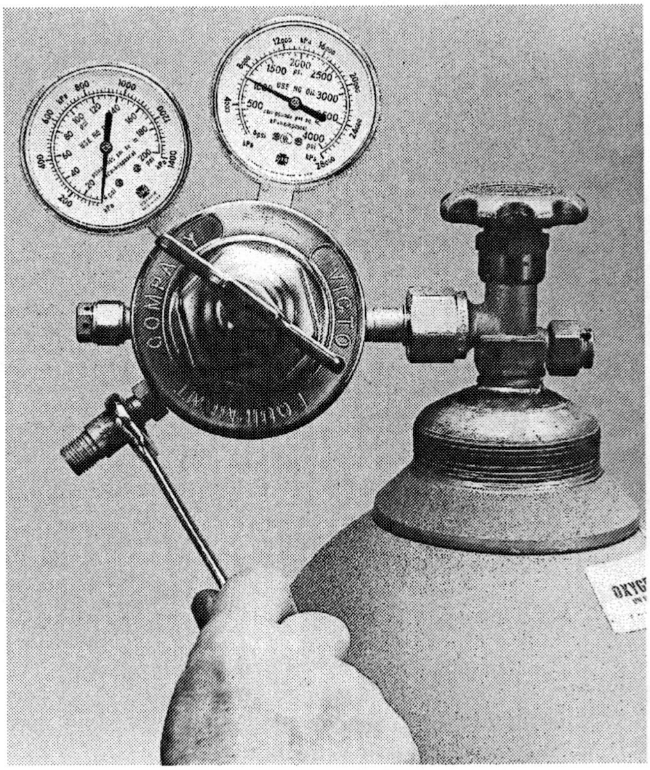

JDPX7233

Fig. 25 — Attaching an Accessory Reverse-Flow Check Valve

Remember:

1. Reverse-flow check valves only prevent the reverse flow of gas and do not stop a flashback!
2. Reverse-flow check valves must be installed and checked according to the valve manufacturer's recommendations.
3. Flash arrestors only stop a flashback. They do not prevent the reverse flow of gas.
4. Reverse-flow check valves and flash arrestors do not replace the need for good safety practices and the proper maintenance of oxyfuel equipment.

Hose and Connections

Special non-porous hose is used to convey oxygen and fuel gases, including acetylene, from the cylinders to the torch. To prevent a mix-up, the oxygen hose is green and the fuel gas hose is red. Also, the hose fittings are different. Fittings on the fuel gas hose have left-hand threads and grooved nuts, and fittings on the oxygen hose have right-hand threads and nuts without grooves (Fig. 26).

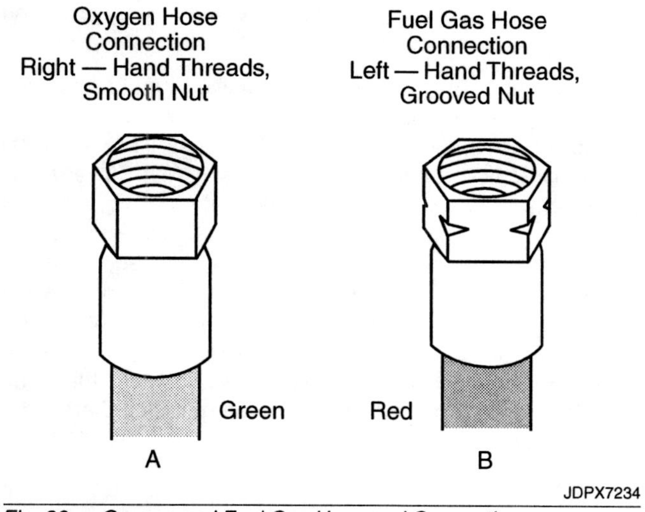

Fig. 26 — Oxygen and Fuel Gas Hose and Connections

Not all welding hose is the same. There are several sizes and grades with different compositions to consider when making a purchase.

When replacing or extending gas hoses be sure to get a hose with a diameter large enough to supply the torch with adequate gas flow and pressure requirements for the distance the gas has to travel. Using a hose with too small a diameter can make it difficult to get the proper flow to the torch. Manufacturers of gas hose provide charts for determining the proper hose size for a given distance to provide adequate delivery pressure and flow.

Grade "R" hose is designed for use with acetylene only. Grade "T" is designed for use with all fuel gases, including acetylene. Hose used for a fuel gas that it was not designed for can deteriorate on the inside and plug passages to and including the torch.

Inspect and maintain gas hoses on a regular basis. Replace hoses that are showing deep weather cracks and repair damaged spots properly on otherwise good hose.

CAUTION: Don't make repairs using electrical tape and hose clamps — it is dangerous.

Welder's Clothing

Persons using and persons observing the use of oxyfuel equipment should wear safety gear and clothing that protects their eyes, ears, and body from potential hazards. Goggles with a dark filter lens or a dark face shield should be worn to protect everyone's eyes from the glare of the flame, from heat (infrared rays), and from bits of flying hot metal and sparks.

A shade 4 or 5 lens is recommended (ANSI Safety Standard Z49.1) for brazing, heating, cutting (to 6 inch/152 mm), and for light to medium welding (to 1/2 inch/12.7 mm). Each person must choose a shade of lens that is not too dark to see the work area and not too light to cause eye strain. Sun glasses are not dark enough to watch oxyfuel operations.

Fire-resistant gloves should also be used for welding. Heavy clothing or aprons are also suggested for skin and body protection from sparks and hot metal.

Steps in Setting Up an Oxyfuel System

CAUTION: Before setting up an oxyfuel system, extinguish open flames that could ignite the fuel gas or be accelerated from oxygen. (This includes smoking materials such as cigarettes.) Also, stop other nearby operations such as grinding that could be a possible source for igniting fuel gas when the cylinder is "cracked" in step 3. There should also be adequate ventilation for breathing and to prevent the accumulation of an explosive gas mixture.

1. Secure Cylinders:

   Place the oxygen and fuel gas cylinders where they are to be used and secure them in the upright position.

IMPORTANT: Always secure cylinders in a vertical position on a cart or in a cylinder rack. When transporting them, always secure the cylinders to a suitable support with a chain or other holder.

2. Remove Cylinder Caps:

   Unscrew the valve protection caps. Put the caps away as they must be replaced on the empty cylinder.

3. Crack Cylinder Valves:

   Slightly open (called cracking) the oxygen and fuel gas cylinder valves to make sure they do not stick, and blow out any dirt or dust lodged in the valve. Close cylinder valve.

CAUTION: Welding cylinders contain gas under high pressure. If damaged, a cylinder can explode. Since gas cylinders are normally part of the welding process, be sure to treat them carefully.

CAUTION: Do not stand directly in front of the valve; stand behind (Fig. 27) or to one side.

NOTE: *Never use force on a cylinder valve to either open or close it. If the valve will not open or close by hand, notify the supplier and have the cylinder replaced.*

JDPX7235

Fig. 27 — Stand in Back when Cracking a Cylinder Valve

4. Connect Regulators to Cylinder Valves:

IMPORTANT: Some old style acetylene regulators and cylinder valves may have right-hand threads.

NOTE: *Before connecting regulators, examine the cylinder valve and regulator threads for damage (Fig. 28). If damaged, do not use. Instead, replace parts. Connect regulator to oxygen cylinder valve using the proper wrench. Do not use excessive force, but tighten the regulator nut firmly. Connect regulator to acetylene cylinder valve using the proper wrench. Again, do not use excessive force, but tighten the regulator nut firmly.*

JDPX7236

Fig. 28 — Examine Threads Before Making Connections

5. Open Cylinder Valves:

IMPORTANT: Be sure the regulator adjusting screw is in the out position. The adjusting screw is turned counterclockwise to release the tension.

   Open the oxygen cylinder valve slowly so the high-pressure regulator gauge needle will move up slowly until approximately 2000 psi (13,800 kPa) pressure is registered (if the cylinder is full). Never stand directly in front or in back of a regulator when opening an oxygen cylinder valve. Always stand to the side of the cylinder valve. After full pressure is reached, open the valve completely, seating the valve open. Opening the fuel gas cylinder valve only one-half to three-fourths of a turn will normally allow an adequate flow. For safety, never open the fuel gas cylinder valve more than one turn — it may need to be turned off quickly if there is an emergency.

6. Connect Hoses to Regulators:

   Connect oxygen hose (green) to the outlet of the oxygen regulator. The oxygen hose nut has right-hand threads. To purge the hose, turn the pressure adjusting screw on the regulator clockwise until a reading of 5 psi (35 kPa) shows on the low-pressure gauge. Allow the oxygen to escape until you are sure the hose is clean on the inside. When clean, turn the adjusting screw counterclockwise until working pressure reads 0 (zero). New hoses may have talcum powder in them to protect the hose lining while in storage. Dust, dirt, and especially talcum powder, if left in the hose, may enter the small gas passages in the torch and plug them or cause a flashback. Connect the fuel hose (red) to the fuel gas regulator outlet. The fuel nut has left-hand threads, with a notch in the connecting nut. Blow out the hose in the same way as you did the oxygen hose.

   Remember that fuel gas will burn. Keep it away from open flames as you are blowing out the hose.

7. Connect Hoses to Torch Body:

   Connect the free end of the hoses to the torch body. The green hose connects to the oxygen. The oxygen connecting nut will have right-hand threads. The fuel gas connection will have a left-hand thread with a notch in the connecting nut and torch body connection. The torch body may have "Oxy" and "Fuel" marked on it.

8. Connect Welding Tip to Torch body:

   Select the tip size suitable for the job you are to do. Refer to the welding tip selection charts furnished by the manufacturers. Tighten the welding tip into the torch. Be sure that seating rings are in place. Tighten the tip only hand tight.

9. Set Working Pressures:

   Partially open the torch body oxygen valve by turning it in a counterclockwise direction. Adjust the oxygen regulator until the working pressure corresponds to that listed for the tip selected (Fig. 29). Close the torch body oxygen valve by turning clockwise. Repeat the procedure for adjusting the fuel gas regulator working pressure.

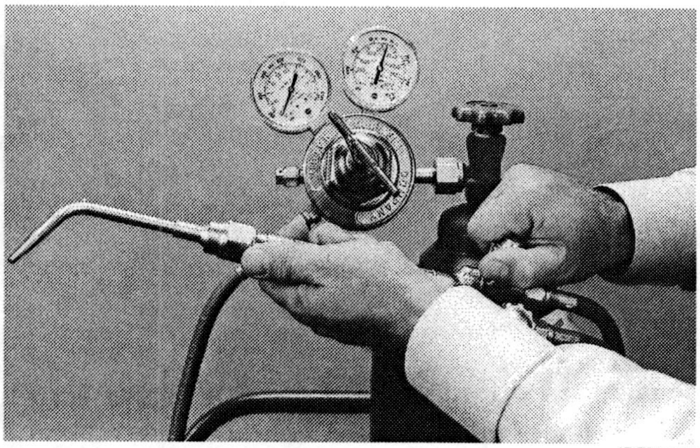

JDPX7237

Fig. 29 — Setting the Working Pressure

NOTE: As mentioned earlier in the section on regulators, observe the amount of rise in the pressure on the working pressures gauges immediately after the torch body valves are closed. This slight change in pressure is the reason for setting the working pressure with each torch valve open.

After completing step 9, check the system for leaks.

Checking the System for Leaks

CAUTION: Before beginning a system check for leaks, extinguish open flames that could ignite the fuel gas or be accelerated from oxygen. (This includes smoking materials such as cigarettes.) Also, stop other nearby operations such as grinding that could be a possible source for igniting fuel gas from a leak.

A simple way to begin the process of checking for leaks is to pressurize the system just as if preparing to do a project. Next, close both cylinder valves and watch each cylinder pressure gauge for at least 2 minutes. Also, listen for any whispering or hissing sounds that could indicate a leak. If the cylinder pressures remain exactly the same after 2 minutes with the valves closed, there are no leaks and the system is ready for use.

If the pressure drops the slightest amount after 2 minutes, back the tension off the regulator adjusting screws by turning them in a counterclockwise direction. Then watch both the cylinder pressure and working pressure gauges again for more pressure drop. If the pressure drops on the cylinder pressure gauge, there is a leak between the cylinder and the the regulator. If the pressure drops on the working pressure gauge, there is a leak between the regulator and the end of the tip on the torch.

Leaking torch connections, torch valves that don't seat, crimped hose connections that leak, and leaks coming from the regulators should be taken for repair to a qualified oxyfuel repair and service center for the equipment you are using.

Repeat the procedures for "Checking the System for Leaks" and for "Stopping Leaks" until the system is leak-free. Also, in the future, use these procedures at the beginning of each day to locate and correct leaks instead of waiting for an accident to happen.

Locating Leaks

Key points to check for leaks in the system:

1. Cylinder valve stems
2. Cylinder regulator connections
3. Regulator hose connections
4. Hose to torch body connections
5. Torch body welding tip connections
6. Torch body valves (check at tip orifice)

Return the system to full pressure by reopening the cylinder valves slowly in a counterclockwise direction and then turning the pressure adjusting screws clockwise to restore working pressure on the hoses. With the equipment under normal operating pressures, begin locating leaks by covering each connection (Fig. 30), from the cylinder to the end of the torch, with a leak detection liquid approved for use with oxygen. Gas bubbles will indicate the location of leaks. Mark each location that leaks and shut the system down before beginning to correct the leaks.

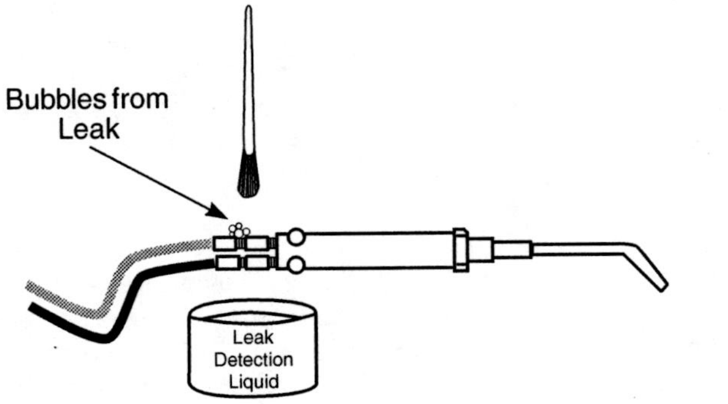

Fig. 30 — Locating a Leak

IMPORTANT: Soap or liquid detergents should not be used for making leak detecting solutions. Obtain leak detection solutions from a local welding gas supplier.

Stopping Leaks

To begin correcting the problems that are causing the leaks, first check the tightness of each threaded connection that leaks, including the packing nuts on the torch valves. If the connections seem to be tight, shut the system down.

Shutting Down the System

1. Close the cylinder valves in a clockwise direction.
2. Open the torch valves in a counterclockwise direction until the pressure is released (the high- and low-pressure gauge on each regulator reads "0").
3. Close the torch valves in a clockwise direction.
4. Release the tension on the regulator adjusting screws by turning them in a counterclockwise direction.

With the system shut down, loosen leaking connections and inspect them carefully for damage and/or foreign material such as dirt. Replace damaged fittings on anything but the cylinder valve connections. Return cylinders with damaged valve connections to the welding supplier for replacement. Clean the fittings carefully and protect them from more contamination until reattached. When repairing holes in the hose, only use splice connections approved for oxyfuel equipment.

Procedures for Lighting the Torch and Adjusting the Flame Properly

1. Hold the torch in one hand and the flint striker (also called a flint lighter and friction striker) in the other hand (Fig. 31).

CAUTION: Don't use matches; any type of lighter, including butane; or anything else that could make the procedure dangerous!

2. Point the torch away from other persons (including yourself), the cylinders, and any flammable materials. Open the torch fuel valve by turning it in a counterclockwise direction approximately 1/4 of a turn and ignite with a flint striker.

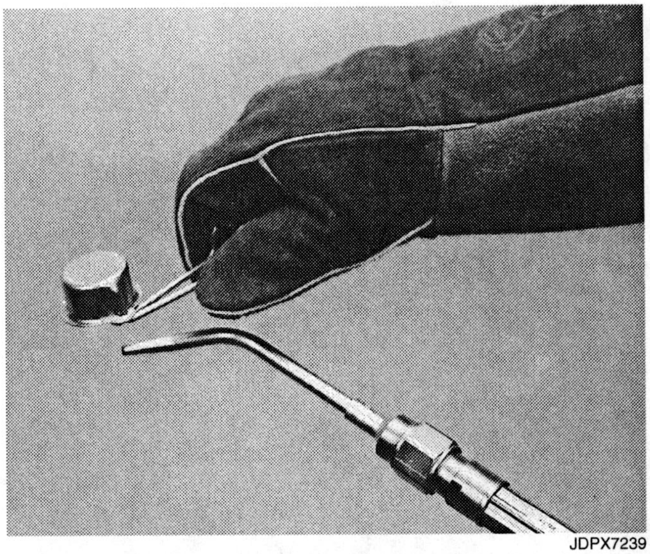

Fig. 31 — Lighting the Torch with a Flint Striker

3. Keep opening the torch fuel valve until there is no soot (Fig. 32), (Fig. 34A) and the flame has stopped excessive smoking. If the flame leaves the end of the tip (Fig. 33B), close the fuel valve by turning it in a clockwise direction until the flame returns to (touches) the end of the tip (Fig. 33A). If the flame won't stop smoking, the pressure is not correct for the tip or there is a significant pressure drop from the regulator to the torch. To correct the problem, increase working pressure or obtain a hose with a larger diameter that will provide adequate flow and pressure to the torch.

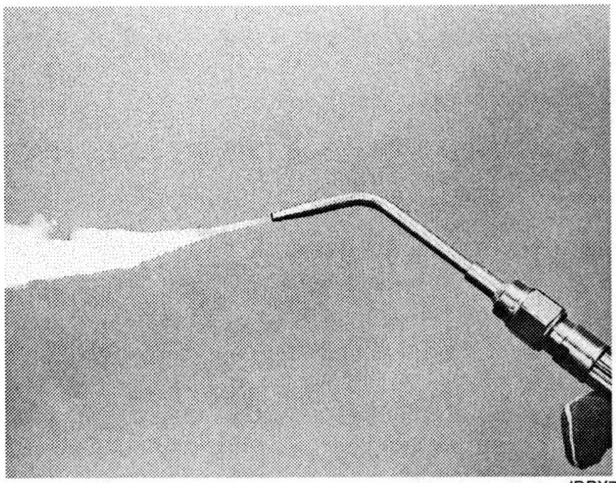

Fig. 32 — Setting the Flame

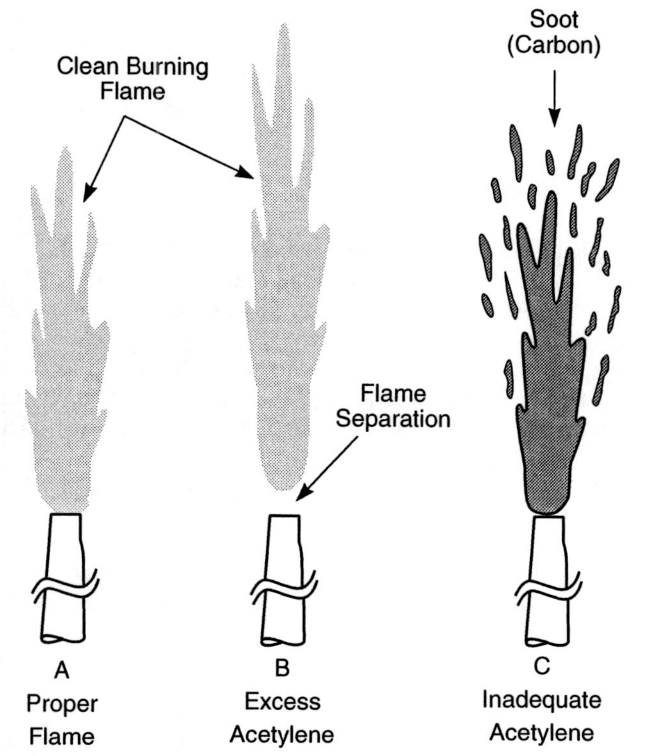

Fig. 33 — Adjusting the Flame Size

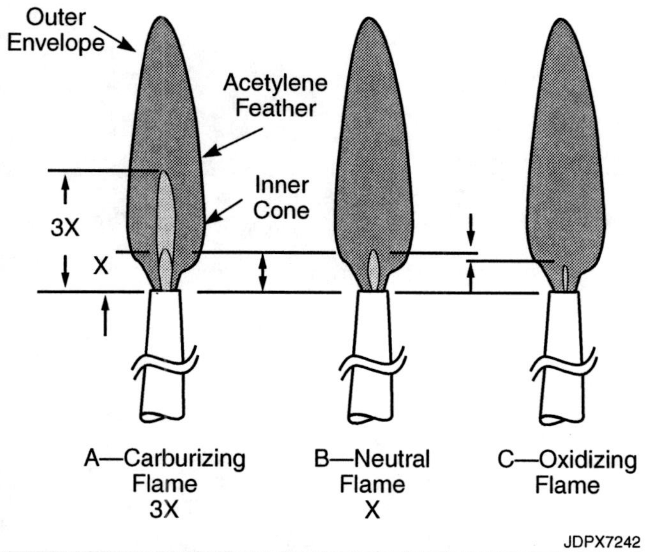

Fig. 34 — Types of Acetylene Flames

4. Open the torch oxygen valve by turning it counterclockwise until a bright inner cone and acetylene feather appears in the flame (Fig. 34A). The point at which the acetylene feather merges with or joins the inner cone is called a neutral flame. Adjust the torch oxygen valve back and forth until you can recognize each flame.

A Neutral Flame is used for brazing, heating, soldering, and cutting. The name "neutral" comes from the fact that there are equal amounts of oxygen and acetylene burning (combining), and there is no chemical activity taking place in the outer envelope of the flame. This neutral outer envelope provides a shield for the weld zone from the air in the same way the shielding gas protects the weld zone with either the GMAW (gas metal arc welding) or GTAW (gas tungsten arc welding) processes.

A Carburizing Flame is the first flame formed when the torch is lit (Fig. 34A). The acetylene feather is 3 times the length of the inner cone. The name carburizing comes from the fact not all of the carbon in the acetylene burns (combines) with the oxygen (there isn't enough oxygen being made available), and an acetylene feather or area of free carbon is formed in the flame. This free carbon will combine with very hot and molten steel to increase the carbon content enough to change the properties of the steel. A carburizing flame is used for hardfacing, welding aluminum, and sometimes soldering.

An Oxidizing Flame (Fig. 34C) is formed when oxygen is added to a neutral flame. This excess or free oxygen will combine with or oxidize carbon in liquid steel and make the steel weaker. An oxidizing flame is something that should be avoided for most applications. A slightly oxidizing flame can be used for brazing.

Tips for identifying the different types of flames:

1. The easiest way to distinguish the difference between a carburizing flame and an other type of flame is that the carburizing flame has an acetylene feather.
2. As oxygen is added to a carburizing flame to make it a neutral flame, the inner cone remains the same size. At the moment the inner cone becomes smaller, the flame is changing from neutral to oxidizing. The inner cone on an oxidizing flame is smaller and appears to be in sharper focus than the inner cone on a neutral flame.

Preparing to Weld

Before starting to weld, learn and follow the safety rules for welding and guidelines for metal preparation.

Safety

IMPORTANT: For complete safety rules, see Chapter 13, on Welding Safety. Observe all safety precautions when you are preparing to weld.

1. Be sure the work area has a concrete or masonry floor.
2. Keep all combustible materials at a safe distance (35 feet/11 Meters).

3. Do not use gloves or other clothing that contains oil or grease.
4. Keep an ABC-type fire extinguisher and a fire blanket within easy reach.

Metal Preparation

NOTE: Any foreign material which gets into the molten metal can weaken the weld. Clean the metal joints to be welded of all rust, scale, paint, dirt, and especially oil. Also be sure the metal is dry.

Weld joints that require maximum strength will need to have the edges beveled on thick pieces. Space will also need to be left between the pieces to ensure the entire depth of the joint melts and flows together. (Fig. 35) illustrates guidelines for beveling and spacing.

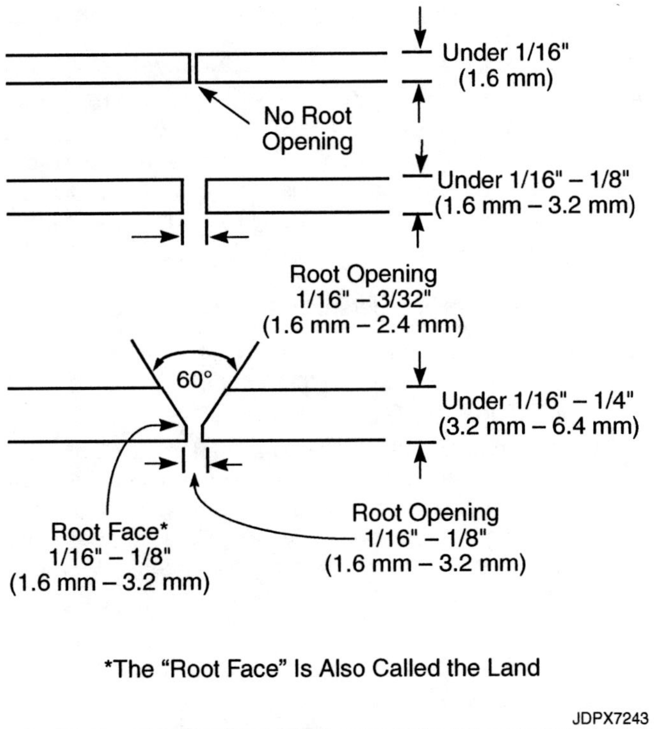

Fig. 35 — Beveling and Spacing Weld Joints

Oxyacetylene Welding (OAW)

Five key points controlling the weld puddle:

1. Heat (volume of heat per tip size)
2. Distance (inner cone from weld puddle and work)
3. Angles (work and travel angles of torch and rod)
4. Travel speed (movement along the metal)
5. Manipulation (movement of torch and rod)

The heat from a neutral flame positioned precisely between the two pieces causes the actual fusion of the metal by forming a liquid puddle as the torch is moved along the weld joint. Torch positioning, torch manipulation, filler metal application, and travel speed are the key actions which should be learned after the proper size tip has been selected according to the equipment manufacturer's recommendations.

Torch Positioning

The flame from the welding torch must be positioned the proper distance from the metal as well as maintained at the proper work angle and travel angles as illustrated in (Fig. 36). Welding can be accomplished by using either the forehand technique or backhand technique (Fig. 37). Melting of the metal occurs when the torch is at the low point of the vertical movement. The forehand technique is the most commonly used.

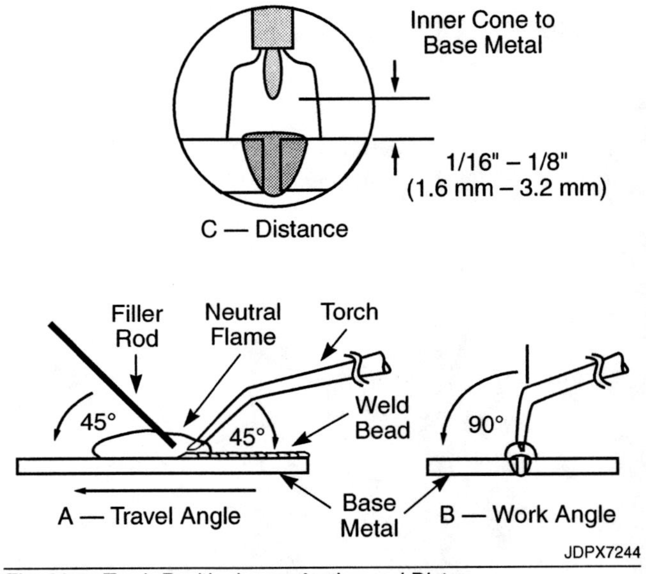

Fig. 36 — Torch Positioning — Angles and Distance

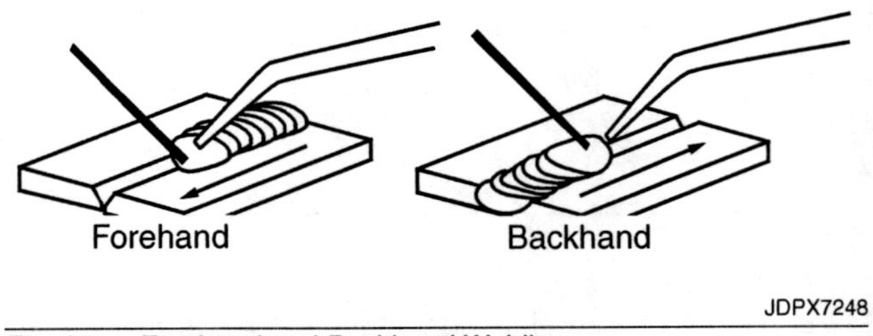

Fig. 37 — Forehand and Backhand Welding

Travel Speed

A travel speed must be maintained so that the edges of the metal melt at the front of the weld puddle to form a "keyhole" that indicates proper penetration and fusion. Movement along the weld is determined by the base metal thickness, tip size, and torch motion.

Torch Manipulation

A motion is necessary to make the flame form the desired size weld puddle (liquid metal) and to make the puddle advance along the edges of both pieces of metal to form a weld. Three examples of torch manipulations are illustrated in (Fig. 38). The "dipping" method minimizes the overheating of thin metal (1/16 inch/1.6 mm), reduces the tendency to melt holes in the metal, and creates the smallest bead size. No matter which manipulation is used, the filler metal must be melted in the liquid puddle on the base metal. If the filler rod is melted with the flame and/or there is not a liquid puddle on the base metal, the deposited metal will lie on top of the base instead of fusing or combining with it.

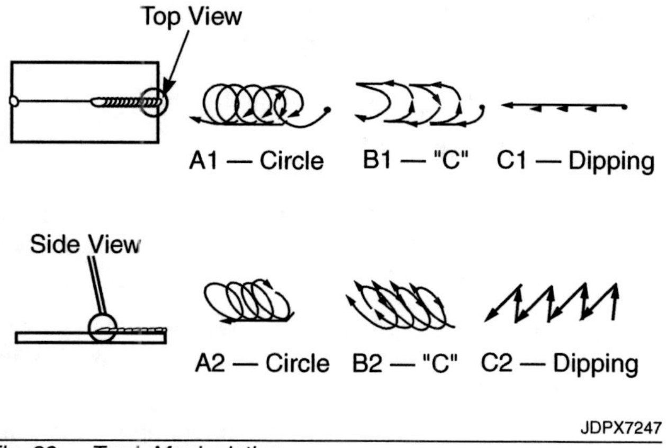

Fig. 38 — Torch Manipulation

Welding Technique for a Square Groove

When the weld is started, the first thing to look for is the forming of a keyhole (Fig. 39). After the keyhole is formed, dip the filler metal rod in the back of the keyhole (Fig. 39). Do not dip the rod directly in the keyhole or it will fall through. As you dip the rod in and out of the puddle, it will partly fill the keyhole and a new keyhole will be formed as the torch moves forward.

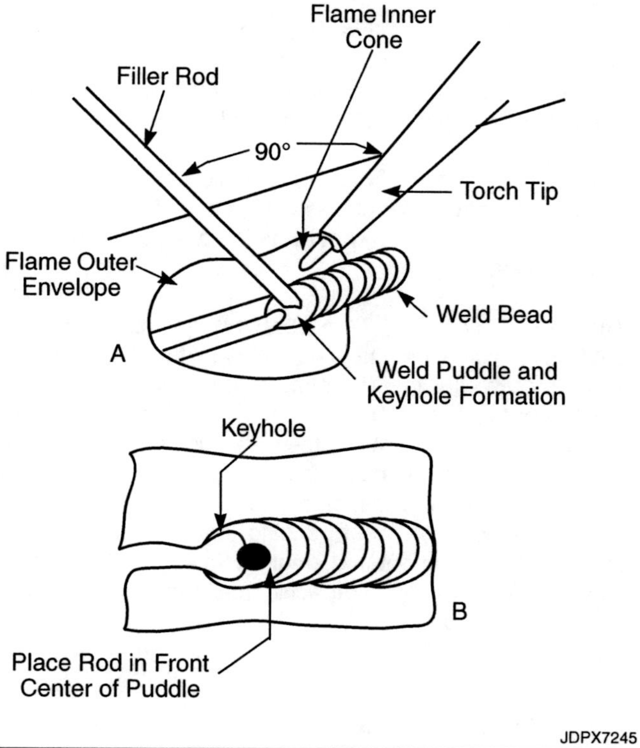

Fig. 39 — Forming a Keyhole and Making a Weld

Types of Filler Metal and Applications

The filler rod should match the base metal composition. Two common oxyacetylene filler metals have AWS (American Welding Society) classification RG60 and RG65. The rod is held at about the same angle as the torch, but slanted away from the torch as in (Fig. 40). It takes good hand/eye coordination to synchronize the torch/rod motions and a lot of practice to make welds that are strong and have a nice appearance.

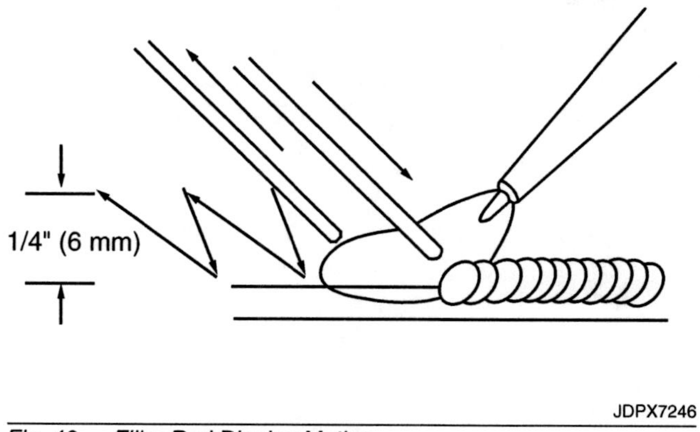

Fig. 40 — Filler Rod Dipping Motion

The diameter of the rod should not be much greater than the thickness of the base metal. The most popular diameters are 1/16 in., 3/32 in. and 1/8 in. (1.6 mm, 2.4 mm, and 3.2 mm). You can tell if the rod size is too large because the rod will keep freezing to the weld bead. (There is insufficient heat in the molten weld puddle to melt the rod.)

Another way to tell is if the weld puddle has to be made so hot to melt the rod, that holes melt in the metal. Welding with too small a rod will work, but it takes a lot of extra time to complete the weld.

The filler rod comes in 36 in. (0.91 m) lengths and is much easier to handle if it is cut in half. A slight bend can be made in the rod about 6 in. (152 mm) from the end. This permits the welder to position the hand to one side or the other from the most intense heat above the weld puddle. When the 6 in. (152 mm) section has been deposited, another bend can be made in the rod. The last small piece can be welded to the next new piece instead of bending the end. Also, the long end of rod can be bent in a small circle so it is less likely to cause an injury if someone accidentally runs into it or gets poked with the rod.

Welding Suggestions

1. Be sure the torch is centered over the groove and make both edges of the metal melt to form a keyhole.
2. Make the weld penetrate both sides of the joint completely.
3. Increase the travel angle to 90° to put more heat on the metal when starting a weld, and then decrease it to 45° for welding.
4. Dip the rod directly into the center of the puddle and try not to touch the sides or it will stick.
5. Hold the inner cone close to the metal when heating the puddle and do not move it too far away when dipping the rod (Fig. 40).
6. Do not melt the rod with the flame.
7. Keep the weld puddle small. The weld face should be 1/16 in. (1.6 mm) wider than the groove.
8. Decrease the travel angle to less than 45° to put less heat on the metal when completing a weld or when having trouble melting holes in the metal.

Acceptable Welds

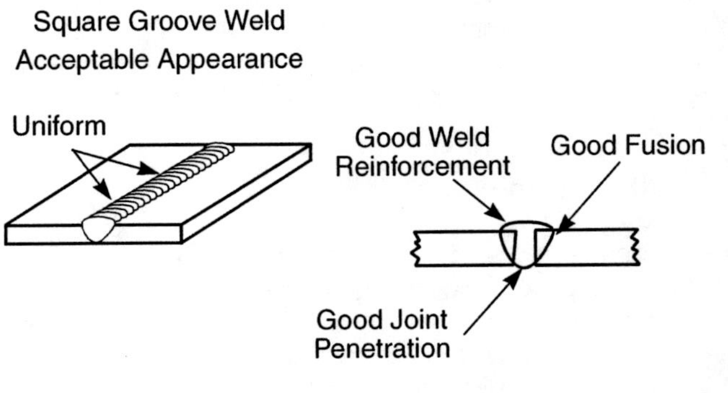

Fig. 41 — Acceptable Groove Weld on 1/8 inch (3.2 mm) Metal

To be acceptable, a groove weld must meet both visual inspection, structural strength requirements, and show evidence of good workmanship, including uniformity. Visual inspection includes measuring weld size, penetration of the weld joint, reinforcement above the surface, and depth of undercut. Weld strength and fusion is checked by bending the piece in the weld area. These requirements are specified by the welding code which applies to the function of the finished part. More detailed information is given in ANSI/AWS "Qualifying Welding Procedures and Welders."

Unacceptable Welds

(Fig. 42) illustrates common weld defects. Generally, a weld is considered unacceptable under any one of the following conditions:

1. The weld doesn't completely fill the weld joint because of a lack of penetration and/or lack of reinforcement above the top of metal.
2. Weld penetration and/or reinforcement above the top of metal is greater than 1/8 in. (3.2 mm).
3. Undercut is greater than 1/32 in. (0.83 mm).
4. The weld breaks or the base metal breaks in the weld area during a bend test.
5. The weld has not fused to the metal.

Troubleshooting Oxyacetylene Weld (OWA) Defects

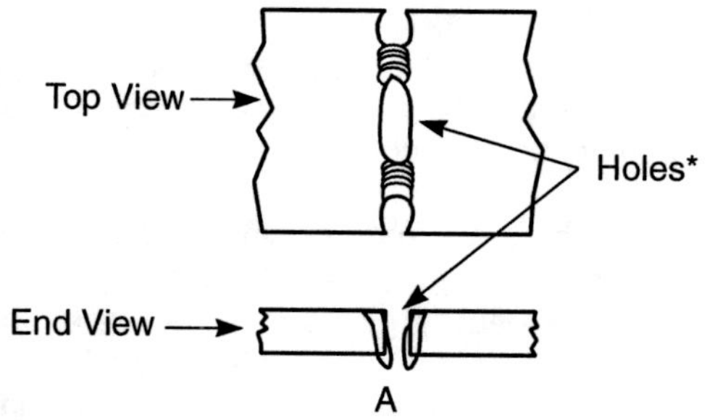

A - Intermittent Holes Instead of a Complete Weld
Corrective actions:

1. Increase travel speed
2. Use a smaller tip
3. Add more filler metal
4. Decrease travel angle to less than 45° from metal

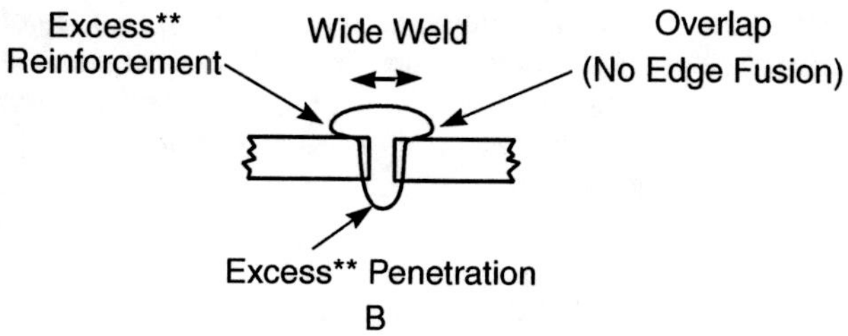

B - Oversized Weld, Excess Reinforcement and Penetration
Corrective actions:

1. Increase travel speed
2. Do not use as much filler metal

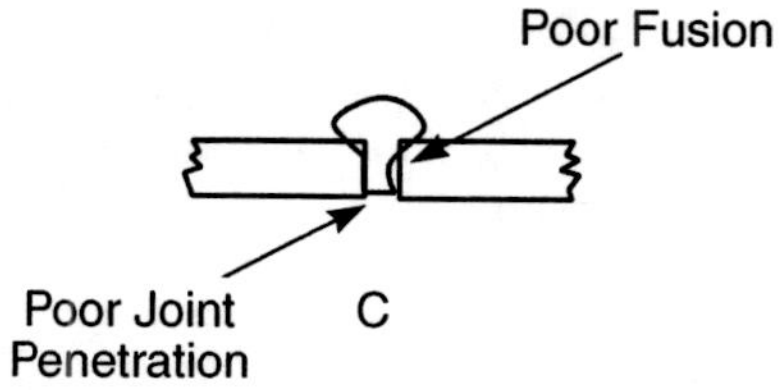

C - Poor Fusion and Penetration into Base Metal
Corrective actions:

1. Do not add filler metal until metal forms a puddle
2. Do not melt rod with flame; melt rod in puddle
3. Use a larger tip if the base metal will not melt

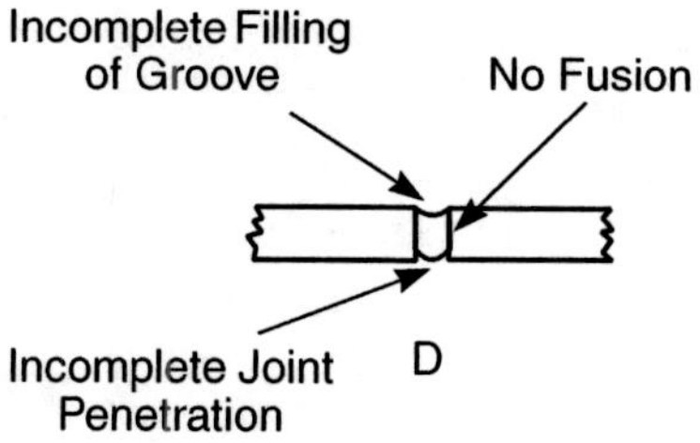

D - No Base Metal Fusion, Incomplete Joint Penetration and Incomplete Filling of Grooves
Corrective actions:

1. Do not melt rod with flame and then try to melt it in with the flame
2. Fill the groove by making a second pass
3. Add more filler metal
4. Form keyhole
5. Large tip size

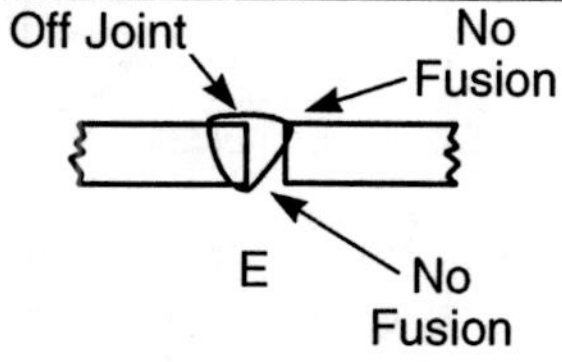

E - No Fusion on One Piece
Corrective actions:

1. Point torch work angle equal by on both pieces
2. Center the flame over both pieces

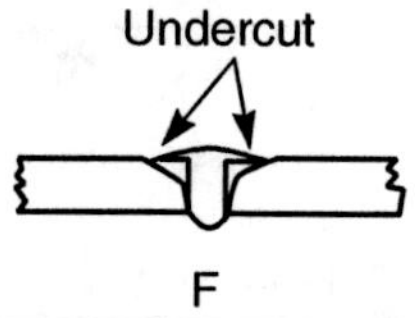

F - Undercut
Corrective actions:

1. Decrease tip size so the weld is not overheated
2. Decrease travel speed and/or add more filler metal

*Examples are for single-pass welds on 1/8 in. (3.2 mm) metal. Beveling is optional on this thickness, but it should be done on thicker metal.
**Over 1/8 in. (3.2 mm).

JDPX7250

Fig. 42 — Common Oxyacetylene Weld (OAW) Defects

Steps for Shutting Down the System

1. Extinguish the flame (shut off the torch).
2. Close the cylinder valves.
3. Release pressure by opening torch body valves (bleed both cylinder and working pressure gauges to zero).
4. Close torch body valves.
5. Release both oxygen and acetylene regulator pressure adjusting screws.

Shutting Off the Torch

To extinguish the flame, shut off the torch by closing the oxygen valve in a clockwise direction, and then close the acetylene valve. If the acetylene valve is closed first, the torch will make a loud popping noise. This pop can cause carbon to be pulled inside the torch and later interfere with the operation of the torch valve.

Turning off the oxygen first eliminates the pop and also provides a check of the operation of the torch valves. If a small flame continues to burn with a blue color with the oxygen valve turned off, the oxygen valve is not closing. If a small yellow flame continues to burn when the acetylene valve is turned off, the acetylene valve is not closing. If a small blue flame continues to burn after both valves are turned off, both valves are not closing. If either valve is not closing completely, the torch should be taken out of use and sent to a person qualified to service oxyfuel equipment.

Stopping Work

When work is to be stopped for considerable periods of time, such as during lunch hour or overnight, the cylinder valves should be closed and all gas pressure released from the regulators. Follow the steps listed below.

1. Close the oxygen cylinder valve by turning it in a clockwise direction.
2. Open the torch oxygen valve in a counterclockwise direction to release all pressure from the hose and regulator.
3. When both pressure gauges read "0," close the torch oxygen valve by turning it in a clockwise direction.
4. Turn out the pressure adjusting screw on the oxygen regulator by turning it in a counterclockwise direction until it turns freely.
5. Close the acetylene cylinder valve by turning it in a clockwise direction.
6. Open the torch acetylene valve in a counterclockwise direction to release all pressure from the hose and regulator.
7. When both pressure gauges read "0," close the torch acetylene valve by turning it in a clockwise direction.
8. Turn out the pressure adjusting screw on the acetylene regulator by turning it in a counterclockwise direction until it turns freely.
9. Check the high-pressure gauges after a few minutes (5 to 10) to be sure the cylinder valves are closed completely and are not leaking.

Closing the cylinder valve and then opening the torch valve relieves all pressures in the regulator and hose line. After the gauge readings (both gauges) have reached zero, the pressure adjusting screw should always be released, since this must be done before the cylinder valve is opened again.

The acetylene and oxygen pressures should not be released simultaneously, and care should be taken that the release of acetylene does not create a fire hazard. If acetylene is supplied through a hydraulic back-pressure valve, first simply close the service valve. After regulators are removed from cylinders, the dust plugs should be screwed in place on the cylinder connection nuts.

When regulators are to be out of service for several weeks or longer, it is good practice to turn the pressure adjusting screw in just enough to relieve the spring pressure on the valve seat. This aids in lengthening the life of the valve seat, especially when the seat is made of relatively soft material. Before putting the regulator into service again, turn the pressure adjusting screw out until tension is released.

Other Types of Fuel Gases

Fuel Gases	Specific Gravity (Air = 1)
Acetylene	0.91
MAPP® Gas	1.48
Propane	1.52
Propylene	1.48
Natural Gas	0.62
Hydrogen	0.07

Table 2 — Specific Gravities for Common Fuel Gases

The common fuel gases, their corresponding flame temperature, and specific gravity is given in Table 2. The gases having a specific gravity greater than 1 are heavier than air. There are special fire and safety regulations in many communities for using standard-size cylinders inside a building. Any of these gases could accumulate in a low area like a basement, pit, or sewer if some part of the welding outfit developed a leak. If the leak was not detected, a static spark or other source of heat could ignite the accumulated gas pocket and cause a serious explosion and fire. Check local fire codes concerning the regulations for using these gases inside a building.

MAPP®, propylene, and propane are available in large cylinders for industrial use and in small cylinders for use by small shops. The 1-lb cylinders for small torch sets are used for soldering, silver brazing small-diameter copper tubing, and light-duty heating projects.

Natural gas is used for the same applications as MAPP®, propylene, and propane. It is delivered through the same natural gas pipeline used for heating buildings.

Acetylene is the only fuel gas that is currently being used for the very limited applications for oxyfuel fusion welding of carbon steel. Acetylene is also the main fuel gas used for flame-hardening and metalizing operations. The other gases are unacceptable for fusion welding carbon steel because they have an oxidizing flame (See "Flame Temperature Comparison of Fuel Gasses" in Table 1). The oxidizing flame creates a weld of unacceptable quality. (MAPP® and propylene can be used for welding sheet metal if a "highly deoxidized" filler rod is used.)

MAPP® and Propylene Gases

MAPP® and propylene-based fuel gases use the same regulators as acetylene. Special tips should be used for cutting and heating. The same tips can be used for brazing and soldering. However, because these gases are more dense it will take a tip one or two sizes larger to do the same job compared to using acetylene as a fuel gas. A grade "T" type hose is recommended for MAPP® and propylene-based fuel gases.

MAPP® (methylacetylene-propadiene) was developed by Dow Chemical Company as a fuel for welding, brazing, cutting, flame hardening, and metalizing operations. It has many of the physical properties of acetylene but lacks its shock sensitivity. Mapp® gas is the result of rearranging the molecular structure of acetylene and propane.

Since MAPP® gas is not sensitive to shock, like acetylene, it can be stored and shipped in lighter cylinders. Acetylene, to be kept safe, must be stored in cylinders filled with a porous filler and acetone. Where empty acetylene cylinders weigh around 220 pounds (100 kg), MAPP® cylinders weigh only 50 pounds (23 kg). Normally a filled cylinder of acetylene weighs 240 pounds (109 kg), while a filled cylinder of MAPP® gas weighs 120 pounds (54 kg).

MAPP® gas has a very distinct odor, so any leakage can readily be detected. For actual welding, the same equipment and operations can be used. However, a slightly larger tip is required because of the greater gas density and slower flame propagation rate. The only significant difference is in the flame appearance. A neutral flame will produce a long inner cone and needs a fuel-to-oxygen ratio of 1 to 2.3.

The explosive limits of the vapor of MAPP® gas in air and oxygen are much narrower than acetylene and about the same as propylene, propane, and natural gas. It can be used at full pressure safely (up to 375 psi/2590 kPa) for jobs that are dangerous with acetylene.

Propylene has properties similar to MAPP® including a very distinct, but different odor. The slightly lower flame temperature is not readily noticeable for most heating and cutting operations.

Propane and Natural Gas

Although propane and natural gas have the desired stability, their limiting factors for welding are their low flame temperature and an oxidizing flame at the maximum flame temperature. Both gases require special regulators.

Hydrogen Gas

Hydrogen can be used for welding thin sections of aluminum, aluminum alloys, and lead, and for brazing operations where lower temperatures are required. Because of its cost and flame temperature, hydrogen has a very limited use in the welding industry compared to the other fuel gases. Presently, the GMAW and GTAW processes are much more efficient for welding aluminum.

One of the unusual characteristics of oxyhydrogen is that the flame is practically nonluminous. Consequently, it is often difficult to adjust for a neutral flame. To avoid welding with what might be an oxidizing flame, adjust the regulator so it provides a greater flow of hydrogen. The excess hydrogen is harmless since the reducing atmosphere does not leave carbon deposits.

The equipment for oxyhydrogen welding is essentially the same as used in oxyacetylene welding, except that a special hydrogen pressure regulator is used.

Air-Acetylene Welding

The air-acetylene flame is generated by burning a mixture of acetylene with air. The torch operates on the same principle as the Bunsen burner. As the acetylene flows to the torch under pressure, it draws in the right amount of air for proper combustion. The temperature of the air-acetylene flame is even lower than that of the oxyhydrogen and so it has restricted uses for welding. Generally, the air-acetylene flame is used for soldering and brazing very light metal sections. It has wider applications in the plumbing industry, where it is used for soldering and brazing of copper tubing.

Test Yourself

Questions

1. What are the two stages of combustion in an oxyacetylene flame?
2. Name the safety device on an acetylene cylinder.
3. List the basic equipment needed for oxyfuel welding (OFW) and oxyfuel cutting (OFC) systems.
4. What is the purpose of the cylinder cap?
5. What is the maximum pressure in an oxygen cylinder at 70°F (20°C)?
6. What type of valve is on the oxygen cylinder?
7. What type of safety device is on the oxygen cylinder valve?
8. What is the maximum pressure in an acetylene cylinder at 70°F (20°C)?
9. What is the procedure for opening the oxygen cylinder valve?
10. Acetylene should never be withdrawn from a cylinder at a rate per hour greater than ____ the cylinder's capacity.
11. Free acetylene should never be used at a pressure above ______________.
12. Fuel gases can be purchased in what two forms?
13. What is the purpose of the oxygen and acetylene regulators?
14. Oxygen regulators are classified under what two types?
15. What are the two main functions of the torch body?
16. What type of material are the oxyacetylene welding tips made of?
17. How is the size of a welding tip determined?
18. When cleaning a welding or cutting tip, you should always start with the _____ size tip cleaner.
19. What shape are the orifices in a fuel gas heating tip (also called rosebud)?
20. When connecting the welding tip to the torch body, how tight should the tip be tightened?
21. List three things that will cause the tip to backfire.
22. When a flashback occurs, what torch body valve is closed first?
23. List two things that will cause a flashback.
24. What is the purpose of a flashback arrestor?
25. List two ways the fuel gas hose connections can be identified.
26. What color is the oxygen hose?
27. What is the recommended shade number of lens for brazing, heating, welding, and cutting?
28. When setting up the oxyacetylene cylinder system, what position should the cylinders be secured in?
29. All hose connecting nuts, regulator connections, and cylinder connections are made of what kind of metal?
30. List six areas to check for leaks on the oxyacetylene cylinder system.
31. List the types of oxyacetylene flames.
32. List five key points to control the welding puddle in oxyacetylene welding (OAW).
33. What fuel gas produces the hottest flame in the oxyfuel process?
34. List four other gases used besides acetylene for oxyfuel heating and cutting.

Shielded Metal Arc Welding (SMAW)

3

JDPX7161

Fig. 1 — Shielded Metal Arc Welding (SMAW) of Carbon Steel Plates

Introduction

Shielded metal arc welding (SMAW) is also known as stick welding and arc welding. Coated electrodes (nicknamed sticks) are melted by an electric "arc" to form a weld. The basic welding circuit is illustrated in Fig 2. The formation of a shielded metal arc weld is illustrated in Fig. 3.

Shielded metal arc welding is used on construction projects, pipe fabrication, and maintenance applications that are not suited for the GMAW and GTAW processes. Arc welding can be used for joining the commonly welded metals that include carbon steels, stainless steel, alloy steels, and cast iron. The type of metal to be joined determines the type of electrode to be used. The current required to operate the electrode will, in turn, determine whether an alternating or direct current welding power source is needed for the project.

The equipment is relatively inexpensive and easy to set up at any job site. The electricity needed to operate the welding equipment can be supplied from either a single-phase or a three-phase electric circuit. Power sources in the 180-amp to 225-amp category usually operate on 230-volt, single-phase electricity. Gasoline or diesel engine power generators (or alternators) supply portable welding power sources with the necessary electricity for welding, auxiliary lighting, and for using power tools.

Good safety practices are essential for the person welding, as well as for nearby workers. Anyone in the welding area must be protected from the arc's rays, welding fumes, or sparks and from coming in contact with hot metal. The rays will sunburn unprotected skin and cause serious injury to a person's eyes if exposed to the direct rays from a welding arc. Concentrations of welding fumes can also be harmful to anyone in the area. Everyone should be informed of the proper safety precautions for avoiding potential hazards from shielded metal arc welding (see Chapter 13, Welding Safety for details).

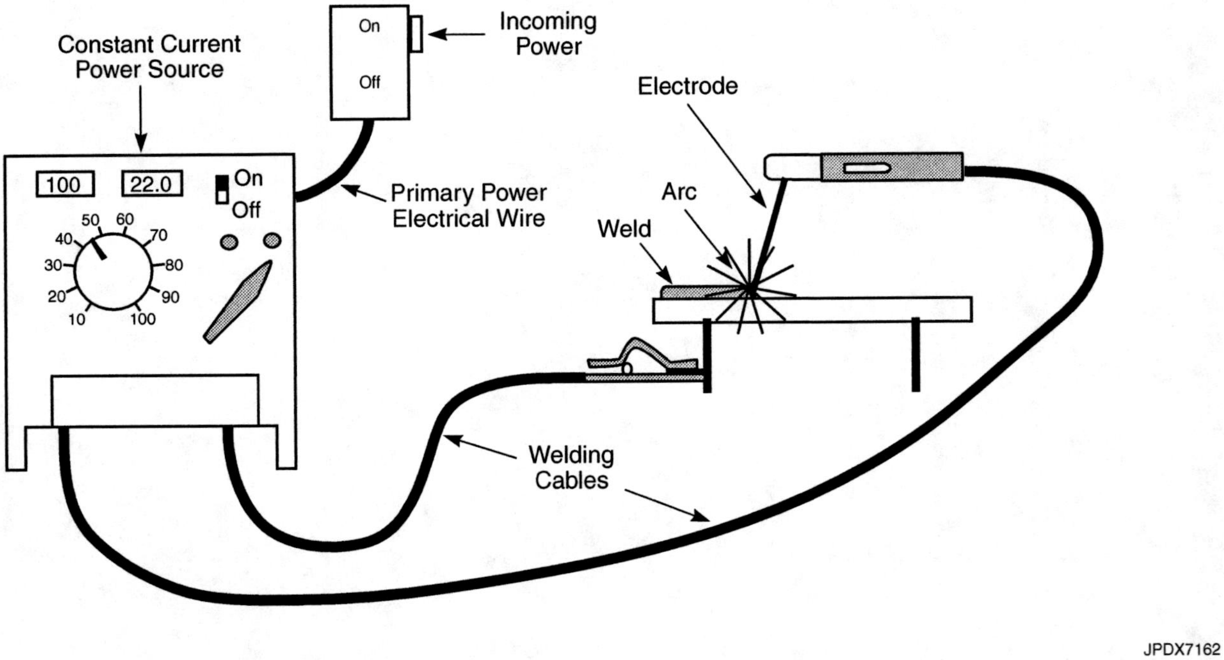

Fig. 2 — Basic Shielded Metal ARC Welding (SMAW) Circuit

Shielded Metal Arc (SMAW) Welding Process

The shielded metal arc welding process is similar to the electric furnace process used for making steel. The tip of a coated electrode is melted from a 1/8-inch (3-mm) long arc that is struck and maintained between the metal to be welded and the electrode (Fig. 3).

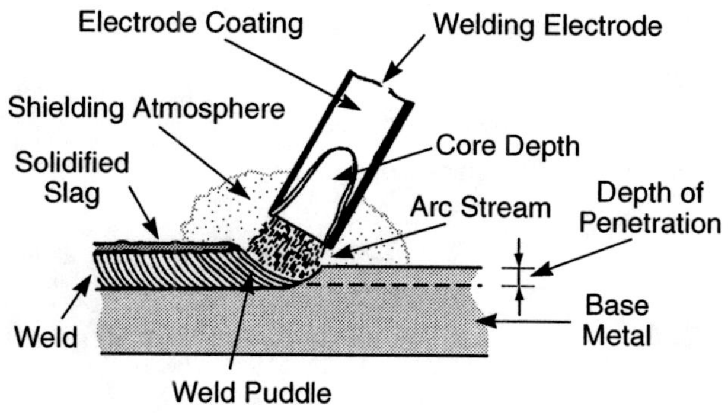

Fig. 3 — Formation of a Shielded Metal Arc Weld (SMAW)

A temperature of over 9000°F (4980°C) produced by the arc at the tip of the coated electrode melts the base metal, the coating on the electrode, and the core wire to form a weld. A slag forms on top of the weld puddle to help shape the weld bead and to protect the weld from the surrounding atmospheric air until the weld has cooled below a temperature at which it will not be contaminated. The high temperature from the electric arc also causes some ingredients in the flux to form a gaseous shield that protects the molten droplets of metal as they are propelled across the arc stream at the end of the electrode into the weld puddle.

The electrical energy necessary to start and maintain the arc is provided by a special power source capable of furnishing the proper combination of high amperage and low voltage safe for welding. Power sources are available that can produce alternating and/or direct current that is adjustable for use with a wide range of electrode sizes and corresponding metal thicknesses.

Shielded Metal Arc Welding (SMAW) Equipment

The following is a list of basic shielded metal arc welding and safety equipment see Fig. 4:

- SMAW power source
- Electrode holder, work clamp, and connecting cables
- Welding screen
- Fume collector
- Weld cleaning and finishing tools
- Welder's clothing and safety gear

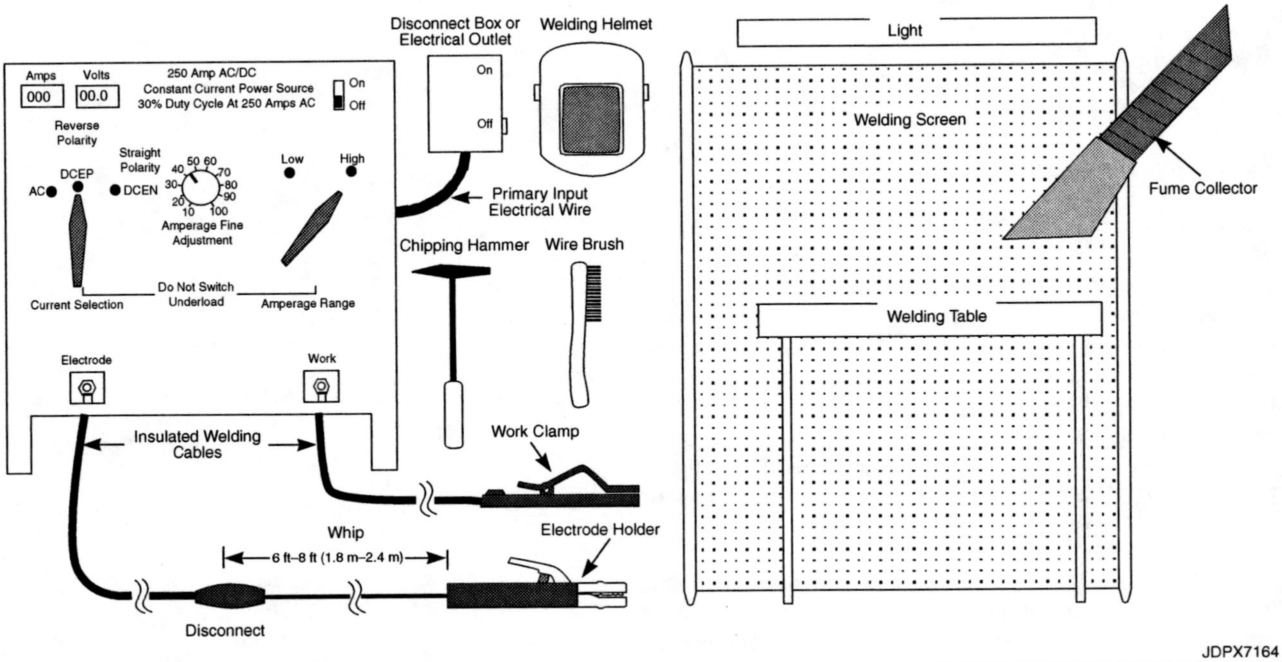

Fig. 4 — Basic Shielded Metal Arc Welding (SMAW) Equipment

Shielded Metal Arc Welding (SMAW) Power Sources

Shielded metal arc welding requires a power source that provides a constant (non-fluctuating) controlled amperage output of either alternating current (AC) and/or direct current (DC) to the welding arc. Amperage is the key factor that determines melting of the electrode and base metal, and each person who makes a weld depends on the power source to maintain a constant amperage output so that he/she can make welds that are uniform, have the proper appearance, and are structurally strong.

Constant current or CC is the name given to welding power sources that have a controllable amperage output. Because there are power sources that are capable of providing constant current for both shielded metal arc welding and gas tungsten arc welding, and constant voltage (CV) for gas metal arc welding, it is no longer completely correct to identify a power supply as being classified as CC or CV. Power sources with this dual capability are classified as multi-process power sources.

The main point to remember about these two descriptions for power sources is that amperage is adjustable on CC output and voltage is adjustable on CV output. This means that a power source designed for CC output can be used for shielded metal arc and gas tungsten arc welding, and a power source designed for CV output can be used for gas metal arc welding.

Factors to consider in choosing an SMAW power source include:

1. Amperage output and duty cycle
2. Amperage ranges and adjustments
3. Input power requirements — voltage and amperage
4. Initial cost and operating cost
5. Size and portability

The amperage output and duty cycle will be determined by the thickness of metal to be welded and the actual continuous welding time on the project at the amperage for the electrode selected. This duty cycle information is found on the name plate as the power source and in the owner's manual. Fig. 5 illustrates the output rating and corresponding duty cycle for a 225-amp CC power source and Fig. 6 illustrates the output rating and corresponding duty cycle for a 250-amp CC power source.

The duty cycle is the percentage of a ten-minute period of time that a welding power source can be safely operated at a given output. The diagonal lines in Fig. 5 and Fig. 6 illustrate that as the amperage increases the duty cycle decreases and, conversely, amperage decreases as the duty cycle increases. Point "A" in Fig. 5 means that the 225 amp power source can be safely operated at 200 amps for 2 1/2 minutes in a 10-minute period of time. Point "A" in Fig. 6 means that the 250-amp power source can be safely operated at 200-amps for 5 minutes in a 10-minute period of time.

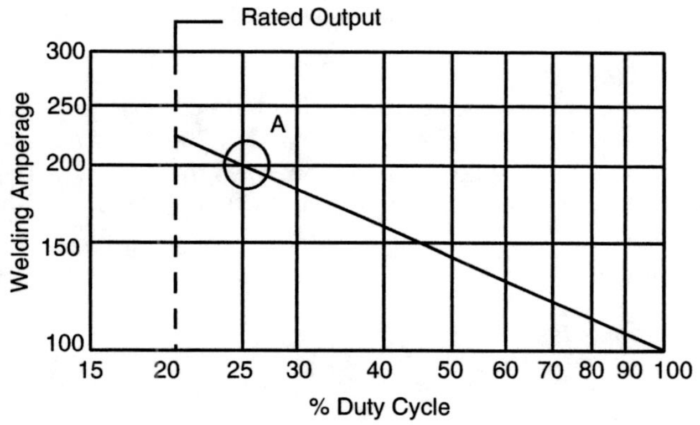

Fig. 5 — Typical Duty Cycle Chart for a 225-Amp Constant Current (CC) Power Source with a 20% Duty Cycle

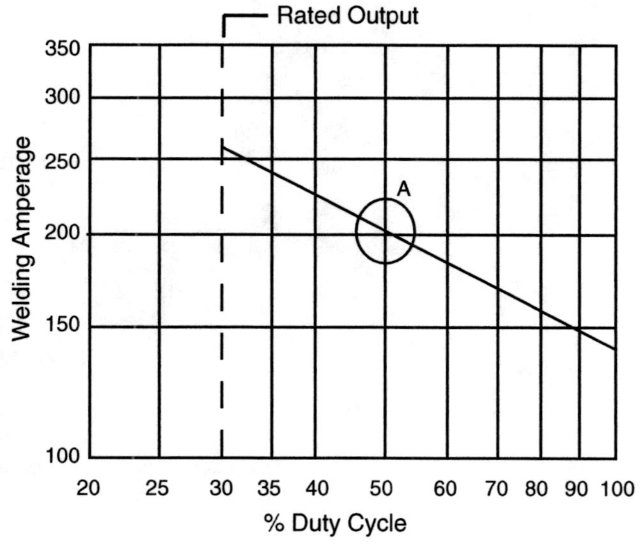

Fig. 6 — Typical Duty Cycle Chart for a 250-Amp Constant Current (CC) Power Source with a 30% Duty Cycle

If either power source is operated at a lower amperage, the duty cycle increases above point "A." If the amperage is increased above point "A," the duty cycle decreases.

CAUTION: Exceeding the duty cycle will cause the power source to overheat and cause damage to the power source.

Calculating the Duty Cycle of a Power Source

To determine what the power source's duty cycle must be for a project, first select the type and diameter of electrodes that will be used. Next find the amperage range for the electrode and estimate the percent of a 10-minute period for actual welding. This number will be the duty cycle.

For example, if a project called for using 1/4-inch (6-mm) metal and 1/8-inch (3-mm) electrodes were selected for the project, locate the amperage range on the manufacturer's chart for the electrode. A realistic amperage range would be 100 to 150 amps. If it was estimated that the welding time would be 5-minutes out of 10-minutes, a power source would be needed that had a 50% duty cycle at a maximum of 150 amps.

Match the 150 amps to the duty cycle chart for the power source that is being considered and you can tell if it has adequate capacity for the project. (Basic duty cycle information is also found on the name plate of the power source.) To make a final purchasing decision, compare the cost to corresponding features of power sources that have adequate capacity (including duty cycle) for your needs.

Reminder: When figuring the overall welding time for a project and the corresponding duty cycle needed for the power source, don't forget to take into consideration the time that will be taken for part setup, electrode changes, and weld cleaning. These activities will typically take 50% of a person's work time, so that the power source is automatically given a 50% duty cycle — there is a five-minute period of time when no welding is taking place and the power source can cool properly. It is reasonable to take into account the possibility of using a larger electrode requiring a higher amperage when determining the amperage capacity needed for a power source. However, not many people are going to exceed this 50% work time when shielded metal arc welding and do not need a power source with a duty cycle much above 50% at the amperage they will be operating.

Check the input power (voltage and amperage) available to the home shop or industrial location before purchasing an SMAW power source. Single-phase welding power sources will operate from 3-phase power, but special converters are required to make single-phase power usable for a 3-phase power source. Also, some power sources have internal electrical connections that allow them to be changed for operation on either 208, 230, and 460 volts. Other power sources are made to operate on a single voltage. Check the input voltage requirements before purchasing a power source.

The amperage input requirement is another very important item that should be considered when selecting a power source. Power sources in the 180-amp to 225-amp range operating from 230 volts typically require a 60-amp service. A 250-amp power source can easily draw 100 to 120 amps from a 200-volt line running to a shop. If there is only a 100-amp service available, or if the power lines to the shop do not have the capacity for operating a welding power source, it can be more expensive to get electrical service to operate a power source than it will be to purchase it.

The amount of electrical power required during welding and while the power source is just on can vary considerably between power sources. The inverter-type power sources use electricity more efficiently than conventional power sources.

However, they are also much more expensive to purchase. Therefore, before making a purchasing decision, compare the savings in operating cost over several years to the difference in the purchase price of power sources having the same capacity and features. Always check the amperage input requirements before purchasing a power source.

Physical size and portability are also important considerations in selecting a power source. Inverter power sources are becoming very popular for in-plant maintenance because of their size and portability. A comparison of power sources having CC output follows:

The conventional (not an inverter) 225-amp transformer-type power source in Fig. 7 has a 25% duty cycle at 200 amps and weighs 85 lb (39 kg). The output is only alternating current (AC).

JDPX7167

Fig. 7 — 225-Amp Alternating Current (AC) Transformer CC Power Source

The conventional 300-amp transformer/rectifier type power source in Fig. 8 has a 20% duty cycle at 300 amps AC and 20% duty cycle at 200 amps DC weighs 107 lb (49 kg). The output is both alternating current (AC) and direct current (DC).

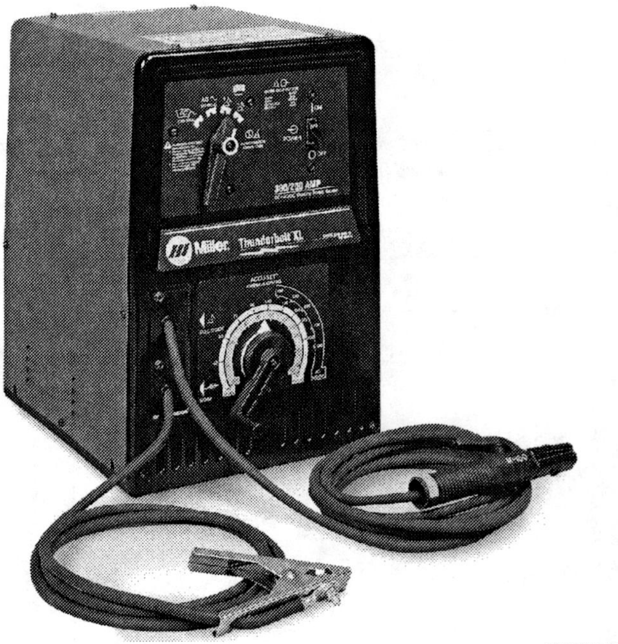

JDPX7168

Fig. 8 — 300-Amp Alternating Current (AC)/200-Amp Direct Current (DC) Transformer/Rectifier Constant Current (CC) Power Source

The portable 225-amp engine-driven power source in Fig. 9 has a 100% duty cycle at 200 amps, is 31 inches (532 mm) high, 18-3/4 inches (476 mm) wide, and 26 inches (1164 mm) deep and weighs 532 lb (241 kg). The output is both alternating current (AC) and direct current (DC). In addition, the power source can provide 5000 watts of 115/230 volt 60-cycle alternating current for auxiliary power when in the auxiliary power mode of operation. Adequate electrical power is also available in the welding mode for operating power tools, such as a grinder or a drill, and for the operation of work lights.

JDPX7169

Fig. 9 — 200-Amp Alternating Current/Direct Current (AC/DC) Portable Engine-Driven Power Source

The 400-amp inverter power source in Fig. 10 has a 60% duty cycle at 300 amps, is about 17" (432 mm) high, 13 inches (318 mm) wide, and 24 inches (610 mm) deep and weighs 76 lbs (34.5 kg). The output is direct current (DC) only. However, by purchasing additional accessories, it can also provide FCAW and GTAW-DC capabilities.

JDPX7170

Fig. 10 — 400-Amp Direct Current (DC) Inverter Power Source

The type of amperage control or setting is another important feature to consider in choosing a power source. Power sources that have plugs or a switch (Fig. 11) only allow the amperage to be changed in steps of 10 to 20 amps. Power supplies that have amperage controls that are infinitely adjustable may have a crank a wheel , a dial/rheostat, or a combination switch and dial/rheostat (Fig. 12).

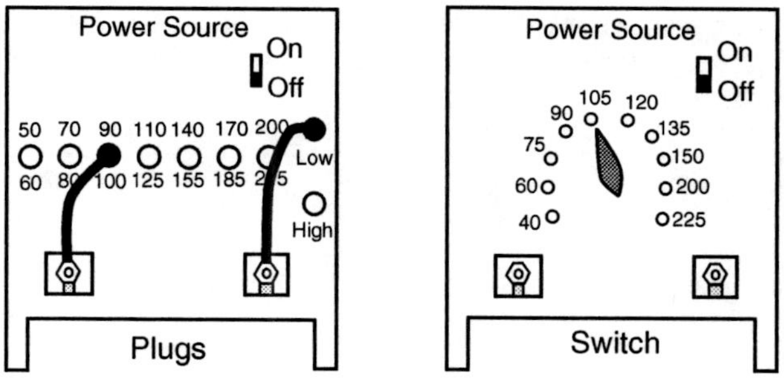

JDPX7171

Fig. 11 — Step Types of Amperage Controls

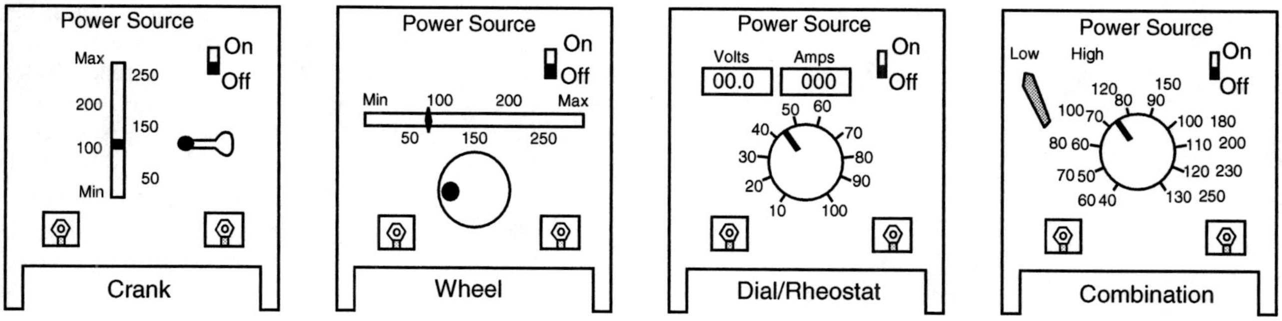

JDPX7172

Fig. 12 — Step Types of Amperage Controls

The ability to make "fine adjustments" to the amperage is most important when using small-diameter (3/32 inch/2.4 mm) electrodes on thin metal and for out-of-position welding using 1/8 inch (3.2-mm) diameter electrodes. This feature may be slightly more expensive to purchase, but it allows a person with welding experience to set the amperage precisely where it is needed for the best control of the weld puddle.

A power source that has a step-type amperage control is easier for a person not having welding experience to set, since there is a limited number of choices to make in finding a setting that is suitable for the project at hand.

The main components that produce the proper output current for shielded metal arc welding are listed below:

1. Transformer (AC)—(Fig. 7), (Fig. 13)
2. Rectifier (DC)—(Fig. 14)
3. Transformer/rectifier (AC/DC)—(Fig. 8)
4. Generator (DC/AC)—(Fig. 9)
5. Inverter (AC/DC)—(Fig. 10)

A transformer inside the power source converts incoming high voltage, low-amperage alternating current to a low-voltage, high-amperage alternating current suitable for shielded metal arc welding. A simple AC welding transformer and welding circuit is illustrated in Fig. 13. Note that the primary side of the transformer has a large number of wraps (windings) of small-diameter wire (copper or aluminum) around a metal core (that is actually laminated and not of solid construction). The secondary side has a few wraps of large-diameter wire (copper or aluminum) around the same metal core.

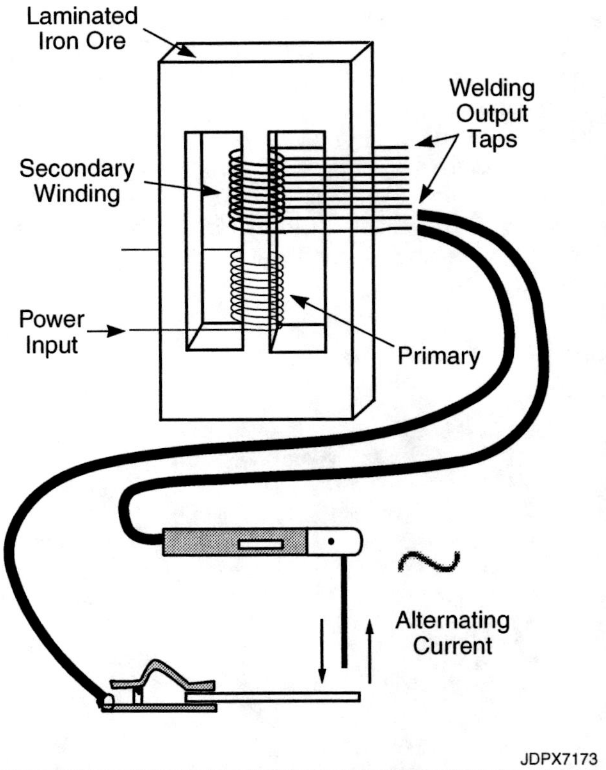

Fig. 13 — Simple Alternating Current (AC) Welder Circuit Tapped Output

When alternating current flows through the primary winding it creates a magnetic field in the metal core of the transformer. This magnetic field induces alternating current in the secondary winding which is the output for welding. The lowest amperage output is obtained by connecting the electrode cable to the end of the secondary winding and the work cable to the beginning of the secondary winding. The highest amperage output is obtained by moving the electrode connection to the first winding next to the work cable connection. Amperages in between these two connections can be obtained by moving the electrode cable connection progressively closer to the last winding.

A rectifier is an electrical device that changes alternating current to direct current. The most commonly used type used in welding power sources is silicon diodes. The diodes are placed in the output portion of the secondary winding's circuit as illustrated in Fig. 14. The resulting flow of direct current (flow of electrons) can be made to flow from the electrode to the work (DCEN—direct current electrode negative) or can be made to flow from the work to the electrode (DCEP—direct current electrode positive) by changing the position of a switch or by changing the location of the electrode and work connections on the outside of the power source.

A transformer/rectifier is capable of producing both alternating (AC) and direct current (DC) for welding. A switch is located on the output of the secondary winding to bypass the rectifier for alternating welding current or to allow alternating current to go through the rectifier for direct welding current.

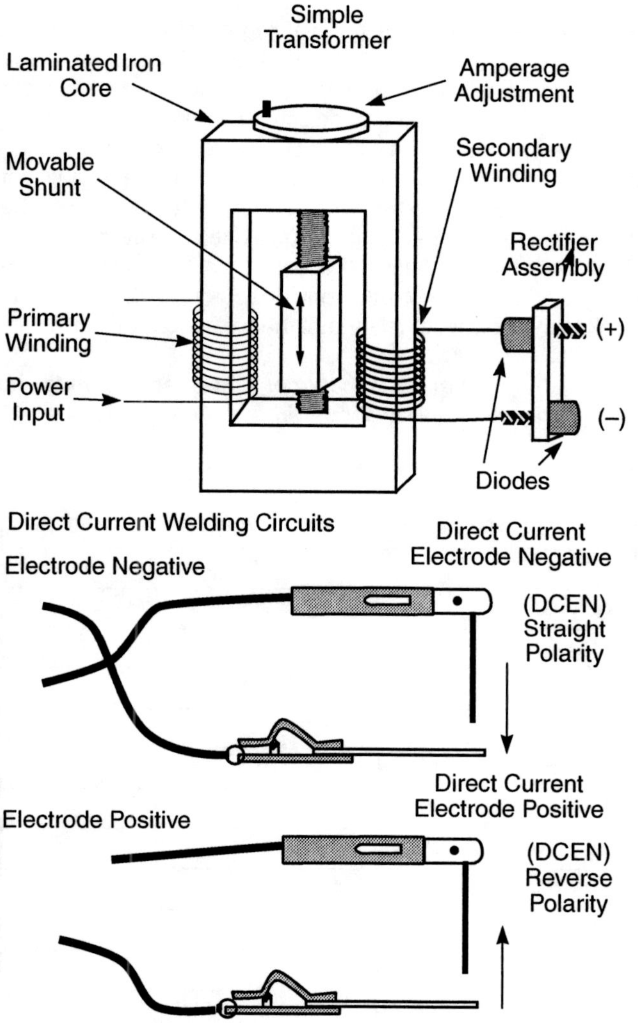

Fig. 14 — Simple Direct Current (DC) Welder Variable Adjustable Output and Welding Circuit

Engine-driven generators are of two basic types. One type produces direct current only and the other type produces alternating current. The second type is technically an alternator even though it is called a generator. For the welding output from the alternator-type generator to be direct current, a rectifier has to be added to the welding circuit. Both types of generators produce 115-volt electricity for operating power tools and lights. This output current may be either AC or DC.

IMPORTANT: Some electrical devices require alternating current (AC) and will either not operate or will be ruined if operated on direct current (DC).

The alternator-type generator typically can also be used for producing electricity for appliances, power tools, and lighting in remote areas or in case there is an electrical power outage.

The auxiliary power output ranges from 3000 to 10,000 watts of 50- or 60-cycle 115- and 230-volt current.

Inverters for welding power sources are much lighter weight, smaller size, and have a higher capacity than conventional power sources. As an example, the conventional 250-amp (30% duty cycle @ 200 amps) power source similar to Fig. 8 weighs 345 lb (156 kg) and the 400-amp (60% duty cycle @ 300 amps) inverter power source in Fig 10 weighs 76 lbs (34.5 kg). The compactness and efficiency of inverter-type power sources is made possible by the use of solid-state electronic devices instead of the mechanical devices found in conventional power sources.

Most inverter power sources produce only DC welding current (as technology improves there will be more inverters that can produce AC welding current). These inverters are being used for SMAW, GMAW, GTAW, and plasma arc cutting (PAC). Many of the inverter power sources have multiprocess capability when equipped with the proper accessories and only direct current is needed for the application.

Electrode Holder, Work Clamp, and Interconnecting Cables

The electrode holder (Fig. 15) is used to hold the electrode and to make an efficient transfer of electric current from the welding cable to the electrode. It must also protect the welder from the electric current and keep his/her hands a safe distance from the heat of the arc.

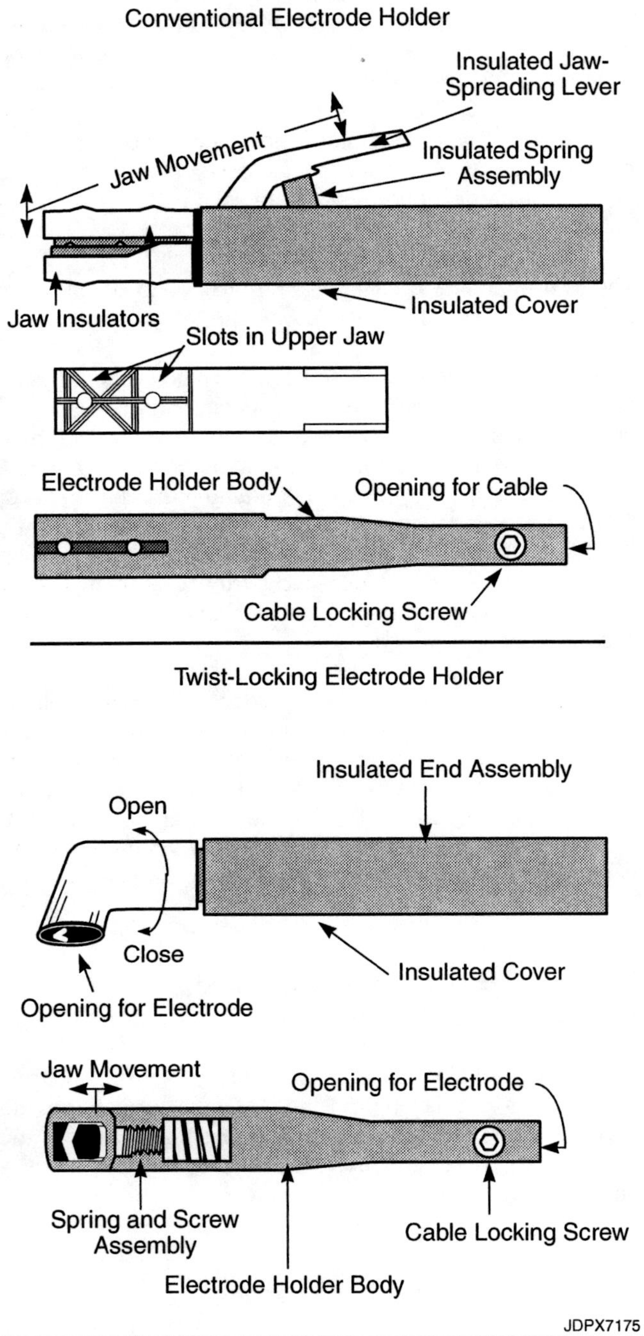

Fig. 15 — Common Types of Electrode Holders for Arc Welding

CAUTION: Do not touch non-insulated parts of the holder with bare hands while the machine is on because there is danger of getting an electrical shock. This is why the electrode holder has been nicknamed "the stinger." Always replace insulated parts when they are worn to the point of not providing cover to the current-carrying parts of the holder.

Electrode holders range in capacity from 100 to 500 amps. The current-carrying capacity of the electrode holder, electrode cable (also called lead), and work cable size should be matched to the amount of current and length to be used for welding. For example, an electrode holder rated at 200 to 250 amps has adequate capacity for most jobs and is lightweight enough for easy handling.

Specially designed insulated welding cables are needed to complete the welding circuit from the power source to the electrode holder, and from the power source to the work connection (also called ground clamp) as illustrated in Fig. 4. The cables must be of sufficient size to carry the welding current without overheating and/or causing a significant voltage drop. Table 1 gives recommendations for the current-carrying capacity of different sizes of welding cables. The size of cable should be based on the total distance (length of cables) from the power source to the work and back to the power source — not the distance one way.

NOTE: Remember distance from power source must be doubled to allow for power and ground leads.

Distance		**Amperes**								
ft	**m**	**100**	**150**	**200**	**250**	**300**	**350**	**400**	**450**	**500**
50	15	2	2	2	2	1	1/0	1/0	2/0	2/0
75	23	2	2	1	1/0	2/0	2/0	3/0	3/0	4/0
100	30	2	1	1/0	2/0	3/0	4/0	4/0		
125	38	2	1/0	2/0	3/0	4/0				
150	46	1	2/0	3/0	4/0					
175	53	1/0	3/0	4/0						
200	61	1/0	3/0	4/0						
250	76	2/0	4/0							
300	91	3/0								
350	107	3/0								
400	122	4/0								

Note: Remember, distance from power source must be doubled to allow for power and ground leads.

Table 1 — Arc Welding Cable Chart

The electrode holder can be attached directly to the electrode cable or attached to a short cable 6 to 10 feet (1.8-3 m) long called a whip (Fig. 16) of a smaller diameter but of sufficient capacity for the short distance. A whip is easier to maneuver and lighter to handle than a full-sized cable.

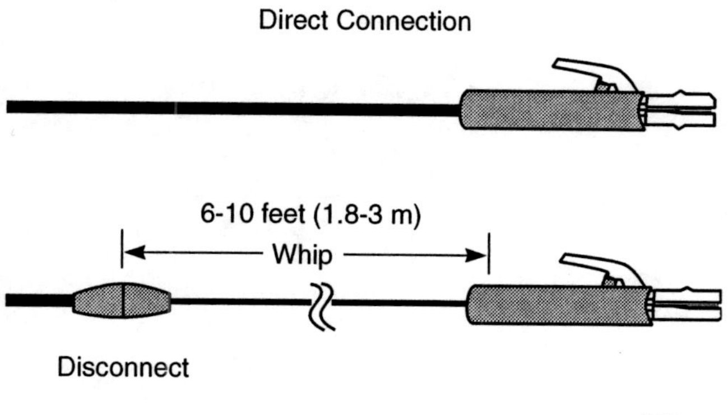

Fig. 16 — Electrode Holder-to-Cable Connection Options

Follow the equipment manufacturer's recommendations when purchasing welding cables and primary electrical wire from the source of electricity to the welding power source. Using a welding cable that is too small for current and distance will make the arc hard to start and the arc will seem to be cold for what should be a the proper setting. It will also cause the welding leads to overheat.

The primary electrical wire size will be determined by the input voltage options of the power source and the voltage supplied at the location of the installation. A relatively small primary wire will be needed to operate a power source on 460 volts compared to the size needed to operate the same power source on 230 volts.

IMPORTANT: Always follow local, state, and national regulations regarding the installation of electrical wiring and other devices necessary for the connection of welding equipment. This includes complying with the provisions in the latest National Electrical Code (NEC).

The work clamp (previously called the ground, but should be referred to as the work clamp to avoid confusion with electrical system ground) completes the welding circuit for both AC and DC welding. The location for the attachment of the work clamp or connection should be free of rust, paint, or other material that prevents a good electrical connection.

CAUTION: Always attach the work connection directly to the part that is to be welded.

Electrical equipment ranging from a computer in the shop office to the radio on a farm tractor can be damaged by improper work connections from welding. Bearings and gears will be ruined if the welding current passes through them and pits the load-bearing surfaces. If damage from arcing through a gear or bearing is suspected, the part should be disassembled and thoroughly inspected. Serious damage could be done to the equipment if the bearing or gear fails during service.

A poor work connection can be identified by checking the temperature of the connection after welding for 5 to 10 minutes or by checking for evidence of arcing. The connection will be much warmer or even hot at the point of contact when there is not a good connection. (A slight amount of warming when welding at high amperages is normal.)

Electricity will take the path of least resistance, which is not necessarily the shortest distance between two points. If at all possible, attach the work connection directly to the workpiece. Improper attachment of the work connection can result to the welding current traveling through electrical systems, bearings, and gears.

Types of work connections are illustrated in Fig. 17. Any of these connections may become warm after 30 minutes of continuous welding. However, there should not be enough heat to cause steam if a drop of water touches the contact surface. Also, the fused surface of the tack weld must be at least equal to the cross-sectional area of the work cable.

CAUTION: Approach the connection carefully — don't just grab the connection with your bare hands when checking for a poor connection. The connection may be hot enough to burn your skin!

Also, there may be arcing from the connection to the work. It is important to ground (attach) the power source directly to the parts being welded because electrical equipment can be damaged by becoming part of the current path.

Arcing can be identified by any or all of the following:

1. Flashes of light at the point of contact.
2. Burned marks on the metal at the point of contact.
3. An arc that is hard to start or that goes out for no reason.
4. An off-and-on (intermittent) hissing or frying sound.

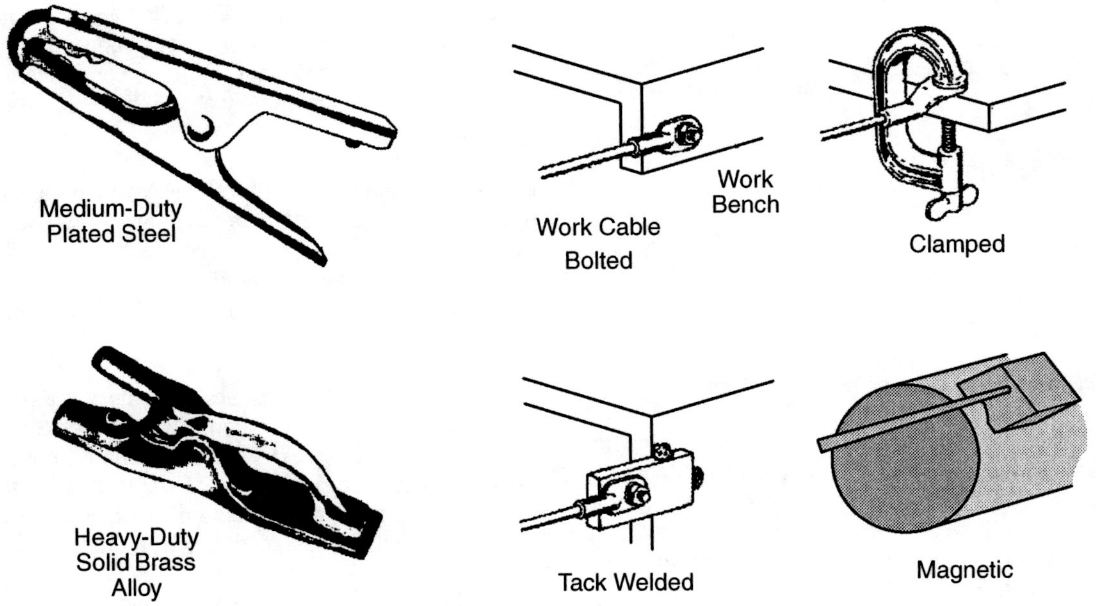

Fig. 17 — Common Types of Work Clamps and Connections

Welding Screen

A welding screen, curtain, or solid barrier should always be used to protect other persons, pets, livestock, or other animals from the welding rays within a 75-foot (23-m) radius of a visible welding arc. Welding screens that block the harmful welding rays, while allowing visible light into the weld area, are preferred over ones that block all light to and from the welding area.

"Homemade" welding screens should be of a material that is both heat- and fire-resistant. A non-reflective paint should be used to cover shiny surfaces like galvanized metal that are sometimes used to separate welding booths. The reason is that the arc rays can be reflected and can cause the same type of injuries as the direct rays. Damaged screens should be repaired or replaced when they contain holes that allow the arc rays to escape the weld area. The welding screens should also contain or stop hot sparks from leaving the weld area.

Fume Collector

A welding fume collector should be used to remove welding fumes from the welding area. Current EPA (Environmental Protection Agency) and OSHA (Occupational Safety and Health Act) or any other regulatory bodies regulations should be followed when selecting the type, size, and location for a fume collector in the area used for welding.

Weld Cleaning and Finishing Tools

Shielded metal arc welds normally require cleaning and finishing after they have cooled to room temperature. A chipping hammer is used to remove the slag covering on the weld. A wire brush is used to remove small pieces of weld spatter and dust around the weld area. Also, a grinder is sometimes used to make the surface of the weld blend evenly with the base metal before the welded part is painted.

Welder's Clothing and Safety Gear

The person in Fig. 18 is properly dressed for welding and is protected from the light rays, hot sparks, and heat produced by the electric arc. A welder's basic work clothes should include the following:

1. Long-sleeved shirt with pockets covered or removed.
2. Trousers (without cuffs) that cover shoe tops.
3. Boots or high-topped leather shoes with safety toes. (Tennis shoes are not acceptable — welding sparks can burn through a good pair of tennis shoes very quickly!)

CAUTION: Avoid fabrics that are so thin or loosely woven that they do not block the arc rays from reaching your skin. Avoid fabrics that contain synthetics that could melt to your skin if contacted by hot metal or from being close to the arc. Avoid fabrics that are highly flammable (nylon and polyester) that could start on fire or melt to your skin if contacted by hot metal or a spark

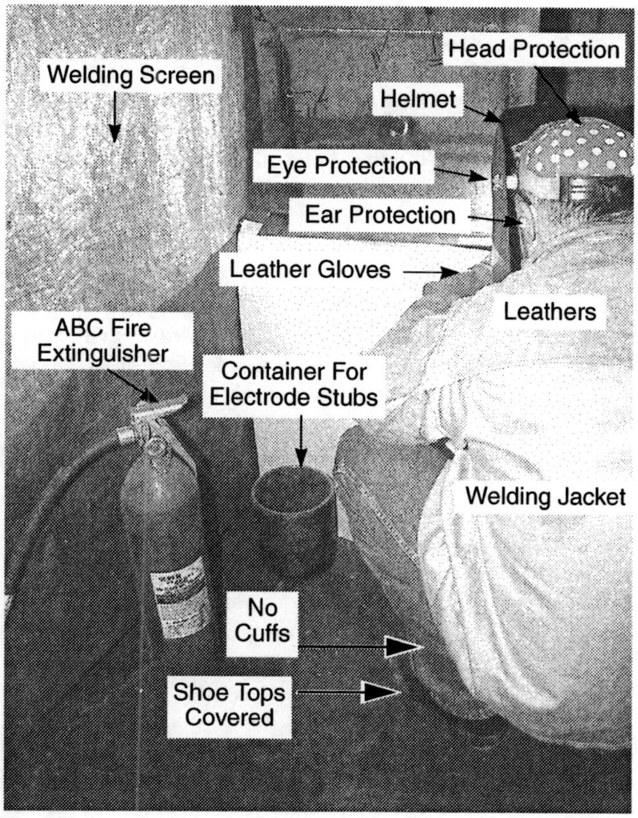

JDPX7178

Fig. 18 — Protective Clothing and Safety Gear for Arc Welding (Always Wear Proper Eye and Ear Protection)

Safety gear especially needed for welding includes the following:

1. Safety glasses with side shields (in addition to the welding helmet) to protect your eyes from flying sparks and slag while cleaning the weld.

CAUTION: Safety glasses with side shields may be uncomfortable to people who do not normally wear glasses, but the discomfort is extremely minor compared to the pain from hot slag getting into an eye and/or the handicap of being blinded in one eye forever.

2. Leather gauntlet gloves. (Use pliers or clamps to pick up hot objects. Only rely on the gloves to protect your hands if you accidentally touch something hot.)
3. A cap or beanie to cover the head and hair from hot sparks.
4. Ear plugs should always be worn to protect a welder's ears from hot sparks going inside.
5. Jackets and/or aprons made of leather or specially-treated fabrics that protect a welder's clothing and body in any situation (like overhead welding) where sparks could penetrate normal work clothing.
6. A welding helmet (Fig. 19 and Fig. 20) is the piece of equipment that protects a welder's eyes and face from hot welding sparks, from the heat, and from the light produced by the electric arc. The helmet has a window in the front containing a special dark filter lens and protective cover lens (some are glass and some are plastic). The lense blocks the harmful infrared and ultraviolet rays from sunburning the face and eyes, and reduces the brightness of the light to a comfortable level for viewing the weld area.

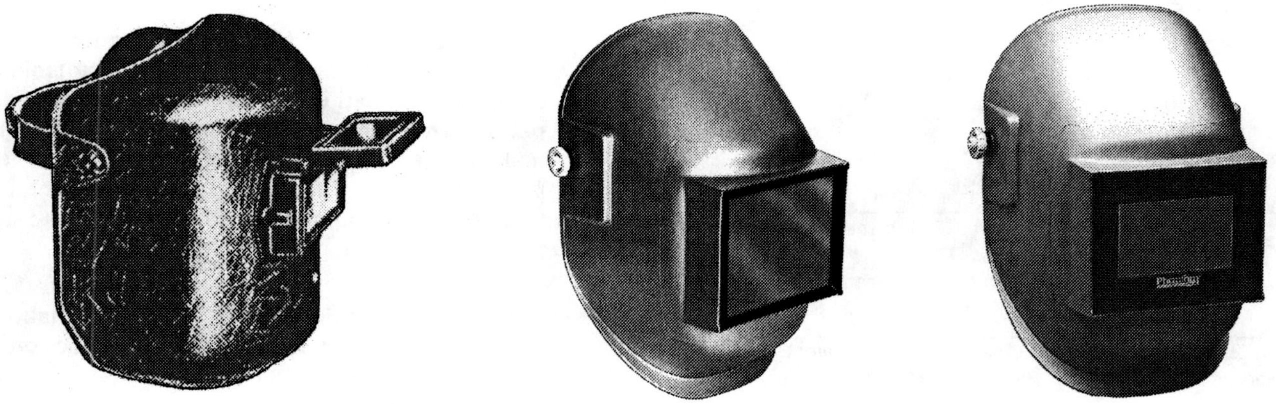

JDPX7180

Fig. 19 — Types of Helments for Arc Welding

Rays from arc welding can cause painful sunburn to the unprotected skin in a short time (less than 5 minutes) and to the eyes is a few seconds. Viewing the arc without the proper eye protection is called getting a flash. It may take as long as 5 or 6 hours to get the full effects, which feels like someone put hot sand in your eyes. The symptoms from a flash can last for several hours and should be treated with special medication from a doctor.

Protect the dark filter lens from smoke, spatter, and other types of damage with a clear cover lens in the front and back (Fig. 20). Replace cover lens whenever cleaning doesn't restore clear vision to the weld area.

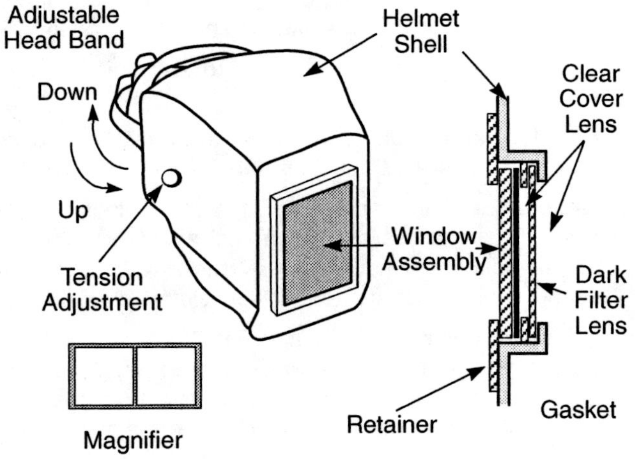

Fig. 20 — Typical Welding Helmet and Accessories

Use Table 2 to match the dark filter plate to the welding amperage for the job.

Current	Minimum Shade	Suggested Shade*
<60	7	----
60–160	8	10
160–250	10	12
250–550	11	14

** Start with a shade that is too dark to see the outline of the work with the arc lit. Then try the next lighter shade until the work can be seen clearly with the arc lit.*

Table 2 — Shielded Metal Arc Welding (SMAW) Filter Selection Guidelines for Helments

Auto-darkening filters allow good visibility of the weld zone with the helmet in the down position until the arc is struck. When the arc is struck, a battery-powered sensor causes the filter to darken instantly (1/25,000 of a second) to the preset shade for welding and return to a visible shade a few seconds after the arc is stopped to allow easy setup and alignment. (Deluxe models have a variable adjustable for both shade darkness and time delay before returning to a visible shade.)

Any damage to the welding helmet and/or the dark filter lens can reduce the protection from the arc's rays. Check the helmet for light leaks at least once a day by holding it up to a trouble light or fluorescent light. Replace dark filters and/or helmets that show any light leaks. Never look at the arc with the naked eye within a distance of 75 feet (23 m).

A magnifier (or "cheater" as it is sometimes called) can be secured behind the inside cover plate to make it easier to see very fine detail when welding small intricate pieces. It can also be very helpful for improving the close vision of people who have difficulty seeing clearly at close distances for tasks like welding and reading. Magnifiers are available from 1.00 to 2.50 diopters, but come in only one size – 2 x 4-1/4.

NOTE: Give a welding helmet reasonable care and it will provide a lifetime of service. Be careful about leaving any of the lightweight helmets on hot objects as the helmets can melt. Don't substitute a welding helmet for a grinding face shield — it is not designed to withstand the impact of a grinding wheel if it should happen to break.

Setting Up Shielded Metal Arc Welding (SMAW) Equipment

CAUTION: The installation of an electrical outlet and disconnect switch for operating welding equipment should be performed only by qualified people. The same qualifications are necessary for making internal electrical connections on the power source for the primary power cable connection and for changing voltage connections to correspond to the shop voltage available.

IMPORTANT: Always follow local, state, and national regulations regarding the installation of electrical wiring and other devices necessary for the connection of welding equipment. This includes complying with the provisions in the latest National Electrical Code (NEC).

Attach the electrode holder, work clamp, welding cable plugs (or lugs), and welding cable to a power source as follows:

1. Turn off the power to the electrical outlet.

2. Turn off the on/off switch on the power source. (On engine-driven power sources, turn off the engine if it is running.)
3. Attach the electrode holder, work clamp, and cable plugs (or lugs) to the welding cable per the equipment manufacturer's instructions.
4. Attach the cables to the designated locations on the power source to obtain the proper polarity for operating on direct current.
5. Attach the work clamp to the welding table or, when possible, directly to the workpiece.

NOTE: Attaching the work clamp directly to the workpiece prevents possible damage to the work from arcing between the table and work during welding.

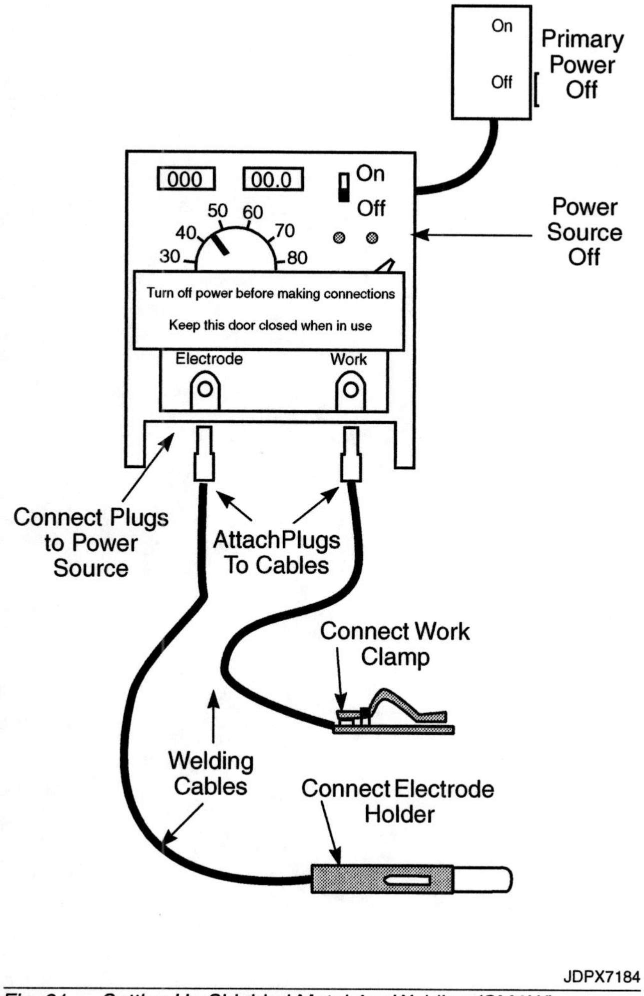

Fig. 21 — Setting Up Shielded Metal Arc Welding (SMAW) Equipment

Safety Precautions Before Welding

Take the necessary precautions to protect yourself, others in the area, the work area and equipment from the potential hazards created by welding operations.

Before starting to weld, learn and follow the safety precautions given by the equipment manufacturer. Also, follow the safety rules listed below and the safety rules listed in Chapter 13, welding safety. Welding is safe when the proper precautions are taken, but it can also be hazardous if the proper safety precautions are not followed.

1. Be sure all power wiring is correctly installed and maintained.
2. When possible, remove all combustible materials from the work area. When combustibles cannot be removed, protect them with a fire-proof covering and have fire watchers present with the proper type and size fire extinguisher to extinguish any type of fire that could occur. Also, have a fire watcher present for at least one hour after welding has stopped to ensure that a fire does not start from something that has been smoldering.
3. Wear the proper clothing and gear to protect yourself from the arc rays, hot sparks, and heat produced by the electric arc.
4. Do not use gloves or wear other clothing that contains oil, grease, or other combustible substances.
5. Protect others with a welding curtain and warn them not to enter the welding area without the proper protective clothing and safety gear. Keep children away from welding operations. Pacemaker wearers should keep away from welding until after consulting with their doctor.
6. Protect equipment in the area from damage from hot welding sparks. Cover painted surfaces and glass windows from pitting by hot sparks landing on them.
7. Use a welding fume collector to remove welding fumes from the welding area.
8. Never work in a damp or wet area — it increases the hazard of an electrical shock.
9. Shut off the welding power source before making repairs or adjustments that could be an electrical shock hazard if power was on.
10. Don't overload the welding cables or use cables with damaged insulation.

Metal Cleaning and Preparation for Welding

Remove as much of the rust, scale, paint, dirt, and foreign material including oil and grease, as possible, from the metal before starting to weld. Remove any slag that maybe left over from a cutting torch. Remember that the dust from using a grinder to remove coatings on the metal such as paint and galvanize can be just as hazardous to breathe as the fumes from welding on these coated surfaces.

The examples in Fig. 22 provide guidelines for preparing the edges and for spacing to obtain complete weld joint penetration on common types of groove welds. The shape of the edges that form a groove between the two pieces actually determines the name of the weld used to make the weld joint. The examples provide guidelines for preparing the edges of the metal and for spacing the pieces to allow for full joint penetration on square groove, bevel groove, and "V" groove welds ("V" is also called "vee.")

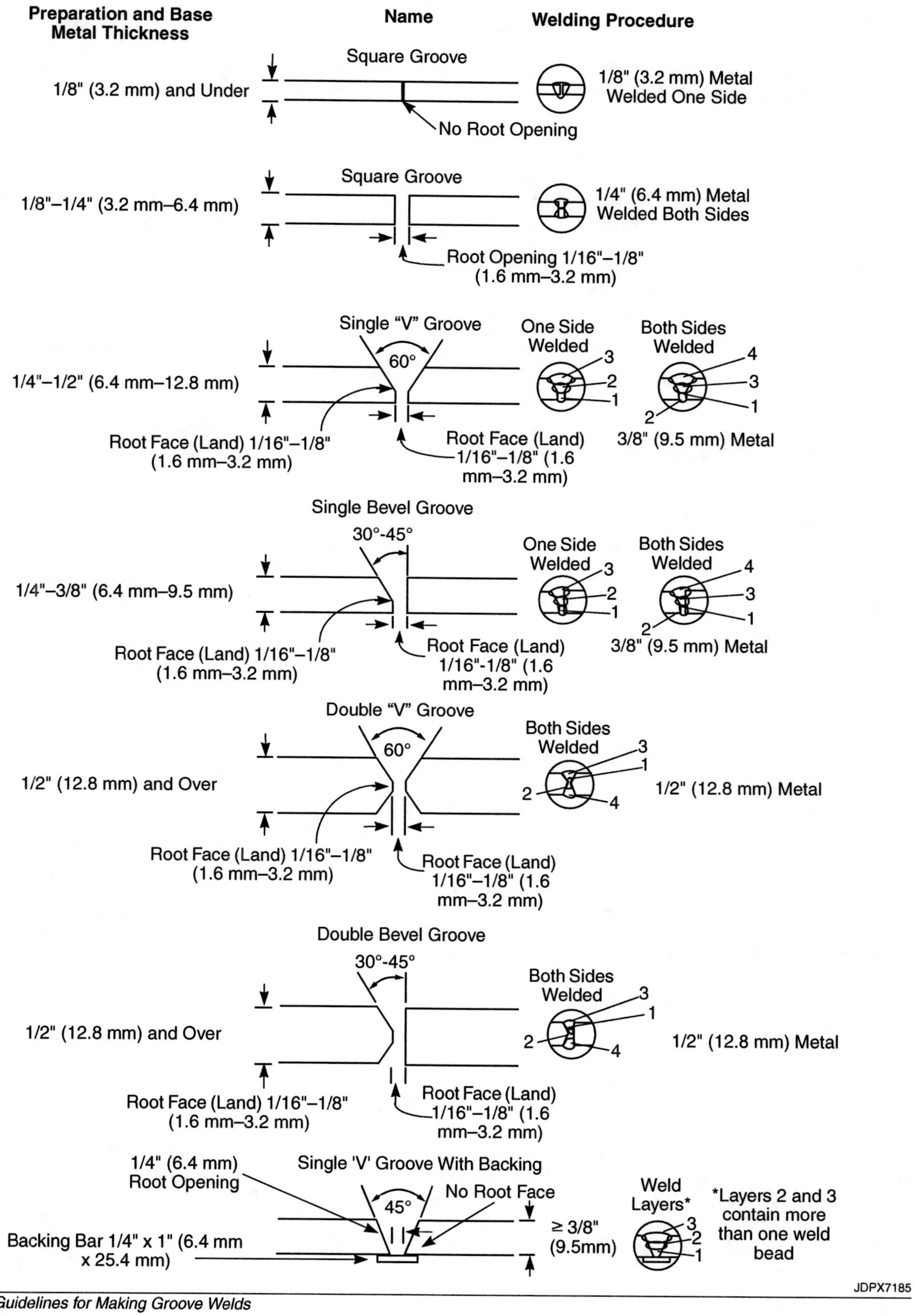

Fig. 22 — Guidelines for Making Groove Welds

NOTE: The weld should penetrate the entire depth of the metal if it is going to have as much strength as the base metal. Joint design is discussed more fully in Chapter 12, Design of Welded Units.

When metal is removed from the edge of a piece (as in the examples that do not have square edges) it is called beveling. The measurement of the angle for metal removal is called bevel angle. The angle formed when two beveled pieces are placed end to end is called included angle.

The need for beveling the edge of one plate, or both edges of both plates, will depend on the thickness of the metal. It will also depend on whether the plates are accessible for welding from both the top and bottom or are limited to welding from one side. The amount removed during beveling is normally between 30° and 35° from each piece.

Electrodes for Shielded Metal Arc Welding (SMAW)

Shielded metal arc electrodes are grouped according to their base metal applications. The common groups of electrodes are the following:

1. Mild steel
2. Low-alloy, high-strength steel
3. Stainless steel
4. Hardsurfacing
5. Cast iron
6. Aluminum
7. Copper alloys

Selecting the Type and Diameter of Electrode

The type of base metal and the kind of welding current available from the power source are two of the main factors that determine the type of electrode best suited for a project. Charts and manuals that provide up-to-date information for matching an electrode to a specific base metal are available from electrode manufacturers. This information will also specify the amperage range for the electrode and whether the current should be AC, direct current electrode positive (DCEP), and/or direct current electrode Negative (DCEN) when there is a difference in arc performance with one type of current or polarity.

Base metal thickness and welding position are also important factors that must be considered in selecting the diameter of an electrode for a project. A 3/32" or 1/8" (2.4 to 3.2 mm) diameter electrode can be used for welding metal ranging from 1/8" to 1/4" (3.2 to 6.4 mm) in thickness. These sizes also allow good control of the weld puddle when welding in the horizontal, vertical, and overhead positions on all thicknesses of metals.

As a rule, never use an electrode that has a diameter larger than the thickness of the base metal. Some welders prefer larger electrodes because they permit faster travel along the weld joint and thus speed up welding, but this takes a lot of skill.

Weld bead location in relation to the weld joint is also an important factor in determining the electrode diameter. For example, on a thick metal section with a narrow "V," a small diameter electrode should be used to make the first weld, which is called the root pass. This procedure ensures full penetration at the root of the weld. Successive passes are then made with larger diameter electrodes.

Electrode Identification, Grouping, and Classification

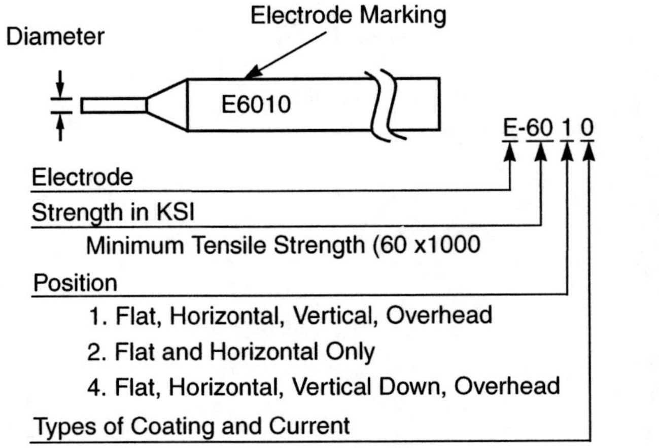

Digit.	Type of Coating	Welding Current
0	Cellulose Sodium	DCEP
1	Cellulose Potassium	AC or DCEP or DCEN
2	Titania Sodium	AC or DCEN
3	Titania Potassium	AC or DCEP
4	Iron Powder Titania	AC or DCEP or DCEN
5	Low-Hydrogen Sodium	DCEP
6	Low-Hydrogen Potassium	AC or DCEP
7	Iron Powder Iron Oxide	AC or DCEP
8	Iron Powder Low-Hydrogen	AC or DCEP
E6020	Iron Powder Sodium Oxide	AC or DCEP or DCEN

JDPX7186

Fig. 23 — American Welding Society (AWS) Identification System for Arc Welding Electrodes

Electrodes are identified by each manufacturer's trade name and/or by a combination of letters and numbers from the American Welding Society's (AWS) classification system. Each electrode manufacturer has a unique method for identifying and naming their electrodes. This makes it difficult to compare electrodes from different manufacturers when the electrodes do not have an AWS classification.

The AWS system requires that an electrode within each of its classifications must comply with a set of minimum chemical and physical standards. This means an electrode with a specific classification manufactured by one company meets the same minimum standards of an electrode with that classification manufactured by a different company. Strict manufacturing standards must be met and maintained before an electrode receives an AWS classification. Table 3 illustrates the AWS letter/number designations of shielded metal arc welding electrodes for mild steel and other types.

The freezing and filling characteristics of mild steel electrodes can be helpful in selecting the electrode most suited for a project.

Fast-freeze is the name given to a group of electrodes for which the weld puddle and slag solidify or "freeze" quickly. Compared to other electrodes for mild steel, they have the least amount of flux coating and smallest amount of slag covering the completed weld. Sometimes the slag does not completely cover the weld and is difficult to remove.

These electrodes have a forceful arc that gives the deepest penetration of any of the mild steel electrodes. They are well suited for out-of-position welding and for welding metal when the dirt, rust, and/or paint cannot be completely removed. They are also used for making welds that need deep base metal penetration.

Fast-fill is the name given to a group of electrodes that produce the fastest fill or highest deposit rate (the most weld per inch/millimeter of electrode used). They are also known as drag rods or iron power electrodes. Once the arc has been started, the electrode can be placed on the weld joint and dragged across the piece. The slag is easy to remove and the weld doesn't have much spatter to remove.

The thick flux coating contains iron powder that also melts in the weld puddle to give a high rate of deposit and produces a soft arc and a shallow penetrating weld. Compared to other electrodes for mild steel, they have the thickest flux coating and most slag covering on the completed weld.

Fast-fill electrodes are limited to uses like making groove welds in the flat position and fillet welds in the flat and horizontal positions. The weld puddle cools relative slowly compared to the fast-freeze and fill-freeze electrodes. The heavy slag acts as an insulating blanket that slows the solidification of the weld puddle. The slow cooling of the weld makes it impossible to keep the liquid metal and slag from running and dripping when out-of-position welding. The liquid slag and weld metal are pulled from the weld joint by the force of gravity before the weld puddle can solidify.

Fill-freeze is the name given to a group of electrodes that have some of the freezing characteristics of the fast-freeze electrodes and fast filling characteristics of the fast-fill electrodes. They are typically called general purpose electrodes and can be used for welding in all positions. They are relatively easy to strike an arc with and produce a clean weld that contains a minimum amount of spatter. They can be used for welding in all positions. However, the weld puddle is much easier to control when welding in the horizontal, vertical, and overhead positions than with electrodes in the fast-freeze group.

Electrodes in a group called low-hydrogen electrodes do not contain much hydrogen in the form of moisture in the flux coating or in chemical form. They have many of the freezing and filling characteristics of the fill-freeze electrodes but produce welds that have outstanding resistance to cracking under high stresses and vibration-type loading conditions. The use of low-hydrogen electrodes can reduce or eliminate the need for preheating medium carbon steels and thick weld sections.

Type of Electrode	AWS Class	Current Type	Welding Position	Weld Results
Mild Steel	E6010	DECEP	F, V, OH, H	Fast-freeze, deep penetrating, flat beads, all-purpose welding
	E6011	DCEP, AC	F, V, OH, H	
	E6012	DCEP, AC	F, V, OH, H	Fill-freeze, low penetration, for poor fit-up, good bead contour, minimum splatter
	E6013	DCEP, DCEN, AC	F, V, OH, H	
	E6014	DCEP, AC	F, V, OH, H	
				Fast-fill, high deposition, deep groove welds, single pass
	E6020	DCEP, DCEN, AC	F, H	
	E6024	DCEP, DCEN, AC	F, H	Iron powder, high deposition, deep penetration
			F, H	
	E6027	DCEP, DCEN, AC		Iron powder, low penetration, high speed
	E7014	DCEP, DCEN, AC	F, V, OH, H	Iron powder, high deposition, single and multiple pass
	E7024	DCEP, DCEN, AC	F, H	
Low Hydrogen	E6015	DCEP	F, V, OH, H	Reduce or eliminate the need for preheating on heavy sections of mild steel. Reduces or eliminates weld failures from vibration and high stress.
	E6016	DCEP, AC	F, V, OH, H	
	E6018	DCEP, AC	F, V, OH, H	
	E7016	DCEP, AC	F, V, OH, H	
	E7018	DCEP, AC	F, V, OH, H	
	E7028	DCEP, AC	F, H	Welding of high-sulphur and high-carbon steels that tend to develop porosity and crack under weld deposit
Stainless Steel	E308-15, 16	DC, AC	F, V, OH, H	Welding stainless steel 301, 302, 303, 304, and 308
	E309-15, 16	DC, AC	F, V, OH, H	Welding 309 alloy at elevated temperature application and dissimilar metals
	E310-15, 16	DC, AC	F, V, OH, H	Welding type 310 and 314 stainless steel where high corrosion resistance and elevated temperatures are required
	E316-15, 16	DC, AC	F, V, OH, H	Welding type 316 stainless steel and welds of highest quality. Contains less carbon to minimize carbon transfer in the weld. Type 316 reduces pitting corrosion.
	E347-15, 16	DC, AC	F, V, OH, H	For welding all grades of stainless steels
Low Alloys	E7011-A1	DCEP, AC	F, V, OH, H	For welding carbon moly steels
	E7020-A1	DCEP, DCEN, AC	F	
	E8018-C3	DCEP, AC	F, V, OH, H	For low alloy, high-tensile strength
	E10013-G	DCEP, AC	F, V, OH, H	For low alloy, high-tensile steels

DCEP — Direct Current Electrode Positive (Reverse Polarity)

DCEN — Direct Current Electrode Negative (Straight Polarity)

AC — Alternating Current

F — Flat

V — Vertical

OH — Overhead

H — Horizontal

Table 3 — Electrode Comparison Chart for Arc Welding

Storing of Electrodes

Keep electrodes in the sealed container until ready for use, and store the containers in a dry area away from moisture and condensation from extreme changes in temperatures. Air and moisture in the air will combine with chemicals in the flux coating and damage or even ruin the electrodes.

Special heated storage containers are required for open containers of low-hydrogen electrodes and for unused low-hydrogen electrodes. The electrodes should be stored at a temperature of at least 215°F (102°C) and not more than 450°F (232°C) in specially vented rod ovens to prevent the flux coating from absorbing moisture from the air. If the storage ovens are not vented properly, they can actually bake moisture in the electrodes. In as little as two hours, low-hydrogen electrodes can absorb enough moisture (the main source for hydrogen) from the air to lose their low-hydrogen characteristics. Special procedures can be used for reconditioning low-hydrogen electrodes that have absorbed moisture.

Fast-freeze electrodes should be kept cooler at 10° to 20°F (5° to 10°C) above room temperature. Fill-freeze, fast-fill, and most other types of electrodes, including stainless steel electrodes, should be stored at warmer temperatures, between 215° and 250°F (102° and 121° C).

In summary, keep the electrodes dry.

Functions of the Core Wire and Flux Coating of Electrodes

The core wire of the electrode is responsible for conducting current from the electrode holder to the end of the electrode and for adding filler metal to the weld. The diameter of core wire is what determines the size designation of the electrode, not the total diameter of the core wire and flux combined. The most common sizes of electrodes are 3/32", 1/8", 5/32" and 3/16 inch (2.4, 3.2, , 4.0 and 4.8 mm).

The flux coating on the electrode is very important and may perform all of the functions listed below.

1. Protects the droplets of metal transferred from the electrode to the base metal, protects the weld puddle and surrounding base metal from the atmospheric air as heat from the arc turns the flux into gaseous and liquid components.
2. Forms a slag covering on the weld, which determines the solidification rate of the weld; helps shape the weld bead; and protects the hot weld metal from the atmosphere until it has cooled to a safe temperature.
3. Combines with impurities in the weld puddle and floats them to the surface of the weld where a slag is formed and then removed by chipping and wire brushing.
4. Contains ionizing elements necessary to maintain the arc when operating an electrode on alternating current.
5. Contains iron powder and/or special alloys that are added to the weld.

Shielded Metal Arc Welding (SMAW) Techniques

Five Key Points for Controlling the Weld

1. Heat (amperage)
2. Distance (arc length)
3. Angles (work and travel angles of the electrode)
4. Travel speed (movement along the metal)
5. Manipulation (movement of the electrode to give additional control of the weld puddle)

The heat necessary for shielded metal arc welding is controlled by the amperage setting on the power source. The amperage range for a particular type and size of electrode can be obtained by referring to a chart from the electrode manufacturer and/or by looking on the electrode container. When this information has been obtained, begin by setting the amperage in the middle of the range given for the electrode. The examples in Fig. 24 represent typical bead shapes to expect from welds made at amperages in the high, middle, and low portion of the range for a given electrode.

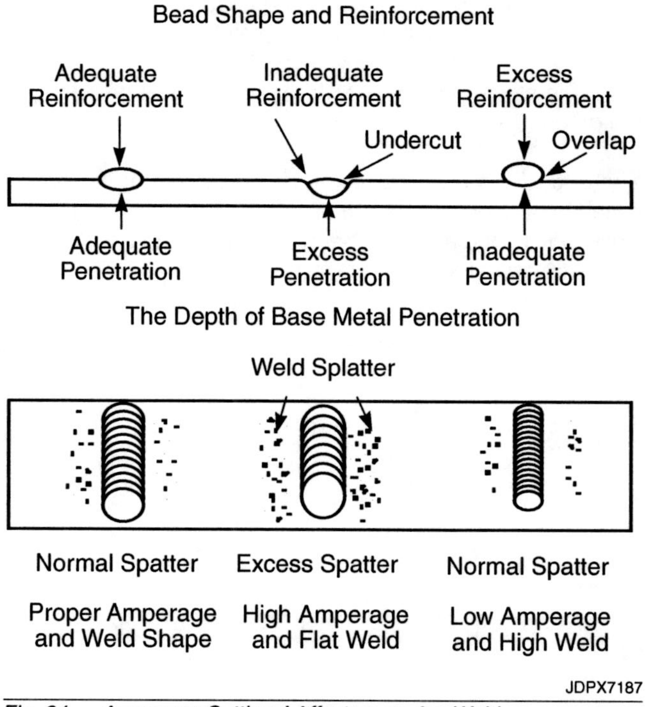

Fig. 24 — Amperage Settings' Affect on an Arc Weld

The amperage needed for a welding project depends on the following:

1. Base metal thickness
2. Position of welding
3. Electrode type and diameter

The base metal thickness, electrode size, and the welding position should be the first factors to consider in choosing an electrode for a project. For most flat-position welding, the electrode diameter should be as large as the base metal thickness will allow. Electrode diameters used for out-of-position welding should be small enough to keep the weld puddle small, from dripping, and from running out of the weld joint. The most common sizes used for-out-of position welding are 3/32 and 1/8 inch (2.4 and 3.2 mm).

As a rule, amperages at the high end of the recommended range for an electrode are used for welding in flat position. Amperages at the low to middle part of the range are used to make welds out-of position (vertical, horizontal, and overhead positions).

Whenever the amperage is too high, the flux coating and the electrode will turn a red color and sometimes start to burn over its entire length. Also, if the weld is a red color for a distance of 3/4 inch (19 mm) or more immediately after the arc is stopped, the amperage is either set too high or the travel speed is too slow.

When the amperage is too low there will not be enough heat to melt the base metal, slag will be trapped in the weld,and the weld deposit will not fuse completely to the base metal — it will just be lying on top. Also, the arc may be hard to start, the electrode will stick to the base metal, and the arc will go out.

The end of the electrode must be positioned at the proper distance from the base metal and directly over the weld joint, while maintained at the proper work angles and travel angles as illustrated in Fig. 25.

Electrode Positioning and Travel

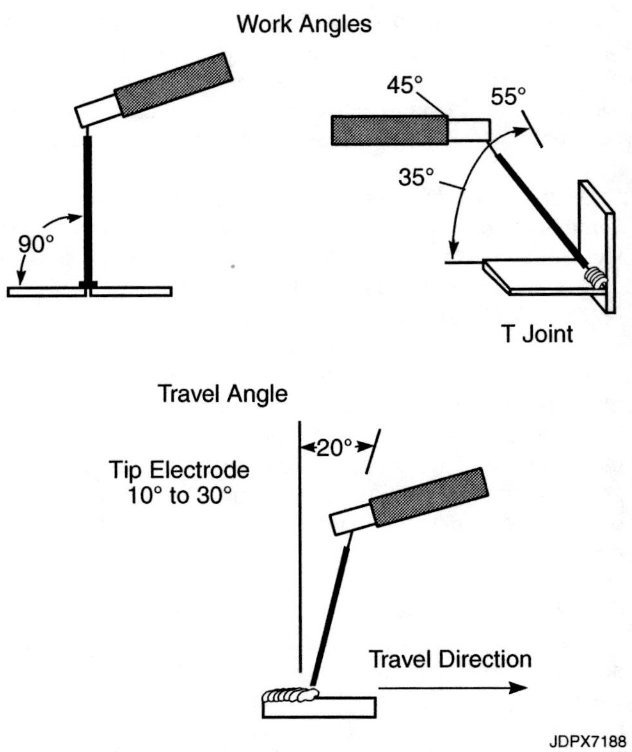

Fig. 25 — Shielded Metal Arc Welding (SMAW) Work and Travel Angle Examples

The arc length or distance is determined by the diameter of the electrode. The effect of arc length on weld appearance is illustrated in Fig. 26.

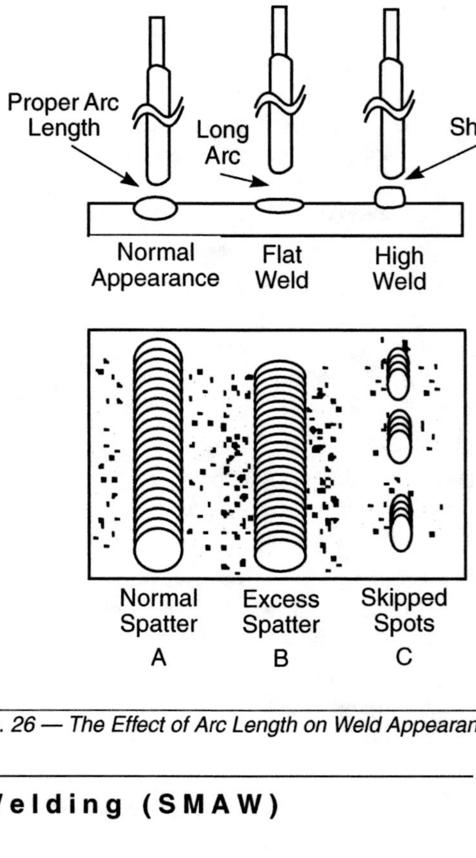

Fig. 26 — The Effect of Arc Length on Weld Appearance

Generally, speaking, the arc length should be approximately equal to the diameter of the electrode core wire and will produce a frying crackling sound when the proper arc length is maintained. For example, the arc length should be about 1/8 inch (3.4 mm) when welding in flat position with a 1/8 inch (3.4 mm) diameter electrode. A shorter arc is used for out-of-position welding as a means for helping hold the liquid weld puddle in place until it can solidify.

The travel angle and work angle affect the depth of base metal penetration, the placement of the weld bead in relation to the weld joint, and the formation of the slag at the back of the weld puddle (Fig. 27). The travel angle is the angle of the electrode in relation to the direction of travel. For welding in the flat, horizontal vertical down, and overhead positions, the electrode is tipped 10° to 30° in the direction of travel. For vertical welding, the electrode is normally pointed from a 90° angle to the direction of travel, to being tipped 10° to 30° away from the direction of travel.

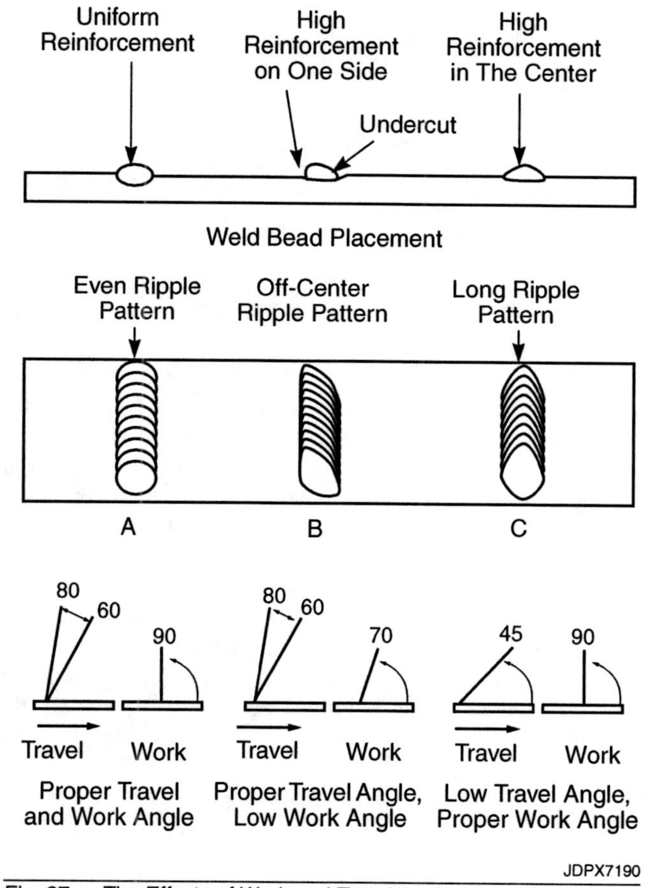

Fig. 27 — The Effects of Work and Travel Angle on Arc Weld

The work angle controls the distribution of the weld metal on the weld joint. The work angle should be positioned so that equal amounts of weld form on each piece. As illustrated in Fig. 27 the work angle for a butt joint is 90° and for a fillet weld ("T" and lap joints) the work angle will vary from 35° to 55°.

Improper work and travel angles will cause shallow base metal penetration Fig. 27, more weld to form on one side than the another Fig. 27, and slag to be trapped in the weld.

Travel speed must be maintained so that the base metal is melted by heat from the arc. Thus the weld penetrates to the desired depth into the base metal and forms to the desired size. A welder must learn to watch the weld puddle and judge how fast to travel for the amperage setting and for the type of electrode being used on the weld joint.

One way to judge the travel speed is by comparing the size of the weld to the diameter of the electrode core wire as illustrated in Fig. 28. The weld (after the slag has been removed) should be 1-1/2 to 2 times the diameter of the core wire. When this rule is followed and the weld is still too narrow in relation to the metal thickness, try decreasing travel speed, using a larger diameter electrode, manipulating the electrode by using a weave, or consider making two weld passes instead of one to obtain the necessary weld build-up. When the weld is too wide, try increasing travel speed or use a smaller diameter electrode.

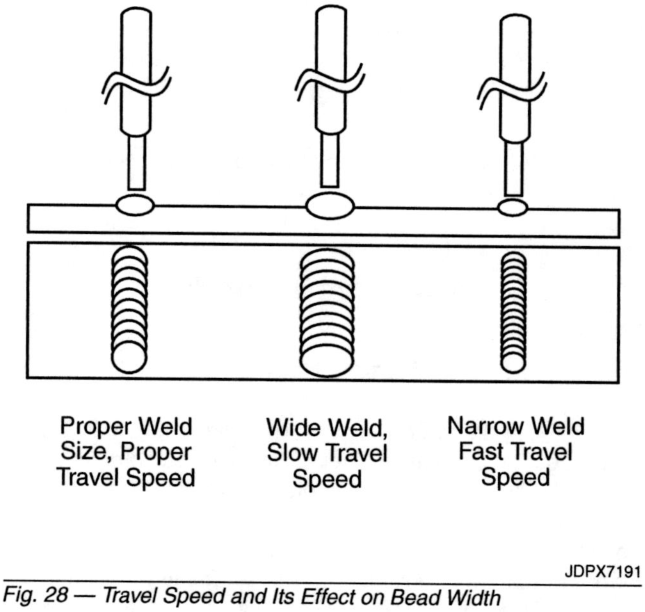

Fig. 28 — Travel Speed and Its Effect on Bead Width

Electrode Manipulation

Normally, when making flat position welds, a welder should not have to manipulate or make movements with the electrode to obtain a weld that is the proper size and that has an acceptable appearance. However, a movement called a "weave" is sometimes used to obtain a wide bead. This wide bead is called a cover pass when it is placed on the top of a multiple pass weld. Electrode manipulations are necessary for making welds in the horizontal, vertical, and overhead positions.

Unnecessary and improper movements can cause slag to be trapped in the weld. Also if the movement is not controlled properly the edges of the weld will not be even (Fig. 29) and the depth of penetration can vary from too much to not enough.

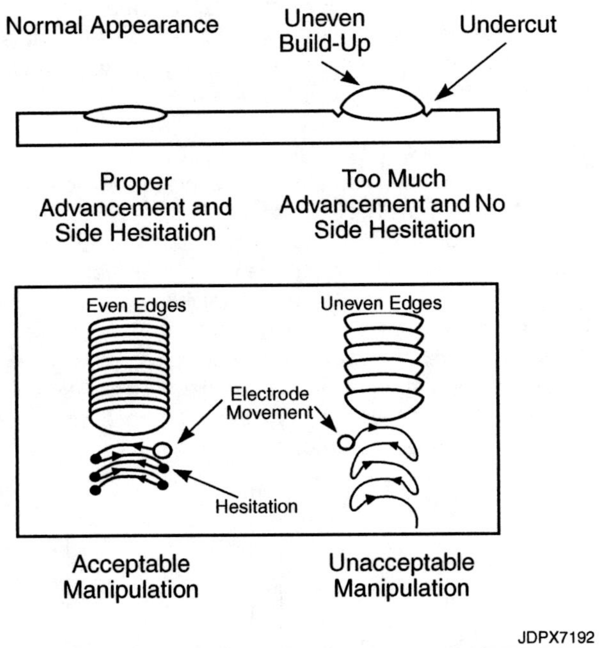

Fig. 29 — The Effect of Electrode Manipulation on Appearance of Arc Weld

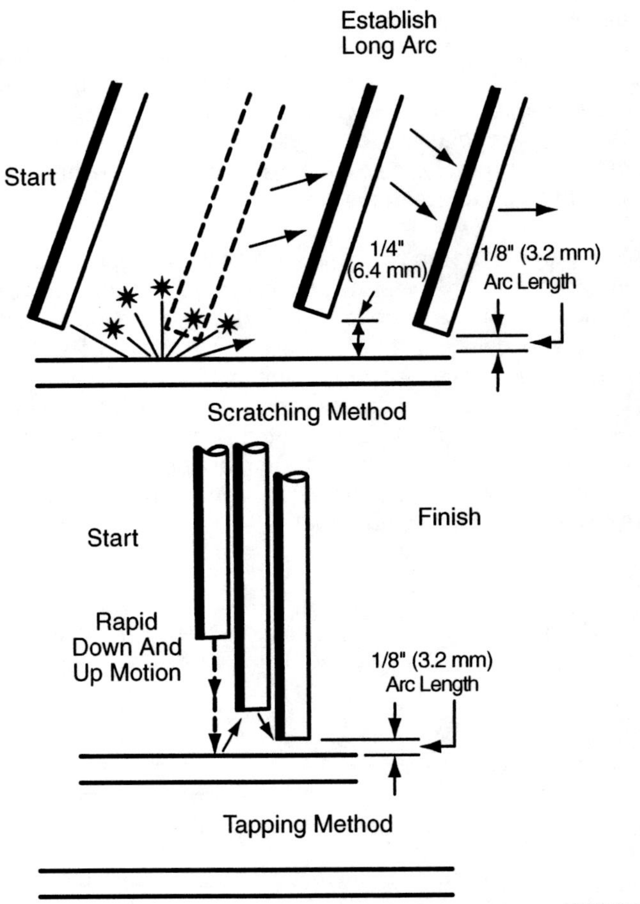

Fig. 30 — Starting the Arc — Two Methods

Flat-Position Welding Techniques

Striking and Maintaining an Arc

The first thing to do before trying to strike an arc is to set the amperage (and polarity when using DC) according to the type and diameter of the electrode selected for the project.

To start or strike an arc, scratch or tap the electrode on the work as illustrated in (Fig. 30). The scratching motion is the easiest to learn because the electrode isn't as likely to stick. Drag the electrode across the work in the same fashion as lighting a match.

With either motion, as soon as the arc starts, raise the electrode to a distance of 2 to 3 electrode diameters for a few seconds until the electrode begins to burn evenly with a steady sound. (If you raise too high the arc will go out, and if you don't raise high enough the electrode will stick.)

When an electrode does stick, try freeing it by moving it back and forth with a twisting motion (Fig. 31). If this doesn't work, release the electrode from the holder (Fig. 32). The electrode must be freed or released from the holder within a few seconds or it will become red hot and probably be ruined.

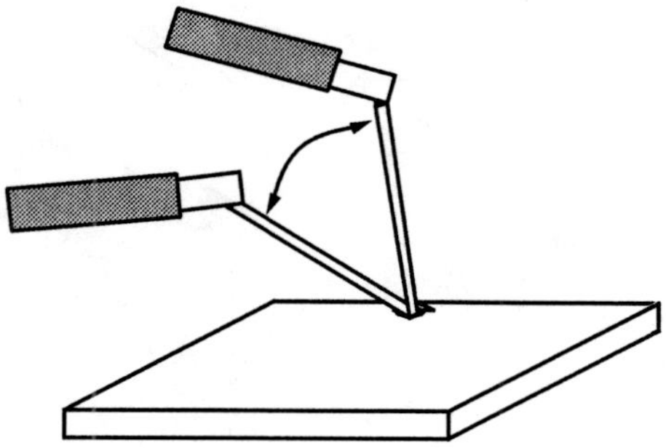

JDPX7194

Fig. 31 — Freeing a Stuck Electrode

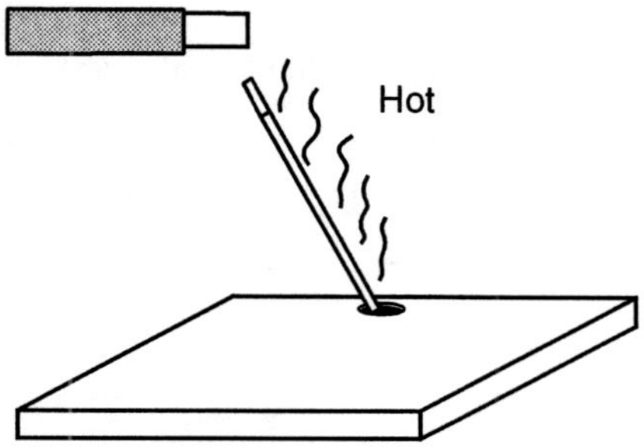

JDPX7195

Fig. 32 — A Stuck Electrode Will be Hot

CAUTION: Keep your welding helmet down and warn others not to look until the electrode has been freed. The rays from the short flash of light when electrical contact is broken can hurt your eyes.

CAUTION: Never turn off the on/off switch on the power source while current is flowing through a electrode that has stuck or while the arc is lit. The power switch could be ruined, and there could be enough arcing in the switch to cause a burn injury to your hand.

CAUTION: Don't touch an electrode that has been stuck until it has had a chance to cool for a few minutes — the glove and your hand can be burned if the electrode is hot when you touch it!

Making Weld Beads, Restarting Welds, and Filling Craters

Making weld beads is the next task to learn after mastering setting the amperage and striking an arc. Make lines on a piece of metal using a soapstone or center punch. Place a mark at about the center of each line to use for practicing stopping and restarting each weld as illustrated in Fig. 33 Hesitate at the end of the weld before completely reaching the edge of the metal to allow the crater to fill.

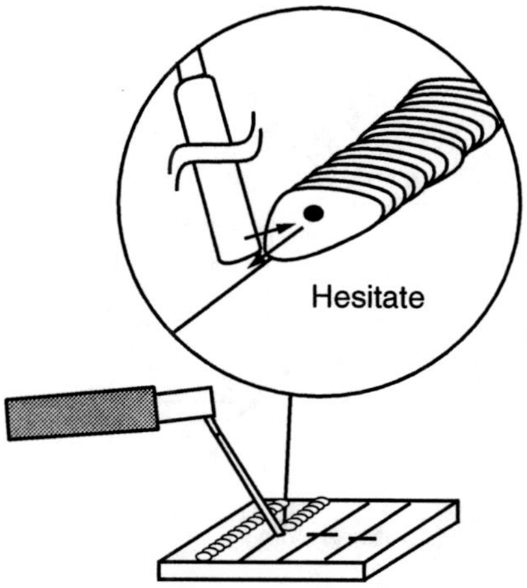

JDPX7196

Fig. 33 — Restarting and Tieing Into an Existing Arc Weld

Making Single-Pass Welds

Practice making square groove welds (butt joints) on one side of two pieces of metal about 1/4 inch (6.4 mm) thick. Welds made with the proper amperage and spacing should penetrate at least halfway to the other side (Fig. 34).

Next, practice making fillet welds (either "T" joints or lap joints) on one side of pieces of metal about 1/4 inch (6.4 mm) thick. Welds made with the proper amperage and travel speed should extend out on each piece the same distance as the metal thickness (Fig. 34).

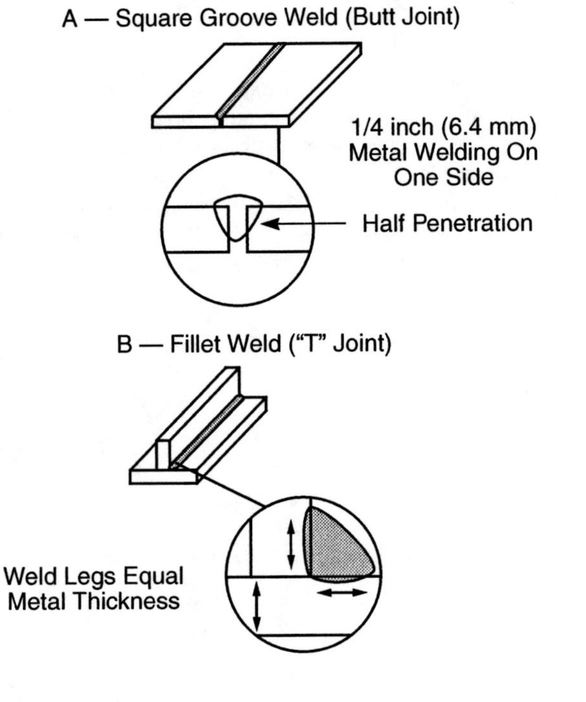

Fig. 34 — Single-Pass Arc Welding Procedures

Making Multiple-Pass Welds

To make a multiple-pass "V" groove weld, bevel two pieces of 3/8 inch (9.5 mm) metal with a grinder or cutting torch as indicated in Fig. 35 . Plan to complete the weld by making at least three passes. A keyhole will form at the front edge of the first weld, indicating complete base metal penetration. Be sure to remove the slag and weld spatter from each weld before making the next weld.

IMPORTANT: If the weld is not cleaned properly, slag will be trapped and the next weld will not fuse completely to the base metal, creating slag holes in the weld and areas with no fusion.

A multiple-pass fillet weld can also be made on of 3/8 inch (9.5 mm) metal. Also, plan to complete the weld by making three passes as indicated in Fig. 35. Be sure to clean each weld before making the next weld.

Safety Precautions

Proper safety precautions should be observed while welding. Make certain there are no flammable objects in or around the work area. Also, lighters cared in a pocket can pose potential life threatening danger. Sparks landing in or on a pocket containing a lighterer can create a serious safety hazard.

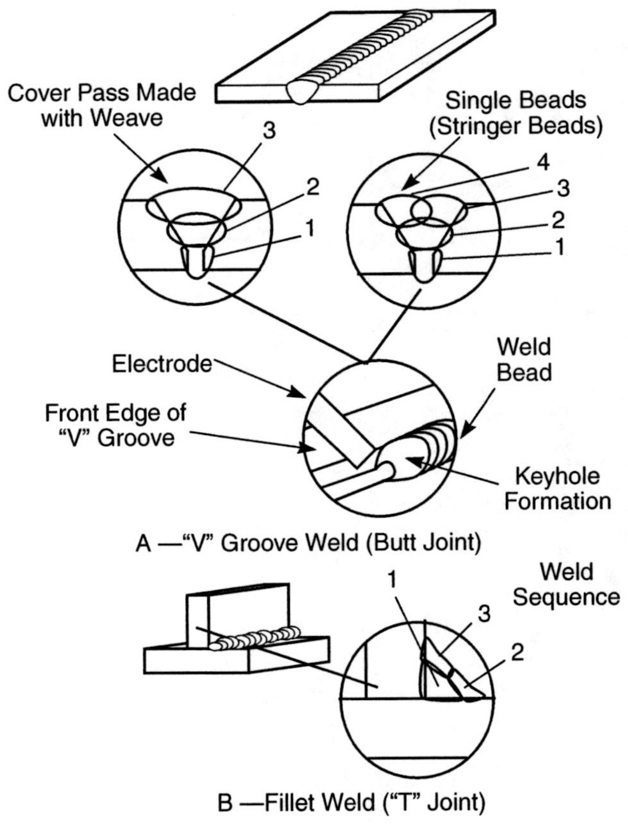

Fig. 35 — Multiple-Pass Arc Welding Procedures

Out-of-Position Welding Techniques

More skill is needed to make welds out-of-position than is needed for flat position welding. Choosing the proper electrode type, size, and corresponding amperage setting have a greater impact on the degree of acceptability of the finished weld. Metal preparation and fit-up also are more important. In short, to make welds that are acceptable when out-of-position welding, a welder must pay close attention to the detail of each aspect of making the weld.

Safety Precautions

Make certain that all parts of your body are protected from hot sparks from above — your head, ears, eyes, neck, hands, arms, legs, and feet. Also, make certain there are no holes in your clothing or open pockets that can catch a spark. Out-of-position welding generally creates larger sparks and more of them — don't learn the hard way and get your skin burned. Otherwise, follow all the safety rules for welding in flat position.

Electrode Selection and Amperage Settings

The fast-freeze electrodes (E6010 and E6011) and low-hydrogen (E7018) are the most common mild steel electrodes chosen for out-of-position welding. Small diameter electrodes (3/32 inch /2.4 mm and 1/8 inch /3.2 mm) are the easiest to control and to learn to use.

The amperage should be set between the bottom and the middle of the range recommended by the electrode manufacturer. There will be a much narrower window of amperages that will give acceptable welds compared to flat position welding.

Electrode Manipulation

Electrode manipulation is necessary for maintaining control of the weld puddle when making welds in the horizontal, vertical, and overhead positions. The most common manipulations for making welds in the vertical up position are illustrated in Fig. 36 for using a E6010 and E6011 electrode and in Fig. 37 for using a low-hydrogen electrode.

IMPORTANT: Do not interchange these manipulations of the electrode. You will not be able to control the weld puddle if you use the wrong motion.

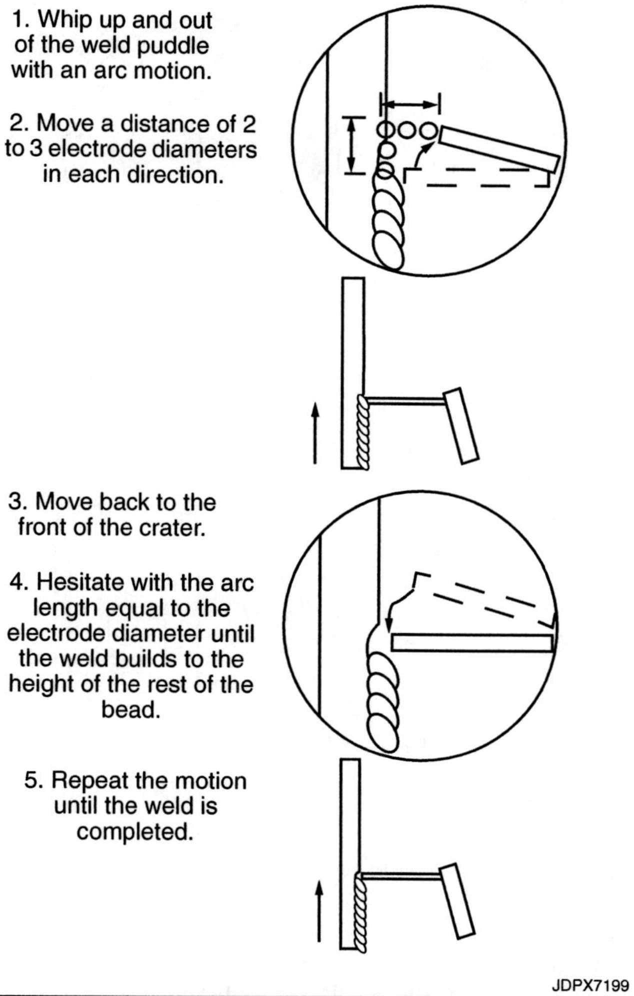

Fig. 36 — Whipping Motion Used for Out-of-Position Arc Welding with E6010 and E6011 Electrodes

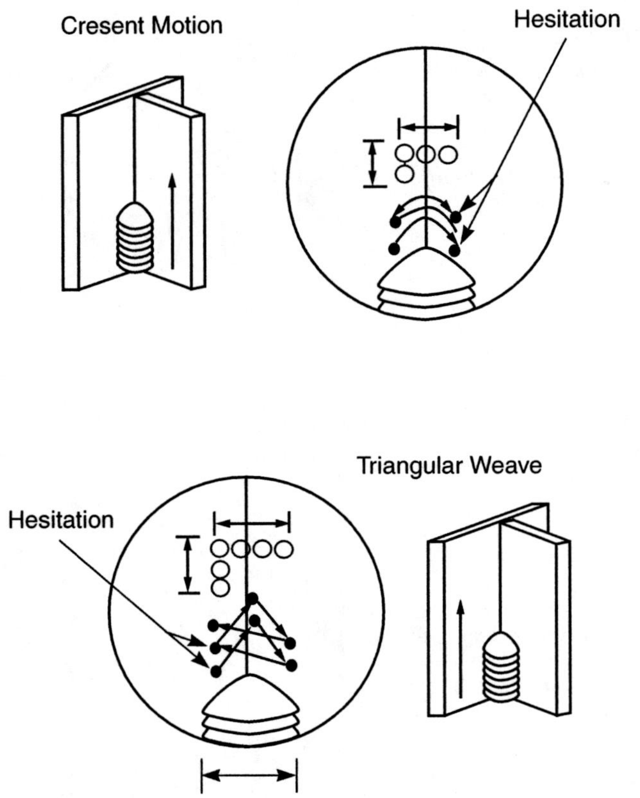

Fig. 37 — Crescent And Triangular Weave Used for Vertical Up Arc Welding with E7018 and E6013

IMPORTANT: Do not interchange these manipulations of the electrode. You will not be able to control the weld puddle if you use the wrong motion.

Strong welds with an acceptable appearance can be made when welding in vertical down position if a few simple facts are understood and followed.

1. Don't weld vertically down with low-hydrogen electrodes unless the manufacturer has specifically indicated the electrodes can be used for making welds in the vertical down position. Otherwise the weld will contain defects — porosity and slag inclusions.
2. The amperage should be set toward the maximum for the electrode.
3. Travel fast enough to keep ahead of the slag.
4. Keep the weld size small and do not make wide weaves or other types of manipulations.
5. Limit vertical down welding to metal thickness of 1/4 inch (3.8 mm) or less. Base metal penetration is much less compared to vertical up welding the same thickness of metal. Welding thicker metals could result in not enough base metal penetration for the weld to hold when it is put into service. Vertical down welding should be used to keep from melting holes in sheet metal.

Electrode Positioning

Hold as close an arc as possible without causing the electrode to stick when making welds out-of-position. Usually, this is about one half the length used for flat position.

Work and travel angles for the vertical up, vertical down, horizontal and overhead methods of welding are illustrated in Fig. 38, Fig. 39, Fig. 40, and Fig. 41.

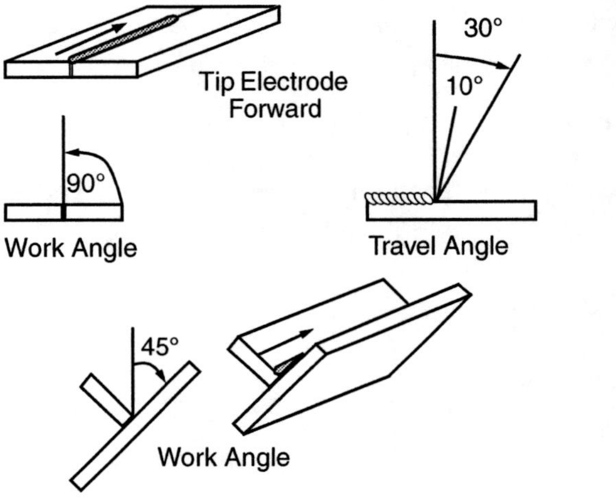

Fig. 38 — Flat Position Arc Welding

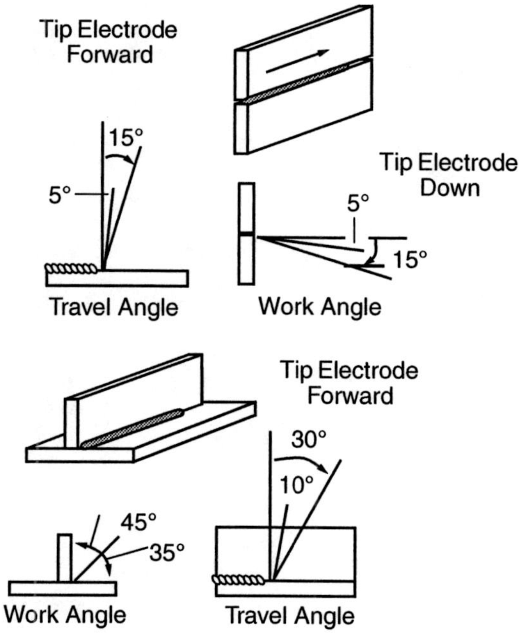

JDPX7202

Fig. 39 — Horizontal Arc Welding

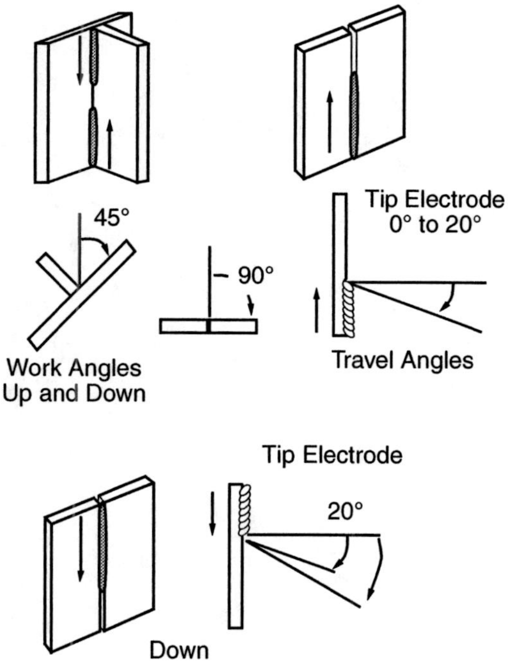

JDPX7203

Fig. 40 — Vertical Up and Down Arc Welding

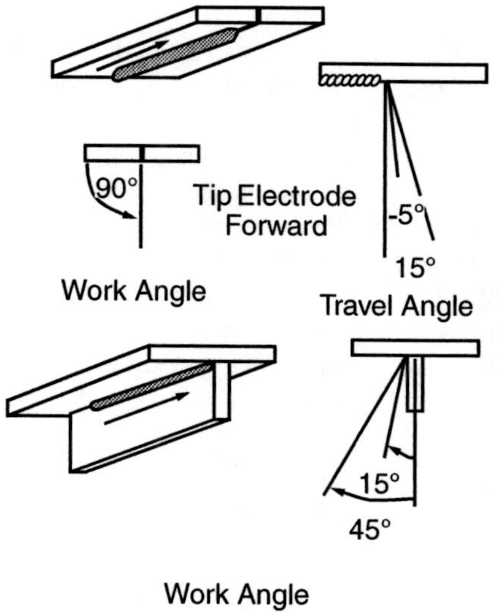

JDPX7204

Fig. 41 — Overhead Arc Welding

Welding Suggestions

1. Follow all safety precautions.
2. Keep the welding area clean and free of combustibles.
3. When chipping slag, always wear safety glasses with side shields, and when possible, let the metal cool before removing the slag.
4. Use charts and manuals for selecting the electrode and type of current, and for setting the amperage.
5. Always prepare the metal properly — clean the metal before starting to weld, space the metal to allow for penetration, and bevel the edges when necessary to obtain full penetration.
6. Use the proper arc length, work angle, and travel angle for the weld joint and welding position.
7. Clean and inspect completed welds carefully.

Steps for Shutting Off the Power Sources and for Stopping Work

1. Remove the electrode from the holder.
2. Turn off the power source and disconnect switch.
3. Wrap up the welding cables.
4. Mark all hot metal "HOT."
5. Returned unused electrodes to their container or to a heated storage oven.
6. Properly dispose of electrode stubs, slag, and scrap.

Acceptable Arc Welds

To be acceptable, groove and fillet welds must meet both visual inspection and structural strength requirements, and show evidence of good workmanship, including uniformity. Refer to Fig. 42 and Fig. 43. Visual inspection includes measuring weld size, penetration of the weld joint, reinforcement above the surface, and depth of undercut. Weld strength and fusion are checked by bending the piece in the weld area. These requirements are specified by the welding code that applies to the function of the finished part. More detailed information is given in Chapter 1.

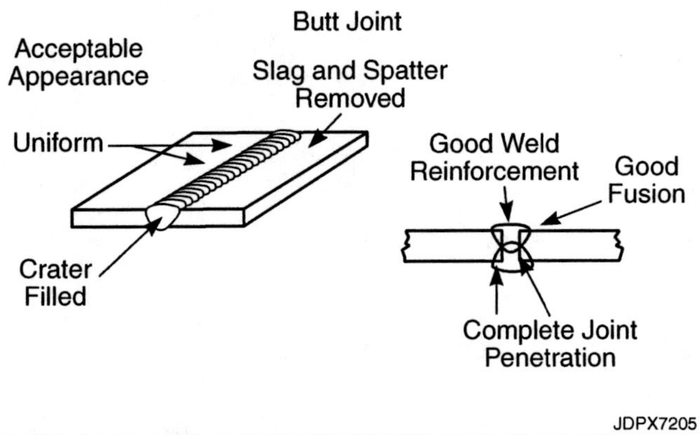

Fig. 42 — Acceptable Square Groove Arc Weld on 1/4 inch (6.4 mm) Metal

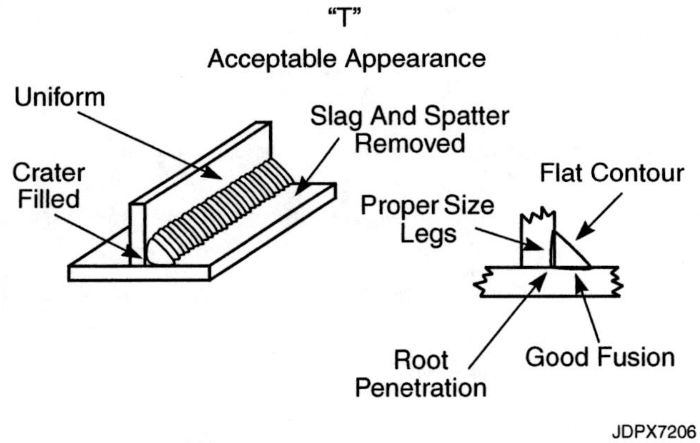

Fig. 43 — Acceptable Fillet Arc Weld on 1/4 inch (6.4 mm) Metal

Unacceptable Arc Welds

Generally speaking, a weld is considered to be unacceptable under any one of the following conditions:

1. A groove weld that doesn't completely fill the weld joint because of a lack of penetration and/or lack of reinforcement above the top of metal.
2. A fillet weld that does not have the proper leg size, throat measurement, and/or penetration at the root.
3. Weld penetration and/or reinforcement above the top of metal is greater than 1/8 inch (3.2 mm).
4. Undercut is greater than 1/32 inch (.83 mm).
5. The weld breaks or the base metal breaks in the weld area during a bend test.
6. The weld has not fused to the metal and/or contains slag holes.

Defective arc welds are illustrated in Fig. 44.

Troubleshooting Shielded Metal Arc Weld (SMAW) Defects

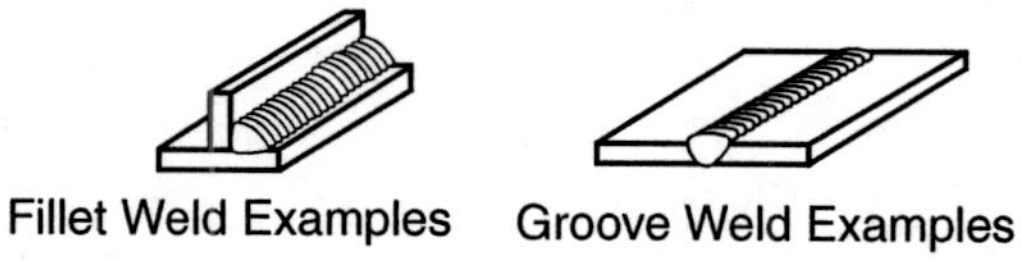

Defect	Corrective Action

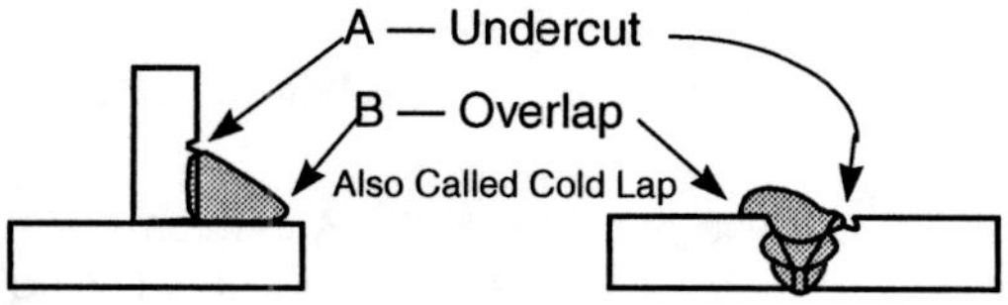

A — Undercut
1. Use proper work angle (fillet and groove weld)
2. Decrease amps (fillet weld)
3. Increase speed and/or decrease weld size
4. No hesitation or not enough hesitation on sides (groove weld)

B — Overlap
1. Decrease weld size per pass by increasing travel speed, using a smaller electrode or making smaller weld beads (fillet weld)
2. Use proper work angle (groove and fillet weld)

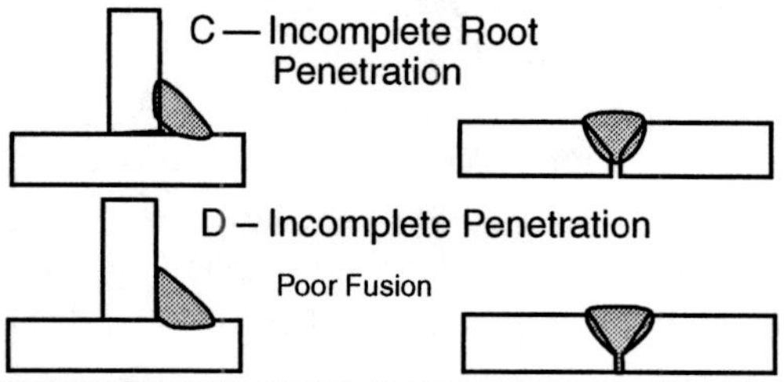

C and D — Poor base metal fusion and joint penetration
1. Increase amps and/or decrease weld size per pass (fillet and groove weld)
2. Do not weld vertical down (fillet and groove weld)
3. Increase root opening (groove weld)

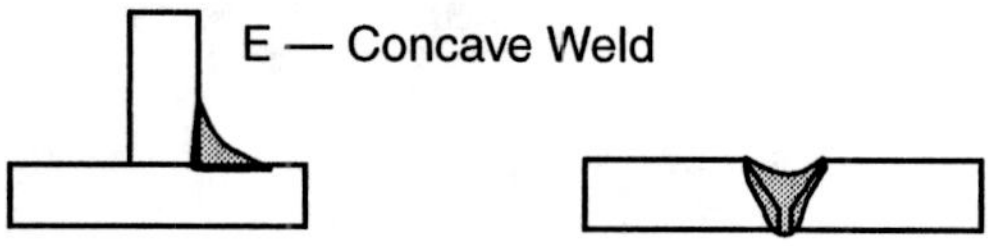

E — Concave weld (also called underfill on groove weld)
1. Do not weld vertical down (fillet and groove weld)
2. Make another weld pass (groove weld)
3. Decrease amperage (fillet weld)

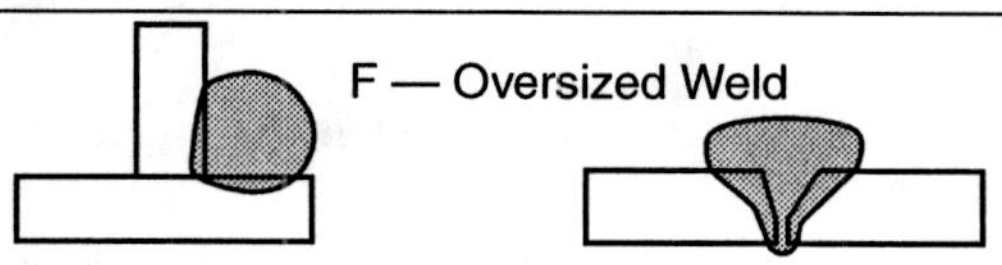

F — Oversized weld and excess reinforcement
1. Increase travel speed (fillet and groove weld)
2. Use smaller electrode (fillet and groove weld)
3. Follow proper welding procedures;do not make welds oversized

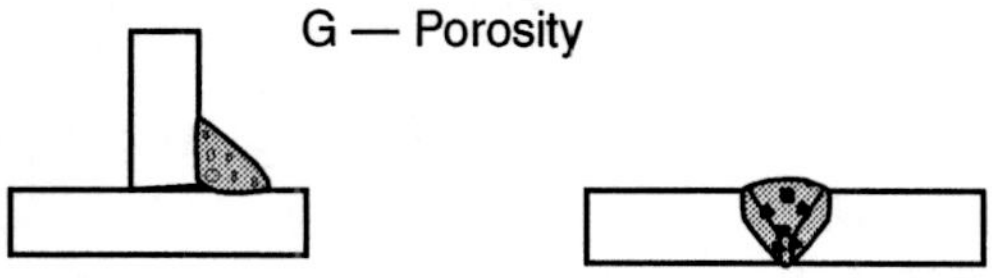

G — Porosity (fillet and groove weld)
1. Contamination on base metal
2. Replace wet low-hydrogen electrodes with dry or new electrodes
3. Hold proper arc length with low-hydrogen electrodes
4. Increase speed and/or decrease weld size per pass

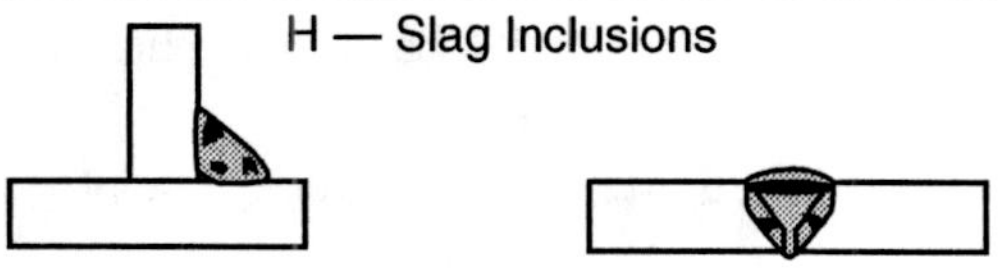

H — Slag holes and inclusions (fillet and groove weld)
1. Remove slag between making the next weld
2. Increase amperage
3. Increase speed and/or decrease weld size per pass

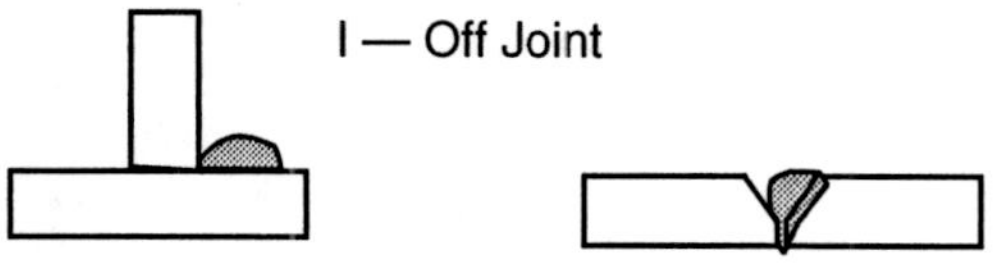

I — Off joint (fillet and groove weld)
1. Use proper work angle
2. Center end of electrode over both pieces
3. Correct vision problem — clean or replace cover glass, use lighter shade lens, or get eyes checked

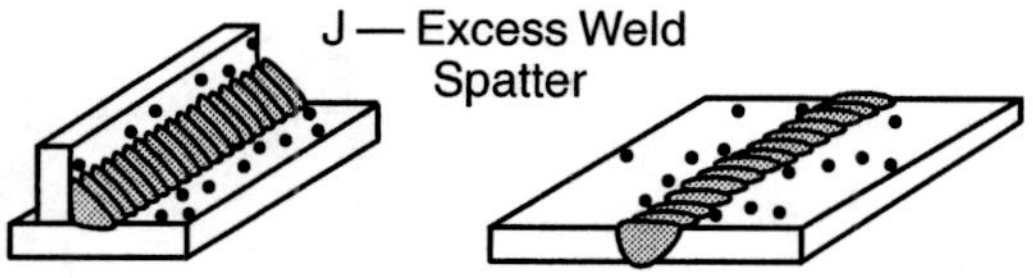

J — Excess weld spatter (fillet and groove weld)
1. Use proper arc length (holding too long an arc)
2. Decrease amperage
3. Replace E6010 or E6011 with another type electrode

JDPX7207

Fig. 44 — Common Defects in Shielded Metal Arc Welds

Test Yourself

Questions

1. Shielded metal arc welding is also known as _____ welding.
2. What is the main factor that determines the type of electrode to be used?
3. The input electrical power to operate the welding power source can be __________or _________ phase electric current.
4. During arc welding, the welder and all persons around must be protected from ___________ of the welding arc.
5. List two functions of the shielded metal arc welding (SMAW) electrode coating.
6. What is the purpose of the shielded metal arc welding (SMAW) power source?
7. What type of power source is needed for shielded metal arc welding (SMAW)?
8. What is the difference between a constant current (CC) welding power source and a constant voltage (CV) power source?
9. What two types of welding processes is a constant current (CC) welding power source designed to perform?
10. Referring to (Fig. 6), what would be the approximate amperage output for the 250 amp constant current (CC) welding power source at 60% duty cycle?
11. Referring to (Fig. 5), how long would you be able to sustain an arc using the 225 amp constant current (CC) power source, and using a 40% duty cycle, what amperage would be used?
12. Before purchasing a constant current (CC) welding power source, what are four items to check?
13. What type of current does a transformer welding power source produce?
14. What type of current does a transformer/rectifier welding power source produce?
15. What determines the size of the electrode holder, ground clamp and welding cable to be used?
16. Referring to Table 1, what size welding cable is recommended for 125 feet at 150 amps?
17. What are two reasons that will cause cable connections, the electrode holder, and cable to become hot?
18. What types of shoes (footwear) are not recommended when welding?
19. What type of eye protection should be worn when welding or working in a welding area?
20. Referring to Table 2, what shade lens is recommended for welding at 150 amps?
21. What first safety step should be taken before setting up shielded metal arc welding (SMAW) equipment?
22. Referring to Fig. 22, what is the recommended root opening for a single "V" groove with a backup strip?
23. Referring to Fig. 22, what is the recommended bevel angle for a single V groove with a open root?
24. Referring to Fig. 22, what is the recommended root face (land) for a single bevel groove?
25. What two items will determine the type of electrode that is recommended for the welding project?
26. What two identification numbers or letters do electrodes have?
27. What do the first two numbers in the American Welding Society (AWS) electrode identification system indicate?
28. When the American Welding Society (AWS) electrode identification number ends in 8, what type of current (polarity) is recommended?
29. Referring to Table 3, what type of stainless steel electrode is recommended for welding dissimilar metals?
30. What shielded metal arc welding (SMAW) electrodes will give you a fast freeze, deep penetration, flat bead appearance, and all-purpose welding results? (Hint... AWS#)
31. Shielded metal arc welding (SMAW) electrode size is measured by the _____ of the _____ _____?
32. What electrodes need special heated storage containers?
33. List five key points to control the weld puddle in shielded metal arc welding (SMAW).
34. Before setting the amperage, the welder must know what three items?
35. What work angle is recommended for "T" and lap joints?
36. What is the approximate bead width after the slag has been removed?
37. Name two ways to start the arc in shielded metal arc welding (SMAW)?
38. What is the maximum size of reinforcement for a groove weld?
39. When welding a groove joint, what is the purpose of forming a keyhole on the root pass?
40. Sketch the bead sequence on a multi-pass "T" joint.

Gas Metal Arc Welding and Flux-Cored Arc Welding (GMAW and FCAW)

4

JDPX7107

Fig. 1 — Semiautomatic Gas Metal Arc Welding (GMAW) Equipment at Work in a Farm Shop

Introduction

Gas metal arc welding (GMAW), dipmatic, microwire, CO_2, and MIG are some of the more common names given to the gas metal arc welding process. Fig. 2 illustrates the basic GMAW equipment and circuit.

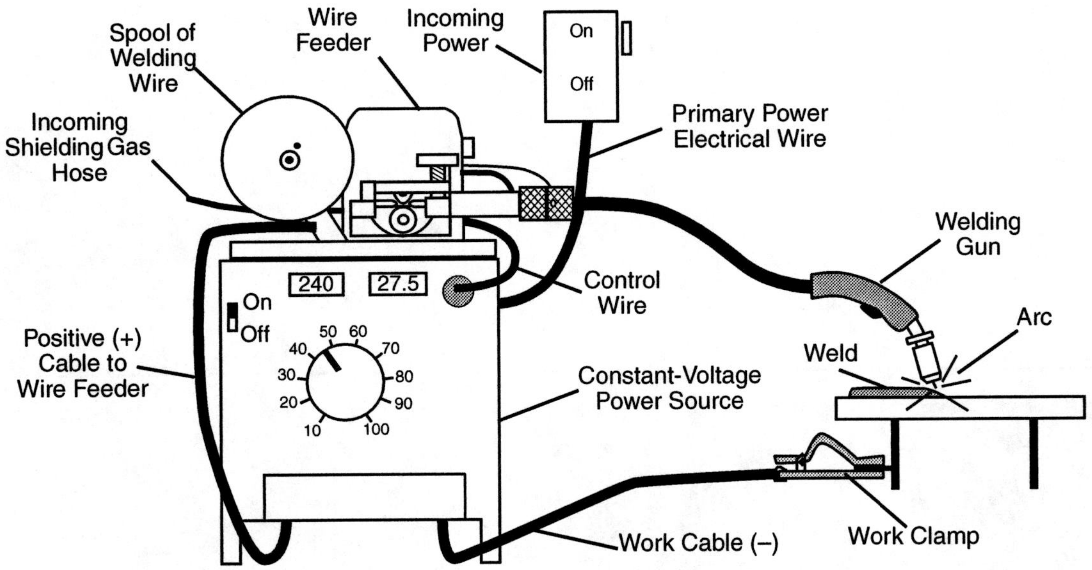

Fig. 2 — Basic Gas Metal Arc Welding (GMAW) Circuit and Equipment

Flux-cored arc welding (FCAW) and gas metal arc welding (GMAW) are two groups of processes within what is called the "continuous wire" welding process. A continuous wire from a spool or coil is driven by wire feeder through a flexible conduit to a welding gun. The welding wire exits the gun through a copper part called a contact tip (or tube) into an electric arc produced from a constant-voltage power source. The wire melts and combines with the base metal to form a weld (Fig. 3).

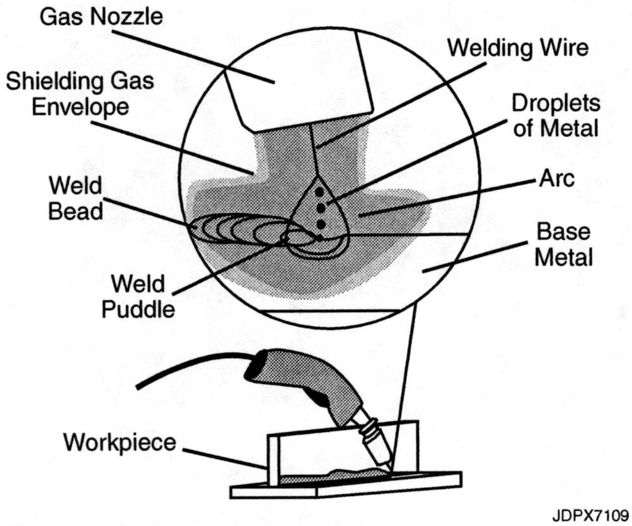

Fig. 3 — Formation of a Gas Metal Arc Weld (GMAW) Process

The molten metal is protected from the surrounding atmospheric air by a gas shield. The shielding gas may be supplied from a nearby high-pressure cylinder or a manifold system. A shielding gas can also be formed from elements in the flux located in the core of the wire electrode as it melts. In some instances, weld shielding is achieved from both elements in the flux core and from an externally supplied shielding gas.

Methods of Controlling the Weld

Welding with these processes can be done using semiautomatic, automatic, or mechanized welding equipment.

Semiautomatic welding equipment controls the feeding of wire electrode to a welding gun held by the person doing the welding. Once the proper wire speed has been set, welding is started and stopped by activating the trigger on the gun. All the welder has to do is position the gun and control the travel speed along the weld joint. The welding amperage output of the power source is determined by setting the wire speed on the wire feeder. A variably-adjustable voltage control on the power source must be set to match the type of metal transfer for the size and type of wire electrode (and for the shielding gas if one is required). The voltage from the power source actually controls the arc length necessary for welding. However, the welder must maintain the proper distance between the gun and the work in somewhat the same manner as he/she maintains the proper arc length when shielded metal arc welding (SMAW) (refer to Chapter 3). This semiautomatic welding is considered hand welding because the welders position the gun with their hands (Fig. 4).

JDPX7110

Fig. 4 — Building Construction Using the Gas Metal Arc Welding (GMAW) Process

Automatic welding on a part is done without any assistance from a person other than the initial setup and programming of all the equipment. A central controller operates mechanized equipment that loads components of a part into a fixture for proper alignment. Then robotic equipment positions the welding gun along the areas to be joined and makes the necessary welds (Fig. 4). When welding has been completed, mechanized equipment removes the part from the fixture and the operation starts over again. Once the welding and parts-handling equipment has been programmed properly, the entire operation can function without any assistance from a person, unless something goes wrong.

JDPX7111

Fig. 5 — Automatic Gas Metal Arc Welding (GMAW) Equipment

Mechanized welding of a part is accomplished by attaching a welding gun to a device that can be controlled or manipulated into the proper position by a person as the part moves under the welding gun. Or the gun can be mounted on a carriage mechanism to move the arc along the joint. Welders, or "welding operators" as they are sometimes referred to, are responsible for operating the controls that start and stop the weld, and for keeping the welding gun properly aligned from start to finish.

Mechanized welding is used for joining parts that have too much variation in size and fit-up to be fully automated, but can be mechanized to the extent that a welder doesn't have to hold the gun in their hand during welding or control the travel speed.

Table 1 shows these application relationships of the various types of welding we have just reviewed.

A	M	S	H	Variables
E	E	E	W	Electrode Feeding
E	E	E	W	Arc Length
E	O	W	W	Electrode Angle
E	O	W	W	Joint Alignment
E	E	W	W	Travel Speed
A = Automatic			E = Equipment	
M = Mechanized			O = Operator	
S = Semiautomatic			W = Welder	
H = Hand				

Table 1 — Welding Applications and Variables to Control

Fabrication and repair shops now use variations within the continuous wire welding process for most metal joining applications. Welds that were once very time consuming or that were difficult to make on carbon steel, stainless steel, aluminum, and many other metals with the oxyacetylene flame or shielded metal arc welding electrodes are now routine procedures with the GMAW and FCAW processes.

Thin metals like those found in automobiles and household appliances, and thick metals like those found in farm machinery and earth moving equipment, are easily joined with one of the continuous wire processes. Welding may be done manually with a hand-held gun as illustrated in Fig. 1 and Fig. 4, or with fully automatic equipment as illustrated in Fig. 5.

Shielded metal arc welding safety practices also apply to the continuous wire welding processes. A darker shade of welding lens may be necessary because the shielding gas for gas metal arc welding does not block the arc's rays like the shielding from shielded metal arc welding. Darker lenses are also necessary because higher amperages are used for spray transfer, pulsed-spray transfer, and FCAW applications, compared to shielded metal arc welding applications.

Gas Metal Arc Welding (GMAW) Processes

Names for variations within the gas metal arc welding process are determined by the melting of the wire electrode and by the way the metal is transferred to the weld puddle to form a weld. The following descriptions provide guidelines for understanding each variation within the gas metal arc welding processes. Fig. 3 illustrates the melting of the wire as a weld is formed.

Gas Metal Arc Welding—Short Circuiting (GMAW-S)

GMAW-S, "S" for short circuiting (often called MIG short arc, dipmatic, and microwire), is a method of metal transfer recognized by the American Welding Society as short circuiting. With short-circuiting transfer, the end of a small diameter (0.023–0.045 inch [0.6–1.1 mm]) solid wire electrode is melted by a series of electrical short circuits (Fig. 6-A) occurring from 20 to 250 times per second at the point the wire makes contact with the base metal.

When the switch on the welding gun is activated to start welding, shielding gas flows through the nozzle on the welding gun, the power source energizes the welding wire, and wire feeds through the welding gun. When the wire contacts the base metal, current flows through the wire, causing the end at the point of contact to get hot (Fig. 6-A-2). The magnetic force around the portion of wire that has almost reached the melting point is pinched off (Fig. 6-A-3). An electric arc is formed at the end of the wire for a small fraction of a second (Fig. 6-A-4). Heat from the arc and heat from the droplet of metal form a weld puddle. As the wire continues to feed through the gun, the end of the wire is driven into the weld puddle. The arc is extinguished and creates another short circuit—thus the short-circuiting process begins all over (Fig. 6-A-5).

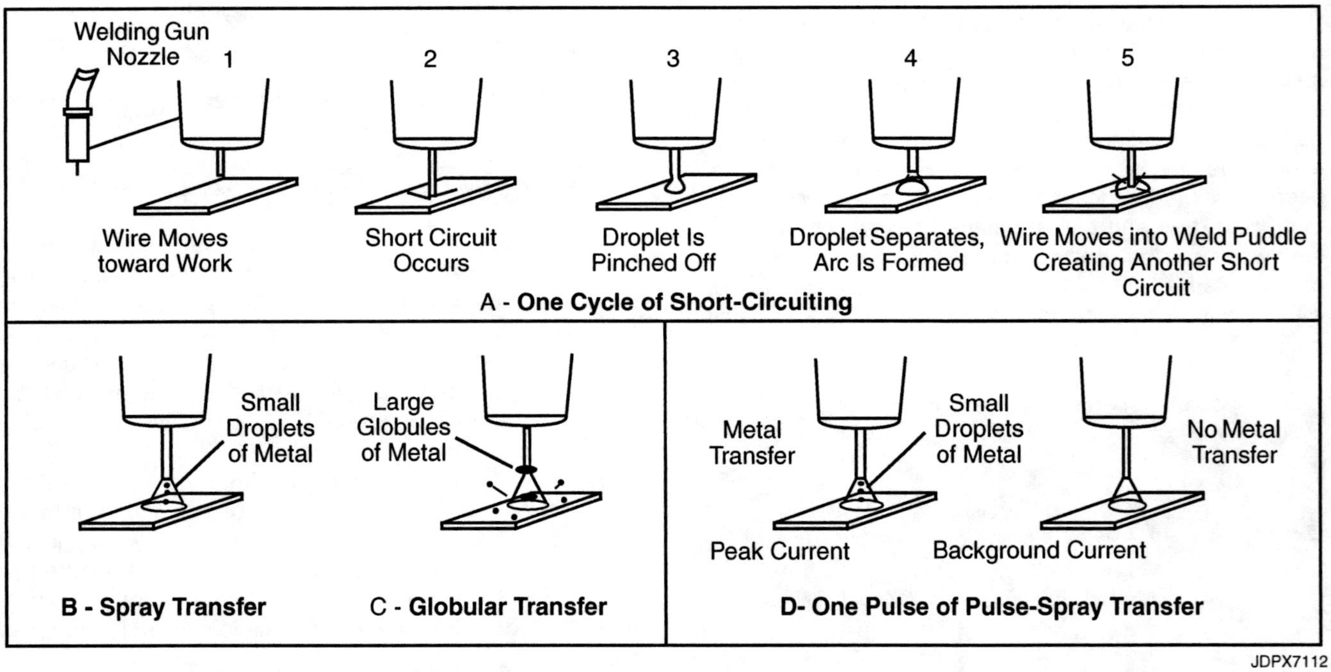

Fig. 6 — Type of Gas Metal Arc Welding (GMAW) (Metal Transfer Variations)

When the proper technique and parameters are being used, short-circuiting transfer can be recognized by a constant buzzing or sharp crackling sound. The weld will be almost completely free of spatter and smoke. The surface of the weld will contain small gray to brown colored scales called tenacious isles of silicon (Fig. 7). The surface will not be completely covered with these isles of silicon like a shielded metal arc weld is covered with slag.

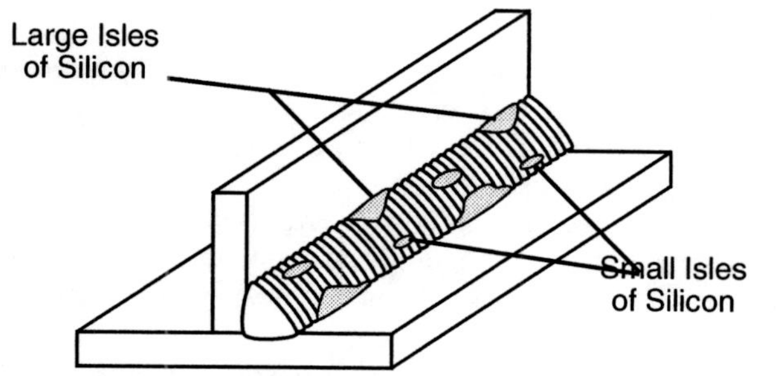

Fig. 7 — Tenacious Isles of Silicon on Weld

CAUTION: Hot flying isles of silicon can cause serious burn injuries to unprotected skin, especially the eyes and face. As the weld cools to room temperature, some of the isles of silicon pop off the surface of the metal and can strike anyone nearby. Always wear the proper safety equipment, especially safety glasses with side shields, until out of the welding area.

IMPORTANT: In some areas of the weld these isles of silicon are large enough to be hard to cover with paint (Fig. 7) and in other areas they are small enough not to affect painting(Fig. 7)). The weld may require wire brushing or special cleaning to remove these isles of silicon before painting or the weld may require a second coat of paint.

Short-circuiting transfer allows a welder to have maximum control in keeping the weld puddle size small and penetration shallow. Control over these variables make short circuiting ideal for welding ferrous and nonferrous metals ranging in thickness from 0.030—0.25 inch (0.015—6.4 mm). Short circuiting is also well-suited for applications requiring out-of-position welding, for joints with poor fit-ups (wide gaps between pieces that must be filled with weld), and for controlling warpage.

Following the proper welding procedures and using the proper welding parameters for short circuiting will ensure strong welds that have an acceptable appearance. Weld defects, such as a lack of base metal fusion and excess weld spatter, are common problems caused by not following the proper welding procedures and/or welding parameters for short circuiting. When starting a new project using short circuiting, always inspect and test a practice weld on scrap material to ensure there is proper weld fusion to the base metals.

Gas Metal Arc Welding (GMAW) Spray Transfer

GMAW spray transfer (often called MIG spray) is the name recognized by the American Welding Society for the next variation in the GMAW processes. With spray transfer the arc is established with the same sequence of events as with the first short-circuit cycle. However, this is where the similarity between the two methods of metal transfer stops. Once the end of the wire is pinched off and an arc is formed, there continues to be an arc between the end of the wire and the base metal as illustrated in Fig. 6-B. The constant arc causes the end of the wire to melt into small droplets that are propelled to the weld puddle. The weld puddle is larger and more fluid, and penetration of the weld into the base metal is much deeper than with short-circuiting transfer.

When properly used for making a weld, the arc will produce a light hissing or whispering sound. The weld will also be free of smoke and spatter. The surface of the weld will also contain isles of silicon. The surface covering will be similar to that of a weld made with short circuiting; however the isles of silicon will be larger, sometimes as large as a pencil eraser. The weld surface will have patches of silicon and will not be completely covered, as in shielded metal arc welding where slag covers the entire weld.

In many applications using spray transfer, it has been necessary to use automated equipment in order to take advantage of the high travel speeds possible. The relatively high voltages and wire speeds produce a deeper, wider, and very fluid weld puddle that is difficult for the welding operator to control. In either automated or manual applications, spray transfer is best suited for making fillet and groove welds in flat positions on ferrous and nonferrous metals over 1/4 inch (6.4 mm) in thickness. Fillet welds can be made in the horizontal position; however, there is a tendency for the vertical piece to be undercut.

Gas Metal Arc Welding—Pulse-Spray (GMAW-P)

GMAW-P is the abbreviation used by the American Welding Society for pulse-spray transfer. The arc starts when the wire contacts the base metal, just like with short-circuiting and spray transfer. However, after the arc has started, the welding current is varied from a peak or high setting to a background or low setting. This pulsation from a high to a low current (Fig. 8) produces a heating and cooling of the weld puddle similar to that which occurs with a short-circuiting cycle. Fig. 6-D illustrates melting of the wire electrode and metal deposition that occurs when the current reaches the peaks of the pulse cycle.

Depending on equipment adjustments, the transfer of the molten metal can occur during peak (driving, digging, deep penetrating arc) or during background (soft, shallow penetrating arc).

Descriptions for each part of the pulse wave form are shown below (Fig. 8).

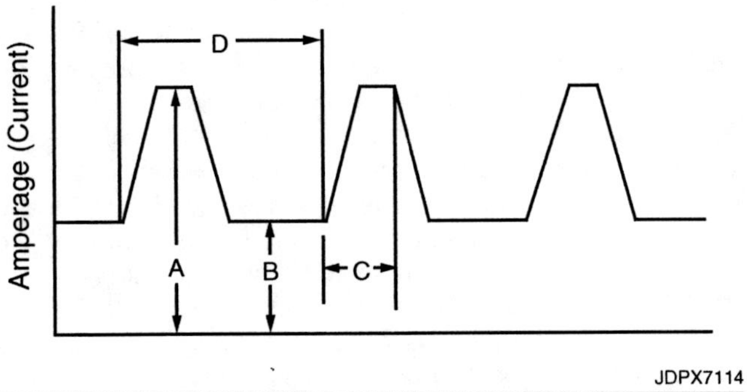

Fig. 8 — Description of Pulse Wave Form with Gas Metal Arc Welding-Pulse Spray (GMAW-P)

A. Peak Current

- The highest current levels in the pulsing waveform
- Current level where spray transition current is achieved

B. Background Current

- The lowest levels in the pulsing waveform
- Current level low enough to maintain an arc but allow weld pool control

C. Peak Time (or Pulse Width)

- The time peak current is applied
- It is the current rise to the beginning of current fall time during a pulse
- Longer time produces a wide arc column; short time produces a narrow arc column

D. Pulses Per Second (PPS or Pulse Frequency)

- A peak and a background equals a pulse; this represents the number of these per second
- Adjusting pulse frequency affects the background current time
- This control has a great effect on arc length

A special type of constant-voltage power source and wire feeder is needed for pulse-spray transfer. The power source is called an inverter and the wire feeder must contain a microprocessor. Some equipment can be interfaced with a computer and the pulse waveform can be displayed and manipulated on screen. Welding programs are developed using the microprocessor for controlling the peak welding current, the background welding current, the duration of time at the peak, the duration of time at the background current, and the number of pulses per second. Several programs (eight or more) of welding parameters for commonly used welding applications can be stored in the memory of the microprocessor.

Synergic and Adaptive Controls

Once the proper combination of peak current, background current, pulses per second, and wire speed have been programmed into memory for a particular project, a change in wire speed will make the microprocessor change everything else proportionally. This is called a synergic control and saves reprogramming every time the wire speed is changed for a different metal thickness or weld joint.

Inverter power sources can also have an adaptive control that automatically regulates the output for changing weld conditions such as electrode extension (stick-out or contact tube-to-work distance). A welder only needs to select the electrode wire size and wire speed for a project when using a preprogrammed power source and wire feeder that has an adaptive synergic pulse control.

When the proper parameters are being used, there will be a low-pitched humming sound from arc with pulse-spray. The surface of weld will look similar to a weld made with spray transfer, including the isles of silicon.

Pulse-spray transfer Fig. 6-C can be used in applications where it is difficult to obtain acceptable welds with either short-circuiting or spray transfer without having a person with a very high level of welding skills. It is possible for a person with average welding skills to make spatter-free welds that have adequate base metal penetration on both ferrous and nonferrous metals with pulse-spray transfer, even when neither short-circuiting or spray transfer welds were acceptable.

Typical applications for pulse-spray transfer include out-of-position welding, joining thin metals (1/16 inch [1.6 mm]) to thick metals (1/4 inch [6.4 mm]) and for welding on joints that have gaps to fill that are not more than 3/16 inch (4.8 mm) wide and for eliminating weld spatter. For pulse welding, joints must have consistent root opening, root face, joint alignment, and no gaps greater than 3/16 inch (4.8 mm) wide. Pulse-spray can also be used for welding thicknesses of hot-rolled steel and galvanized steel on which adequate weld penetration is difficult to obtain with short circuiting or that are too thin to weld with spray transfer. Pulse-spray transfer can sometimes reduce warpage on applications using spray transfer and increase warpage on applications using short circuiting.

Globular transfer as illustrated in Fig. 6-C occurs when the wire speed and voltage are not balanced properly when trying to achieve either a short-circuiting or spray-type transfer. The arc produces an intermittent popping or dripping sound, the weld contains large droplets of weld spatter, and base metal penetration can be unacceptable. Welding in the globular transfer range does not produce welds that have a nice appearance. Base metal penetration is usually not adequate with globular transfer when trying to weld metals in the spray-transfer thickness range.

Flux-Cored Arc Welding (FCAW) Process

The name flux-cored arc welding was given to the process because a weld is formed by melting a flux-cored tubular steel wire in an electric arc on the base metal. Names for variations within this welding process are determined by the type of shielding required and by the composition of the flux core of the electrode wire.

Flux-cored arc welding wires can be divided into two groups:

- Externally shielded wires (Fig. 9)
- Self-shielded wires (Fig. 9)

Both types of wire produce welds that have a slag covering similar to that formed on the top of a shielded-metal arc weld. The slag should be removed between weld passes and before the part is painted.

In general, FCAW is an easy, forgiving process to use. Learning to produce high-quality production welds is quickly done. It produces good fusion and penetration abilities. Electrodes are generally in a range of 0.030—0.125 inch (0.76—3.2 mm) in diameter.

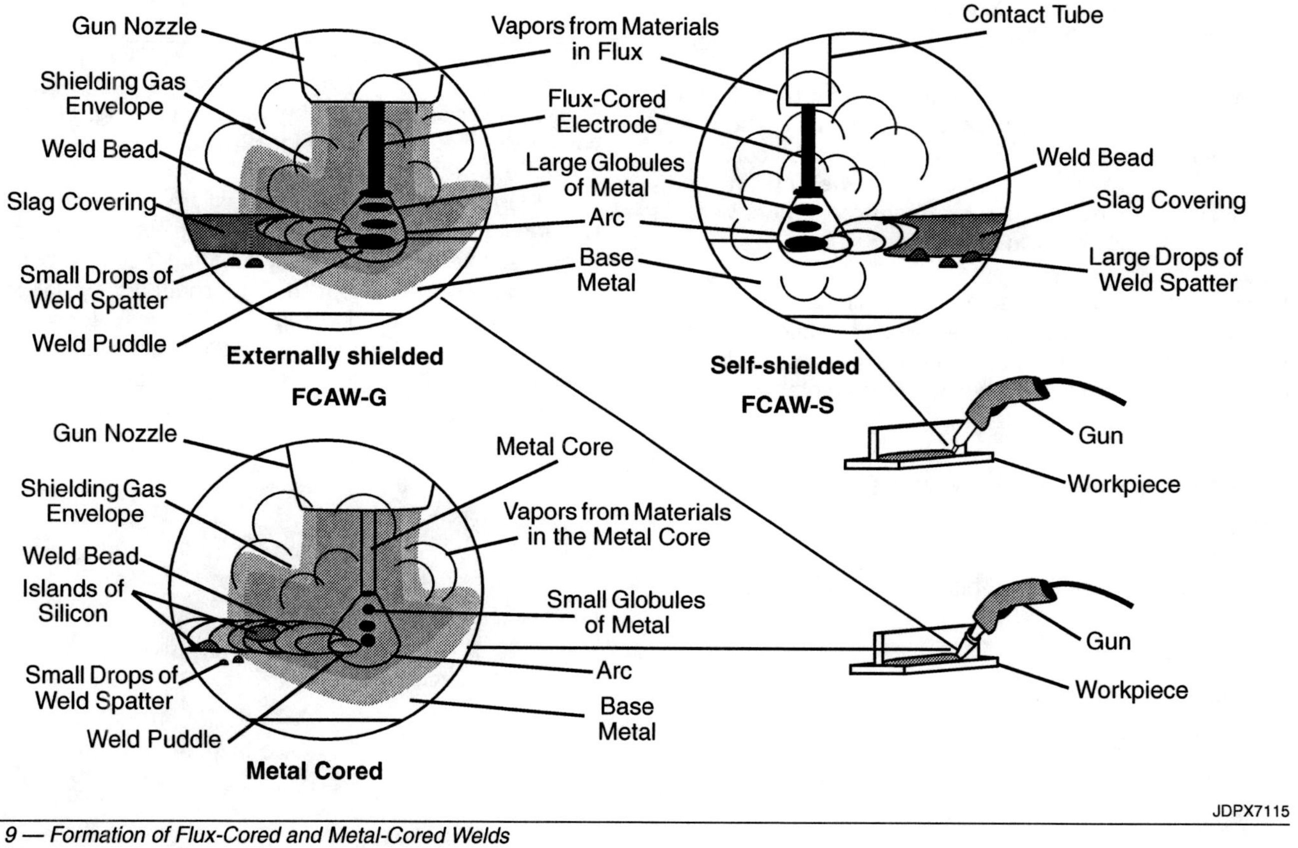

Fig. 9 — Formation of Flux-Cored and Metal-Cored Welds

Self-shielded flux-cored wires obtain their shielding from gases produced as the flux inside the core wire decomposes in the welding arc. In addition to removing slag, cleaning is necessary to remove weld spatter and smoke from the weld before the part is painted.

Proper ventilation is needed in the weld area because the volume of fumes is equal to or greater than that produced from shielded metal arc welding. These wires are typically used outdoors on construction sites where a build-up of fumes in an enclosed area is less likely. Self-shielded wires can also handle more wind or drafts than a gas-shielded wire.

As the name implies, externally shielded flux-cored wires require an external shielding like gas metal arc welding with a solid wire. Typically, less cleaning is required to remove weld spatter and smoke on welds made with externally shielded flux-cored wires compared to welds made with self-shielded flux-cored wires.

Most of the wires are limited to flat position groove and fillet welds and are of a diameter greater than 5/64 inch (1.9 mm). Some wires can be used out-of-position and are a diameter of 5/64 inch (1.9 mm) and less.

These wires also produce a large volume of fumes and require good ventilation in the weld area. However, the ventilation system cannot interfere with the shielding process. The weld must be protected from air currents or drafts that could cause the gas shield to be blown or drawn away.

Metal-Cored Arc Welding (MCAW)

Metal-cored wire is a type of tubular wire that requires an external shielding gas (Fig. 9). The weld is not covered by slag. However, there are intermittent brown scales on the surface of the weld similar to those found on a weld done by spray transfer.

Because of the ability to deposit large amounts of weld with metal-cored wire, good ventilation is important. The use of pulse is beneficial in reducing fumes with metal-cored wire. When using ventilation systems or welding in drafty conditions, the weld must be protected. Otherwise, its shielding gas will be blown away.

Using metal-cored wire is somewhat similar to gas metal arc welding in that it uses a tubular wire containing powdered metal and other materials that melt to form a weld, similar to the way a solid wire forms a weld with gas metal arc welding.

Gas Metal Arc Welding (GMAW) Equipment

The following is a list of basic gas metal arc welding equipment:

- Power source
- Wire feeder and controls
- Welding gun and components
- Shielding gas delivery system
- Weld cleaning and finishing tools

Gas Metal Arc Welding (GMAW) and Flux-Cored Arc Welding (FCAW) Power Sources

Power sources for GMAW and FCAW are of the direct current constant voltage type. This is unlike the constant current type you learned about in Chapter 3 on SMAW. The constant voltage (CV) machine is designed to deliver a very steady voltage. This voltage maintains a very consistent arc length. As the wire feeder is adjusted to feed more or less wire, the power source outputs more or less amperage to burn the wire off and maintain the arc length. Fig. 10 is a volt-amp curve for a CV machine.

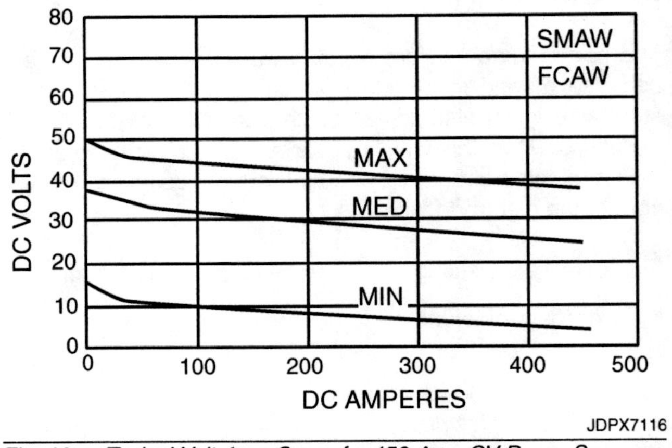

Fig. 10 — Typical Volt-Amp Curve for 450-Amp CV Power Source

These power sources can be very simple (Fig. 11) or they can be very sophisticated (Fig. 12).

An all in one unit (Fig. 11) is actually a power source and wire feeder combination enclosed in one sheet metal case. It has a simple heat control switch that controls both the voltage and wire feed speed. This control is similar to the synergic control described in the GMAW-P section. This heat switch only allows for a limited number of settings and does not allow the welder many options for fine-tuning the welding arc.

Fig. 11 — Wire Feeder and Power Source (All in One Unit)

JDPX7118

Fig. 12 — Industrial Inverter Power Source

An industrial inverter (Fig. 12) has very sophisticated controls for fine tuning the welding arc. This machine has a control system that allows the voltage to be adjusted infinitely from minimum to maximum of its range. It also has an inductor control that can be used to reduce weld spatter and make the weld puddle more or less fluid on stainless steel or virtually any type of material. (The inductor actually controls the power source's response time.) It requires a separate feeder/control to function.

Most industrial GMAW and FCAW power sources are rated at 100 percent duty cycle. Since these power sources are being used with a continuous wire process, the "arc on time" can be very high. There is no need to stop and change electrodes as in shielded-metal arc welding. This is why most industrial machines are rated at 100 percent duty cycle. The 100 percent duty cycle rating is at a given amperage. If the machine is capable of putting out higher amperages than the rating, the duty cycle must be lowered.

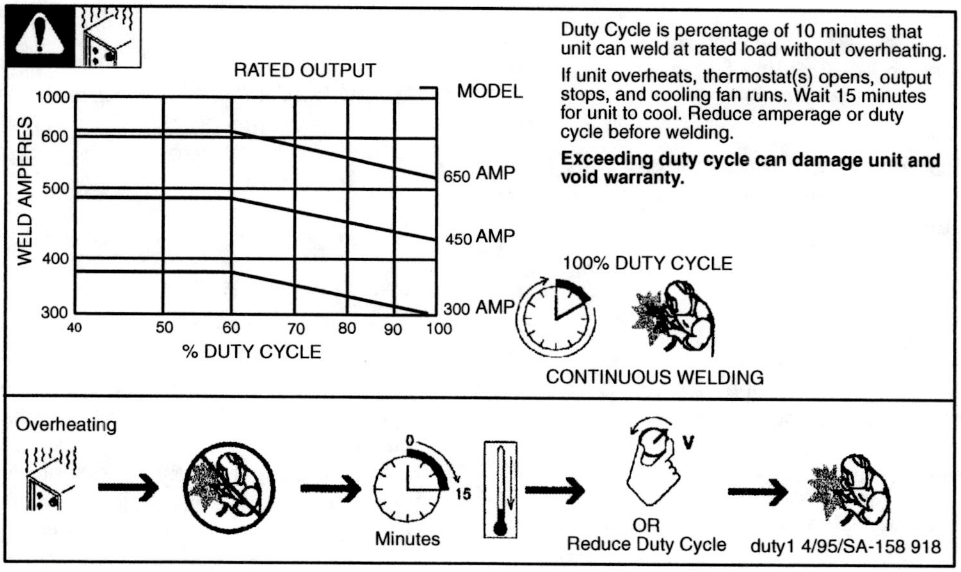

JDPX7119

Fig. 13 — Duty Cycle for Power Source

Tips for keeping a power source functioning properly:

- Follow the duty cycle capability of the power source. The owner's manual should contain a duty chart (Fig. 13). Some manufacturers put heat sensors in their machines to shut them down and prevent overheating.

IMPORTANT: A power source can be damaged if it is not equipped with a heat sensor and the duty cycle is exceeded.

- Install the power source where the air is not restricted going in and out of its louvers. As a rule of thumb, the distance from the nearest obstruction should be at least equal to the fan opening size (Fig. 14).

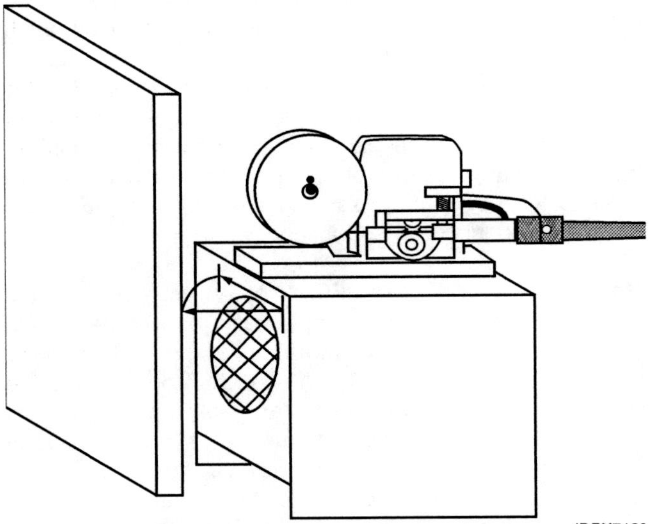

JDPX7120

Fig. 14 — Power Source Location Away from Obstructions — Minimum Distance

- The power source should be periodically cleaned out to remove any dust and dirt the fan has drawn inside the machine.

IMPORTANT: Cleaning should be performed only by a person qualified to service welding equipment.

- Don't use filters over the air inlets unless the manufacturer recommends them.

In selecting a GMAW or FCAW power source, the cost must be compared to the features and limitations of the equipment. The items listed below are important factors to consider when selecting a power source for one of the continuous wire processes:

- Type of metal: carbon steel, stainless steel, aluminum, etc.
- Metal thickness range: sheet metal or plate, pipe wall thickness, etc.
- Type of metal transfer: short circuiting, spray, pulse spray.
- Wire diameter range: 0.023 inch (0.58 mm), 0.030 inch (0.76 mm), 0.035 inch (0.89 mm), 0.045 inch (1.14 mm), etc.
- Primary power available: voltage, amperage, single or three-phase.
- Capacity of power source: voltage and amperage output range must be adequate for the type of metal transfer needed for the type and thickness of the metal to be welded.
- User friendly controls and adjustments: the people (welders, technicians, engineers) responsible for using the power source must have the ability to understand and use the features.

Wire Feeders and Controls

The wire feeder controls the delivery or feeding of wire from a spool or coil to the welding gun.

Feeding of the wire may be done by pushing the wire from the wire feeder to the gun (Fig. 15-A and C), by pulling the wire to the gun with a small drive motor in the gun (Fig. 15-B), or by using two drive motors at different locations to simultaneously push and pull the wire to its destination (Fig. 15-D).

As a weld is being made, controls on the wire feeder continuously interact with the power source, the wire drive motor, and the source of shielding gas. These controls are activated by the welder from a switch located on the welding gun.

The wire feeder and controls deliver the wire to the arc at a very consistent rate. If the wire cannot be delivered at a steady rate, the welding arc and resulting weld will not be very consistent.

NOTE: Welding will not be consistent if the interconnecting power source/wire feeder cables, work cable, and work clamp size do not meet manufacturer's recommendations.

There are basically two types of wire feeders:

- Constant-speed feeders
- Voltage-sensing feeders

The constant-speed feeder delivers the wire at whatever rate you set on the control. This is independent of what is happening in the welding arc. This feeder is the most common and gives the most consistent wire feed, regardless of the type of metal transfer being used. The main drawback to this type feeder is that it must be supplied with electrical power to operate (anything from 24 VAC to 115 VAC). It also needs a cord to energize the power source contactor. These additional cords can be damaged when used in rough work environments.

The voltage-sensing feeder operates off the welding voltage supplied by the power source through the electrode lead. These feeders have a switch that allows them to work off of either the constant voltage or constant current power source. When used on a CV machine, voltage-sensing feeders offer fairly consistent feed rates, due to the machine supplying a stable voltage to the drive motor. They still are generally not recommended for short circuit or pulse transfer.

When the voltage-sensing feeder is used on a constant current (CC) machine, the wire feed speed can fluctuate more since these machines don't maintain a constant voltage. They can be used with certain flux-cored wires and spray arcs operating at higher voltage levels. They cannot be used for small diameter wires and the short-circuiting transfer method. Their main advantage is they do not have any power or contactor cords going to them. They have only the large, well-insulated electrode weld cable that is not as easily damaged. For rough construction or field use, these feeders are very popular.

If the feeder is not marked as constant feed or voltage sensing, you can tell by looking at it how it is connected to the power source. The voltage-sensing feeder will have only the heavy welding cable going to it and a small lead coming out of the feeder to clip to the work. The constant speed feeder will have additional leads going to it for power and contactor control.

Additional variations in feeders are how they are packaged. They can be set up to feed a single wire or two wires. The feed motor can be separated from the control with the motor supported on the end of a boom. They can be set up on an open base that makes it easy to access the drive rolls and gun connection. It also eases wire supply changes. These feeders are usually large and bulky (Fig. 15-A). Generally, they are used in a shop environment and are called bench feeders.

Other enclosed feeders are called suitcase feeders (Fig. 15-C). They are set up for portable work. It can be more difficult to access the drive rolls and feed wire into the gun. They generally accept only the 8- or 12-inch (203- or 305-mm) spools of wire. The enclosed feeder's reduced size and weight are more important than the drawbacks.

The feeder can also be built directly into the power source, making it fairly compact and less likely to be damaged by rough usage. Typically, these are small units for the home hobbyist with low current and operating on 115 VAC or 230 VAC. They have duty cycles in the 20 percent area.

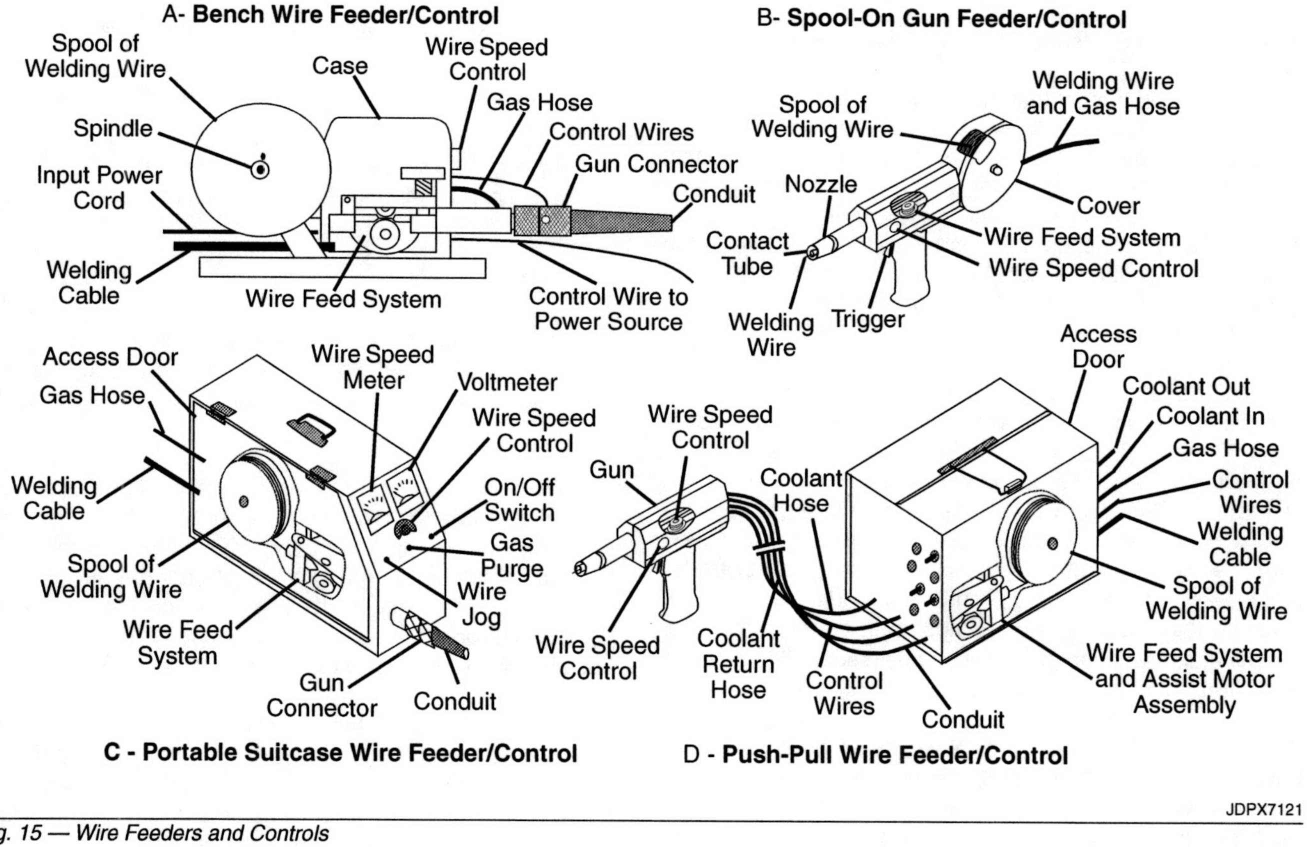

Fig. 15 — Wire Feeders and Controls

The power source/wire feeder (Fig. 16) is more of an industrial all-in-one unit that has high current output at 60 percent duty cycle and can operate on 460 VAC three-phase power as preferred in most industrial locations.

When welding aluminum, spool-on gun feeders (Fig. 15-B) are very popular as are push-pull feeders (Fig. 15-D).

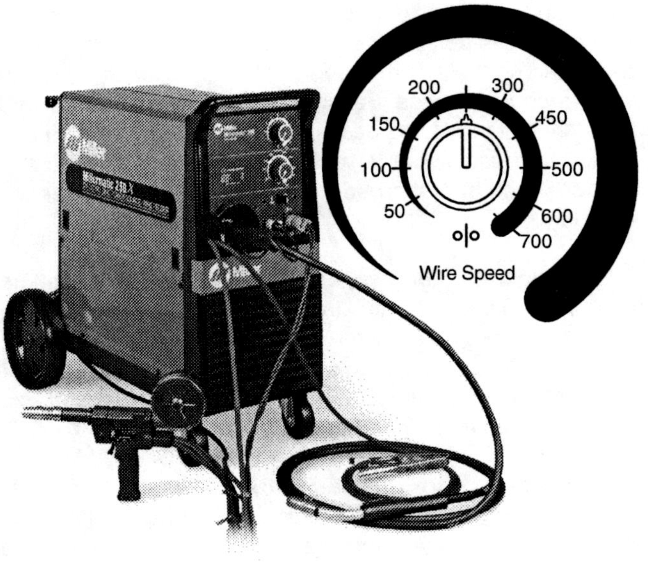

JDPX7122

Fig. 16 — Industrial GMAW Power Source and Wire Feeder Combined

The spool-on gun feeder uses a small 4-inch (102-mm) diameter spool. The gun has a drive motor built into it, as well as a speed control that makes it very convenient for the welder to use. The soft aluminum wire is fed directly from the spool straight into the drive rolls and directly into the contact tube area of the gun. Feeding problems are minimal with such a direct feeding path and short distance of approximately 10 inches (254 mm) from the drive rolls to the end of the contact tube. Note the spool fits into the gun—thus its name. This does make the gun heavy and difficult to get into tight places.

The push-pull feeder (Fig. 15-D) has a drive motor in the gun, but the 12-inch (305-mm) spool of wire may be located 25 to 30 feet (7.6 to 9 meters) away. This spool canister has an additional assist motor to make feeding the soft aluminum wire through the long wire liner easier. The larger spool of wire makes it more economical in a high-usage situation over the more expensive 4-inch (102-mm) spool. Both the spool-on gun feeder and the push-pull feeder can be used for more than just aluminum. Check with the manufacturer, but generally spool-on gun or push-pull feeders can be used with steel, stainless steel, and cored wires.

Wire feeders also come with a variety of controls. Some have one-knob approximate speed adjustment (Fig. 16). Others have digital adjustment with readout to 1 inch (25 mm) per minute accuracy with closed loop feedback to maintain speed, or computer-controlled microprocessor technology (Fig. 17).

JDPX7123

Fig. 17 — Dual Wire Microprocessor Wire Feeder/Control

The feeder controls other things in a weld sequence other than just wire feed speed. Following are some of the additional features found on some feeder/controls:

- When the shielding gas should flow (preflow is the time the gas flows before the arc starts and post-flow is the time the gas flows after the arc stops)
- The speed of the wire at arc starting
- Voltage and time for arc starting at the beginning of the weld
- Voltage and time for filling the crater at the end of the weld
- Burning the wire free of the weld pool at the end of the weld, called melt-back or burn-back time

Wire Drive System

The drive system on the wire feeder pulls the wire from a spool or a coil to the drive rolls. It then pushes the wire from the drive rolls to the end of the welding gun where it exits the gun.

The main components of a drive system (Fig. 18) are the wire guides to and from the drive rolls, the drive rolls, the drive roll tension adjustment, the drive motor assembly, and wire speed control. For the drive system to be effective, the wire must feed evenly at the speed set on the wire speed control. Uneven wire feeding can cause the arc to be unstable and weld quality to be inconsistent.

The wire guides keep the wire aligned with the drive rolls and minimize the buckling of the wire as it is pushed into the torch liner. Buckling, flexing, or anything that interferes with the wire feeding creates an unstable arc, thus making it difficult to obtain uniform welds with consistent quality.

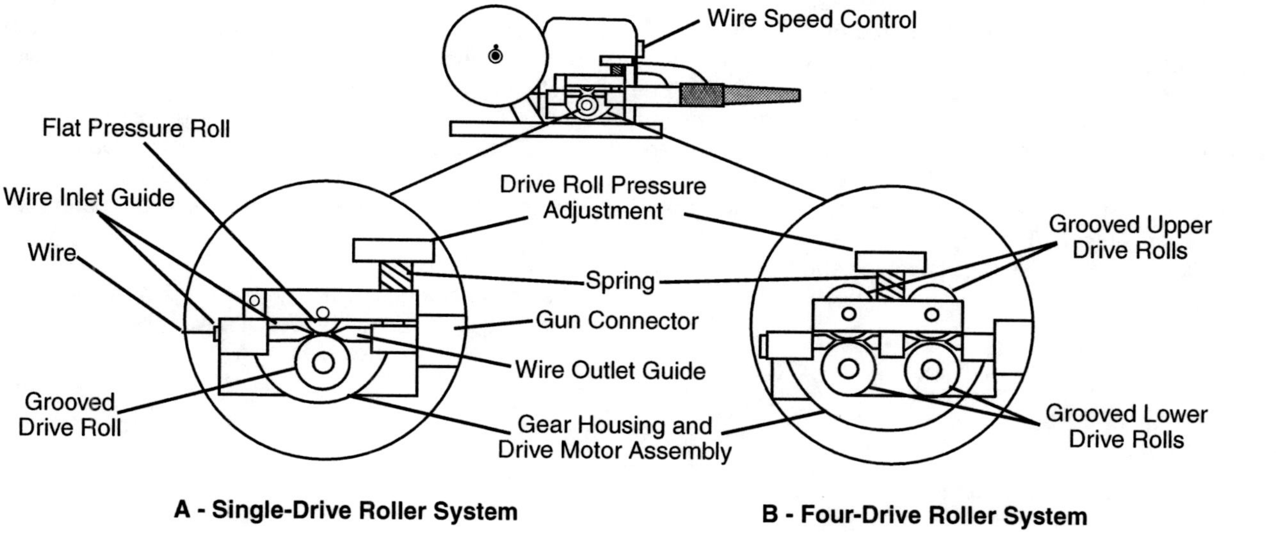

Fig. 18 — Wire Drive Systems

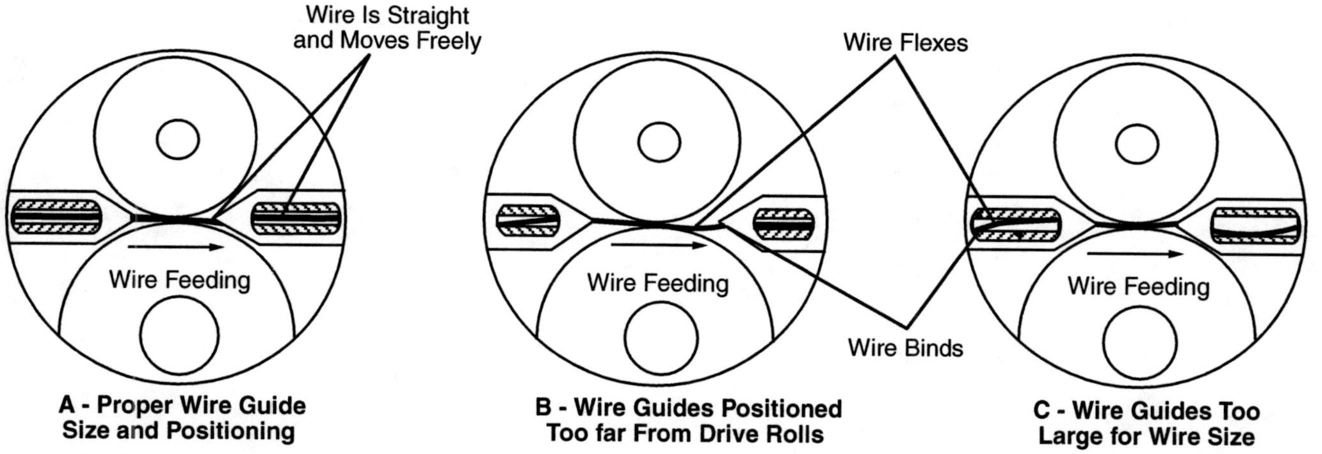

Fig. 19 — Wire Guide Size and Positioning on a Single-Drive Roll System

The wire guides must be the proper size (Fig. 19-A), positioned properly, and in alignment (Fig. 20-A) with the drive rolls. Oversized wire guides (Fig. 19-C) will cause the wire to flex or buckle as it travels through the drive system. Too much space (Fig. 19-B) between the drive rolls and wire, and drive rolls and wire guides that are out-of-alignment, will also cause the wire to flex or buckle (Fig. 20-B). When the wire buckles, it stops feeding into the outlet guide and forms a tangled coil called a birds nest as the wire is pushed between the drive roll and the outside of outlet guide. The wire will not go into the opening when the wire guides are too small.

Small filings caused by excess rubbing of the wire (Fig. 19-B, 19-C and Fig. 20-B) will plug the opening in the wire guides and cause the wire to stop feeding. Properly sized, positioned and aligned wire guides ensure a smooth movement of wire and a consistent welding arc.

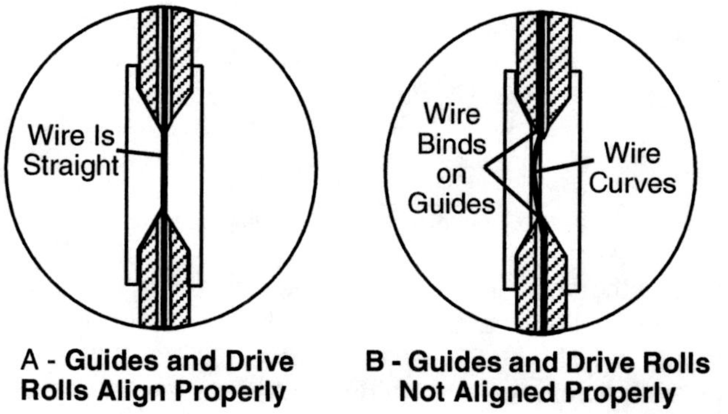

Fig. 20 — Wire Guide Alignment

A spring-type pressure adjustment keeps adequate pressure against the feed rolls for gripping the welding wire. There must be just enough pressure exerted on the drive rolls to grip the wire to prevent slipping and wearing flat spots on the wire. There cannot be too much tension or the wire will be deformed into an egg shape. If the wire is deformed the slightest bit, it will bind as it moves through the liner and contact tube. This is especially true with soft wires like aluminum and flux-cored wires. There are several types of spring arrangements (Fig. 18), and some even have a scale that indicates the relative tension on the spring.

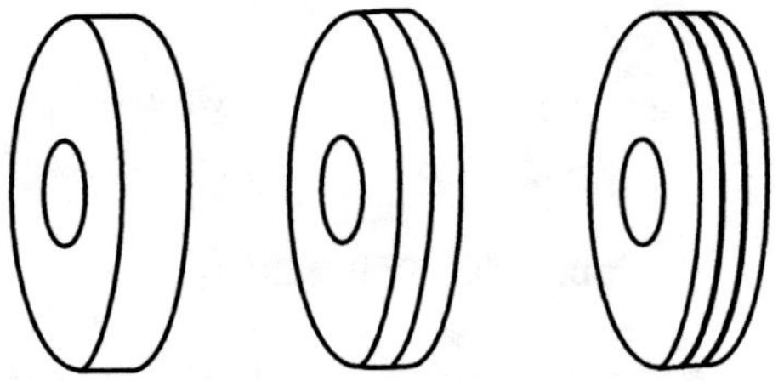

Fig. 21 — Types of Feed Rolls

There are single (Fig. 18-A), two, and four drive (Fig. 18-B) roll systems available. Each drive roll may contain one or two grooves (Fig. 21). A flat roll (Fig. 21-A) is used to apply the pressure necessary for the wire to be gripped by a wire feed system having only one drive roll. Drive rolls containing two grooves may have the same size groove so that the roll can be turned around when the first groove wears out. Some rolls are manufactured with two different-sized grooves so the roll can be turned and used for two different sizes of wires.

The proper shape and size groove (Fig. 22) in the drive roll is more critical than the number of drive rolls in getting a good grip on the wire without deforming it.

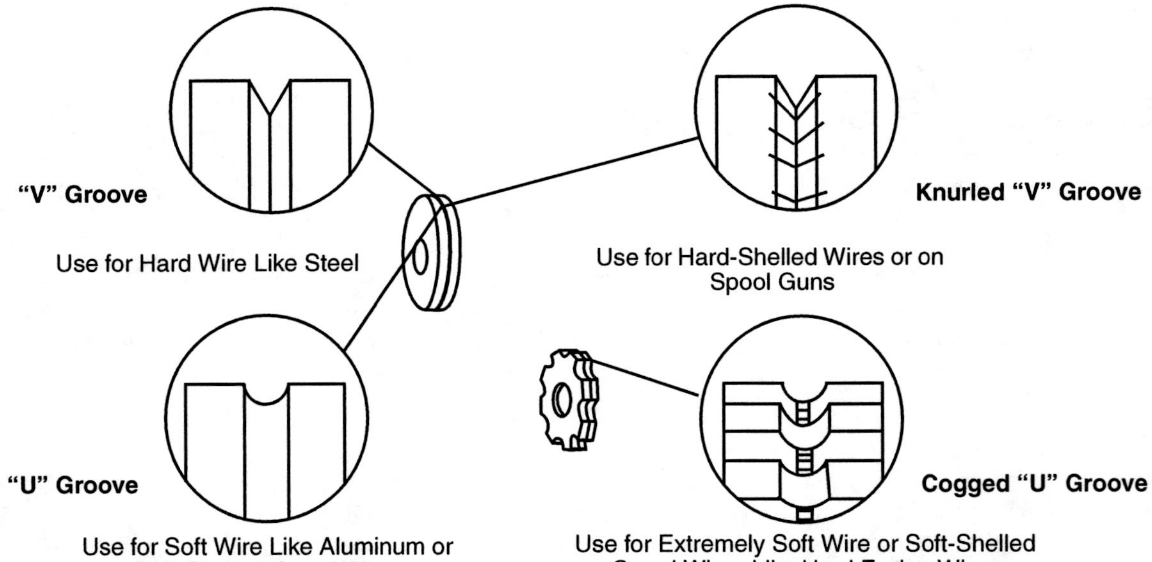

Fig. 22 — Drive Roll Groove Shapes and Applications

One of the first things to check when there is a wire-feeding problem is the clearance between the drive rolls. Normally, there will be 0.005—0.010 inch (0.123—0.254 mm) clearance between the two drive rolls (Fig. 23-A) when they are gripping the proper-size wire. An easy way to check for adequate clearance is to slide a piece of paper into the open space between the drive rolls as they turn. If the rolls grab the paper, the rolls are too large for the wire (Fig. 23-B), the rolls are worn (Fig. 23-C), or there is dirt between the two rolls that is keeping them from coming together and gripping the wire.

CAUTION: Keep fingers, hands, and clothing a safe distance from the drive rolls and other moving parts when checking drive roll clearance.

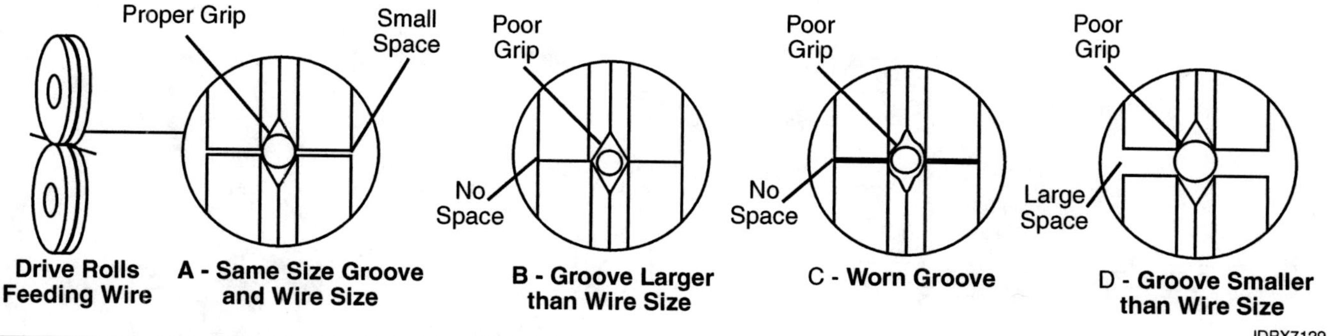

Fig. 23 — Examples of Drive Roll Groove Size and Wire Grip

Selecting the wire feeder depends upon whether you want the convenience of a bench-type feeder or the portability of a suitcase feeder; if you plan to use a difficult-to-feed wire like aluminum or titanium; or if a power source with a built-in wire feeder suits your short- and long-term welding needs. Once that decision is made, the wire feeder must have the proper wire feed speed ranges for the work being done. Slow wire feed speed is used for large diameter wires, while high wire feed speed is needed for small diameter wires. A typical wire feeder will have a wire feed range of 70—750 inches (178—1905 cm) per minute. Check a burn-off chart for the electrode wire you are using to make certain you have the correct speed range.

Welding Guns and Components

A welding gun (Fig. 24) is the equipment connected to several other components needed to make welds with the continuous wire processes. The welding current, the wire electrode, and in many instances a shielding gas come together in the welding gun.

Welding guns are manufactured in a variety of shapes, sizes, and lengths for a given amperage and corresponding duty cycle. Each welding gun is either designed to be hand held by a person or attached to mechanized welding equipment like a robot.

Gun shapes are usually of the goose neck (Fig. 24-A and B) variety, but for some applications the pistol grip (Fig. 24-C) is preferred.

Gun size is determined by the way it is cooled and by the amperage at which it is designed to operate. Gun cooling is achieved by room air contacting all exposed surfaces or by circulating a coolant like water through passageways in the gun and welding cable. Guns operated intermittently or in a low-to-medium amperage range are air-cooled. Guns that operate at high duty cycles and high amperages are water-cooled. An air-cooled gun must have a lot of surface area for cooling and is larger than a water-cooled gun with the same amperage rating.

Welding guns are generally rated in amperage, for a specific gas, and at a given duty cycle. They are also rated for the diameter of wire they are designed to feed. Conduit length is based on the length of the liner that directs the wire from the wire feeder connector to the entrance of the gas diffuser. The shorter the conduit length, the better. There will be fewer feeding problems and more welding time between liner replacement with conduit lengths of 10 feet (305 cm), 12 feet (366 cm) and 15 feet (457 cm).

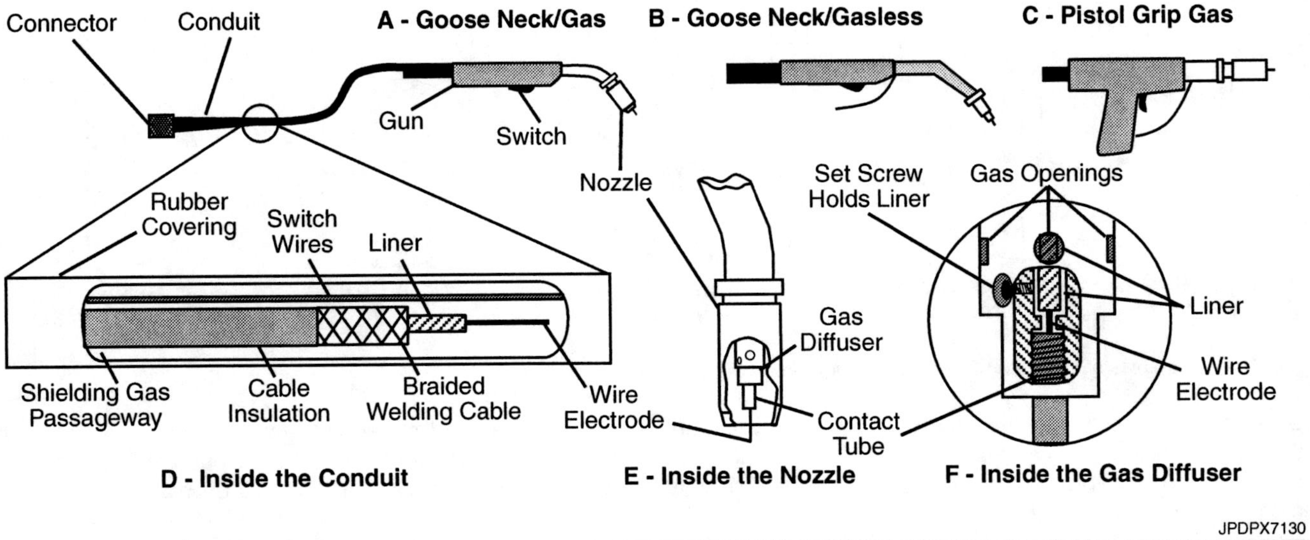

Fig. 24 — Welding Guns and Components

Most guns rated under 400 amps are designed for use with CO_2 gas and are air-cooled. Guns over 400 amps can be used with gas mixtures containing high percentages of argon (90 percent plus) and are water-cooled. The amperage capability of a gun rated with CO_2 must be reduced in half to remain at the same duty cycle as when used with gas mixtures containing high percentages of argon. For example, a 300-amp, 100 percent duty cycle CO_2 gun should only be used at 150-amp, 100 percent duty cycle with gas mixture containing 90 percent argon and 10 percent CO_2. Otherwise it will overheat and lead to early failure of its parts. When doing pulse welding, the shielding gas is primarily argon (90 percent to 98 percent for carbon steel and 100 percent for aluminum). High peak amperages are 300 to 500 amps — these are the amperages the gun should be sized for, not the average amperage. The proper gun to use for pulse welding should be water-cooled.

The gun liner and contact tube must be appropriate for the type and diameter of the wire being fed or wire feeding will not be consistent. If the inside diameter of the liner or contact tube is too close to the outside diameter of the wire, friction will cause wire feeding to be erratic or to stop. If the inside diameter of the liner is too large, the wire will tend to buckle up in the liner and then spring forward, producing a very erratic wire delivery at the arc. Soft wires like aluminum feed best when the inside of the liner is coated with teflon.

The liner must also support the wire from the point it enters the gun connector at the wire feeder until it reaches the contact tube. If the liner is too long it will prevent the gas diffuser from connecting to the gun. If the liner is too short (Fig. 25-B) it will be difficult to start the wire feeding through the gas diffuser into the contact tube without removing the gas diffuser and/or the contact tube. Once the wire has gone into the contact tube, feeding will be erratic, as the wire flexes and rubs against the inside of the gas diffuser between the liner and the contact tube. This erratic wire feeding may also cause the wire to stick to the contact tube when the arc first starts.

Most replacement liners are too long and must be cut to the proper length for the gun being used. For best gun performance, follow the instructions that are provided with a new liner for determining the proper length to cut the liner.

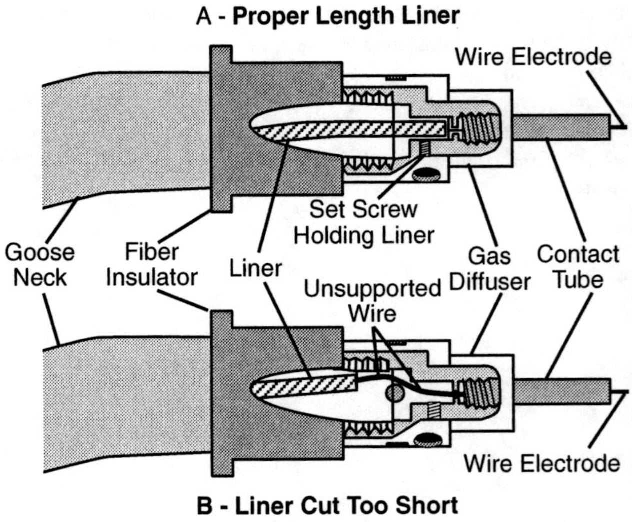

Fig. 25 — Liner Position in the Welding Gun

Proper liner installation includes securely tightening the set screw that holds the liner in place (Fig. 25-A). Do not overtighten the set screw, or the liner will be compressed and restrict the wire as it moves past the area.

If the contact tube is too large, the electric pickup will be poor and the arc will be erratic. Use the proper-sized liner and contact tube the manufacturer recommends for the wire diameter you are using. As the liner and contact tube wear, the arc will also become erratic. An accumulation of foreign material (dirt and small filings from the wire) will also cause the arc to become erratic. Preventive maintenance and troubleshooting suggestions are given later in this chapter.

The gas nozzle and contact tube gas nozzle relationship must be appropriate for the type of metal transfer in the work at hand. A large diameter (5/8 inch/16 mm) heavy-walled nozzle is required for spray transfer, pulse-spray transfer, and for FCAW-G applications using large diameter wires in the 3/32—1/8 inch (2.4—3.2 mm) range.The contact tube should be recessed inside the gas nozzle (Fig. 26-C) approximately 3/16 inch (4.8 mm) for these high-energy applications.

For good short-circuiting control, a smaller-diameter gas nozzle (1/2 inch [13 mm]) (Fig. 26-A) or even a tapered nozzle (Fig. 26-B) is preferred. Also, an acceptable contact tube position range can be from recessed (Fig. 26-B) inside the gas nozzle up to 1/8 inch (3.2 mm) to extended (Fig. 26-A) up to 1/8 inch (3.2 mm) beyond the end of the gas nozzle. This allows the welder to maintain the proper electrode extension while still being able to see and track the joint. When there are obstacles that can arc against the contact tube with it extended, recess the contact tube until the problem is corrected.

IMPORTANT: When welding with short circuiting transfer, never recess the contact tube more than 1/8 inch (3.2 mm). It will be difficult to see the weld joint and maintain a 3/8 inch (10 mm) electrode extension.

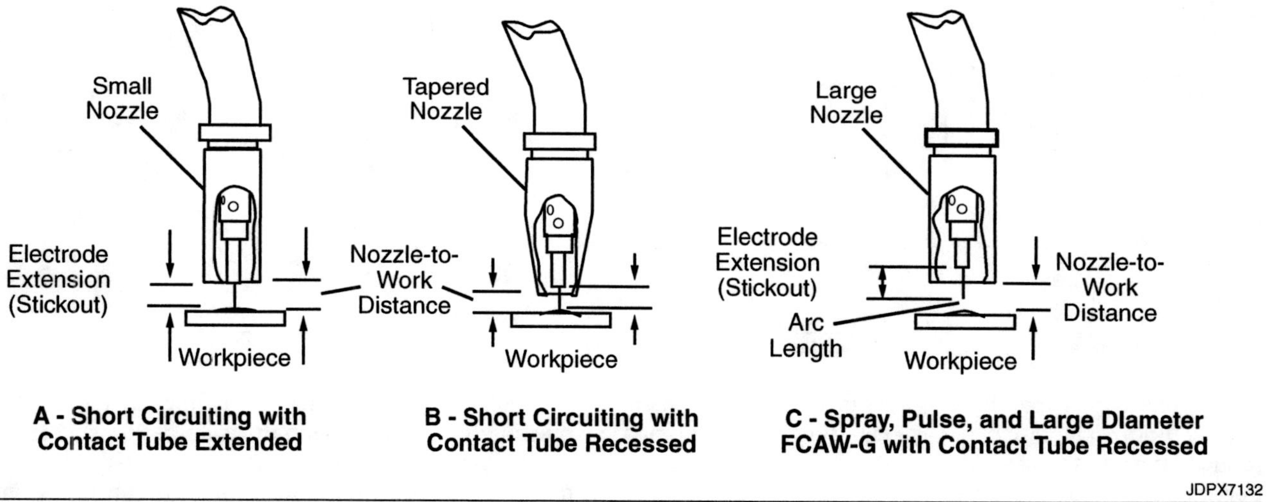

Fig. 26 — Gun Nozzle and Contact Tube Positioning

Shielding Gas Delivery System

The GMAW, metal cored, and FCAW-G processes require an externally supplied shielding gas. The shielding gas is delivered from steel containers to the nozzle at the welding gun in vapor form. As with other types of gases used in the welding industry, the cylinder pressure must be reduced to a safe working pressure and the flow rate controlled according to the use for the gas.

Pressurized containers similar to those used for storing oxygen are used for storing shielding gases (Fig. 27). Some containers store the gas in vapor form at very high pressures that exceed 2,000 psi (13 790 kPa). One of these high-pressure cylinders can deliver about 300 cubic feet (8.5 cubic meters). Other types of specially-insulated steel containers have liquid inside that is vaporized during use to produce pressures slightly over 100 psi (690 kPa). These range from cylinders that contain about 4000 cubic feet (113 cubic meters) to large bulk containers that hold as much as 88,000 cubic feet (2464 cubic meters).

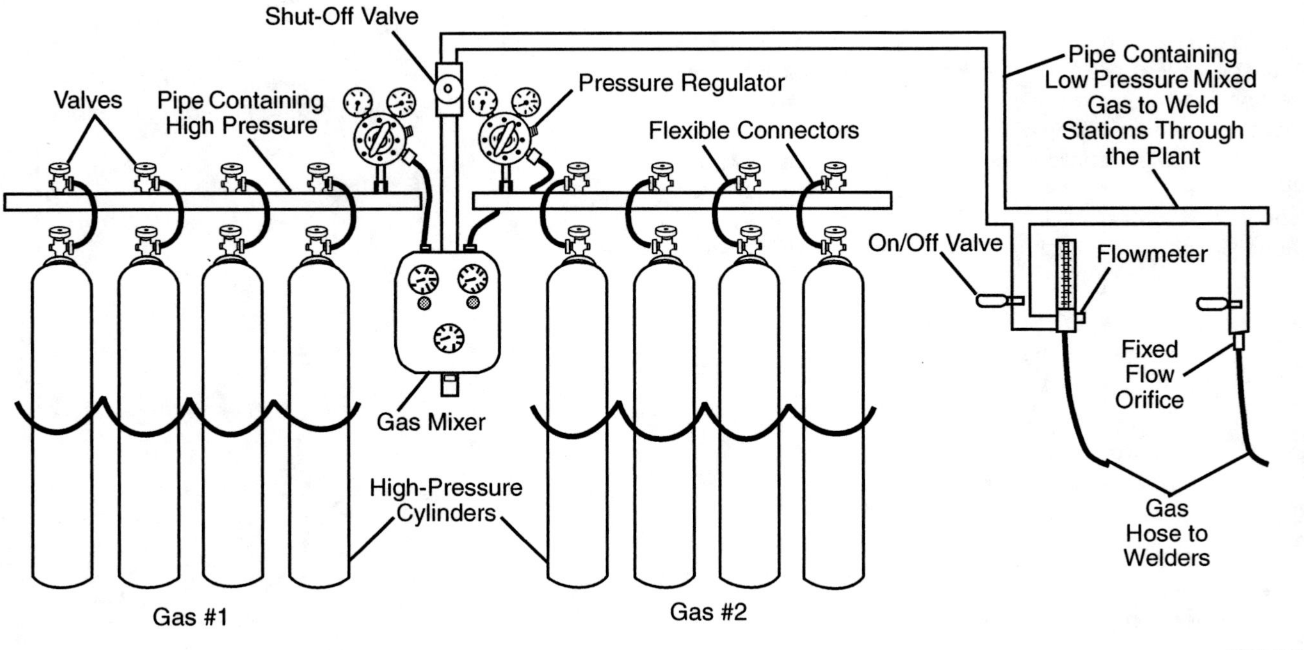

Fig. 27 — Components of a Simple Shielding Gas Manifold System and Mixer

The shielding gas may be supplied to welding equipment directly from one of these cylinders (Fig. 28) or from a reduced-pressure manifold system. The gas may be delivered to the manifold system from high-pressure cylinders or from a liquid storage container. Large users of shielding gas mixtures purchase each individual gas in liquid form, mix them according to their needs, and use a manifold system to transport the gas to each welding station. Some companies have a back-up manifold system of high-pressure mixed shielding gas that can be activated in case there is a problem with the gas mixer or one of the liquid containers.

IMPORTANT: Always secure cylinders in a vertical position on a cart or in a cylinder rack. When transporting, always secure cylinder to a suitable support with a chain or other holder.

NOTE: High-pressure cylinders and low-pressure liquid cylinders should never be attached to the same manifold system at the same time.

A regulator is always needed to reduce the high source pressure to a safe working pressure of 50 psi (345 kPa) for use by the welding equipment. Final delivery to the welding gun is controlled by a flowmeter or by an orifice inserted between the regulator and the gas hose to the welding gun.

A gas mixer is needed to combine the shielding gases in the desired proportions once the gases have been reduced to a safe working pressures for the welding equipment.

A flowmeter is a device that controls gas flow. The flow rate is variably adjustable within the range of the flowmeter (normally 10 CFH/0.28 CMH to 80 CFH/2.3 CMH). Some flowmeters are designed for use with a single gas and others calibrated for use with three to four different gases. (Fig. 28) illustrates a flowmeter that can be used with four different gases. The flow rate is read from the top of the ball.

NOTE: Not all flowmeter manufacturers calibrate their equipment for reading the top of the ball; some require reading the middle of the ball and others require reading the bottom of the ball. When the flow rate must be exact, follow the equipment manufacturer's instructions for the proper place to read the gas flow.

A fixed flow orifice, (Fig. 28-B) has a fixed opening that allows only a predetermined gas flow at a given pressure. On a manifold system supplying several pieces of welding equipment, a fixed flow orifice offers considerable cost savings compared to installing a flowmeter at each piece of equipment. The fixed flow orifices also eliminate the problem of wasting gas that occurs when flowmeter rates are set higher than job specifications.

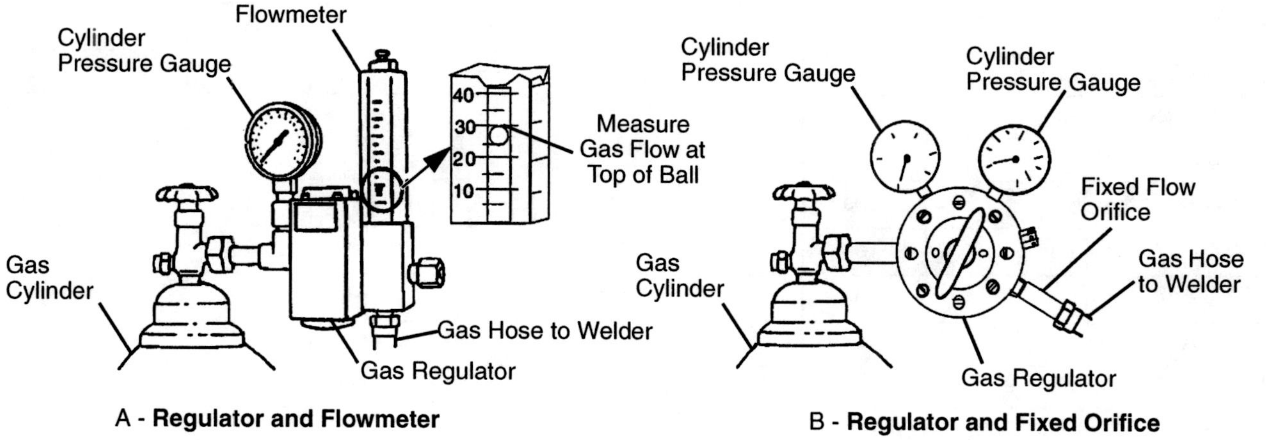

JDPX7134

Fig. 28 — Single-Cylinder High-Pressure Shielding Gas Delivery System

Consumables

Shielding gases are used with the GMAW, metal cored, and the FCAW-G processes to prevent atmospheric contamination of the arc, weld pool, and solidifying weld. The air we breathe contains contaminants to the weld such as oxygen and nitrogen. These elements can make a weld hard and brittle and very porous.

The type and mixture of shielding gas affects how smooth the arc runs, the amount of spatter produced, and the profile of the deposited weld metal (Fig. 29).

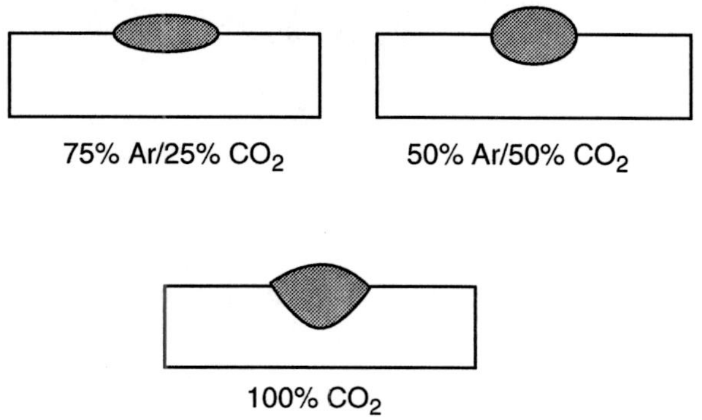

JDPX7135

Fig. 29 — Weld Cross Sections and Profiles with Short Circuiting

The shielding gases are broken down into several groups: Inert (argon, helium), reactive (oxygen, carbon dioxide) and mixed gases (combinations of the inert and reactive gases).

Table 2 shows some of the common gases used with these processes on various materials.

Gas	Base Metals							
	Carbon Steel Process			Stainless Steel Process			Aluminum Process	
	Short Circuit	Spray Pulse	FCAW-G Metal Cored	Short Circuit	Spray Pulse	FCAW-G Metal Cored	Short Circuit	Spray Pulse
100% Carbon Dioxide (CO_2)	Yes	----	Yes	----	----	Yes	----	----
100% Argon (Ar)	----	----	----	----	----	----	Yes	Yes
100% Helium (He)	----	----	----	----	----	----	----	Yes
Gas Mixture*								
75% Ar/25% CO_2	Yes	----	Yes	Fair**	----	Yes	----	----
90% Ar/10% CO_2	Yes	Yes	Yes	----	----	----	----	----
92% Ar/8% CO_2	Yes	Yes	Yes	----	----	----	----	----
95% Ar/5% Oxygen	----	Fair***	----	----	Yes	----	----	----
98% Ar/2% Oxygen	----	Fair***	----	----	Yes	----	----	----
90% He/7.5% Ar/2.5% CO_2	----	----	----	Yes	----	----	----	----

** Common Mixtures*

*** This mixture produces considerable weld spatter and causes carbide perception on stainless steel.*

**** Argon/oxygen mixtures have a narrow parameter range and have limited use in pulse welding applications.*

Table 2 — Shield Gas Selection Chart

The gas flow rate is determined by the joint type, method of metal transfer and corresponding power level, travel speed, air flow in the general work area, and the type of shielding gas. Table 3 gives some general flow settings of the shielding gas for variations within the continuous wire process. In applications using helium or helium mixtures, the flow rate will need to be increased above the values in the chart. Check with your gas supplier for the proper flow rate when using helium or helium mixtures.

Metal Transfer	Gas Flow Rate Per Hour	
	Cubic Feet	Cubic Meters
Short Circuiting	20 to 30	0.57 to 0.85
Spray	30 to 50	0.85 to 1.4
Pulse Spray	30 to 50	0.85 to 1.4
Flux Cored	20 to 50	0.57 to 1.4
Metal Cored	30 to 50	0.85 to 1.4

Table 3 — Shielding Gas Flow Rates

The welding wire (electrode) is used both to conduct the electrical power to the arc and to melt and become part of the weld. The wire electrode may be of a solid or tubular construction (Fig. 30). Both types of wire require the base metal be free of foreign materials that include paint, dirt, oil, grease, moisture, and rust. Some tubular wires will tolerate small amounts of foreign material and rust, but not moisture.

Welding wires are manufactured for joining most metals, including steel, stainless steel, aluminum, magnesium, and cast iron. Small diameter wires are used to join thin metals and for welding out-of-position. Large diameter wires are used for welding heavy plates that can be placed in the flat and downhand position.

The wire is supplied in spools (4 inch/10.2 cm, 8 inch/20.3 cm and 12 inch/30.5 cm in diameter), or large-capacity drums or pay-off spools. The small spools can weigh as little as one pound (0.45 kg) and the large pay-off systems can weigh over 1000 pounds (454 kg). Generally, the larger the spool capacity, the less expensive the wire is per unit (pound/kilogram).

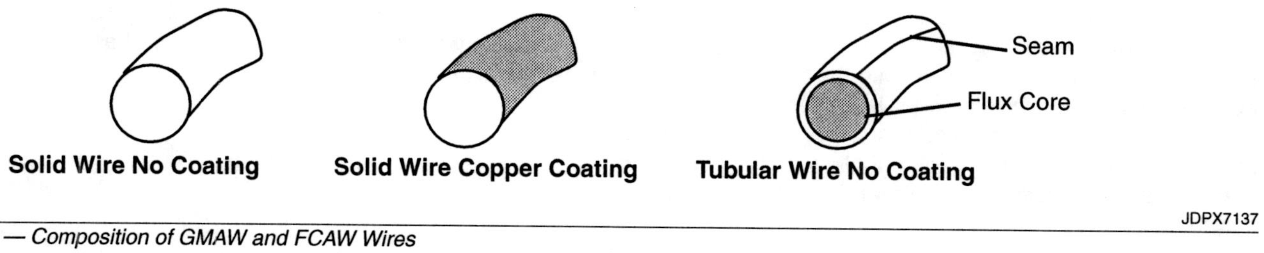

Fig. 30 — Composition of GMAW and FCAW Wires

IMPORTANT: If the wire will not be used up in a reasonable amount of time (two to three months), smaller spools are recommended. This is especially true when the wire is not used on a regular basis. Over an extended period of time the wire can corrode or otherwise deteriorate until it will not feed properly and/or contain enough rust to contaminate the weld. When this happens, the wire should be thrown in the scrap. (Depending on contamination, it may be possible to salvage part of the spool by removing the top layers of contaminated wire down to good wire.

When removing contaminated wire, unwind it directly from the spool. Do not run the contaminated wire through the feeder — the wire guides, liner, and welding tip will be plugged with contamination and have to be replaced.

Welding wires are produced in a variety of diameters. The smaller diameter wires are more expensive per unit (pound/kilogram) because they are more expensive to manufacture. For good short-circuit transfer, never use wire over 0.035 inch (0.9 mm) diameter. When welding with spray or pulse-spray transfer, use the largest diameter wire practical. Larger diameter wires cost less per unit (pound/kilogram); are easier to feed, especially aluminum; and surface contamination on the wire has less effect in the arc.

Common wire diameters and packaging are listed in Table 4.

Sizes		Types and Common Packaging		
Inches	mm	Solid	FCAW	Metal Core
0.023	0.6	X	----	----
0.030	0.8	X	X	----
0.035	0.9	X	X	X
0.045	1.1	X	X	X
3/64 (0.047)	1.2	----	X	----
0.052	1.4	X	X	X
1/16	1.6	X	X	X
5/64	2.0	----	X	X
3/32	2.4	----	X	X
7/64	2.8	----	X	----
1/8	3.2	----	X	----

Table 4 — Common GMAW and FCAW Wires

The wire must be properly alloyed and fabricated so that it has good feeding ability. It must not be too hard or too soft, and its surface finish must be smooth and free from scale, rust, scratches, or slivers. It must have proper helix and cast (Fig. 31).

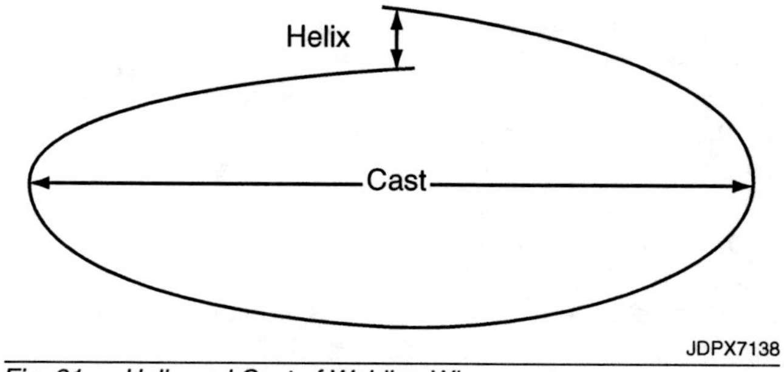

Fig. 31 — Helix and Cast of Welding Wire

The straighter the wire, the easier it is to feed. Since this wire is coiled up, it wants to loop back up and twist at the same time as it exits the contact tube. This wears on the contact tube and makes the arc oscillate with the twisting wire. One advantage a four-drive roll feeder has over a two-drive roll feeder is that it will straighten the wire if it is adjusted properly. If too much pressure is applied to the drive rolls, they can actually deform the wire and make the helix and cast worse.

Welding wire classification by the American Welding Society is by a series of letters and numbers. Solid wires are manufactured for a larger number of base metals than tubular wires.

Solid Wires	**-------------Tubular Wires-------------**		**Metal Core Wires**
All external shielded	Self-shielded	Externally shielded	Externally shielded
ER70S-3	E70T-4	E70T-1	E70C-6M
ER70S-6	E71T-11	E71T-1	E70C-3M
--------Letter and number designations--------			
ERXXS-X	EXXT-X	EXXT-X	EXXC-X
E=Electrode----------			
R=Rod			
XX=Tensile strength x 1000	X=Tensile strength x 1000----------		
S=Solid	T=Tubular----------		C=Metal Core
	0=Flat & horizontal position----------		
	1=All position----------		
X=Usability and performance capabilities such as wire chemistry, gas shielding, and polarity			

Table 5 — GMAW and FCAW Wires — Their AWS Classifications

Table 5 illustrates the basic explanation of the system for solid and tubular wires.

Solid wire means exactly what the name implies — solid in construction. It does not contain flux on the outside like SMAW electrodes or a flux on the inside like the flux-cored wires. However, some manufacturers of wire for carbon steel put a thin copper coating on the outside of the wire, while other manufacturers leave the wire uncoated. Solid wire selection should be matched to the base metal, mode of metal transfer, and the type of shielding gas used. Common types of wire include those for carbon steel, stainless steel, aluminum, magnesium, nickel, and copper alloys. Silicon bronze wires are manufactured for joining galvanized metal or thin pieces of carbon steel like those used in the automotive industry and for joining copper alloys.

Two examples of common solid wires used on carbon steel are ER70S-6 and ER70S-3. The “E” indicates the wire can function as an electrode and the “R” indicates the wire can also function as a filler rod. The “70” indicates the wire produces 70,000 pounds per square inch of tensile strength (31750 kg/mm). The “S” indicates the wire is of a solid construction. The “-6” is one of several numbers (2 to 7) used to indicate wire chemistry, special characteristics, and applications.

An ER70S-6 wire can be used for most carbon steel welding applications. When welding in the short-circuiting mode of metal transfer using 100 percent CO_2 (with the proper combination of wire speed and voltage), there will be less weld spatter and better weld pool wetting than with an ER70S-3 wire.

When using CO_2, the arc can be harsh. There is some degree of spatter due to this arc harshness. A weld made using a ER70S-6 wire will have less weld spatter and flow into the base metal more evenly that a weld made using a ER70S-3 wire.

An ER70S-3 wire is commonly used with the argon/CO_2 mixtures in the modes of metal transfer given in Table 2. If an ER70S-6 is used with an argon/CO_2 mixture, there may be more isles of silicone. These isles of silicone may in turn require additional weld cleaning in order to obtain an acceptable coat of paint.

Tubular wires can be grouped according to the type of shielding provided to the weld puddle and by the composition of the flux core. The groups include self-shielded, externally shielded flux-cored, and externally shielded metal-cored.

Self-shielded flux-cored wires obtain their shielding from gases produced as the flux core inside the wire decomposes in the welding arc. The wires are commonly used in outdoor construction locations that require welding out-of-position and that are difficult to supply with shielding gas. They are also used in multipass welds on thick metals (1 inch [25 mm]) and thicker).

Two examples of self-shielded flux-cored wires used on carbon steel are E70T-4 and E71T-11. The "E" indicates the wire can function only as an electrode. It produces 70,000 pounds per square inch of tensile strength (31 750 kg/mm). The "T" indicates the wire is of a tubular construction. The "-11" is one of several numbers (1 to 11) used to indicate wire chemistry, special characteristics, and applications.

Good ventilation is needed in the weld area when using flux-cored wires. The volume of fumes is equal to or greater than that produced when using SMAW electrodes.

Slag is formed on top of the weld. It must be removed between passes and before the part receives a final finish such as painting.

Externally-shielded flux-cored wires require an external shielding gas like the solid wires — in addition to the limited shielding from the flux in the core of the wire. Most of the wires are limited to flat position groove and fillet welds. Some wires can be used out-of-position. These wires are also called dual-shielded wires because they get a small amount of shielding from the flux core.

These wires also produce a large volume of fumes and require good ventilation in the weld area. However, the ventilation system cannot interfere with the shielding process. The weld must be protected from air currents or drafts that can cause the gas shield to be displaced around the weld by room air, and thus cause the weld to be porous.

Slag is also formed on top of the weld as with the self-shielded wires. It must be removed between passes and before the part is painted.

Examples of flux-cored wires are E70T-1 and E71T-1. When only an "E" is given, it means the wire can be used only as an electrode. Both wires have 70,000 pounds per square inch of tensile strength (31 750 kg/mm). The "0" in "70" indicates welding can be done in the flat or horizontal position on fillet welds. The "1" in "71" indicates welding can be done in all positions (flat, horizontal, vertical, and overhead). The "T" indicates the wire is a hollow tube filled with flux, made up of deoxidizers, alloys, and slag formers.

The T-1 electrodes are designed for use with CO_2 shielding. They operate on DCEP. Impact properties can be improved by adding argon to the CO_2. Single- or multi-pass welds can be made with this electrode. The metal transfer is in the form of a spray with low spatter level, flat to slightly convex weld beads, and moderate slag that covers the entire weld bead. For more detailed information, refer to the most current manufacturer's information or AWS A5 booklets.

Metal core wires are a type of tubular wire that requires an external-shielding gas and good ventilation to remove the welding fumes. The weld must be protected from air currents or drafts, including those from the ventilation system, that could remove shielding gas from the weld zone.

Two examples of metal-cored wires used on carbon steel are E70C-6M and E70C-3M. The "E" indicates the wire can only function as an electrode. It produces 70,000 pounds per square inch of tensile strength (31 750 kg/mm). The "C" indicates the wire is a metal-cored electrode. The "-6M" indicates the wire is to be used with an argon/CO_2 mixture, while the "-3M" wire is to be used with 100 percent CO_2.

The weld is not covered by slag. However, there are intermittent brown scales on the weld's surface similar to those found on a weld done by spray transfer.

Weld Cleaning and Finishing Tools

The same tools used to clean welds made with SMAW are used for cleaning welds made with GMAW and FCAW processes. Gas metal arc welds made with the proper parameters and welding techniques generally contain very little if any weld spatter and smoke. In fact, the weld may not require cleaning. However, when weld appearance is very important or when the proper techniques are not followed, post-weld cleaning may be necessary to remove weld spatter, smoke, and dust from the weld bead and surrounding area. Also, as mentioned earlier, it may be necessary to remove the isles of silicone in order to obtain full paint coverage of the weld with the first coat. Weld cleaning is also necessary to remove the slag between passes and before painting with the FCAW processes. The large isles of silicon on welds made with metal-cored wire also need to be removed before painting.

CAUTION: To avoid skin burns from touching the weld, or from flying pieces of silicon, slag, and spatter, allow the welds to cool to room temperature before cleaning.

Special Tools

A wire cutter is needed to trim the wire to the proper shape and length prior to arc starting. A tool to remove spatter build-up on the nozzle and contact tube will also be required. Pliers are needed to replace contact tubes, as well as keep them tight while in use. A welding pliers (Fig. 32) can be utilized for all these tasks.

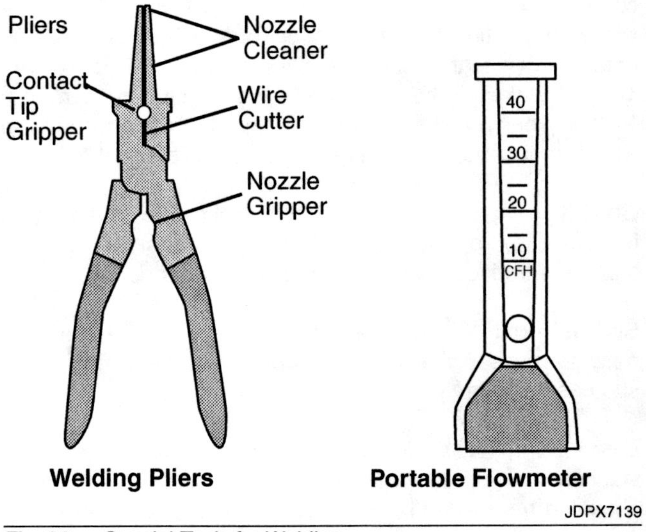

Fig. 32 — Special Tools for Welding

Setting Up Gas Metal Arc Welding (GMAW) and Flux-Cored Arc Welding (FCAW) Equipment

CAUTION: Before beginning work, read and understand warning labels (Fig. 33) on the equipment, the safety precautions, and the operating instructions in the owner's manual for the specific equipment to be used.

Some setup is necessary with new equipment, and preventive maintenance should be performed on existing equipment before starting work. Since there are several pieces of equipment with a continuous wire welding system, the setup, operating procedure, and maintenance is more complicated than with SMAW equipment.

Read and Understand Warning Labels

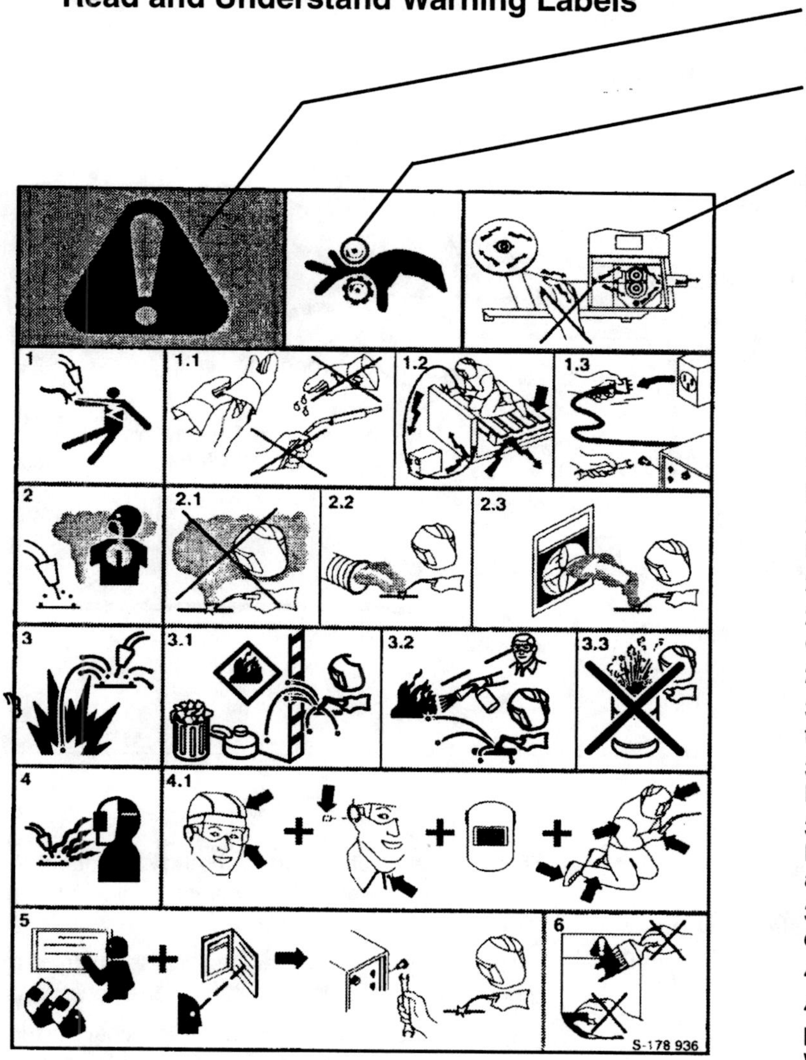

Warning! Watch Out! There are possible hazards as shown by the symbols.

Drive rolls can injure fingers.

Welding wire and drive parts are at welding voltage during operation — keep hands and metal objects clear.

1. Electric shock can kill.

1.1. Wear dry insulating gloves. Do not touch electrode with bare hand. Do not wear wet or damaged gloves.

1.2. Protect yourself from electric shock by insulating yourself from work and ground.

1.3. Disconnect input plug or power before working on machine.

2. Breathing welding fumes can be hazardous to your health.

2.1. Keep your head out of the fumes.

2.2. Use forced ventilation or local exhaust to remove the fumes.

2.3. Use ventilating fan to remove fumes.

3. Welding sparks can cause explosion or fire.

3.1. Keep flammables away from welding. Don't weld near flammables.

3.2. Welding sparks can cause fires. Have a fire extinguisher nearby and have a watch person ready to use it.

3.3. Do not weld on drum or any closed containers.

4. Arc rays can burn eyes and injure skin.

4.1. Wear hat and safety glasses. Use ear protection and button shirt collar. Use welding helmet with correct shade of filter. Wear complete body protection.

5. Become trained and read the instructions before working on the machine or welding.

6. Do not remove or paint over (cover) the label.

JDPX7140

Fig. 33 — Examples of Welding Equipment Warning Label and Definitions

A working understanding of how everything connects, the function of all the welder controls, and basic maintenance/troubleshooting procedures plays a very important part in working efficiently and in meeting weld quality standards. When selecting equipment, it must function as a system and work together. The power source, wire feeder, and accessories, like remote controls, interconnecting cables and wires, gas shielding, or cooling system, must all function together.

Fig. 34 illustrates the main points that should be set on a new system or checked on an existing GMAW system.

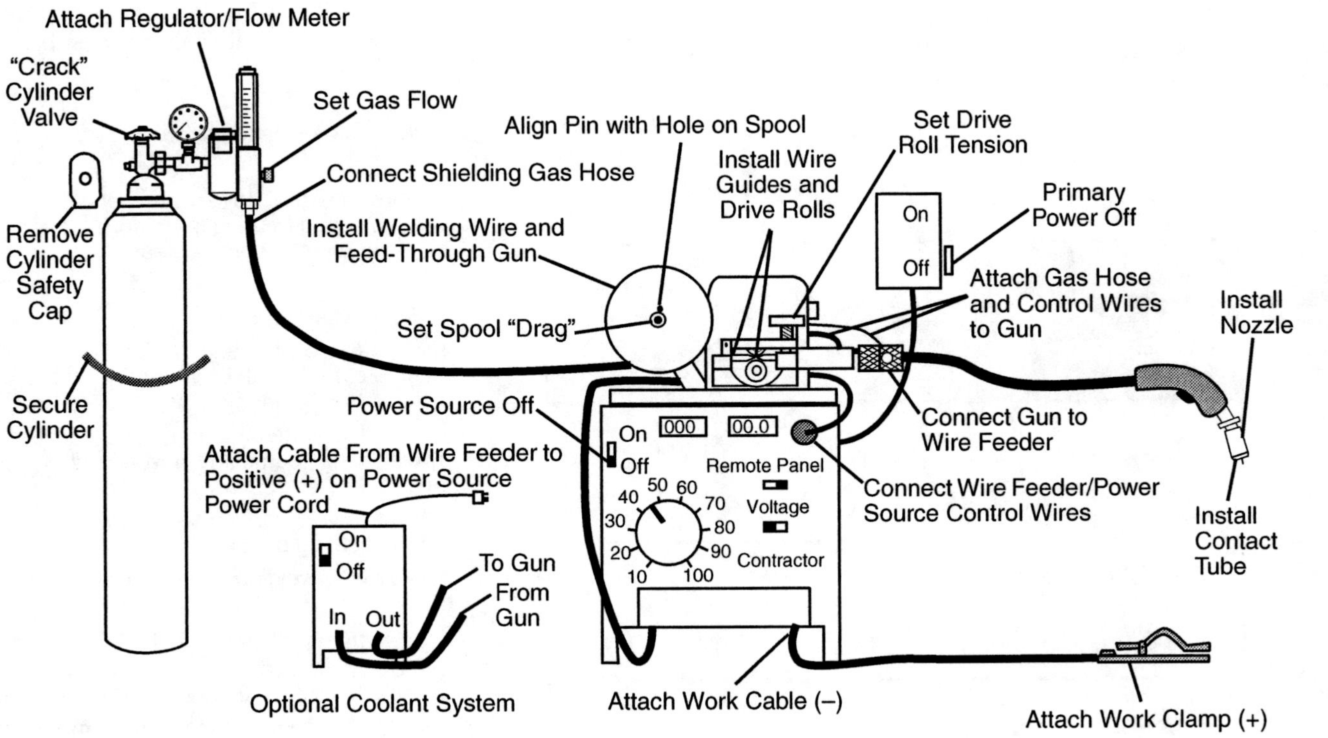

Fig. 34 — Setting Up Gas Metal Arc and Flux-Cored Arc Welding

Power Source — Wire Feeder Connections

CAUTION: The installation of an electrical outlet and disconnect switch for operating welding equipment should be performed only by qualified people. The same qualifications are necessary for making internal electrical connections on the power source for the primary power cable connection and for changing voltage connections to correspond to the shop voltage available.

IMPORTANT: Always follow local, state, and national regulations regarding the installation of electrical wiring and other devices necessary for connecting welding equipment. This includes complying to the provisions in the latest National Electrical Code (NEC).

1. Turn off the electrical power at the disconnect switch or unplug the power source before beginning equipment setup or preventive maintenance.
2. Follow equipment manufacturer's instructions for checking/changing input voltage to the power source.
3. Attach a cable from the positive terminal on the power source to the welding terminal on the wire feeder.

NOTE: Some self-shielded flux-cored wires run on electrode negative. Follow the wire manufacturer's setup instructions.

4. Attach the control wires to the proper receptacle on the power source.
5. Attach a work cable and work clamp to the power source.

Wire Feeder — Gun Connections

1. Attach the gun to the wire feeder. Adapters may be necessary to attach a gun manufactured by one company to a wire feeder manufactured by another company, or between different models of wire feeders from the same manufacturer.
2. Attach the gas hose to the wire feeder.
3. Plug in the control wires from the power source.

Shielding Gas Setup

1. Secure the gas cylinder.
2. Remove the valve protection cap.
3. "Crack" open the cylinder valve.
4. Attach a regulator flowmeter.
5. Set the proper gas flow.
6. Check all connections for leaks.

Wire Feeder — Gas Supply Connections

1. Attach the gas hose.
2. Check all connections for leaks.

Wire Feeder Setup

1. Install the proper size feed rolls for the size and wire type (Fig. 23).
2. Install the proper size wire guides and adjust them as close as possible to the feed rolls without touching (see Fig. 19 and Fig. 20).
3. Install a spool of wire. Make sure the pin on the spool spindle aligns with the hole in the spool and the wire unwinds from the bottom.
4. Feed wire through the gun.

Setting Drive Roll Tension

1. Loosen the tension until the wire stops feeding.
2. Begin tightening until the drive rolls begin to grip the wire and push it through the gun.
3. While feeding wire, place the end of the gun near a non-conductive surface such as a concrete floor or a piece of wood.
4. Increase tension on the feed rolls until a 3—4 inch (76—102 mm) coil of wire forms at the end of the gun (Fig. 45).

NOTE: This procedure works only with solid carbon steel wire. For flux-cored, stainless-steel, and aluminum wires, apply only enough pressure to get the wire to feed without slipping. Too much pressure will make the wire out-of-round until it lodges in the liner and/or contact tube.

Checking the Wire Feed System

To check the condition of the liner and the other parts of the system (when feeding solid carbon steel wire), push the gun against a non-conductive surface as the wire is being fed, until the wire stops feeding. If the drive rolls slip and the wire does not wrap around the feed rolls, the system is functioning properly.

If the wire wraps around the feed rolls and forms what is called a "birds nest," check the following items:

1. Check the condition of the liner for wear, obstructions, and/or damage by feeding wire to the gun, releasing the tension on the feed rolls, and pulling the wire back through the gun to the wire feeder. The wire should pull back through the wire feeder without much force. If a lot of force is required, all or partway, the liner should be replaced.
2. Check the size and position of the wire guides as indicated in (Fig. 19) and (Fig. 20).
3. Check the attachment of the liner to the gas diffuser. It should be attached as illustrated in (Fig. 25).
4. If there is an adapter between the wire feeder and gun, check that it has a jumper liner installed properly and it is of the proper length.
5. Check that the liner is the proper size for the wire.
6. Be sure the gun is not too long for the wire being fed.

Troubleshooting Tips

- As the liner wears and fills with small wire filings and dirt, more tension must be applied on the drive rolls to get the wire to feed properly. When there is too much pressure on the feed rolls to cause the wire to "birds nest" when it is stopped at the gun, the liner needs to be replaced.
- The life of the liner can be lengthened by blowing clean dry air through the gun before putting on a new spool of wire. Remove the unused wire and contact tip from the gun. Start by blowing air through the tip end of the gun. Then remove the conduit from the wire feeder and blow air through it. Alternate blowing air through each end until no more dirt comes out. Reassemble the gun and feed the wire through the system.

Preparing to Weld

When using the GMAW and FCAW processes, follow the same general safety precautions that were followed with the SMAW process — the same hazards are present (Fig. 35) and (Fig. 36).

Welding Wire Can Cause Puncture Wound.

1. Do not press gun trigger until instructed to do so.

2. Do not point gun toward any part of the body, other people, or any metal when threading welding wire.

JDPX7142

Fig. 35 — Label Warning of Puncture Wounds

Do not put feeder where welding wire hits cylinder.

Do not move or operate equipment when it could tip.

JDPX7143

Fig. 36 — Label Warning for Moving Equipment on Cart

Special Precautions for Gas Metal Arc Welding (GMAW) and Flux-Cored Arc Welding (FCAW)

- Greater protection from eye and skin burns from the arc is needed. This means a darker shade of lens and more dense clothing material than for the same amperage used when shielded-metal arc welding. Welding screens should be used to protect other workers from the arc light for a radius of 75 feet (22.9 m). Light from the arc is more intense because there are not as many fumes to block the rays from reaching a person's eyes and skin. Also, higher amperages may be used for spray transfer, pulse-spray transfer and for FCAW compared to similar SMAW operations. Follow the equipment manufacturer's recommendations for the proper lens shade to use.

CAUTION: Caution is required to prevent burns or puncture wounds by the exposed end of the welding wire (Fig. 35).

- Protect yourself from skin burns or puncture wounds from the exposed wire end. A sharp and potentially hot wire is fed out of, or protrudes from, the end of the gun. When shortening or removing the balled end of the wire with a wire cutter, it may cause this small piece of wire to be launched like a projectile and present a hazard. Point the wire down or away from people when trimming the wire.
- Avoid accidentally turning on the gun's trigger. The welder, when working, can easily turn on the large exposed gun trigger. However, if the gun is laid down carelessly, it can be accidentally activated. This in turn may cause arcing, a fire, and/or skin punctures.
- Properly handle and care for high-pressure cylinders. Never move a cylinder unless the valve protection cap is in place. Don't hang the welding gun on the gas cylinder. Never allow an arc strike to occur on the cylinder.
- Never hold the end of the gun close to your head or ears to check (listen) for shielding gas flow. Use a portable flowmeter to check gas flow at the gun nozzle see (Fig. 32).
- Use caution when moving equipment on a welding cart. See the warning labels in (Fig. 36).

Setting Welding Parameters

Welding parameters can be obtained from a written procedure that has been tried, tested, and proven to produce sound welds, or from weld data sheets that contain important settings and techniques. In either case, a range will be given for setting wire speed/amperage, voltage, and gas flow for a given metal thickness range and weld joint. If neither type of information can be obtained, it is a matter of trial and error to establish parameters that produce a weld that meets inspection standards.

If no data is available, the following basic rules can be useful:

First, adjust the wire feed speed to produce the amount of amperage you will need for the thickness of metal being welded. As a rule of thumb, approximately one amp is required for each 0.001 inch (0.025 mm) of metal thickness. See Table 6 for examples.

Gauge Number	Nearest Decimal/Metric Equivalent		Approximate Amperage Needed
	Inch	mm	
18	0.047	1.2	50
16	0.062	1.6	60
14	0.078	2.0	80
12	0.100	2.5	100
10	0.125	3.2	125
8	0.156	4.0	160
6	0.187	4.8	190

Table 6 — Metal Thickness and Corresponding Amperage for Short-Circuiting Transfer on Carbon Steel.

Since wire feed speed controls the amperage, the next step is to set this speed. Electrode manufacturers supply a burn-off chart that shows how many inches per minute of wire is required for a given amperage. The rule of thumb for short-circuiting transfer on carbon steel is 2 inches (51 mm) per amp using 0.030 inch (0.76 mm) wire and 1.6 inch (41 mm) per amp using 0.035 inch (0.89 mm) wire.

To apply these two interrelated rules to making a weld on 10-gauge steel with 0.035 inch (0.89 mm) diameter wire — 0.125 inch (3.2 mm) thick metal would require 125 amps. It takes 1 amp to burn off 1.6 inch (40.6 mm) of 0.035 inch (0.89 mm) diameter wire. A wire feed speed of 0.125 inch (3.2 mm) x 1.6 inch (40.6 mm) = 200 inches (508 cm) per minute. If the feeder has a wire feed speed meter or calibrated setting around its knob, set it for 200 inches (508 cm) per minute.

If the wire feeder has no speed indicators on it, the wire speed will have to be measured. To measure wire feed speed, run out a 6-second amount of wire. Since 6 seconds is 1/10 of a minute, the wire fed in six seconds should measure 20 inches long, as this is 1/10th of the 200 inches (5080 mm) per minute required. Make several trials to be certain of the wire speed.

When the wire feed speed is set, the arc length or voltage can be set. Electrode manufacturers have data sheets with recommendations for the voltage to be used. If this information is not readily available, the arc length can be set by sight, sound, and weld bead appearance. For short-circuiting welding, decrease the voltage to a point where the arc is very short and the wire may stub into the work. This voltage should be recorded or the knob position noted. The voltage should now be raised to where the arc sound is not a crisp crackle or the sound of a bumblebee on the end of the wire, That sound indicates a long arc condition. A high-voltage condition can also be identified when the end of the wire appears to melt in large drops, that results in excess weld spatter. This voltage should be recorded or the knob position noted.

The best voltage for this wire feed speed is the middle setting. If the low voltage was 16 volts and the high voltage was 20 volts, then 18 volts will be the best setting.

When setting the wire speed and voltage for spray transfer, metal-cored and flux-cored wires, use a similar method by watching the arc length and spatter level, by the stable arc sound, and by looking at the weld profile. When the voltage is too low, the end of the wire appears to melt in large drops, that results in excess weld spatter. When the voltage is to high the arc will be long and cause undercut. The longer the arc, the wider and flatter the weld. The shorter the arc, the narrower and more built up the weld will be. (Fig. 37) illustrates this same effect using the beam of light from three flashlights at different distances from a surface. It is easy to see that the greater the flashlight's distance from the surface, the larger the light beam.

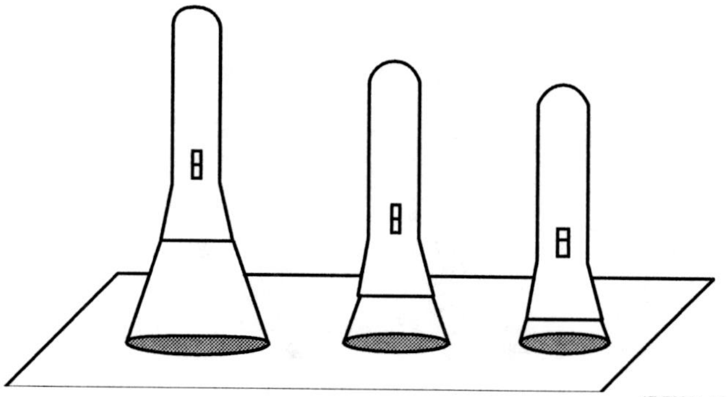

Fig. 37 — Distance and Its Effect on the Size of the Light Beam from a Flashlight

Proper Welding Technique

Flat Position Welding

Welding techniques will be important to master as they affect how the weld metal is deposited. This will affect the appearance and how strong the welds will be. The points to focus on are:

- Electrode extension
- Travel angle
- Work angle
- Travel speed
- Electrode/joint alignment

Electrode extension or "stick out" (Fig. 26) is how much wire is sticking out of the contact tube to the welding arc. It is important for all wire welding processes; however, it is very critical with short-circuiting transfer and especially so when welding stainless steel. Table 7 lists some typical electrode extensions for common applications. Note that these extensions can be varied up to ± 1/8 inch.

Wire Type	Type of Metal Transfer		
	Short Circuiting	Pulse Spray	Globular
Solid	3/8 in.	3/4 in.	NA
Metal Core	NA	5/8 in.	5/8 in.
FCAW-G	NA	3/4 in.	3/4 in.
FCAW-S	NA	NA	3/8—1-1/2 in.*

** Check with the electrode manufacturer for the exact length recommendations.*

Table 7 — Typical Electrode Extensions

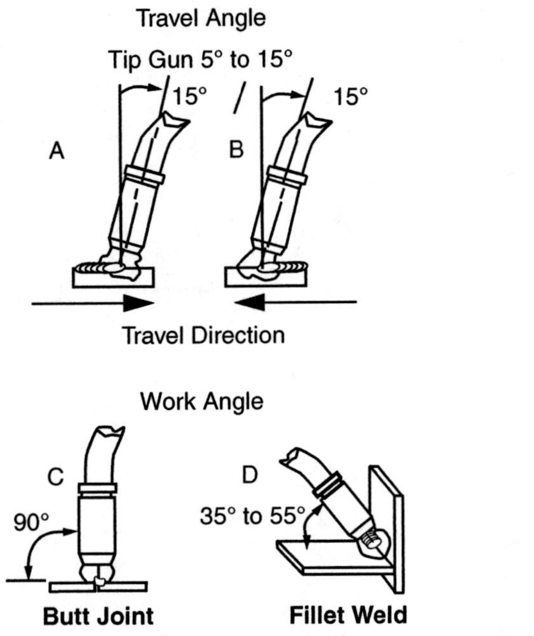

Fig. 38 — Welding Gun Angles

Travel angle is the angle in the direction of travel. The gun is generally held with a 5° to 15° push angle or leading arc (Fig. 38) or drag angle trailing arc (Fig. 38). The welding position, wire type, weld profile, penetration pattern, and slag weld pool formation all impact the travel angle. If the angle is increased to 20° to 30° there will be more spatter, less penetration, and an unstable arc. A push angle will make the weld flatter and less penetrating, while a drag angle will make the weld penetrate deeper with more build-up. The drag angle is used with the flux-cored process, as the arc force tends to keep the molten slag from running ahead of the weld pool.

Work angle is the angle the electrode makes across the weld joint (from the direction of travel). In general, this should split the joint so that the heat of the arc is evenly placed on both sides of the joint. This angle may be varied, depending upon the fit-up and whether the thickness of the joint members is not the same. (Fig. 38) and (Fig. 38) shows gun angles for a butt joint and a fillet weld.

Travel speed is measured in inches per minute. When working to a detailed welding procedure and the travel speed is specified, practice will be required to judge the proper speed. Most semi-automatic welding is done without specifying the travel speed. It is up to the welder to read the weld pool and arc action to achieve the desired results. When doing GMAW short-circuiting transfer (which is, at best, a shallow penetration transfer method), keeping the arc on the leading edge of the weld pool is very important. With spray, flux cored and metal cored, keep the arc more in the center of the weld pool. In some flux-cored wires, the weld pool should flow out well in advance of the arc. This places the arc in the back side of the weld pool. The electrode manufacturer's recommendation should be studied for specific details. (Fig. 39) illustrates the effect of travel speed on the relative depth of base metal penetration.

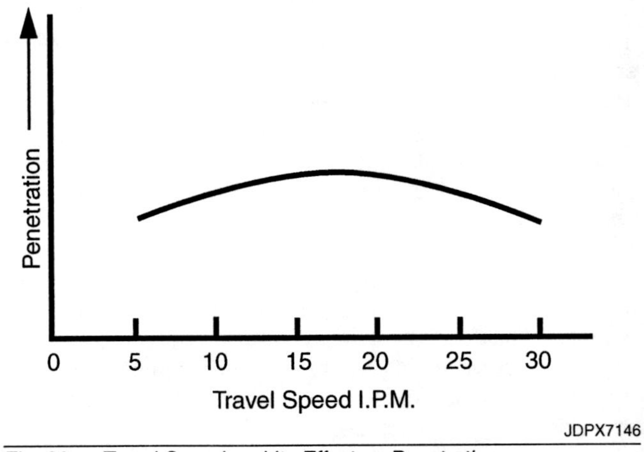

Fig. 39 — Travel Speed and Its Effect on Penetration

Joint alignment describes where on the joint the welding wire is pointed. In most cases, it is pointed directly at the center of the joint. However, when welding at high amperages and voltages, this may not always produce the desired results. Pointing the electrode off the center of the joint will allow the weld metal to flow into the joint, producing a better weld profile and reducing undercut. (Fig. 40) illustrates an example of how this technique is used on a horizontal position tee joint weld.

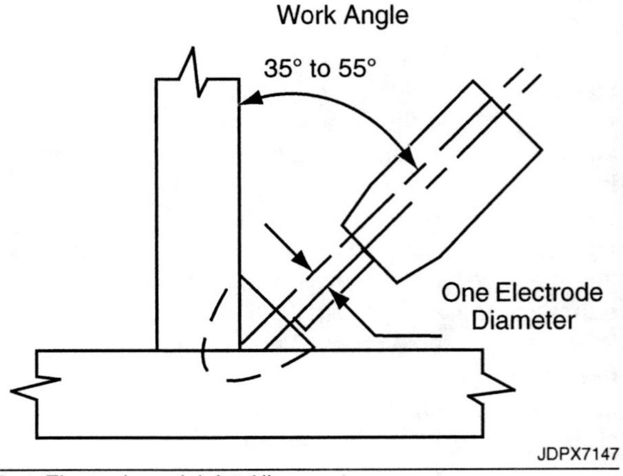

Fig. 40 — Electrode and Joint Alignment

Out-of-Position Welding

Welding out of position can be fairly easy if the proper transfer and electrodes are used. Short-circuiting transfer, pulse-spray transfer, and small diameter flux-cored wires are typically used for out-of-position work. On aluminum and some heavy sections where there is a large amount of heat sink, full spray arcs can be used as well as larger-diameter electrodes.

Various gun angles are shown in (Fig. 41) through (Fig. 43) for out-of-position work. Use wire feed speeds and voltages in the low-to-medium range for the wire being used. Since gravity is working against the weld metal, all other factors must be correct. The joint fit-up must be very good with no contaminants in the weld area. Wear the appropriate protective clothing when welding out of position. Getting under the arc in a position that is comfortable and gives good viewing of the joint is very important.

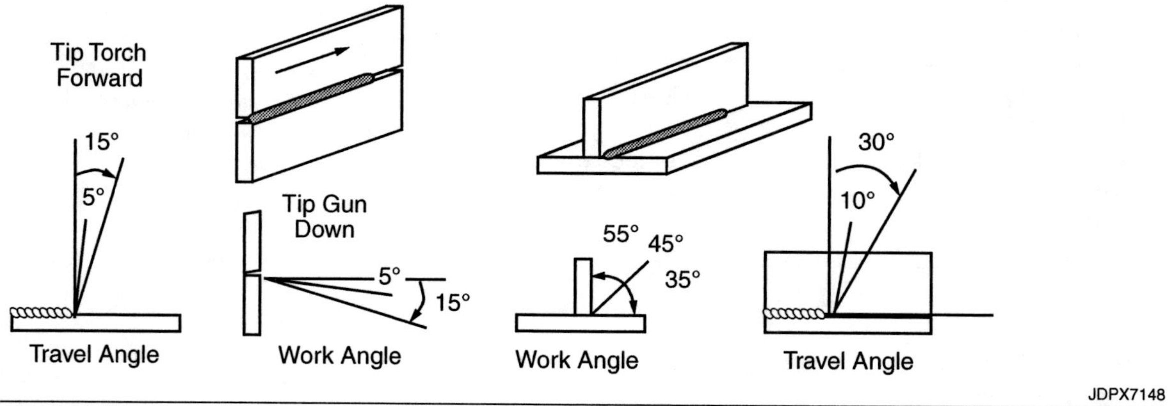

Fig. 41 — Horizontal Welding

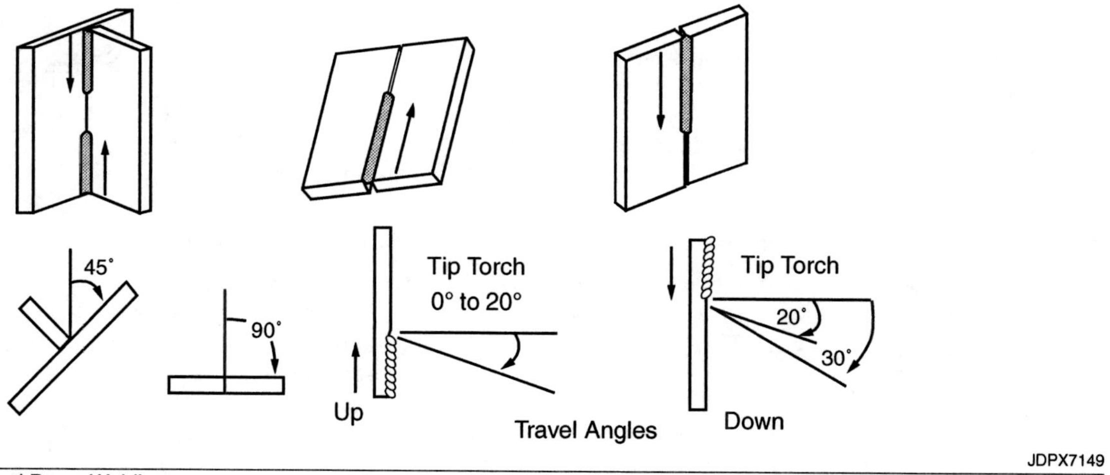

Fig. 42 — Vertical Up and Down Welding

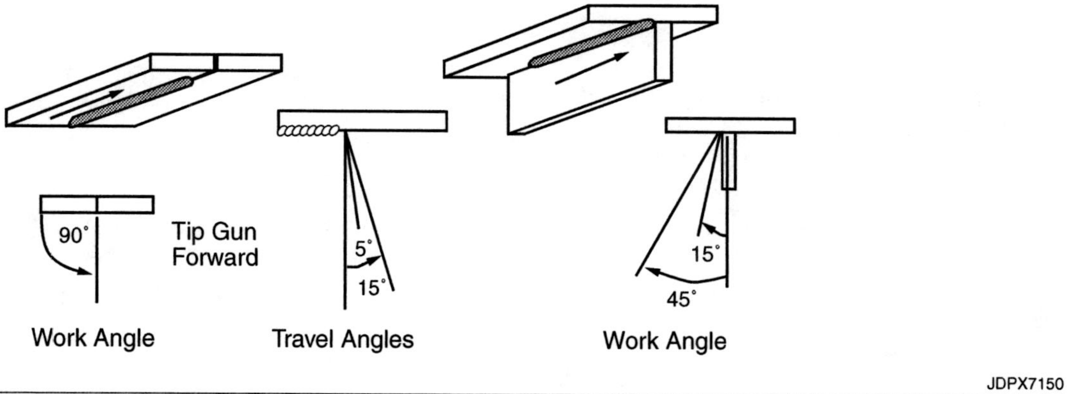

Fig. 43 — Overhead Welding

Welding Suggestions

- Follow all safety precautions.
- Keep the welding area clean and free of combustibles.
- When chipping slag, always wear safety glasses with side shields, and when possible, let the metal cool before removing the slag. If it must be removed hot, deflect the slag away from exposed skin.
- Use charts and manuals for selecting the wire feed speed, amperage, electrode extension, and voltage.
- Always prepare the metal properly — clean the metal before starting to weld, space the metal to allow for penetration, and bevel the edges when necessary to obtain full penetration.
- Use the proper work angle, travel angle, electrode extension, and travel speed for the weld joint and welding position.
- Clean and inspect completed welds carefully.

Shutting Down the Equipment and Stopping Work

1. Turn off cylinder valve or valve to flowmeter, if on a manifold system.
2. Activate purge button on feeder or disengage drive rolls and push gun trigger to open gas solenoid valve. Be careful when doing the latter or the contactor will be energized and the wire will arc if touched to anything grounded. This step drains the shielding gas from the system.
3. Back off (or lower) the flowmeter valve setting.
4. Trim any exposed wire to reduce the possibility of puncture wounds.
5. Turn off the power source and wire feeder.
6. Position the wire feeder for safe storage.
7. Wrap up welding cables and gun. When storing the gun, do not kink the gun liner by coiling it too tightly. Position the gun so that exposed wire cannot be a puncture hazard or come in contact with the shielding gas cylinder.
8. Mark all hot metal as "HOT."
9. Properly dispose of electrode wire trimmings, slag, other residue, and scrap.

Troubleshooting the Semi-Automatic Processes

Troubleshooting can best be done by looking at the specific welding process being used (short-circuiting transfer, spray transfer, pulse-spray transfer, or flux-core). It can be broken down into the process variables, consumables, and the equipment. Most of the problems, once the proper welding variables and consumables are determined, will be with the equipment. Most of these problems will be found starting at the spool of wire and working out to the contact tube at the end of the gun.

(Fig. 44) shows the wire feeder problem areas. (Also see (Fig. 18) through (Fig. 25) for a more detailed view of these parts.)

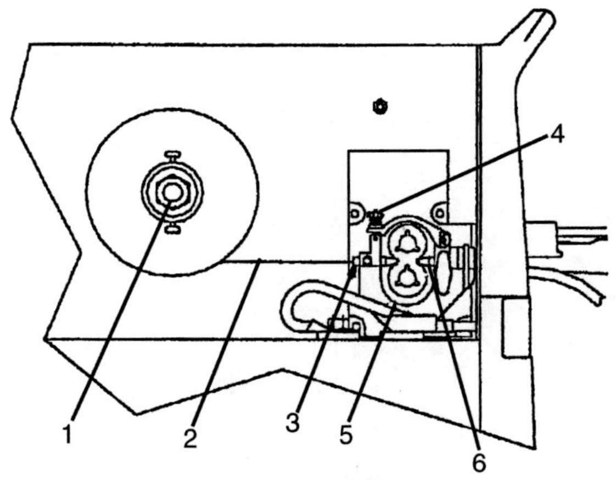

JDPX7151

Fig. 44 — Wire Feeder Problem Areas

1. Adjust the spool brake to the proper tension (Fig. 45). At a high wire feed speed setting, it should coast approximately 1/4–1/2 turn. At any speed, the wire should not be taut between the inlet wire guide and the spool.
2. Be sure the wire is the proper size and type for the work to be done.
3. Be sure the inlet guide is the proper size and not worn.
4. Adjust the drive roll tension by holding the gun several inches from a block of wood and tighten knob enough to prevent slipping (Fig. 45). Don't over-tighten; if slipping still occurs, look for other problems.

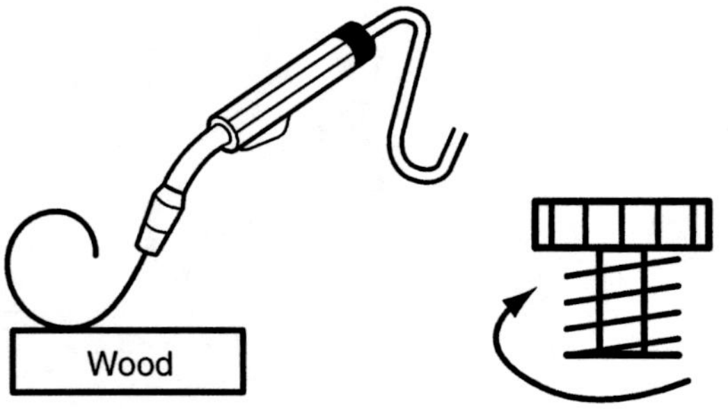

JDPX7152

Fig. 45 — Setting Drive Roll Tension

5. Are the drive rolls of the proper size and shape for the wire being fed? Are they dirty or worn?
6. Is the outlet guide adjusted as close to the drive rolls as possible without touching them?
7. The welding gun is in a rough environment and may get heavy use, so it needs daily inspection. Spatter should be carefully removed, the contact tube should be tight, not worn, and of the proper size. Frequently inspect for

wear of other parts that come into contact with the wire or are in the arc area. (Fig. 46) shows a typical gun contact tube, gas nozzle, head tube, and gun liner that need to be inspected, cleaned and/or replaced.

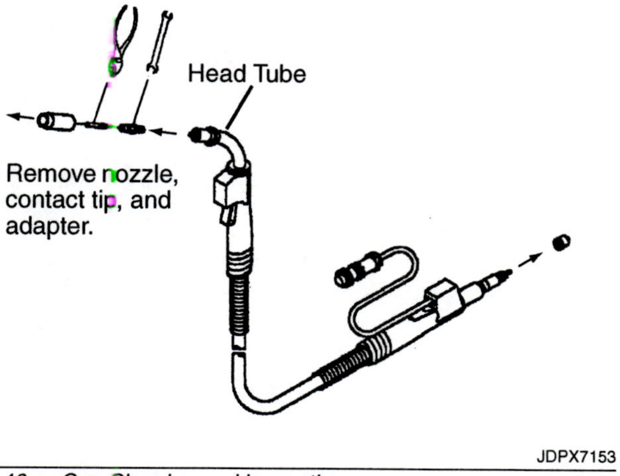

Fig. 46 — Gun Cleaning and Inspection

When removing the gun liner to replace or clean it and the gun casing: Turn equipment off, remove the wire from the gun, and disconnect the gun from the feeder. The gun should be laid out straight. Remove the outlet guide, making sure the liner is not held in place on either end (Fig. 47). Gun manufacturers provide manuals with their guns. Reference the gun's manual for proper gun maintenance and troubleshooting.

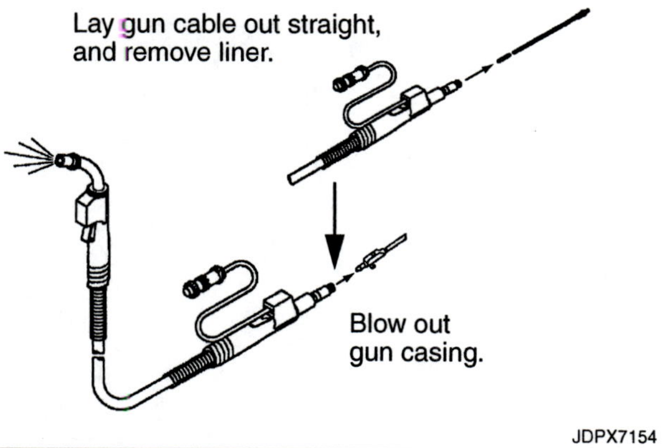

Fig. 47 — Welding Gun Disassembly

(Fig. 48) represents a typical gun with an exploded view illustrating the various parts that need to be inspected.

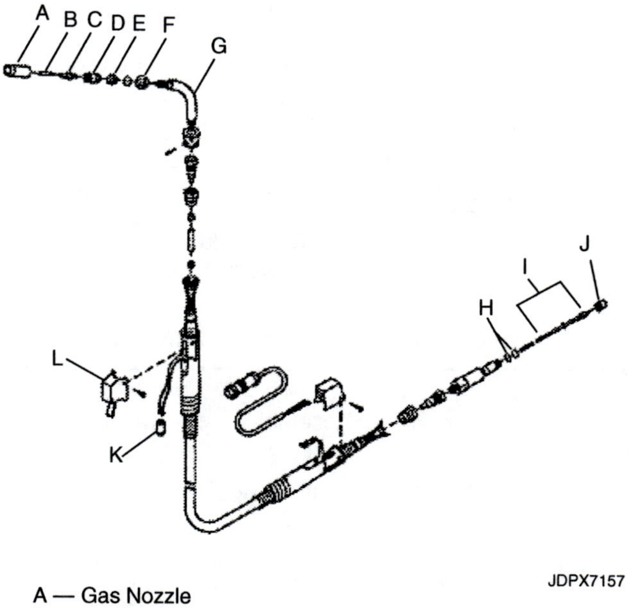

A — Gas Nozzle
B — Contact Tube
C — Contact Tube Adapter
D — Gas Nozzle Adapter
E — Retaining Ring, O-Ring
F — Shock Washer
G — Head Tube
H — O-Rings
I — Gun Liner
J — Outlet Guide
K — Trigger Switch
L — Trigger Housing

Fig. 48 — Exploded View of the Welding Gun (Listing Parts that Must Be Inspected)

Acceptable Welds

To be acceptable, groove or fillet type welds must meet visual inspection and structural strength requirements, and show evidence of good workmanship, including uniformity (Fig. 49).

Visual inspection includes measuring weld size, penetration of the weld joint, reinforcement above the surface, depth of undercut, porosity or other uniformity problems on the weld face, and excessive spatter on surrounding surfaces. Bending the piece in the weld area checks weld strength and fusion. These requirements are specified by the welding code that applies to the function of the finished part. More detailed information is given in Chapter 1 on "Introduction to Joining and Cutting Metals."

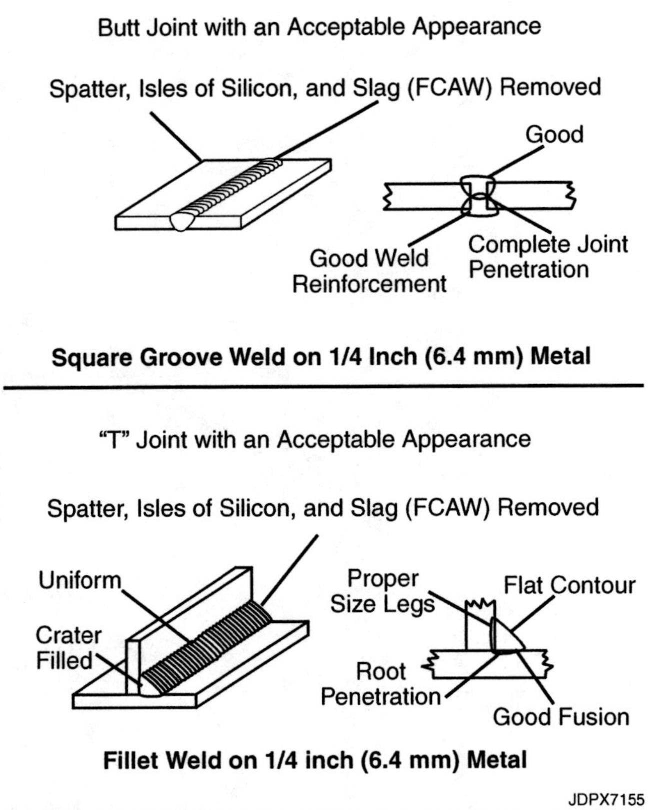

Fig. 49 — Visually Acceptable Welds

Unacceptable Welds

Generally speaking, a weld is considered to be unacceptable under any one of the following conditions:

1. A groove weld that doesn't completely fill the weld joint because of a lack of penetration and/or lack of reinforcement above the top of metal.
2. A fillet weld that does not have the proper leg size, throat measurement, and/or penetration at the root.
3. Weld penetration and/or reinforcement above the top of metal is greater than 1/8 inch (3.2 mm).
4. Undercut is greater than 1/32 inch (0.83 mm).
5. The weld breaks or the base metal breaks in the weld area during a bend test.
6. The weld has not fused to the metal and/or contains slag holes. The weld face contains excess gas holes (porosity) or other gas-caused depressions (worm tracks). Check the welding code or other acceptance requirements for what is excessive.

(Fig. 50) shows typical semi-automatic weld defects with corrective action.

Typical Semi-Automatic Weld Defects

Fillet Weld Examples **Groove Weld Examples**

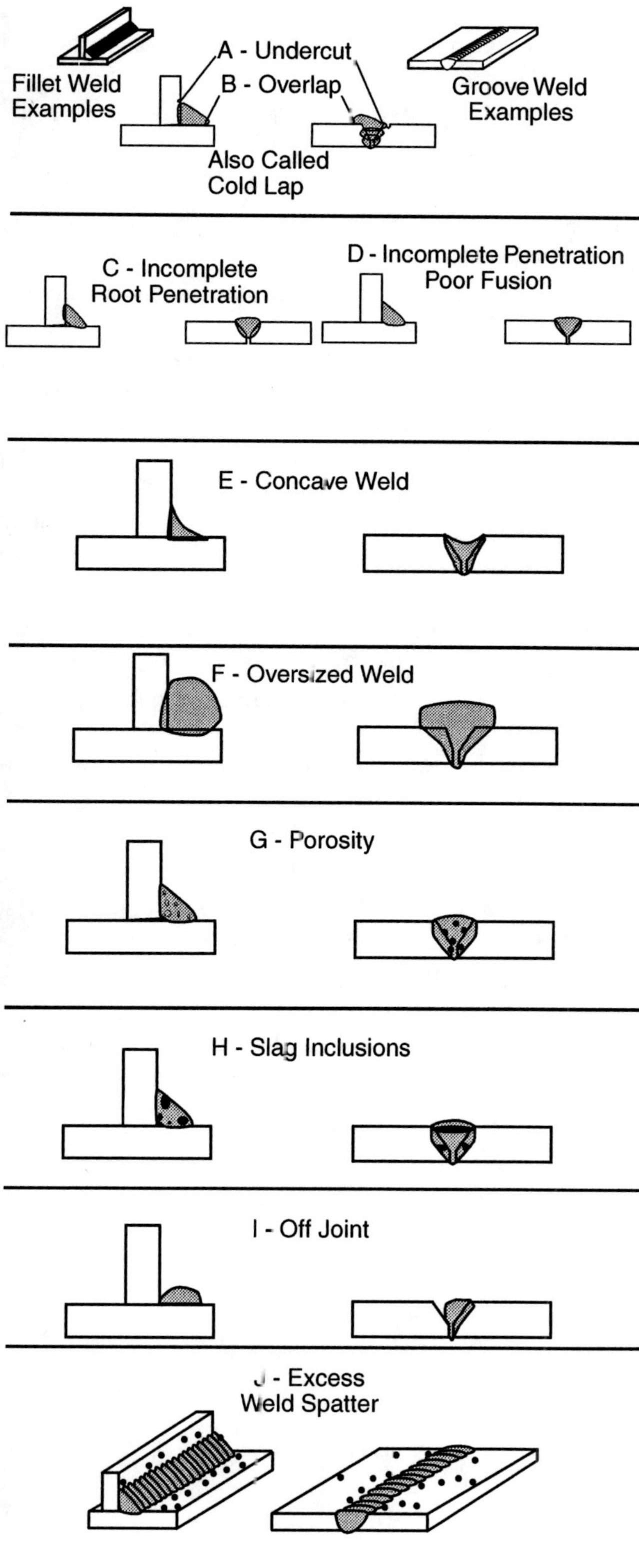

Defect	Corrective Action	Fillet Weld	Groove Weld
A	Undercut Weld		
	1. Use proper work angle	X	X
	2. Decrease voltage (arc length)	X	
	3. Check joint alignment	X	
B	Overlap Weld		
	4. Decrease weld size per pass by increasing travel speed	X	
	5. Use proper work angle		X
C, D	Poor Base Metal Fusion and Joint Penetration		
	1. Increase WFS (amps) and volts (arc length) only if required	X	X
	2. Check electrode extension	X	X
	3. Do not weld vertical down	X	X
	4. Increase root opening		X
	5. Decrease weld size per pass	X	X
E	Concave Weld (Also Called Underfill on Groove Weld)		
	1. Do not weld vertical down	X	X
	2. Make another weld pass		X
	3. Decrease voltage (arc length)	X	
	4. Increase WFS (amps)	X	
	5. Decrease travel speed	X	X
F	Oversized Weld and Excess Reinforcement		
	1. Increase travel speed	X	X
	2. Use smaller wire, less amps/volts	X	X
	3. Follow proper welding procedures; do not make welds oversized	X	X
G	Porosity		
	1. Contamination on base metal	X	X
	2. Not enough gas flow or leaks	X	X
	3. Wind/draft in welding area	X	X
	4. Moisture in flux-cored wire	X	X
	5. Electrode extension too long	X	X
H	Slag Holes and Inclusions		
	1. Remove slag or islands of silicone before making next weld	X	X
	2. Increase WFS (amperage) and voltage length (arc length)	X	X
	3. Use proper travel angle (usually a drag angle)	X	X
I	Off-Joint Welds		
	1. Use proper work angle	X	X
	2. Position end of electrode for proper joint alignment	X	X
	3. Correct vision problem — clean or replace cover glass, use lighter shade lens, or get eyes checked	X	X
J	Excess Weld Spatter		
	1. Use proper voltage (keep arc length as short as possible)	X	X
	2. Decrease WFS (amperage) not below spray condition; make sure gas, wire, amps, volts, and techniques work together	X	X
	3. Incorrect volt and amperage setting	X	X

JDPX7156

Fig. 50 — Typical Semi-Automatic Weld Defects

Test Yourself

Questions

1. (Fill in blank.) Gas metal arc welding is also known as ____________ welding.
2. When doing semi-automatic welding, what two things does the equipment control?
3. GMAW and FCAW have been replaced by what two other processes for many fabrication and repair jobs?
4. What does GMAW-S stand for?
5. A buzzing or sharp crackling sound best describes what type of metal transfer?
6. Eyes, face, and exposed skin must be protected from what, when GMAW welds cool down?
7. (Fill in blank.) The maximum thickness that is typically welded with GMAW-S is ___ inches.
8. (Fill in blank.) GMAW spray welding is generally not done on materials thinner than ___ inch.
9. According to the American Welding Society (AWS), how is the pulse variation of the gas metal arc process abbreviated?
10. In Table 1, the arc length can best be controlled by which variable for pulse welding?
11. What type of power source must be used for pulse welding?
12. A hissing or whispering sound best describes what type of metal transfer?
13. List the two AWS abbreviations for the arc welding processes that melt a tubular electrode with alloys, deoxidizers, and slag formers in their cores.
14. Of the two processes listed in the prior question, which would be best used outdoors?
15. What type of power source is used for all the processes and metal transfer methods listed in Chapter 4?
16. (Fill in the blank.) Because these processes use a continuous wire, the duty cycle for industrial power sources should be ______% duty cycle.
17. (Fill in blank.) The power source should be located no closer than the size of its ________ opening from the nearest obstruction.
18. (Fill in blanks.) The two common types of wire feeders are ________ speed and _______ sensing.
19. What feeder configuration would be used for feeding soft wires like aluminum?
20. (Fill in blank.) Which type of drive roll in (Fig. 18) would be used for feeding aluminum wire? __________
21. (Fill in blanks.) Most 300-amp air-cooled welding guns are rated at this amperage at what duty cycle and with what gas? ________% duty cycle, ________ gas.
22. (Fill in blanks.) The cubic feet per hour for spray or pulse spray is _____ to _____.
23. In (Fig. 29), which gas produces the deepest penetration?
24. (Fill in blank.) An ER70S-6 wire has approximately _________ pounds of tensile strength.
25. The wire sticking out of the gun end presents what two hazards?
26. When clipping the wire end, what additional precautions should be taken?
27. In GMAW and FCAW, amperage is set by which control?
28. (Fill in blanks.) The longer the arc the ______ the voltage, and the shorter the arc the _____ the voltage.
29. List the five points to consider for proper welding techniques.
30. (Fill in blank.) The distance from the contact tube to the arc for GMAW-S is ____ inch.
31. (Fill in blank.) (Fig. 39) shows maximum penetration at approximately 18 I.P.M. for this situation. Going slower or faster will cause the penetration to ________.
32. (Fill in blank.) A ______ travel angle of 5° to 15° is used for FCAW.
33. (Fill in blanks.) The push travel angle for vertical up welding is __ to __.
34. When shutting down the equipment and draining the shielding gas by pushing the gun trigger, what two hazards must be guarded against?
35. What four corrective actions will help eliminate porosity in the weld?
36. (Fill in blanks.) When troubleshooting, break the situation down into these three areas: _______, _______, _______.
37. List two gun parts that need to be inspected frequently.
38. (Fill in blanks.) E7OT-4 wire can be used for making fillet welds in the _______ and _______ positions.
39. (Fill in blanks.) When adjusting drive roll tension, hold the end of the gun several inches away from a _______ and adjust the _______ _______ until slippage stops.

Gas Tungsten Arc Welding (GTAW)

5

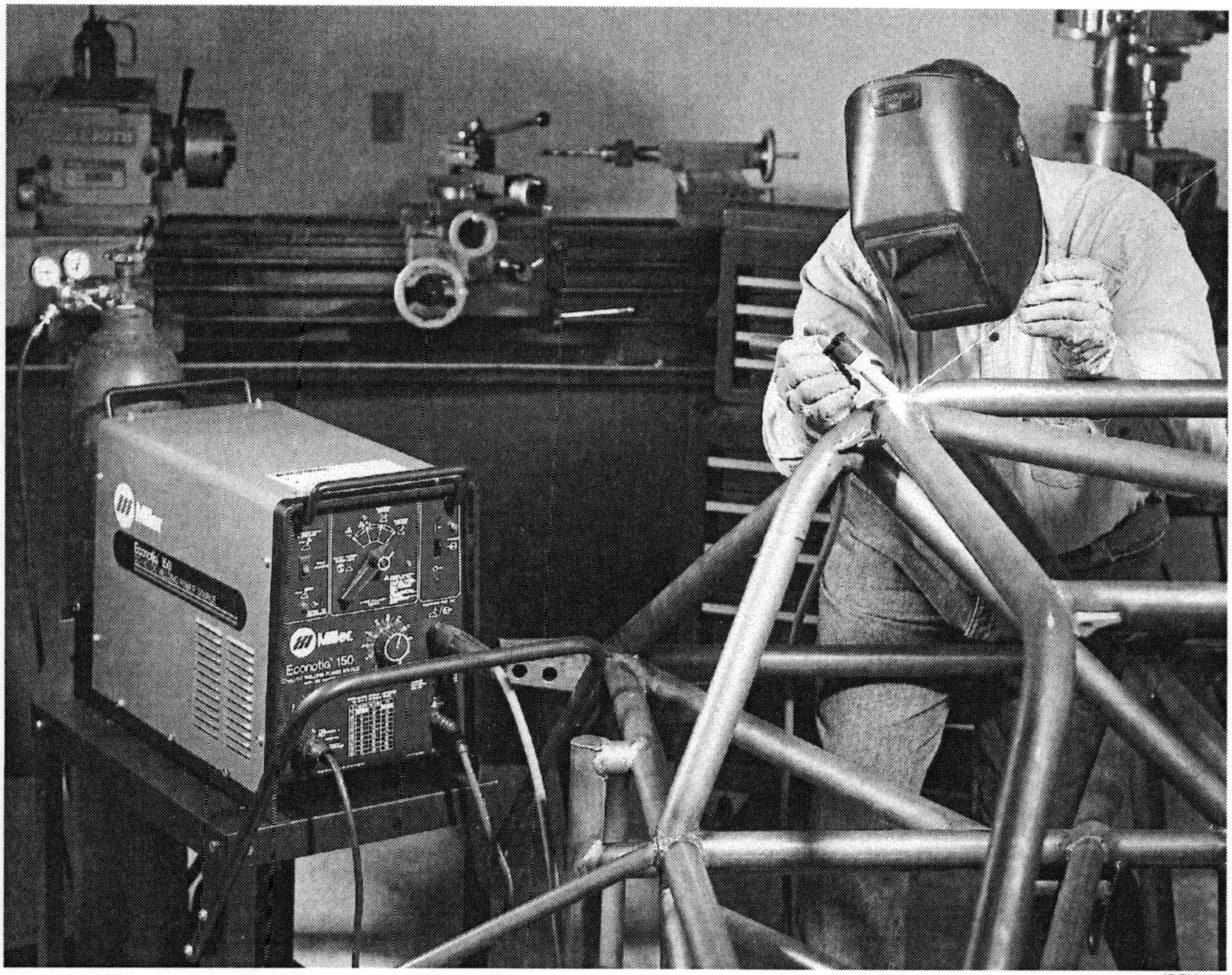

JDPX7054

Fig. 1 — Gas Tungsten Arc Welding (GTAW)

Introduction

GTAW (Gas Tungsten Arc Welding), TIG, heli-arc, or argon-arc all refer to the process (Fig. 1) covered in this chapter.

An electric arc, jumping between a non-consumable tungsten electrode and the workpiece generates the high temperature necessary for welding. An inert gas supplied from a nearby high-pressure cylinder or a manifold system is used to protect the tungsten and weld zone from atmospheric air contamination during welding. Filler metal may be added to obtain the proper weld size.

The basic equipment necessary for gas tungsten arc welding is illustrated in fig. 2, and the formation of a weld is illustrated in fig. 3.

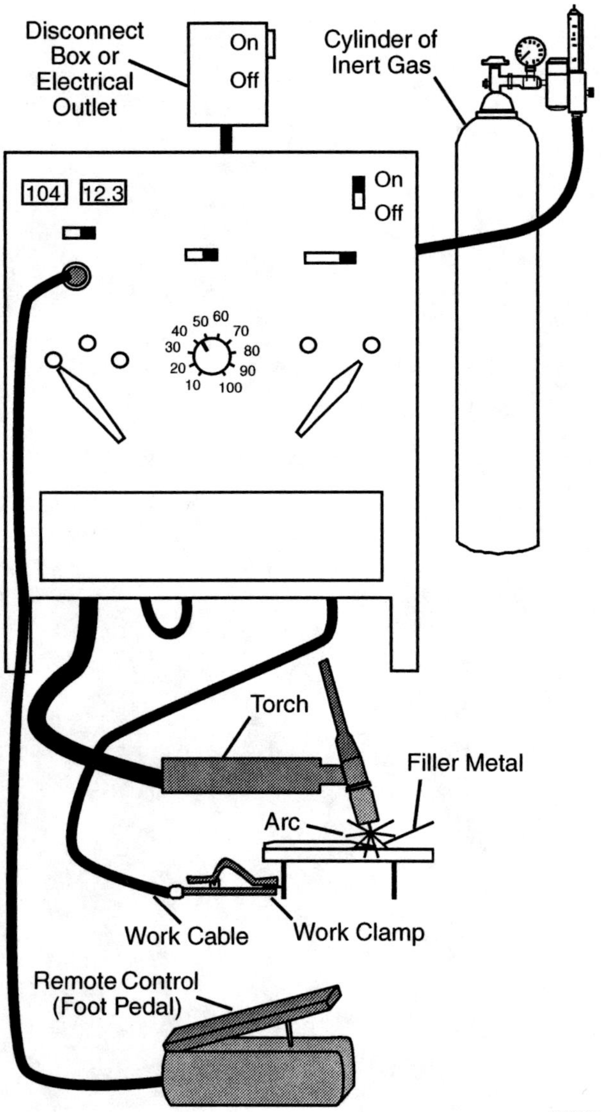

Fig. 2 — Basic Gas Tungsten Arc Welding (GTAW) Circuit and Equipment

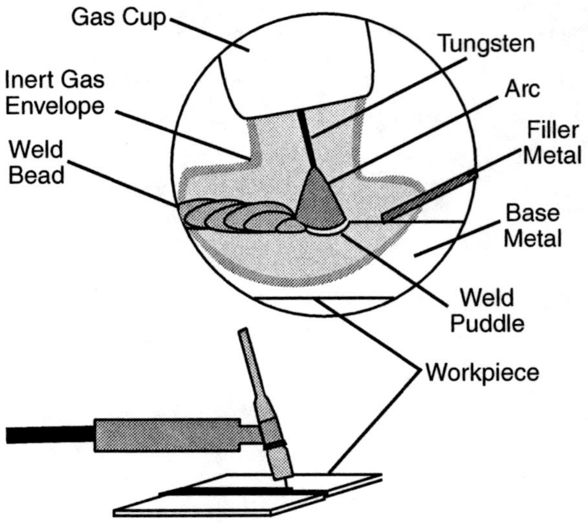

Fig. 3 — Formation of a Gas Tungsten Arc Weld (GTAW)

No weld metal is transferred across the arc as with the shielded metal arc, gas metal arc, or flux cored arc welding processes.

The base metals may be joined by fusion autogenous welding (without filler metal) (Fig. 4-A). Or filler metal may be added (as in oxyacetylene welding) to the weld joint (Fig. 4-B) to ensure the depth of the weld equals the base metal thickness.

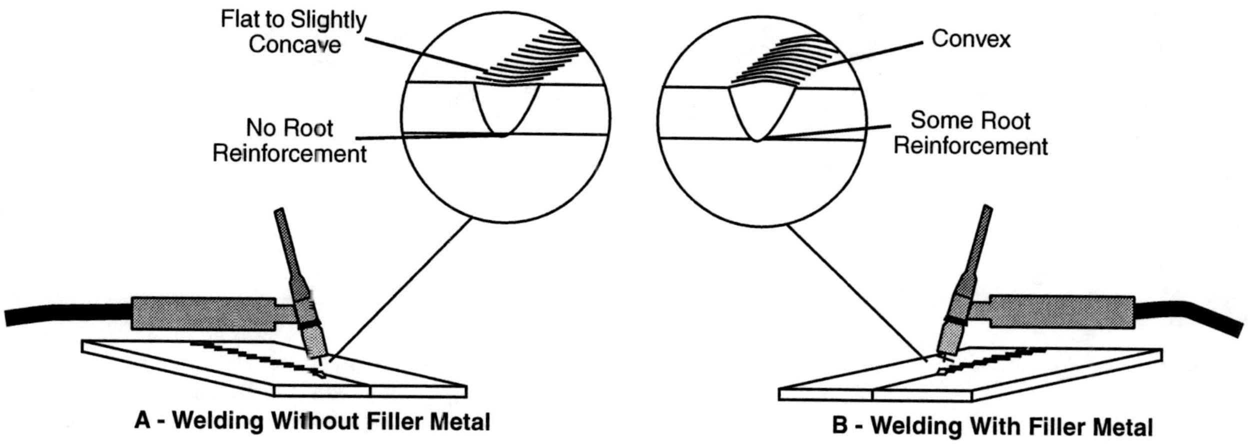

JDPX7057

Fig. 4 — The Effect of Filler Metal on Weld Shape and Reinforcement on Butt Joints

Post-weld cleaning is eliminated because there is no spatter, sparks, smoke, or slag on the weld. These features allow excellent control of the weld puddle and produce welds that are stronger, more ductile, and more corrosion-resistant than welds made with other processes.

Fabrication and repair shops use the GTAW process for metal-joining applications that are difficult with other processes and/or when the job specifications call for a weld of very high quality. These job applications and corresponding equipment requirements range from the very basic to very sophisticated.

Fig. 5 illustrates welding high-strength steel tubing using a basic 150-amp AC/DC power source. In Fig. 5 a stainless steel piping system is being fabricated using a multi-process (GTAW, SMAW, and GMAW) power source. An aluminum engine block is being welded in Fig. 6 using a high-tech 300-amp AC/DC power source that contains the most up-to-date technological features for controlling the arc.

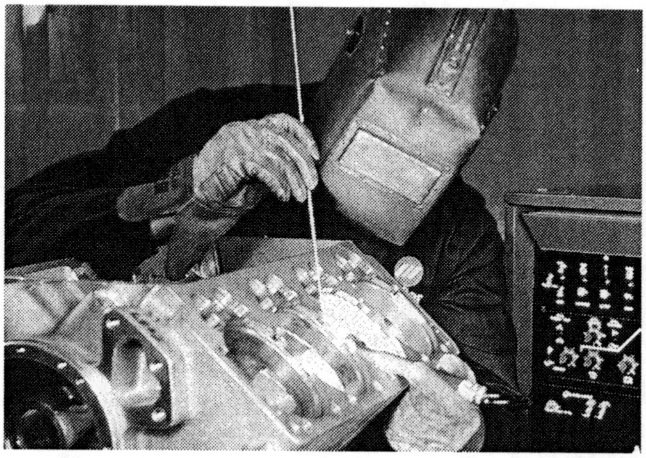

JDPX7059

Fig. 6 — Gas Tungsten Arc Welding (GTAW) an Aluminum Engine Block

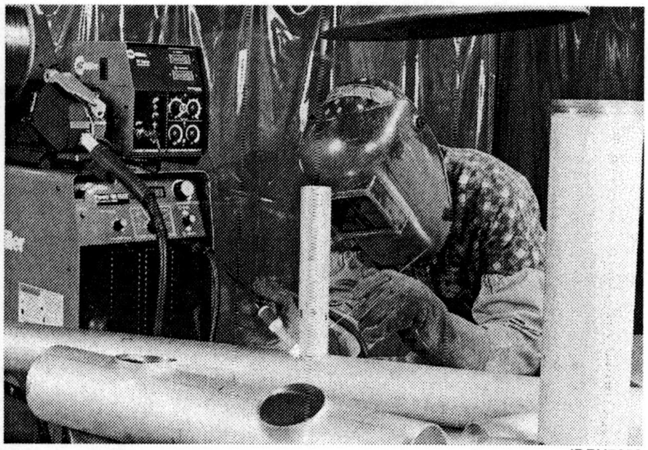

JDPX7058

Fig. 5 — Gas Tungsten Arc Welding (GTAW) of Stainless Steel

The Gas Tungsten Arc Welding (GTAW) Process

The GTAW process is similar to OAW (oxy acetylene welding), in that they both are simply a heat source held in one hand. If filler metal is required, it must be added with the other hand. This two-handed operation (Fig. 7) requires much more skill than typical manual or semi-automatic welding. Beyond this, there is no similarity between GTAW and OAW. GTAW operates at a much higher temperature with a very concentrated arc. So manipulation and control of the weld puddle are much different than for OAW. An OAW-skilled person tends to overheat and overweld when first working with GTAW.

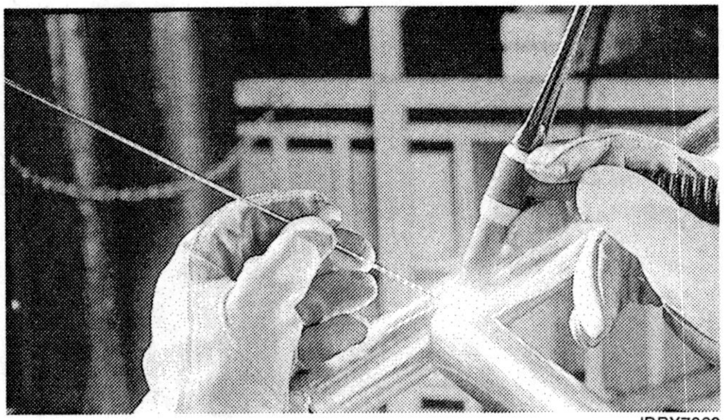

JDPX7060

Fig. 7 — Both Hands Are Usually Needed for Gas Tungsten Arc Welding (GTAW)

The electrode used in the GTAW process is a "non-consumable" tungsten with a very high melting temperature (6,170°F/3,410°C)—the highest melting point of all the metals and the same type material used as the filament in a regular light bulb. Unlike the electrodes used in the SMAW, GMAW, and FCAW processes, this tungsten electrode does not melt to become part of the weld. For this reason, the electrode is classified as a "non-consumable electrode."

The arc can reach a temperature of 11,000°F (6,000°C), which is well above the melting temperature of the tungsten. However, the tungsten is not normally melted because it has the ability to conduct heat well (has good thermal conductivity) and because it is being held in a torch that is a good heat conductor.

IMPORTANT: It is possible to cause the tungsten to melt by exceeding the amperage rating for the tungsten size and type, by trying to weld on direct current electrode positive (DCEP), by extending the tungsten too far beyond the end of the gas cup, and/or by not having an adequate flow of inert gas. Tungsten melting can be identified when the end becomes "white hot" and a yellowish-colored smoke forms in the weld area. Also, a tungsten with a point will become rounded when used at too high an amperage, while a tungsten with a rounded end will become noticeably shorter.

Three potential welding problems are created because of this special tungsten electrode.

One problem is tungsten's incompatibility with the base metal. If the tungsten comes into contact with the base metal (or melts), the tungsten and/or base metal can become contaminated. This contamination can lead to some of the weld defects covered at the end of this chapter.

A second problem is arc starting. How can the arc be started without contaminating either the tungsten electrode or the base metal when first contact is made to start the flow of current? This means, arc starting is more complicated for gas tungsten arc welding because the starting methods used for SMAW, GMAW, and FCAW won't work for GTAW.

The last problem concerns shielding the electrode and the base metal from atmospheric contamination during welding. Since GTAW electrodes are bare, there is no flux coating as with shielded metal arc welding to provide shielding from atmospheric contamination during welding. Shielding must be from an external source similar to that used for GMAW.

The gases used for shielding with GTAW are inert. This means there is no chemical reaction between the gas envelope and the base metal during welding (i.e., the inert gas does not add or subtract anything from the weld puddle). Argon (Ar) is the gas most widely used for GTAW. Helium (He) or combinations of argon and helium are also used for GTAW.

Since there is no flux on the electrode, and the gas is inert, gas tungsten arc welding is a very clean process. There is very little smoke, fumes, sparks, or spatter during welding. The material to be welded must be very clean before welding begins because there is no flux to trap and remove foreign material that gets in the weld puddle. The GTAW process does not work well on dirty, corroded base metal unless it can be cleaned properly before welding.

To obtain optimum weld quality on aluminum and magnesium with the GTAW process, the tough surface oxides must be removed. These oxides do not melt easily, even in the intense heat of the welding arc, because they have a much higher melting temperature than the base metal. For example, aluminum oxide melts at approximately three times the temperature of the aluminum base metal.

There are several ways to remove the surface oxide on aluminum. However, because the oxide begins reforming immediately, welding must be completed before the oxide reformation is great enough to interfere with welding. The surface oxides can be removed mechanically with a stainless steel (not carbon steel) wire brush or with special chemicals before welding. The action of the arc from alternating current also breaks up and removes surface oxides.

The GTAW arc is a very pure heat source that can be used for welding virtually all metals, including aluminum, titanium, stainless steel, carbon steel, and copper. On thin sections (1/8 inch/3.2 mm or less) with the proper joint design, no filler rod is needed. Only fusion (autogenous) welding is done. This makes it a one-handed operation or an extra-steady two-hand method which makes the job easy and fast.

The addition of filler metal is required on heavier sections or with certain joint designs like the intersections on the tubing in Fig. 7. Filler rods come in a variety of alloys, which will be covered under consumables. If no filler rod is available, or if it is difficult to obtain the filler metal for some exotic metal, a piece can be sawed or sheared from the base metal and used as a substitute filler rod.

Since the heat source for gas tungsten arc welding is an electric welding arc, the type of current used is very important. Direct current electrode negative is used for most direct-current applications because it produces deeper penetration and allows faster travel speeds. Direct current electrode positive is seldom used for this process because it produces shallow penetration and slow travel speeds. (This is just the opposite of SMAW, and GMAW which are mostly done with electrode positive.). Alternating current is used for welding metals like aluminum and magnesium, which have tough surface oxides that require removing.

The current flow, gas ion flow, and heating relationship for both types of current are illustrated in Fig. 8. Note that a 1/16 inch (1.6 mm) diameter tungsten electrode operating on DCEN (Fig. 8-A) can accomplish the same welding task as a 1/4 inch (6.4 mm) diameter tungsten electrode operating on DCEP (Fig. 8-B). This is possible because the electrons flowing from the electrode to the work (DCEN) transfer approximately 70 percent of the heat created in the arc to the base metal. Likewise, the electrons flowing from the work to the electrode (DCEN) transfer approximately 70 percent of the heat created in the arc to the electrode. Even by using a large-diameter tungsten, on DCEP it is difficult to obtain the heat necessary on the base metal for efficient welding.

One very important occurrence with DCEP is surface cleaning. Surface oxides on the base metal are broken up and lifted away by the flow of gas ions. However, despite this cleaning action, DCEP is seldom used for GTAW unless it is needed to limit the penetration on thin metal.

Action in the arc and on the base metal is more complex with alternating current. As illustrated in Fig. 8-C, the flow of electrons is constantly changing direction. For example, with 60-cycle alternating current, the electrons flow one direction, stop, and change direction 60 times per second. This constant current reversal creates base metal heating for an instant (with negative electron flow) and then provides base metal cleaning the next instant (with positive electron flow). Also, heat from the arc is more evenly distributed between the tungsten and workplace since the electrons are flowing in both directions. The tungsten size to accomplish a task with AC is only slightly larger than that required to do the same task with DCEN.

The benefits of both types of direct current become advantages for using alternating current for gas tungsten arc welding of aluminum and magnesium. As discussed in the next section on power sources, an even better cleaning and heating action can be achieved with high-tech power sources that have special controls for varying the time and amperage on the negative and positive parts of the alternating current.

Gas Tungsten Arc Welding (GTAW) Equipment

The following is a list of basic gas tungsten arc welding equipment:

- Power source
- Torch assembly, power cables, and work clamp
- Inert gas delivery system

Weld cleaning and finishing tools

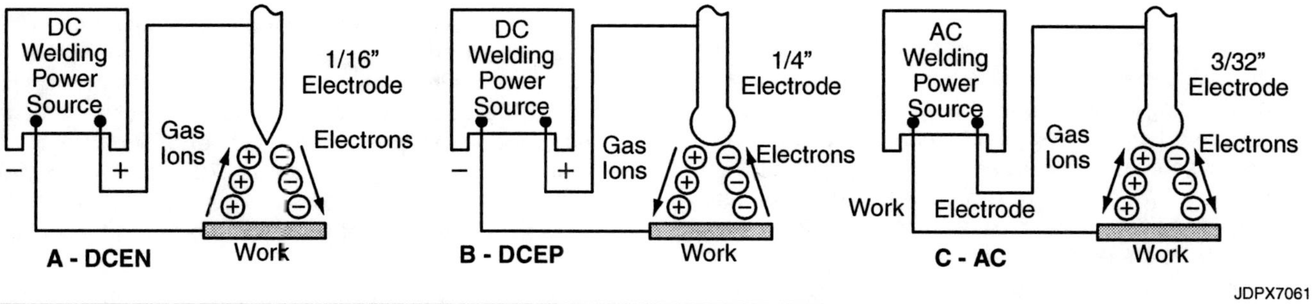

Fig. 8 — Electron and Gas Flow for Gas Tungsten Arc Welding (GTAW) with DC and AC Currents

Power Sources

GTAW power sources have the same type of constant (non-fluctuating) controllable amperage output as a SMAW power source. These power sources are available in two basic types: one that produces both alternating and direct current (AC/DC), and one that produces only direct current (DC). An AC/DC power source is the most versatile. It has the capability of welding any metal that can be joined with the GTAW process. A DC power source is commonly used for joining most metals except magnesium and aluminum.

Even though GTAW power sources are considered to be the constant current type and can be used for SMAW, the capability and performance of the power sources are quite different. Therefore, it is possible, but not always the most practical in the long run, to gas tungsten arc weld with a SMAW power source and vise versa.

The main differences between GTAW and SMAW power sources are listed and explained below.

1. A GTAW power source has the built-in controls necessary for arc starting and stopping, shielding gas control, and remote amperage control. Some or all of these devices are needed to make a SMAW power source usable for GTAW. Many SMAW power sources are not designed for adding remote controls and, therefore, have very limited GTAW capabilities.

2. GTAW power sources must be designed for welding at higher duty cycles and with better cooling than SMAW power sources. Compared to a SMAW power source, a GTAW power source has less of an opportunity to cool between the start and finish of a welding job. With GTAW, there is no need to stop for changing the electrode (it is non-consumable) or for removing slag, smoke, and spatter associated with SMAW. When gas tungsten arc welding aluminum and magnesium with alternating current, a condition called "DC component" is set up, which causes additional internal heat build-up in the power source. These are the reasons a GTAW power source must have a high duty cycle and special cooling.

   A SMAW power source used for AC GTAW must have the duty cycle de-rated (decreased) or risk damage from overheating. Table 1 provides the information used to de-rate a typical 200-amp 20-percent duty cycle SMAW power source for use in an AC GTAW application. The power source's 200 amps are multiplied by 45 percent and equal 90 amps at a 100-percent duty cycle for SMAW. To de-rate it further for AC GTAW, this 90 amps is multiplied by 70 percent. The resulting 63 amps is available at a 100-percent duty cycle for welding on AC.

   An external device for generating high frequency is necessary when using a SMAW power source for AC GTAW applications such as welding aluminum and magnesium. Without the addition of high frequency to the alternating welding current, it is impossible to maintain an arc.

Power Supply Present Duty Cycle	Power Supply Rated Amps Times		Non-GTAW Applications Duty Cycle	De-rated Duty Cycle AC GTAW Welding De-rated Amps Times
60%	75%	=	100%	70%
50%	70%	=	100%	70%
40%	55%	=	100%	70%
30%	50%	=	100%	70%
20%	45%	=	100%	70%

Table 1 — Chart for De-Rating Shielded Metal Arc Welding (SMAW) Power Sources for Gas Tungsten Arc Welding (GTAW)

IMPORTANT: If a SMAW power source is used for AC GTAW according to the SMAW duty cycle rating, the power source can be ruined (burned up) from overheating.

3. Power sources that use high frequency (HF) for arc starting need special protection from potential damage to internal electronic components from the high frequency itself. Manufacturers of GTAW power sources "build in" the necessary protection from the high-frequency current.

   Most SMAW power sources are not equipped for protection against high-frequency current. A SMAW power source should be equipped with the proper protection when a high-frequency generator is added for AC gas tungsten arc welding. The high-frequency current can damage internal components of a power source if they are not properly protected.

4. The minimum output on a DC SMAW power source used for GTAW may not be satisfactory for some applications such as welding thin metal or for crater filling. The output of a power source is rated by the manufacturer on the specification sheet and/or the name plate on the power source. Since SMAW is a higher voltage process than GTAW, the actual minimum amperage output on GTAW may be much higher than indicated. Typically, SMAW voltages are in the range of 25 to 30 with the arc lit, while GTAW voltages are in the 10 to 15 range. This difference in welding voltage could, for example, make the minimum amperage output of 70 amps for SMAW become 100 amps on GTAW. When used for GTAW, the power source may not be capable of going low enough for some jobs.

5. A SMAW power source is not equipped with a solenoid valve for the inert gas used with GTAW. This deficiency can be overcome by using a torch with a manual on-off gas valve.

AC/DC power sources are manufactured with a variety of features and contain controls like those represented on the front control panel in Fig. 9 and the illustration in Fig. 10. A welder must understand the function and be able to set each control properly before beginning a welding job.

Features on a typical GTAW power source:

1. Power (On-Off Switch)
2. Current and Polarity Selector
3. Mode (Also Called Process Selector Switch)
4. Remote Switch (Amperage and/or Output/Contactor)
5. Remote Receptacle (Inside Cover)
6. Amperage Range (Also Called Coarse Adjustment)
7. High-Frequency Control (Off/Start Only/Continuous)
8. High-Frequency Intensity Control (Inside Cover)
9. Amperage Adjustment (Also Called Fine adjustment)
10. Arc Balance Control
11. Post-Flow Timer
12. Preflow Timer
13. Spot Timer
14. Start Control
15. Crater Fill Control
16. Pulse Controls
17. Gas and Water Solenoid Valves
18. Touch and Lift Start Controls*
19. Full Weld Sequencer*

*These features and controls are available on some GTAW power sources that are not illustrated in Fig. 9 or Fig. 10.

Fig. 9 — Front Panel of a Typical AC/DC Power Source Designed for Gas Tungsten Arc Welding (GTAW)

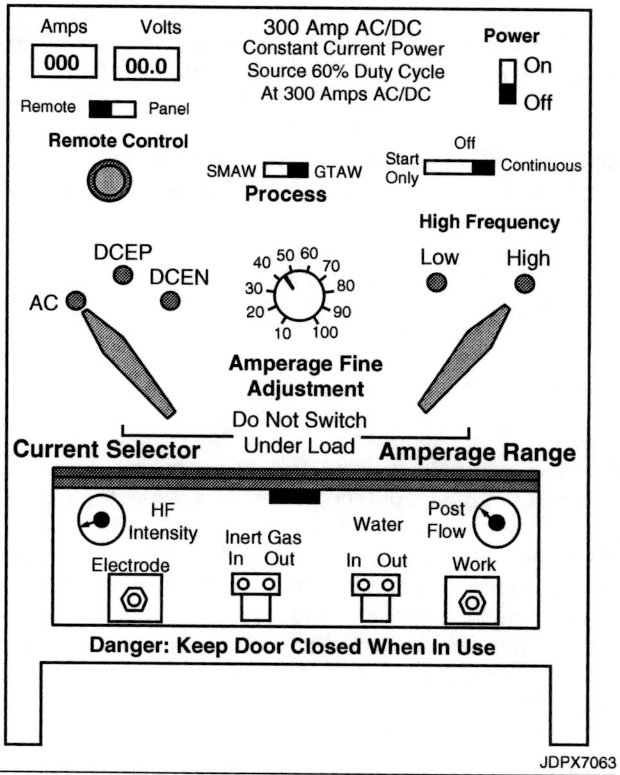

Fig. 10 — Basic AC/DC Gas Tungsten Arc Welding (GTAW) Power Source

CAUTION: Read the owner's manual and follow operating instructions before changing the controls and switches while the power source is in use. Changing the position of most of the switches and controls when current is flowing (if the arc is lit or if the electrode is stuck to the work) could damage the power source and/or cause burn injuries.

Power (On-Off Switch)

The power switch turns the power source on and off. It should not be moved to the OFF position while the arc is lit. The power switch should be in the OFF position before inserting the plug into a primary power receptacle.

Current and Polarity Selector

The current and polarity selector is used for changing back and forth from AC to DC. It is also used to change from DCEN to DCEP. The current selector should not be moved while the arc is lit or if the electrode is stuck.

Mode

The mode or process selector switch must be in the GTAW position for the other controls to function properly when gas tungsten arc welding. A GTAW power source can also be used for shielded metal arc welding when the switch is in the SMAW position. However, as explained in the differences between power sources, a power source designed for GTAW may not perform as well in the SMAW mode as a power source that is specifically designed for SMAW and vise versa. The process switch should not be moved while the arc is lit or if the electrode is stuck.

Remote Switch

The remote or control switch affects the operation of several features and controls on a GTAW power source. The remote amperage and output contactor may operate from one remote switch or each function may have a separate switch. When in the REMOTE position the amperage can be controlled from one of the remote control devices in Fig. 11. If the remote amperage and contactor operate from the same switch, several related functions may be started and stopped simultaneously when a remote control device is operated.

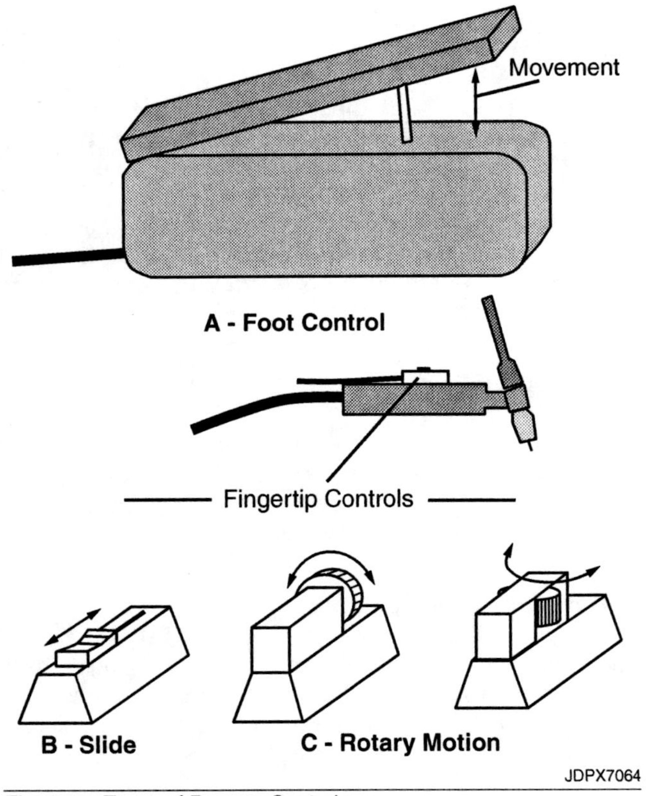

Fig. 11 — Types of Remote Controls

Remote Receptacle

A receptacle located near the remote switch is used for connecting a remote control device. The remote control device in turn may start (and stop) the high frequency, the flow of inert gas, the flow of water (or coolant) to a water-cooled torch, and may close (and open) the circuit that allows welding current to flow. In addition, the amperage can be increased or decreased by moving the position of the remote control mechanism between the minimum and maximum positions.

The advantages of a remote control are listed below:

- Several controls can be started or stopped at the same time and at a distance away from the power source.
- The amperage can be changed (increased or decreased) without stopping a weld and having to go to the power source any time an amperage adjustment is necessary.
- It is much easier (or in some cases possible) to maintain the proper base metal penetration and proper weld puddle size by making immediate amperage adjustments to compensate for changing weld conditions. When a welder recognizes that the temperature of the base metal is beginning to become too hot or too cold, he/she immediately changes the amperage before the weld puddle gets out of control and causes a weld defect.
- It is easier to form a weld puddle quickly and to fill the crater properly on metals like aluminum that conduct heat very well. When welding aluminum, a lot of amperage is needed to generate the heat necessary to form a weld puddle because the whole piece is getting hot. Once a weld puddle has formed, not as much heat is needed to keep the puddle going because the piece has been preheated as the weld puddle formed. When the end of the piece has been reached, the amperage can be decreased and the weld crater can be filled properly. Without a remote amperage control, the setting would have to be at a medium amperage. It would take much longer for the piece to be heated enough for a weld puddle to form and it would be more difficult to fill the crater at the completion of the weld.

The remote control device can be foot controlled as with the foot pedal illustrated in Fig. 11-A. Or a torch mounted fingertip control like one of the examples in Fig. 11-B and Fig. 11-C can be used. A foot control works well when a welder can be in a comfortable standing or sitting position. A fingertip control can be used in any position, and it is almost a necessity when welding in a position where it is not possible to stand or sit to use a foot control. A welder must have a very steady hand to operate a fingertip control and at the same time keep the GTAW torch positioned properly. The slightest unintended movement of the torch can cause the tungsten to touch the workpiece and contaminate both the tungsten and workpiece.

Amperage Range

On power sources with an amperage range (or coarse adjustment as it is sometimes called), set the amperage according to the type and thickness of metal to be welded. (See Table 1 for amperage requirement guidelines.) The amperage range should not be moved while the arc is lit or if the electrode is stuck.

Some power sources do not have a range switch. Unless one of these power sources is equipped with a digital amp meter, the amperage can be difficult to fine-tune using the amperage adjustment on the control panel.

High-Frequency Control

The high frequency is used with the GTAW process when welding on both AC and DC to start the arc without touching the electrode to the work. On direct current, the high-frequency switch should be in the START ONLY position. Once the arc starts, it will keep going without high frequency. On AC, the high frequency should be set to CONTINUOUS. The high frequency is needed to start the arc and also to keep the arc going, because the flow of electricity across the arc stops and changes direction every 1/120 of a second. The high-frequency switch should not be moved while the arc is lit or if the electrode is stuck. The high-frequency controls may be built into the power source or added as an accessory. Fig. 12 illustrates two examples of add-on high-frequency controls.

JDPX7054

Fig. 12 — Types of Accessory High-Frequency Controls

High-Frequency Intensity Control

High-frequency intensity adjustment controls the strength of the high-frequency current necessary for the arc to start without having to touch the electrode to the work. A high-intensity setting will make the high-frequency current jump a larger air gap to start the arc. The intensity may have to be set higher when using long welding leads to achieve the same arc starting that is possible when using short welding leads.

The high-frequency intensity should be set as low as possible without preventing the arc from starting. One reason for using a low setting is to minimize the electrical interference with nearby electronic equipment. High frequency operates in the AM radio frequency range and can cause interference with a variety of electronic devices. A low (high-frequency) intensity setting also reduces the possibility of getting "bit" by the high-frequency current during welding. A hole in a glove or wet clothing can provide a path that makes it easier for high-frequency current to follow, rather than flowing directly to the workpiece. Be careful and always follow the safety rules for working with electricity.

CAUTION: People with electronic health care devices such as heart pacemakers, should consult with their physician before getting near welding equipment that uses high frequency.

Amperage Adjustment

The amperage adjustment (or fine adjustment on power sources that have a range switch) is used to fine-tune or select the amperage that works best for each welding job. When a remote control is used, it works along with the fine adjustment on the power source. The maximum setting on the fine adjustment will be the upper amperage limit on the remote amperage control. For example, if the amperage within a range setting is 50 to 150 amps, and 100 amps is the maximum needed for the job, the fine adjustment can be set at a position that will limit the amperage to 100 amps. This feature allows a welder to set this limit below an amperage that could cause the metal to be overheated and prevents the possibility of melting an unwanted hole in thin material. The position of fine adjustment can be changed while the arc is lit.

Arc Balance Control

The arc balance control illustrated in Fig. 13 is used to vary the percent of time the electrode is negative (DCEN) and positive (DCEP) during each cycle of the alternating current. This means that when conditions are right, the cleaning action can be minimized to increase base metal heating and travel speed, or the cleaning action can be maximized when cleaning is more important than base metal heating and travel speed.

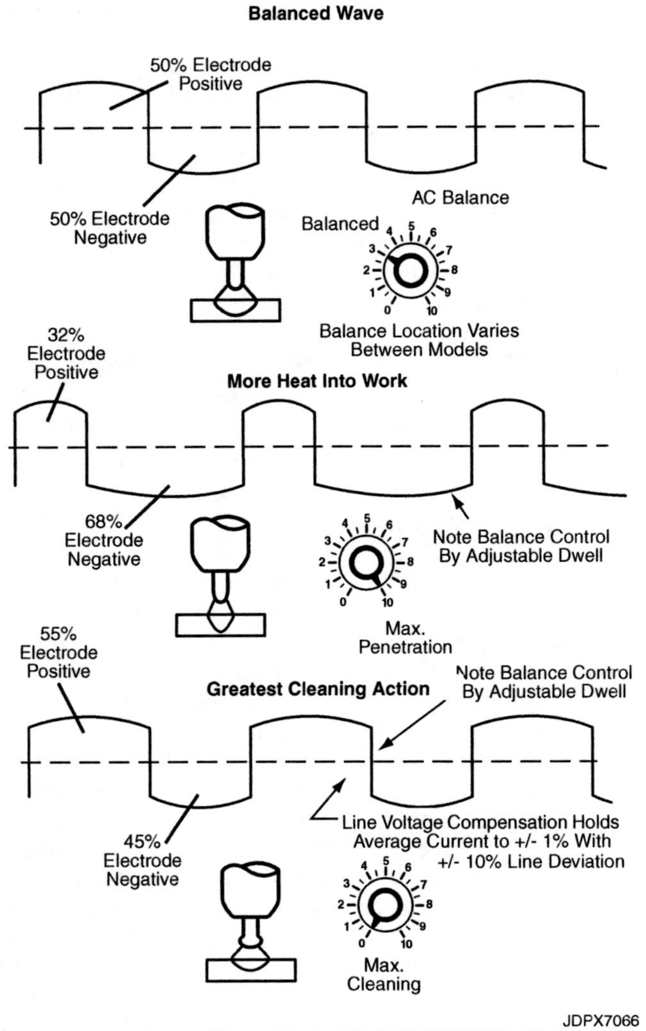

Fig. 13 — Arc Balance and Its Effect on Welding

There are three different types of power sources that produce alternating current for gas tungsten arc welding: one type (the older style shown in Fig. 10) produces a sine wave. Another type produces a square wave (Fig. 14) and the third, more recent type produces an asymmetric square wave (Fig. 15). The more control available for all the parts of the AC cycle, the better the power source will perform when welding aluminum and magnesium. Table 2 compares the controls available on these three different types of AC power sources.

Control	Sine Wave	Square Wave	Asymmetric
Current Overall	X	X	X
Current Electrode (–)			X
Current Electrode (+)			X
Time Electrode (–)		X	
Time Electrode (+)		X	X
Frequency Adjustment			X

Table 2 — Types of AC Controls on GTAW Power Sources

Fig. 14 — Square Wave AC/DC Inverter as GTAW Power Source

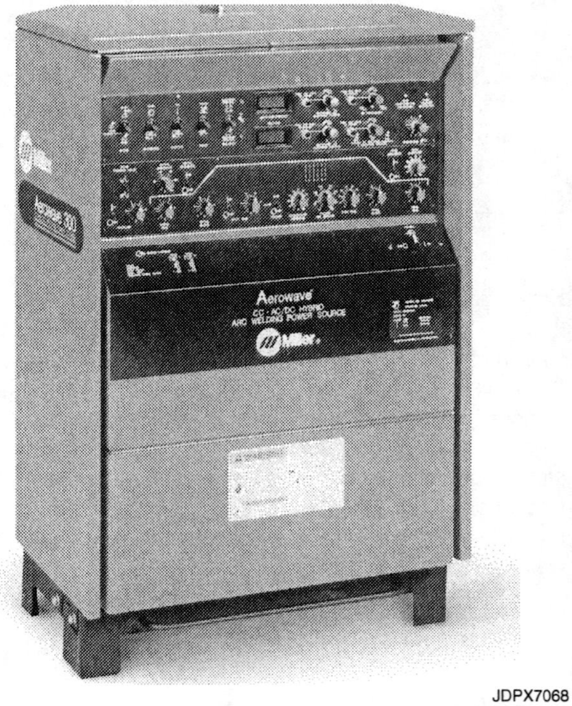

Fig. 15 — Asymmetric Square Wave AC/DC as GTAW Power Source

Since none of the major equipment manufacturers build the AC sine wave GTAW power sources any more, only the square wave and asymmetric wave types will be covered. (Fig. 13) illustrates the way each part of alternating current can be controlled and its effect on base metal penetration and cleaning on a square wave power source.

In a Balanced Wave position (Fig. 13), 50 percent of the time the electrode is negative (DCEN) and 50 percent of the time the electrode is positive (DCEP). This setting provides a corresponding equal combination of base metal heating and cleaning action.

The More Heat Into Work position (Fig. 13) provides more base metal penetration, faster welding speeds, and less cleaning action than the balanced position. At this setting, 68 percent of the time is on electrode negative (heating the work) and 32 percent of the time is on electrode positive (cleaning the work). There is sufficient cleaning when an etched band (Fig. 16) can be seen along the edges of the weld.

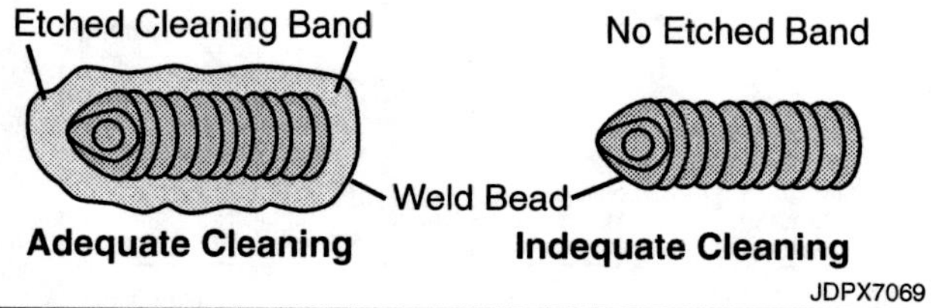

Fig. 16 — Etched Cleaning Band Indicates an Adequate Percent of DCEP on the Arc Balance Control

The Greatest Cleaning Action (Fig. 13) provides more base metal cleaning, shallower penetration, and much slower welding speeds than the balanced position. At this setting, 55 percent of the time is on electrode positive (cleaning the base metal) and 45 percent on electrode negative (heating the work). More time on electrode positive also increases the heat into the tungsten and requires a larger tungsten when compared to welding at the same amperage in the balanced position. This setting should only be used when mechanical and chemical methods are not successful in cleaning the base metal.

Balance, maximum penetration, or maximum cleaning is accomplished by adjusting the percent of time spent on each half cycle of the alternating current on a square wave power source. There is no independent control over the amperage on each half cycle (just an overall amperage setting) and no control over current frequency (not high frequency).

An asymmetric power source (Fig. 15) does have independent controls over half-cycle amperage and frequency. The additional control over the current on electrode positive and negative half cycles, the ability to adjust frequency, and with extended balance control on up to 90% electrode negative gives total control over the AC cycle. This does make the equipment much more complicated and expensive. The controls for an asymmetric square wave and their effect on a weld are illustrated in Fig. 17.

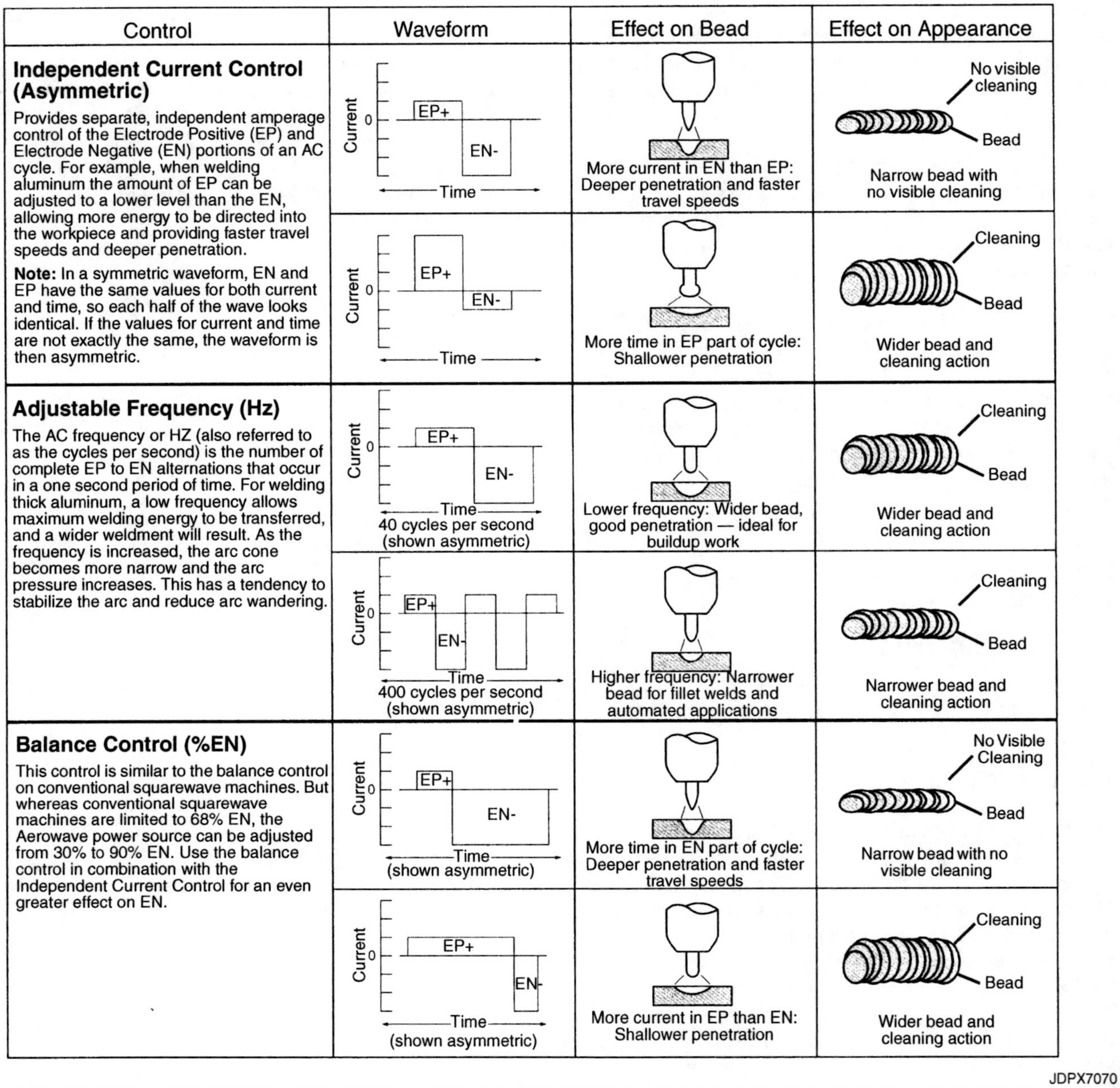

Fig. 17 — Asymmetrical Square Wave Controls and the Effect Each Has on Weld Cleaning and Penetration

Post-Flow Timer

The post-flow timer controls the time the shielding gas and water flow after the arc is stopped. Shielding gas coverage is needed to protect the hot weld metal and tungsten from atmospheric contamination until each has cooled properly after the arc has stopped. The gas flow also provides some cooling for an air-cooled torch. The water (or coolant) should also flow until the torch body (on a water-cooled torch) and power cable have cooled.

The post-flow timer is normally set at 1 second per 10 amps of welding current. For example, when welding at 100 amps, the post-flow should be set for 10 seconds.

Preflow Timer

The preflow timer controls the time the shielding gas flows before the arc starts. Preflow is necessary to ensure the base metal and tungsten are protected from atmospheric contamination when the arc starts. Unless a power source is equipped with accessories for automated welding, the preflow is not generally adjustable. It is preset by the equipment manufacturer. Preflow can be recognized by the sound of gas flowing from the torch and a 3- to 5-second delay each time an arc is started.

Spot Timer

The spot timer is used for setting the time needed to make a tack and/or spot weld. The amperage and time are determined from past experience or by trial and error until the desired results are obtained. The torch is positioned at the proper angle and distance from the work by using some type of torch holder or a specially shaped gas cup that can be placed against the workpiece. A remote control is used to start the arc when the torch is in position, and the timer stops the current at the end of the designated weld time.

Start Control

A start control is used to give the arc a "hot start" or a gradual amperage increase until the welding amperage set on the fine adjustment is reached. The duration of time the start control functions may be varied or it may be preset for 3 to 5 seconds. An amperage adjustment on the control temporarily overrides the main amperage adjustment on the power source.

The purpose for the start control is to either get a weld started quickly by providing a few seconds of high amperage or gradually increase the amperage to the setting for welding without a sudden surge that could melt a hole in a piece of thin metal.

The start control is normally used with a remote control that is only starting and stopping the current. The start control does not serve any purpose if a remote amperage control is also being used.

When troubleshooting a power source that seems to be "starting hot" or "starting cold" and then welding okay after a few seconds of operation, check to see if there is a start control and if it is in the ON position. This condition happens quite often on a power source that is being used interchangeably between SMAW and GTAW. A quick check of the on-off switch on a start control when there is an arc starting problem may save the time and expense of sending the power source in for repair when there is really nothing wrong except the position of the switch.

Crater Fill Control

The crater fill control slowly decreases the amperage over a few seconds until the arc stops. When the contactor is signaled by a remote control to stop the current, the crater fill control is activated. A crater fill control that is on can be mistaken for a remote control that is not shutting off properly. If the arc doesn't want to stop, make sure the crater fill is not turned on before sending the power source and remote control for repair. Crater fill is used in automated welding.

Pulse Controls

GTAW power sources can have built-in controls for pulsing the current or add-on accessory controls (Fig. 18). Pulsing the current makes it possible to get the same weld obtainable from a non-pulse condition, but with less heat input into the work. In other words, by using pulse, it is possible to get good penetration, to obtain good control of the weld puddle when out-of-position welding, to minimize distortion (warping), and to have less difficulty welding thin metals. Power sources with pulse controls are more expensive to purchase and are typically used more in applications that are automated rather than hand-welded.

JDPX7071

Fig. 18 — Accessory Pulse Control

Setting the controls properly for pulsing takes some trial and error. It also takes a person who has a good understanding of welding (how amperage and travel speed control the weld puddle size and base metal penetration) and the effect that changing each control on the power source has on a weld.

There are four controls within a typical pulse control that must be set properly. The first two controls include the current for cycling between a peak (high) and a background (low) value. The peak current is usually set on the amperage fine adjustment on the power source. A separate control (in the group of pulse controls) is used to set the background current level. A third control is used to set the percent of time the current is maintained at the peak and background values. The last control is used to set the number of cycles or pulses per second (PPS). When establishing the pulse parameters for an application, it will be quicker and easier to find the optimum setting for one control at a time, rather than trying to change all the controls at the same time.

Gas and Water Solenoid Valves

The flow of inert gas and water is started and stopped by two separate solenoid valves (see Fig. 10 or Fig. 30). When the contactor on the power source is closed to start the welding current, the solenoid valves open and start the gas and water flowing. The welding current is actually delayed from starting for a few seconds to allow the inert gas to completely cover the tungsten and weld area before the arc starts (preflow). This time delay can be adjusted on power sources with controls for mechanized welding. When the contactor opens and the arc stops, the gas and water continue to flow until the post-flow timer stops. Water is used for cooling the torch and electronics. A portable radiator pack is available if you are not working near a water source.

Touch and Lift Start Controls

A touch and lift start control allows the arc to be started easily on DCEN by touching the tungsten to the workpiece for 2 to 3 seconds. The tungsten is then lifted and the arc starts. There is a minimum amount of contamination and damage to the tungsten and to the workpiece with either type of arc starting. High frequency is not used and the potential danger to other electrical devices from the high-frequency current is eliminated.

Full Weld Sequencer

A full welded sequencer is used to automate welding jobs that can use mechanized equipment for holding and positioning the workpieces during welding. Much of this mechanized welding is done by fusion welding (without adding filler metal); however, it is also possible to have devices for adding filler metal. A weld sequencer may be built into the power source as in Fig. 15 or added as an accessory similar to the unit in Fig. 19.

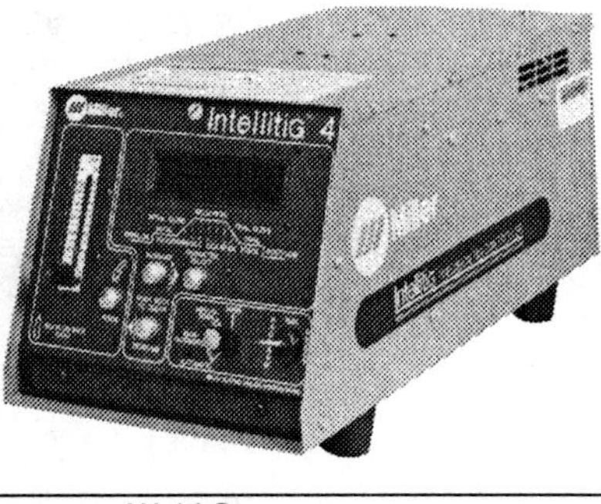

JDPX7072

Fig. 19 — Accessory Weld Sequencer

A weld sequencer has a control for setting all the parameters necessary for making the weld from start to finish. The controls range from the preflow time (before the arc starts) to the rate the amperage decreases for crater filling at the end of the weld. The amperage can be increased or decreased at different intervals during the weld to correspond to changes in metal thickness or joint type. Setting the weld sequence controls properly takes perhaps more knowledge and experience with welding than is needed for setting the controls for pulse welding.

Note on DC Power Sources

DC power sources have limited use in either shop or field construction projects that require the GTAW process. With only a few exceptions, the welding speed and weld cleanliness obtainable with the GMAW process outweigh the very-high-quality welds obtainable with the much slower GTAW process. There are a few specialized applications like welding the thin-wall stainless-steel tubing used in food-processing plants, and for making the root pass weld on the high-pressure pipe found in power plants and chemical plants that need the weld quality obtainable only with DC GTAW.

All that is necessary to do gas tungsten arc welding with a DC power source is to replace the SMAW electrode holder with a properly equipped GTAW torch (including a manual on-off gas valve as illustrated in Fig. 20-D) and to set the polarity to electrode negative. Limitations for this type of a setup include: no aluminum or magnesium welding, and no contactor (arc on-off) for stopping the arc and for scratch starts (dragging the tungsten across the work to start an arc). Also, some DC power sources do not have remote amperage control capabilities and a minimum output (when used for GTAW) low enough to weld thin metal.

Gas Tungsten Arc Welding (GTAW) Torch Assemblies, Power Cables, and Work Clamp

GTAW torches, similar to the ones shown in Fig. 20, are manufactured in a variety of shapes, sizes, and with many options, including cooling.

The torch body acts as a handle for the welder to use as a control for directing the welding current and inert shielding gas to the exact location where the weld is to be placed. As illustrated in Fig. 20-B, the power cable from the power source, the gas hose from the inert gas supply, and the coolant hose, all connect to the torch. The welding current flows from the torch body through a collet and collet body to the tungsten electrode. The inert gas travels through a passageway in the torch to holes in the collet body and exits the torch through a ceramic cup (also called a nozzle).

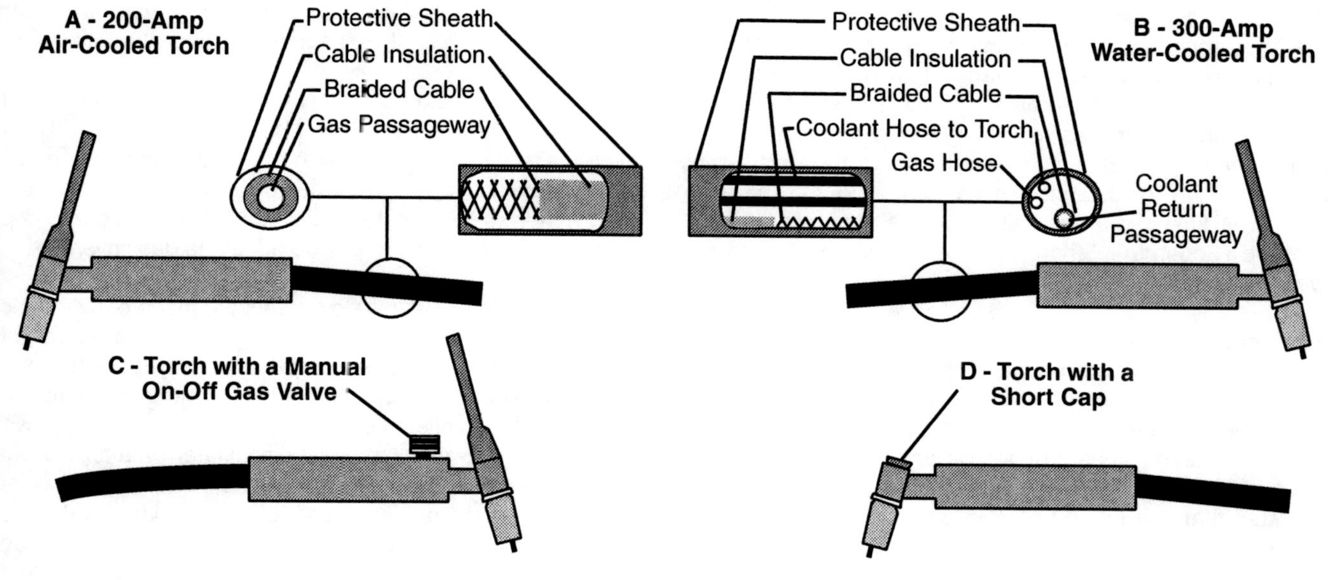

Fig. 20 — GTAW Welding Torches

Torches having a capacity of 200 amps or less may either be air cooled (atmospheric air contacting the torches outer surfaces) as illustrated in Fig. 20-A or water cooled as illustrated in Fig. 20-B. Torches (and power cables) rated much above 200 amps require internal cooling passages that are connected to tap water or a water cooler similar as shown in Fig. 21.

Fig. 21 — Water Circulators and Coolers

Torch size is also a factor that determines the method of cooling. For example, it is possible to build a 100-amp water-cooled torch and power cable that are much smaller than for an air-cooled torch with the same rating. The reason for the size difference is that more surface area per amp is needed for air cooling than for water cooling of both the torch and power cable.

A water circulator or radiator system is a better option than running water from a faucet (also referred to as tap water) to the torch and then out into a floor drain. Water as it comes from most sources contains a lot of minerals that will eventually build up in the small cooling passages. This will cause overheating of the torch and power cable destruction. Cooling equipment manufacturers recommend the proper additives to add to the water to prevent freezing (when shop temperatures go below freezing) and to act as lubricant for the pump. These recommendations should be followed with regard to the type and amount of additive to put in the system. If the wrong additives are used, they could conduct most or all of the high frequency to the cooling system. This transfer or loss (sometimes called bleed off) of high frequency can make arc starting difficult or even impossible.

CAUTION: The power cable will be ruined within a few minutes if used above the manufacturer's amperage rating and duty cycle, or if used without the cooling system operating.

Protect the power cable from hot and/or sharp objects. Puncture damage to the cable and hoses from the power source to the torch will cause a gas or water leak.

The collet and collet body secure the tungsten in the torch and transfer the welding current from the torch to the tungsten electrode. The collet and collet body are made to be used as a set for holding one size of tungsten. This is important because the collet holds the tungsten in place and makes final electrical connection with the torch. If the collet and collet body are not the proper size, the tungsten will not stay in place and/or there will be arcing that could damage the entire torch assembly.

Not all collets and collet holders are the same, even though they are made for holding standard diameters of tungstens. The thread patterns vary between manufacturers, as well as between torch models from the same manufacturer.

Some collet holders are even equipped with a small screen assembly at the bottom called a gas lens. This lens reduces the turbulence in the gas column as it leaves the torch. It also provides better gas coverage (than a standard collet) when the tungsten has to be extended farther than normal past the end of the gas cup to reach limited-access weld joints. The small openings in the screen assembly are easily plugged when the tungsten is contaminated. So, be very careful or don't use a gas lens when the openings become plugged enough to interfere with the flow of gas and cause weld defects.

Torch caps have several functions that determine their size and, sometimes, shape. As the torch cap is tightened, it puts force against the collet, which grips the tungsten to prevent it from falling from the torch or from arcing to the collet body. The torch cap provides a protective enclosure for the top of the tungsten and determines the maximum length for the tungsten that can be used for welding. The torch cap also provides the gas seal for the top of the torch.

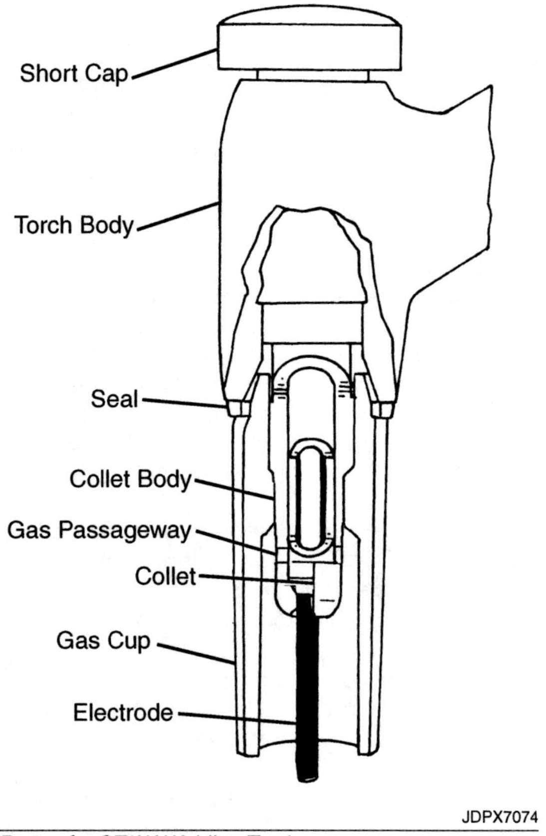

Fig. 22 — Parts of a GTAW Welding Torch

Joint access is usually one of the main factors that determine the torch cap length. When there is plenty of room around the weld joint, a long cap is usually preferred. When access to the weld joint is limited, the smallest cap that will allow access to the weld joint will be used. Be careful; the longer the torch caps, the more easily they break when dropped or misused — always keep a spare just in case.

IMPORTANT: Overtightening the torch cap can deform the collet and possibly cause other internal damage to the torch body.

A gas cup (or nozzle as it is sometimes called) directs inert gas around the tungsten, the arc, and the weld puddle. It must make a gas-tight seal with the torch body and should be of adequate size to allow complete gas coverage of the weld. When welding material like titanium, an oversized cup is used with a gas lens to ensure adequate gas coverage.

Gas cups used on hand-held torches are made of a ceramic material and usually attach to the threads on the outside of the collet body. (Some types of gas cups require the use of a gasket between the cup and the torch.) Although there are some collets, collet holders, and gas cups that will interchange between torches from different manufacturers and between torches from the same manufacturer, there are many parts that won't interchange. This is particularly true with standard collet holders and the collet holders with a gas lens. It takes one style of cup for a standard and another style for a gas lens.

A work clamp (called the ground), should be referred to as the work clamp to avoid confusion with electrical system ground. It is needed to complete the circuit for gas tungsten arc welding in the same manner as for the shielded metal arc process and the continuous wire processes. The location for the attachment of the work clamp or connection should be free of rust, paint, or other material that prevents a good electrical connection. The size of the interconnecting cable and work clamp should be of adequate amperage-carrying capacity for the work to be done.

A poor work connection can be identified by checking the temperature of the connection after welding for 5 to 10 minutes or by checking for evidence of arcing. The connection will be much warmer or even hot at the point of contact when there is not a good connection. (A slight amount of warming when welding at high amperages is normal).

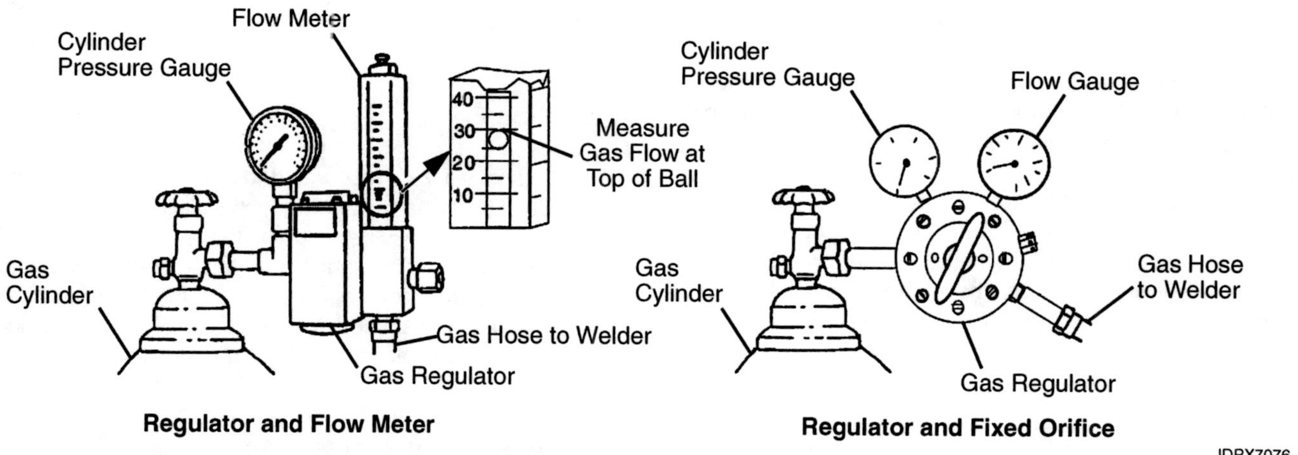

Fig. 23 — Single-Cylinder, High-Pressure Inert Gas Delivery for Gas Tungsten Arc Welding (GTAW)

Inert Gas Delivery System

The inert gas may be supplied to welding equipment from a nearby high-pressure cylinder equipped with a regulator flow meter (Fig. 23), a regulator and flow gauge (Fig. 23), or from a reduced-pressure manifold system similar to the ones used for the GMAW process (see Fig. 27 in Chapter 4). Large users of inert gas purchase the gas in liquid form and use a manifold system to transport the gas to each welding station.

CAUTION: Always secure cylinders to a suitable support with a chain or other holder.

Common inert gases used with the GTAW process are argon and helium. Argon is the most common gas used for GTAW to protect the weld zone from the atmospheric air. Argon is a heavy gas which is one-and-one-half times heavier than air. When it comes out of the gas cup, it tends to blanket the weld zone very well. It has a low ionization potential of 15.7 volts. This makes it quite easy to start an arc in an argon gas shield. Helium is used for some applications where high energy or travel speeds are required. Helium is much lighter than air. As it leaves the gas cup, it rises rapidly away from the arc area. The flow rate for helium must be two or three times the flow rate for argon. It also is much harder to start an arc in helium because its ionization potential is 24.5 volts. Fig. 24 illustrates the difference in these two gases in relation to the arc column and the weld bead profile.

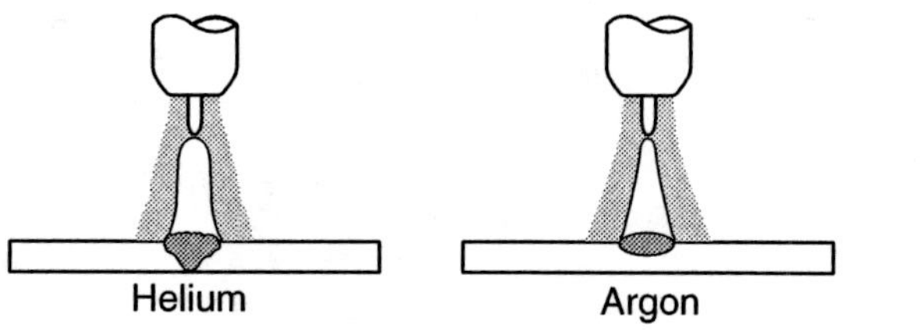

Fig. 24 — Effects of Shielding Gas on Penetration and Bead Width

Nitrogen, which is not an inert gas, but a reactive gas, can be mixed with argon for certain applications. It produces more heat energy into the work than argon alone. When welding on highly conductive materials like copper, this gas mixture can be very helpful. It should not be used as a shielding gas on steel or stainless steel because it makes a weak, porous weld. Nitrogen has been used on some stainless-steel piping installation as a purge gas on the inside of the pipe with good results.

Gas Tungsten Arc Welding (GTAW) Cleaning and Finishing Tools

The cleaning and finishing tools needed for GTAW are essentially the same as the tools needed for the other welding processes covered in previous chapters. Since the GTAW process has a simple heat source and has no flux or scavengers to help clean the base metal, cleaning must be done prior to welding. Grinders, wire brushes, and/or various chemical cleaners are available to prepare the base metal prior to welding. Chemicals containing chlorinated hydrocarbons should not be used to clean metals in a weld area. Grinding wheels and wire brushes should not be used on different metals to avoid contamination. For example, stainless steel will be contaminated with iron (it will rust) when ground or wire brushed with a grinding wheel or brush that has been used on carbon steel.

Welding with dirty or contaminated filler metal may also cause unacceptable weld quality. To obtain high-quality welds, the filler metal must be cleaned prior to using. This is especially important with aluminum filler metals because the dense surface oxides need to be removed. Stainless steel wool or over-the-counter chemicals work well for removing the oxide from the filler metal. The same rule for base metal cleaning applies to filler metal cleaning — don't use the same cleaning tools for different metals. If the tools are used to clean other materials, the filler metal will become contaminated. This may cause unacceptable weld quality.

When possible, keep the hot end of the filler metal in the gas shield. If the hot end is taken out of the gas shield during welding or at the end of the weld joint during the post-flow time, filler metal will be contaminated by the atmospheric air. When the hot end of the filler metal becomes contaminated, it should be cut off or cleaned.

Since there is no slag, spatter or smoke, little post-weld cleaning is required with the GTAW process. Welds can generally be left in the as-welded condition. On stainless steel, the discoloration can be removed with chemicals or mechanically cleaned with a stainless-steel wire brush or buffing wheel.

Supplies for Gas Tungsten Arc Welding (GTAW)

Electrodes

There are several types and diameters of tungstens to choose from when planning a welding job. The type of tungsten will determine whether the electrode will retain a point or balled shape with the welding current. The diameter needed for the job will be determined by how much amperage can be used for welding without the tungsten melting. As illustrated in Table 3, a large-diameter tungsten can carry more amperage than a small-diameter tungsten, while some types of tungstens can carry more amperage than others when used on the same type of current. It is important that the electrode chosen does not melt in the intense heat of the electric arc or spit tungsten (release very small droplets) that cause weld defects or interfere with the welding process.

Typical Current Ranges for Tungsten Electrodes							
		Direct Current (Amps)		Alternating Current (Amps)			
		DCEN	DCEP	Unbalanced Wave AC		Balanced Wave AC	
Electrode Diameter (Inches)	Gas Cup I.D. (Inches)	EWP, EWCe-2 EWLa-1 EWTh-1, EWTh-2	EWP, EWCe-2 EWLa-1 EWTH-1, EWTH-2	EWP	EWCe-2 EWLa-1 EWTh-1, EWTh-2 EWZr-1	EWP	EWCe-2 EWLa-1 EWTh-1, EWTh-2 EWZr-1
0.010	1/4	Up to 15	N/A	Up to 15	Up to 15	Up to 15	Up to 15
0.020	1/4	5–20	N/A	5–15	5–20	10–20	5–20
0.040	3/8	15–80	N/A	10–60	15–80	20–30	20–60
0.060	3/8	70–150	10–20	50–100	70–150	30–80	60–120
0.093	1/2	150–250	15–30	100–160	140–235	60–130	100–180
0.125 (1/8)	1/2	250–400	25–40	150–210	225–325	100–180	160–250
0.156 (5/32)	1/2	400–500	40–55	200–275	300–400	160–240	200–300
0.187 (3/16)	5/8	500–750	55–80	250–350	400–500	190–300	290–390
0.250 (1/4)	3/4	750–1000	80–125	325–450	500–630	250–400	340–525

***All values are based on the use of Argon as a shielding gas. Other current values may be employed depending on the shielding gas, type of equipment, and application.**

DCEN = Direct Current Electrode Negative (Straight Polarity)
DCEP = Direct Current Electrode Positive (Reverse Polarity)

AC = Alternating Current

Table 3 — Amperage Ranges for Tungsten Electrodes and Gas Cups Used for Gas Tungsten Arc Welding (GTAW)

A tungsten electrode may be made of pure tungsten or alloyed (combined) with one or more rare materials to obtain a certain arc when operated on a particular type of current and polarity. Optimum welding results can be expected from a project only when the proper combination of current and polarity, tungsten composition, and tungsten diameter have been chosen. As illustrated in Table 4, tungstens are color coded to make their identification easier.

As illustrated in Table 3, electrodes are manufactured in standard diameters ranging from 0.010 inch (0.025 mm) to 1/4 inch (6.4 mm) and in standard lengths of 3 inches (76.2 mm), 6 inches (152.4 mm) and 12 inches (304.8 mm) with either a ground or cleaned finish. The ground finish is more expensive but makes better electrical and heat transfer contact with the torch collet, especially in applications that are automated. For most hand-welding applications, a tungsten with a cleaned finish performs satisfactorily and provides a cost savings to a project when several packages of tungstens are to be purchased.

AWS Classification	Composition	%	Color
EWP	Pure Tungsten	99.5	Green
EWCe2	Ceria	1.8–2.2	Orange
EWLa1	Lanthanum	0.9–1.2	Black
EWTh1	Thorium	0.8–1.2	Yellow
EWTh2	Thorium	1.7–2.2	Red
EWZr1	Zirconium		Brown
EWG	------Not Specified--------		Gray

Table 4 — Tungsten Electrode Composition and Color Coding

Pure and zirconium electrodes are used for welding aluminum and magnesium using ACHF (alternating current with high frequency). Pure tungstens provide good arc stability on both argon and helium. A zirconium tungsten is preferred when the highest-quality work is necessary. There is less tungsten "spitting" (emission of small droplets of tungsten) and the tungsten is more resistant to contamination when accidentally touched to the filler metal or base metal. As indicated in Table 3, a zirconium tungsten has a higher amperage-carrying capacity than a pure tungsten of the same diameter.

Thoriated electrodes are used in DCEN applications on common metals like carbon steel, stainless steel, nickel alloys, and silicon bronze. They can also be used with AC power sources that have expanded balance controls or asymmetric waveforms. However, when thoriated tungstens are used on power sources other than an asymmetric type, the arc will be harder to start and it will be difficult to maintain a balled end on the tungsten.

NOTE: *The thoria produces a low-level radiation. It is a relatively small amount and has not been found to be a health hazard. Before doing a lot of electrode grinding and welding in confined spaces, check with the appropriate safety personnel. One way of avoiding thoria is to use the EWes-2 or EWLa-1 electrodes.*

Ceriated and lanthanum electrodes can be used in applications which require ACHF or DCEN.

The EWG electrode is a special classification and the manufacturer must identify the specific additions and the amounts.

Tungsten Electrode Preparation

Once the proper size and type of tungsten has been selected for the amperage to be used, the tungsten must be shaped properly. New electrodes come out of the box with a square end that must either be balled(Fig. 25-A) or pointed (Fig. 25-B) before being used for welding. When pure or zirconium electrodes are used, a balled end is preferred. When welding with DCEN, expanded balance-controlled AC square wave, or asymmetric AC square wave power source, the tungsten needs to be pointed.

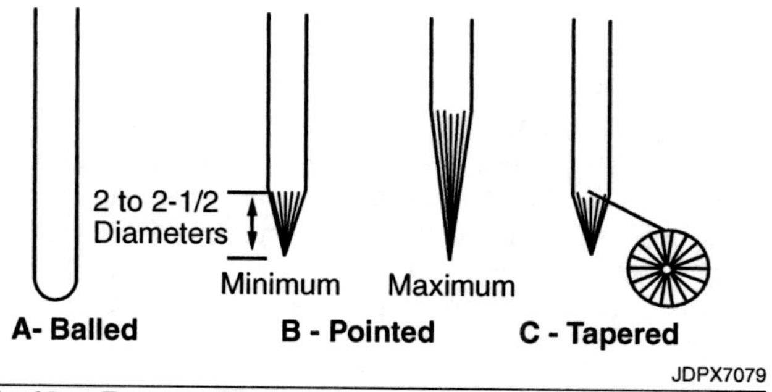

Fig. 25 — Tungsten Electrode Properly Prepared for Welding

There are two methods of forming a ball on the end of a tungsten. With the first method, start an arc on a clean, heavy piece of scrap aluminum and increase the amperage until a ball is formed. With the second method, set the current to DCEP and the amperage to a much lower setting than would be used for welding. Start the arc on a piece of clean copper and slowly increase the amperage until a ball is formed. After a ball is formed, change the current and amperage back to the setting for welding. With either method, the ball should be approximately the same diameter as the electrode.

A point can be put on a tungsten electrode by using a special chemical or by grinding. The grinding wheel should be of a fine grit and used only for grinding tungsten (otherwise the tungsten can be contaminated from other materials ground on the wheel). The proper technique for grinding a point with the tungsten is illustrated in (Fig. 26). Note that the grinding marks run parallel with the tungsten (Fig. 25-B). Improper grinding can cause the arc to wander (jump from one piece to the other). Some welders prefer that the tungsten be tapered (Fig. 25-C) without having a sharp point on the end. After grinding the tungsten to a point, the welder lightly touches the end of the tungsten to the grinding wheel. This removes the sharp point and makes a 0.005 to 0.015 inch (0.13 to 0.38 mm) flat spot on the end.

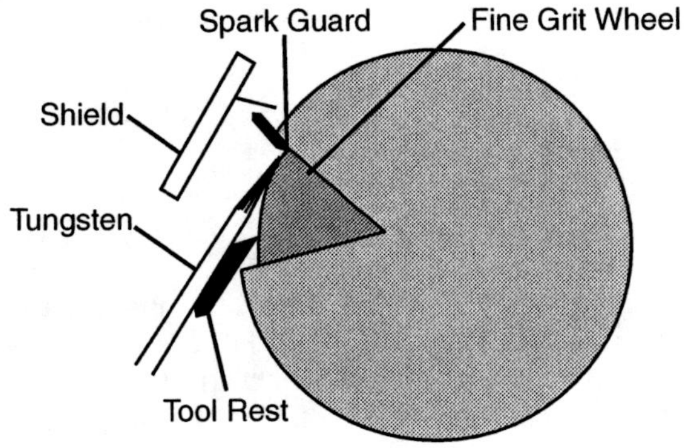

JDPX7080

Fig. 26 — Proper Tungsten Electrode Grinding

Gas Tungsten Arc Welding (GTAW) Filler Metals

Filler metals are manufactured for joining most metals including steel, stainless steel, aluminum, magnesium, and copper alloys. There are also many special filler metals like those used for tool and die repair. The American Welding Society (AWS) has a classification system for filler metals which simplifies the selection process. For detailed information, consult the most recent AWS A5 filler metal documents.

Filler metals for GTAW generally come in 36 inch (914.4 mm) straight bare rods that typically range from 1/16 inch (1.6 mm) to 1/8 inch (3.2 mm) in diameter. Filler metals can be purchased in quantities as small as one pound (0.45 kg) and are normally packaged in standard size containers which weigh from 5 to 50 pounds (2.3 to 23 kg). A 36 inch (914.4 mm) piece of filler metal can be awkward to handle, so many times a welder will cut it in half for easier handling. Also, as illustrated in Fig. 27, it is a good safety practice to bend the exposed end of the filler rod over in a "U" shape. This action reduces the possibility of the end becoming a puncture hazard to the welder's helper or an innocent bystander.

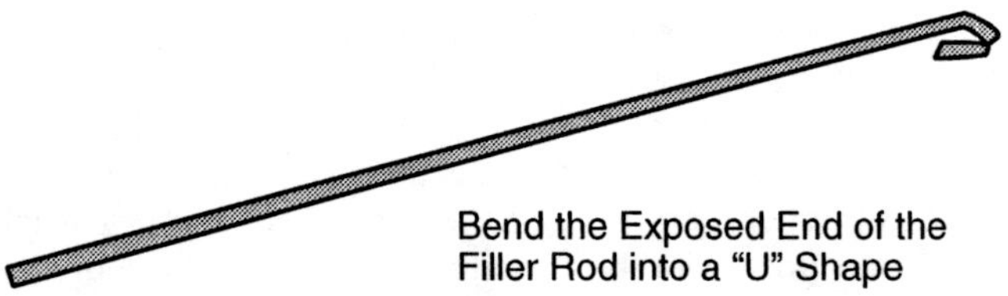

JDPX7081

Fig. 27 — Reduce the Puncture Hazard by Bending one End of the Filler Rod

The method of identification for an individual piece of filler metal varies from very good to none, depending on the manufacturer. As illustrated in Fig. 28, a piece of filler metal may be flag-tagged, flattened and stamped, stamped, or not marked at all.

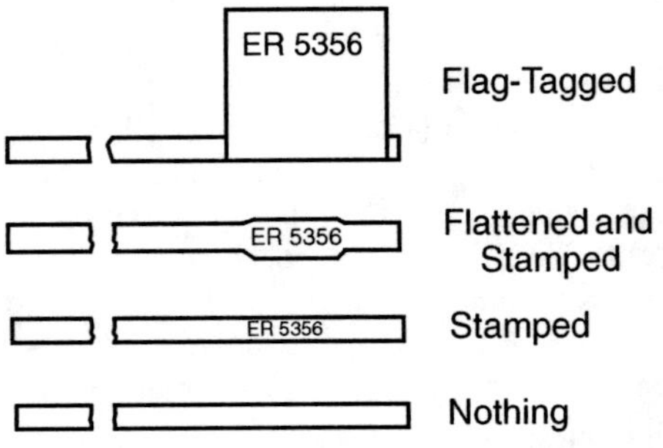

JDPX7082

Fig. 28 — Methods for Identifying Filler Metal

The best practice for selecting a filler metal is to positively identify the base metal(s) and then follow the recommendations of a reputable filler metal manufacturer for welding that base metal. The following filler metals and descriptions provide examples of the letter and number designations within the AWS system.

Carbon Steel

A typical carbon steel filler metal would be ER70S-2. The "E" means it can be used as an electrode for GMAW work or filler rod ("R") for GTAW. The "70" stands for 70,000 pounds per square inch of minimum tensile strength. The "S" means it's a solid wire with no flux coating or no flux in its core. The "2" refers to its chemical make-up and operating characteristics.

Filler metals for steel need a lot of deoxidizers because of the impurities found in the base metal. Coat hangers, bailing wire, and oxyacetylene filler metal don't generally make good filler metal for steel welding. Welds made with these non-GTAW filler metals may be porous and of a poor quality.

Stainless Steel

A typical stainless steel filler metal would be ER309L. The meaning for the "ER" is the same as for steel. The "309" represents a particular stainless steel alloy, and the "L" means the filler metal has a much lower carbon content than a type ER309.

A stainless steel filler metal needs to match the base metal(s) as close by as possible. Not matching the base metal properly could cause serious problems, ranging from premature weld failure to the contamination of a product traveling through a stainless piping system (having welded joints).

The following factors should be considered when selecting stainless filler metals:

- Corrosion resistance
- Service temperature
- Weld ductility
- Resistance to carbide precipitation

Aluminum

Two common aluminum filler metals are ER4043 and ER5356. The meaning for the "ER" is the same as for steel and stainless steel. The explanations for the other numbers are much more complicated and for that reason will not be explained in this text (for more specific information, consult the AWS filler metal specification).

If the designation has been removed from either or both of these two filler metals, they can be re-identified with either of these simple tests.

One test is to drop each of the wires on a hard flat surface and listen to the sound. The ER4043 will make a flat sound compared to a ringing sound from the ER5356 at the time of impact.

A second test is to bend the end of each rod with a pair of pliers or your fingers. The ER4043 will bend relatively easily compared to the ER5356, which has a tendency to spring back more when it is released after bending. The reason for this difference is that ER4043 is somewhat softer and does not have as high a tensile strength as the ER5356.

The following factors should be considered when selecting aluminum filler metals:

- Tensile and/or shear strength of the weld metal
- Weld ductility
- Service temperature
- Freedom from cracks
- Corrosion resistance
- Color match after anodizing

Setting Up Gas Tungsten Arc Welding (GTAW) Equipment

CAUTION: Before beginning work, read and understand warning labels on the equipment, the safety precautions, and the operating instructions in the owner's manual for the specific equipment to be used.

Some setup is necessary with new equipment, while preventive maintenance should be performed on existing equipment before starting work. A working understanding of how everything hooks together, the function of all the controls, and the basic maintenance/troubleshooting procedures play an important part in working efficiently and in meeting weld quality standards.

Since there are several controls and options on a power source and on the torch assembly, the setup, operating procedure, and maintenance of GTAW equipment is somewhat complicated. Fig. 29 illustrates the main connections and controls on a new or existing system that should be checked and/or set before starting a gas tungsten arc welding project.

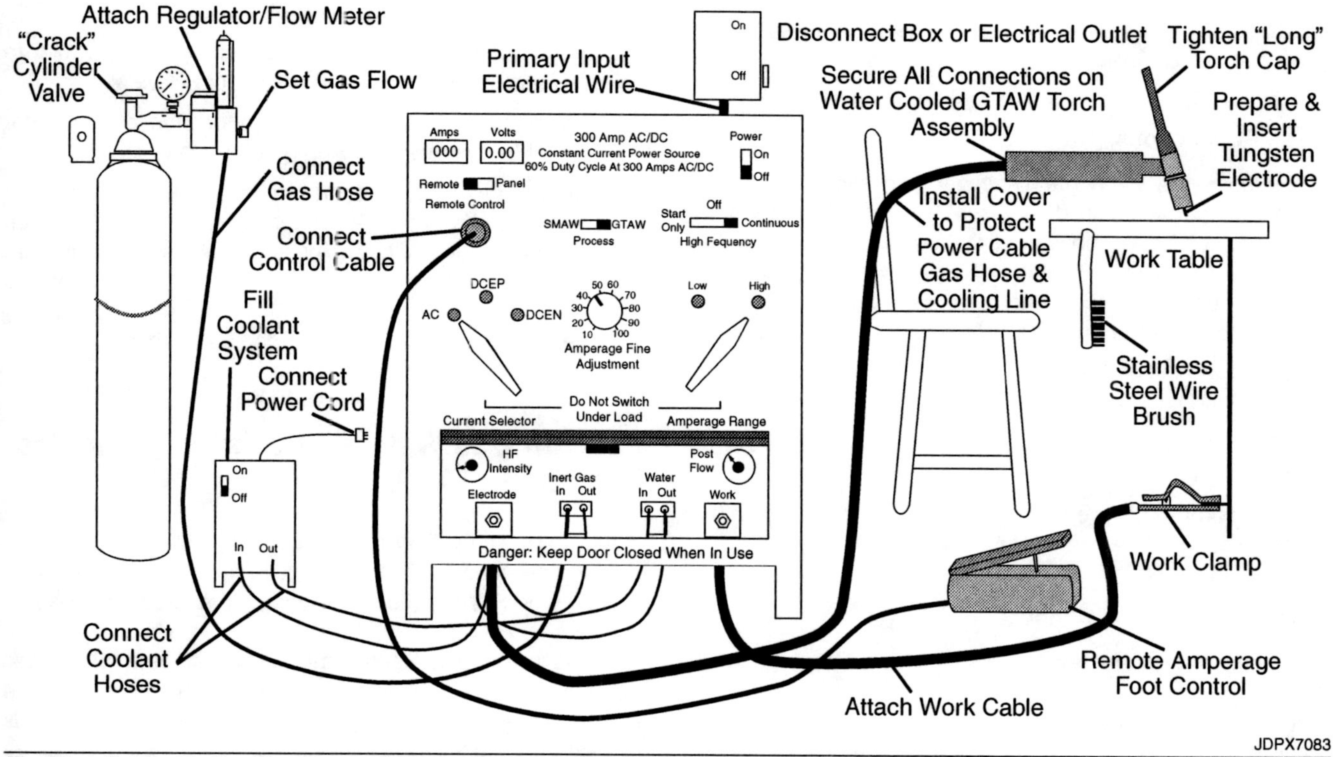

Fig. 29 — Setting Up Gas Tungsten Arc Welding Equipment and Water-Cooled Torch

Power Source, Torch, and Work Cable Connections

CAUTION: The installation of an electrical outlet and disconnect switch for operating welding equipment should be performed only by qualified people. The same qualifications are necessary for making internal electrical connections on the power source for the primary power cable connection and for changing voltage connections to correspond to the shop voltage available.

IMPORTANT: Always follow local, state, and national regulations regarding the installation of electrical wiring and other devices necessary for the connection of welding equipment. This includes complying with the provisions in the latest National Electrical Code (NEC).

To connect equipment, do the following:

1. Turn off the electrical power at the disconnect switch or unplug the power source before beginning equipment setup or preventive maintenance.
2. Follow equipment manufacturer's instructions for checking/changing input voltage to the power source.
3. Attach the torch cable and cable adapter to the terminal on power source labeled "electrode."
4. Plug the remote control power cord to the proper receptacle on the power source.
5. Attach a work clamp to the work cable and make the necessary connection to the terminal labeled "work" on the power source.

Torch Connections

1. Remove the torch handle cover and slide it over the incoming cables and hoses. Then attach the power cable, gas hose, and water line from the power source to the torch. When all other gas and water connections have been completed, check for leaks and fix any leaks. Then replace the torch handle cover.
2. Insert the collet and collet body at the other end of the torch and hand tighten.
3. Attach the proper gas cup (and gasket if required) and hand tighten.
4. Insert a properly prepared tungsten electrode. Attach the torch cap and secure the tungsten.

Inert Gas Setup and Connections

1. Secure the gas cylinder.

CAUTION: Always secure cylinders to a suitable support with a chain or other holder.

2. Remove the valve protection cap.
3. "Crack" the cylinder valve.
4. Attach a regulator flow meter or a regulator and a flow gauge.
5. Set the proper gas flow.
6. Check all connections for leaks.

Power Source-Gas Supply Connections

1. Attach the gas hose from the flowmeter to the in connection of the gas solenoid valve.
2. Connect the gas hose from the torch to the out connection of the gas solenoid valve.
3. Check all connections for leaks and fix any leaks before starting to weld.

Cooler System Setup (Optional)

When a water-cooled torch is used, it may be connected directly to the cooler or connected in series with the water solenoid on the power source. In either case, the coolant should flow to the torch and return through the power cable. The return flow through the power cable allows the use of a smaller electrical conductor and thus makes the torch easier to manipulate. The coolant system should be plugged into an outlet on the power source. This will ensure that the cooler is operating when the power source is on and is off when the power source is not operating.

IMPORTANT: A water-cooled torch can be ruined within a few minutes of operation without the coolant system operating.

1. Connect a hose from the out connection of the cooler to the in connection of the water solenoid valve.
2. Connect a hose from the out connection of the water solenoid valve to the cooling hose to the torch.
3. Connect the water in hose on the cooler to the power cable from the torch. The water goes directly to the torch and returns through the power cable to keep it from overheating.

A cooler is not needed with an air-cooled torch.

Preparing to Weld with Gas Tungsten Arc Welding (GTAW)

When using the GTAW processes, follow the same general safety precautions that were followed with the SMAW, GMAW, and FCAW process. The same hazards are present.

The heavy leather protective gear used with other processes is not necessary. This difference is due to the fact that there is no metal transfer through the arc, so no spatter is produced.

Special Gas Tungsten Arc Welding (GTAW) Safety Precautions

The arc radiation is very intense, and severe arc burn will result on any exposed skin within a few minutes. When using filler rod, bend over the end not being used (Fig. 27).

This will eliminate the possibility of puncture injuries to nearby personnel. Before operating the equipment, be sure to read and understand the manufacturer's recommendation for the following:

- Proper eye, face, head, ear, and body protection from hot metal and from the arc's rays for the welder and persons in the area.
- Proper protection from welding fumes, and methods of removing fumes from the welding area.
- Proper procedures to follow for preventing a fire or explosion during welding.
- Proper procedures for connecting the power source to the shop electrical system.
- Proper procedures for connecting the work clamp directly to the work or work area.
- Proper procedures for operating and maintaining each component in the GTAW welding system.

The warning labels illustrated in Fig. 30 through Fig. 34 represent additional hazards associated with the GTAW process. Precautionary measures should be taken to minimize the potential dangers from inert gas, magnetic fields, high-pressure cylinders, high frequency, and arc noise (produced by some of the fast-switching AC power sources).

Buildup of Shielding Gas Can Harm Health or Kill.

Shut off shielding gas supply when not in use. Always ventilate confined spaces or use approved air-supplied respirator.

JDPX7084

Fig. 30 — Hazards from Inert Gas

Magnetic Fields From High Currents Can Affect Pacemaker Operation.

1. Pacemaker wearers keep away.
2. Wearers should consult their doctor before going near arc welding, gouging, or spot welding operations.

JDPX7085

Fig. 31 — Hazards from Magnetic Fields

High-Frequency Radiation Can Interfere with Radio Navigation, Safety Services, Computers, and Communications Equipment.

Keep high-frequency source doors and panels tightly shut, keep spark gaps at correct setting, and use grounding and shielding to minimize the possibility of interference.

JDPX7086

Fig. 32 — Hazards from High Frequency

Noise Can Damage Hearing.

Noise from some processes or equipment may damage hearing.

Wear approved ear protection if noise level is high.

JDPX7087

Fig. 33 — Hazards from High Noise Levels

Cylinders Can Explode if Damaged.

Welding cylinders contain gas under pressure. If damaged, a cylinder can explode. Since gas cylinders are normally part of the welding process, be sure to treat them carefully.

JDPX6958

Fig. 34 — Hazards from Improper Care of Cylinders

Pre-Welding Checklist

Whether using new or existing GTAW equipment in a familiar or unfamiliar work area, it is a good work practice to make a final check of everything (welding equipment, safety gear, and the work area) before starting a job. Begin by checking the welding equipment at the arc end of the system and work back toward the primary power.

The welding equipment as well as the work could be damaged or ruined if welding is attempted without going through the items on the following list:

1. Tungsten Electrode—Is it the proper type, shape, and size for the work to be done and has it been contaminated?
2. Gas Cup—Is the gas cup the proper size and is the tungsten stick-out adequate for the weld joint?
3. Torch Condition—Is the torch in good working order with no cracked, broken, or burned insulation?
4. Torch Electrical Connections—Is the torch connected to the electrode terminal on the power source and has the connection been tightened properly?

5. Gas Hose Connections—Is the gas hose properly attached to the in/out from the gas solenoid, to the gas supply, and to the torch?
6. Torch Coolant Connections—If the torch is water-cooled, check the flow path. Coolant should travel to the torch head first, then the power cable, power cable adapter, and back to coolant system.
7. Work Cable Connections—Is the work cable and clamp properly connected directly to the work, and not passing welding current through any gears, bearings, or electrical control circuits?
8. High-Frequency Current—HF can damage electronic circuit boards. Disconnect the appropriate components per manufacturer's recommendation prior to welding on anything (vehicles, machinery, etc.) with electronic components.
9. Inert Gas—Is the inert gas of the proper type for the work to be done? Is it securely fastened with the regulator flow meter installed, and is the gas hose connected to the input side of the gas solenoid?
10. On-and-Off Switch—Check for the location of the switch.
11. Arc Starting—Select the starting method to be used: high-frequency-continuous-start only, touch, or lift start.
12. High-Frequency Intensity—If high frequency is used, set the intensity control only high enough to get the arc started.
13. Amperage Control—Set the main amperage control for the maximum amperage that will be needed.
14. Remote Amperage* Control—Is the remote control in position, properly connected, and is the remote amperage switch in the proper position?
15. Remote Contactor* Switch—Is the remote contactor switch in the proper position?
16. Current and Polarity—Select the proper current and polarity: AC for aluminum and magnesium, DCEN (electrode negative) for everything else.
17. Range Switch—If the power source has a range switch, set it in the range that allows the most flexibility in fine tuning the amperage for the job at hand.
18. Balance Control—Set the balance control in the proper position for penetration or for additional cleaning.
19. Post-Flow Timer—Set the post-flow timer to operate approximately 1 second for every 10 welding amps. If the tungsten turns blue or black, more time is needed. If it is set too long, gas will be wasted!
20. Power Source Electrical Connections—Is the power source properly connected to the primary power, and are the cord and connector in good condition and connected to a primary fuse box with a disconnect switch?

*Some power sources will have a mode switch that combines the function of both of these controls.

Other Pre-Welding Checks

It is just as important to check the condition of safety gear and the work area for potential hazards as it is to check the equipment before beginning to weld. Do the following:

1. Before turning on any equipment, go through a complete safety inspection of the personnel protective welding gear and the work area.
2. Mentally start at the top of the head and work toward the feet, making sure the proper protective gear is available.
3. Next, inspect the weld area for proper arc shielding for other personnel in the area.
4. Also inspect the weld area for any flammable or explosive hazards—they must be removed or properly safe-guarded before starting to weld!
5. All electrical connections and conductors should be in good condition. There should not be any frayed connections or conductors with broken insulation.

Final Preparations to Weld Gas Tungsten Arc Welding (GTAW)

Once the area has been determined safe to start work, use the following checklist for making final preparations to weld:

1. Secure the GTAW torch so accidental arcing cannot occur.
2. Turn on the shielding gas.
3. Turn on the power source.
4. Open the gas solenoid valve by pushing the foot pedal or torch switch so the flow rate can be set.
5. When the post flow timer times out, shut off the gas cylinder valve and inspect for leaks if using unfamiliar equipment.

Tips for Welding Common Metals

Table 3 provides general guidelines for selecting the tungsten and gas cup sizes for various amperages. Table 5 through Table 7 provide amperage requirements for different types and thicknesses of base metals.

Aluminum — Manual Welding: Alternating Current, High Frequency Stabilized						
Metal Thickness	Joint Type	Tungsten Electrode Diameter	Filler Rod Diameter (If Required)	Amperage	Gas	
					Type	Flow-CFH
1/16"	Butt	1/16"	1/16"	60–85	Argon	15
	Lap	1/16"	1/16"	70–90	Argon	15
	Corner	1/16"	1/16"	60–85	Argon	15
	Fillet	1/16"	1/16"	75–100	Argon	15
1/8"	Butt	3/32"–1/8"	3/32"	125–150	Argon	20
	Lap	3/32"–1/8"	3/32"	130–160	Argon	20
	Corner	3/32"–1/8"	3/32"	120–140	Argon	20
	Fillet	3/32"–1/8"	3/32"	130–160	Argon	20
3/16"	Butt	1/8"–5/32"	1/8"	180–225	Argon	20
	Lap	1/8"–5/32"	1/8"	190–240	Argon	20
	Corner	1/8"–5/32"	1/8"	180–225	Argon	20
	Fillet	1/8"–5/32"	1/8"	190–240	Argon	20
1/4"	Butt	5/32"–3/16"	3/16"	240–280	Argon	25
	Lap	5/32"–3/16"	3/16"	250–320	Argon	25
	Corner	5/32"–3/16"	3/16"	240–280	Argon	25
	Fillet	5/32"–3/16"	3/16"	250–320	Argon	25

Table 5 — Guidelines for Welding Aluminum

Stainless Steel — Manual Welding: Direct Current Electrode Negative						
Metal Thickness	Joint Type	Tungsten Electrode Diameter	Filler Rod Diameter (If Required)	Amperage	Gas	
					Type	Flow-CFH
1/16"	Butt	1/16"	1/16"	40–60	Argon	15
	Lap	1/16"	1/16"	50–70	Argon	15
	Corner	1/16"	1/16"	40–60	Argon	15
	Fillet	1/16"	1/16"	50–70	Argon	15
1/8"	Butt	3/32"	3/32"	65–85	Argon	15
	Lap	3/32"	3/32"	90–110	Argon	15
	Corner	3/32"	3/32"	65–85	Argon	15
	Fillet	3/32"	3/32"	90–110	Argon	15
3/16"	Butt	3/32"	1/8"	100–125	Argon	20
	Lap	3/32"	1/8"	125–150	Argon	20
	Corner	3/32"	1/8"	100–125	Argon	20
	Fillet	3/32"	1/8"	125–150	Argon	20
1/4"	Butt	1/8"	5/32"	135–160	Argon	20
	Lap	1/8"	5/32"	160–180	Argon	20
	Corner	1/8"	5/32"	135–160	Argon	20
	Fillet	1/8"	5/32"	160–180	Argon	20

Table 6 — Guidelines for Welding Stainless Steel

Mild Steel — Manual Welding: Direct Current, Electrode Negative						
Metal Thickness	Joint Type	Tungsten Electrode Diameter	Filler Rod Diameter (If Required)	Amperage	Gas	
					Type	Flow-CFH
1/16"	Butt	1/16"	1/16"	60–70	Argon	15
	Lap	1/16"	1/16"	70–90	Argon	15
	Corner	1/16"	1/16"	60–70	Argon	15
	Fillet	1/16"	1/16"	70–90	Argon	15
1/8"	Butt	1/16"–3/32"	3/32"	80–100	Argon	15
	Lap	1/16"–3/32"	3/32"	90–115	Argon	15
	Corner	1/16"–3/32"	3/32"	80–100	Argon	15
	Fillet	1/16"–3/32"	3/32"	90–115	Argon	15
3/16"	Butt	3/32"	1/8"	115–135	Argon	20
	Lap	3/32"	1/8"	140–165	Argon	20
	Corner	3/32"	1/8"	115–135	Argon	20
	Fillet	3/32"	1/8"	140–170	Argon	20
1/4"	Butt	1/8"	5/32"	160–175	Argon	20
	Lap	1/8"	5/32"	170–200	Argon	20
	Corner	1/8"	5/32"	160–175	Argon	20
	Fillet	1/8"	5/32"	175–210	Argon	20

Table 7 — Guidelines for Welding Mild (Low Carbon) Steel

Welding aluminum is not difficult, it is just different. Aluminum does not change color as it approaches melting temperature. It also loses most of its strength when it approaches melting temperature. (This property is called "hot shortness.") These two characteristics make it necessary to watch the formation of the weld puddle very carefully. If the welder is not paying careful attention when the weld puddle forms, a hole will develop and/or the piece will collapse before he/she realizes what has happened.

Base metal preparation is very important when welding aluminum. All oil, dirt, paint, or any other contaminants must be removed prior to welding. The AC GTAW arc will clean light oxide coatings, but heavy oxide coatings must be removed. This aluminum oxide melts at 3700°F (2037°C) and the aluminum underneath melts at 1200°F (660°C). If the heavy oxide is not removed before welding, base metal melting will be more difficult and the oxides will reduce weld quality.

Aluminum is a very good conductor of heat. While welding on small parts or weld coupon test pieces, heat builds up until the welding current needs to be reduced. This is why remote foot controls are so popular for welding aluminum with the GTAW process. Large aluminum parts absorb so much heat that preheating is required with an oxyfuel torch or by placing the part in a preheat oven before attempting to weld.

One of the key points in the successful welding of aluminum is learning to judge the size of the weld puddle in relation to the amount of heat being applied to the base metal. If a puddle of metal is not forming, apply more heat by increasing the amperage or preheat the part before starting to weld. When the weld puddle starts to sink below the surface of the base metal and/or a hole forms, the weld puddle has gotten too large and the heat needs to be decreased by decreasing the amperage and/or increasing travel speed.

The information in Table 5 can be used as a guide for making welds on various thicknesses and joint designs of aluminum.

Welding stainless steel is much different from welding aluminum with the GTAW process. The base metal turns a red color just before the melting point is reached, and this makes it much easier to identify the weld puddle as it is forming. Also, stainless steel does not conduct heat nearly as well as aluminum. This means it is easier to start and maintain a weld puddle, and it does not take the high amperages needed for welding a comparable piece of aluminum. Although stainless steel is an alloy of chromium and nickel, the surface oxides that are formed do not cause a problem during welding.

Heat input is one of the main items that must be controlled very carefully when welding stainless steel. The heat input can be kept to a minimum by using low amperages, short arcs, and fast travel speeds. When the proper combination of amperage, arc length, and travel speed has been achieved, the completed weld will be a gold to straw color.

The wrong combination of amperage, arc length, and travel speed will cause the top surface of the weld to have a burned look (dark gray to black color). The back (or bottom) side of the weld will contain dark gray spots the shape of lumps of sugar and is referred to as sugaring. Sugaring will occur on the back (or bottom) side of a weld when the molten stainless steel is exposed to air. Sugaring can be prevented by protecting the back side of the joint with inert gas during welding. Overheating will also cause burning out of the alloys that give stainless steel its' resistance to corrosion.

More elaborate weld fixtures and tack welding procedures are necessary to prevent stainless steel assemblies from distorting (warping) out of shape during welding. Stainless steel distorts easily because it is a poor conductor of heat and has a high rate of thermal expansion. This means stainless steel expands and contracts more per degree of temperature change than most other commonly welded metals.

The information in Table 6 can be used as a guide for making welds on various thicknesses and joint designs of stainless steel.

Welding mild steel (known as low-carbon steel) is somewhat similar to welding stainless steel, but with fewer problems. There are a lot of variations in the manufacture of mild steel which complicate the welding process. Some of these steels are of the "semi-killed" type, which means they are manufactured with some oxygen in them. If bubbles form as the weld is made, this is the likely problem. It can be overcome by using a deoxidized filler wire. Oxyacetylene grade filler metal is not adequate for these steels, and coat hangers and bailing wire are definitely not acceptable. Some of these steels are "killed steel," meaning they have less oxygen in them and are thus easier to weld. These steels must always be mechanically cleaned prior to welding. Any rust, paint, oil, grease, or any other contamination must be removed. Some of these steels are hot-rolled. Mill scale must also be removed in the weld area. Some steel pipes contain a varnish-type coating which must also be removed prior to welding.

Table 7 can be used as a guide for welding various thicknesses and joint designs of mild steel.

Establishing Welding Procedures and Setting Parameters

Weld data sheets or existing welding procedures are both good places to begin when setting the parameters for a GTAW project. Weld data sheets similar to the ones in Table 5 through Table 7 provide enough information to have a starting point to begin a new welding project. However, most projects will still need some trial and error from a skilled welder to find the proper combination of tungsten diameter, filler rod diameter, and amperage to make a weld with an acceptable appearance. When the weld's appearance has reached an acceptable level, the proper testing (see Chapter 1) must be done to determine the strength of the weld and the effect welding has had on the base metal. Record the essential information used for making, inspecting, and testing the weld. This includes an accurate measure of amperage from either the meter on the power source or an amp probe. This written procedure that has been tried, tested, and proven can be used on future welding projects.

The following tips will be helpful in finding the proper combination of tungsten diameter, filler rod diameter, and amperage for a welding job.

Tungsten Electrode Diameter

- A tungsten electrode with a small diameter will provide a more stable arc (than a large-diameter tungsten) when starting an arc and when welding at the lower end of an amperage range.
- When arc starting and arc stability are not problems, use as large a diameter of tungsten as feasible. A large-diameter tungsten does not bend or break as easily as a smaller one when it accidentally touches or gets stuck to the work. Also, it is easier to remove contamination from a large-diameter tungsten without the tungsten breaking.
- A larger-diameter tungsten is needed on AC when the ball becomes more than 1-1/2 times the diameter of the tungsten (Fig. 45) or when the end of the tungsten melts and drops on the base metal.

Filler Metal Diameter

- Use the next LARGER diameter of filler metal when the rod melts before it reaches the weld puddle.
- Use the next SMALLER diameter of filler metal when the rod sticks as it is dipped into the weld puddle.

Amperage

When adjusting the amperage, watch the weld puddle and keep it in proportion to the thickness of the material being welded. Weld puddle size can be controlled with a remote foot control by pushing down to get more heat (increasing amperage) or backing off (decreasing amperage) to get less heat.

The following descriptions can be used to troubleshoot problems caused by too much or too little amperage.

- Increase the amperage when the filler rod sticks as it is dipped into the weld puddle. More amperage will increase the size of the weld puddle and/or the temperature of the weld puddle and the problem will be solved. Decreasing travel speed may also help solve the problem.
- Decrease the amperage when the weld puddle starts to sag (also called excess penetration or melt-through), or when a hole melts in the metal. Increasing travel speed may also help solve either of these problems.

Gas Flow

The gas flow rates given in Table 5 through Table 7 are adequate for most welding projects. Unless there are unusual circumstances, using much higher flow rates will only waste gas and increase the cost of the welding job. When there are weld defects that could be gas related, use a portable flowmeter (see Chapter 4, Fig. 32) to check the actual gas flow from the gas cup on the torch.

Weld coverage and gas flow can be affected by the following conditions.

- Air currents from fans, air conditioners, or open doors can displace the gas and prevent the proper gas coverage of a weld.
- Gas flow may not be set accurately because of faulty equipment like a flow gauge with a broken indicator.
- Something can interrupt or stop the flow of gas between the flowmeter and torch. A sharp object or the hot end of a piece of filler rod can make a puncture hole in the gas hose or torch cable. (There can also be gas contamination from air being drawn into the hose through a puncture hole.) A pinched gas hose or sticking gas solenoid valve can also cause intermittent weld defects by restricting or stopping the flow of gas.

The following descriptions can be used to identify a bad weld that was made without the proper gas coverage.

- Surface Oxidation: When the gas flow is too low, aluminum will not form a clean weld puddle. In addition, the top surface of stainless steel near the weld will be covered with a reddish brown smoke and the weld will be a dark gray to black color. The surface of carbon steel near the weld will be covered with a reddish brown smoke and the weld will be porous as well. Assuming other welding procedures are being followed properly, increase gas flow when any of these conditions are present. Fig. 46 contains examples of surface oxidation.
- If the end of the tungsten electrode melts and turns a dark gray to black color, and/or if there is a greenish white smoke on the weld, increase gas flow.
- Arc starting can become difficult if the gas flow rates are excessive.
- When flow rates are in the range of 4 to 5 times the values in the data sheets, there can be enough turbulence created to pull air into the weld zone and cause weld defects.

Proper Gas Tungsten Arc Welding (GTAW) Techniques

Learning and developing the proper techniques for gas tungsten arc welding is more difficult than with any of the other processes covered in this text. The manipulative skills like hand, eye, and mind coordination are similar to those needed for oxyacetylene welding. However, making welds with the GTAW process requires much more concentration and patience than for oxyacetylene welding. Torch positioning on the weld joint and filler metal application in the weld puddle have to be very exact. A wrong movement can cause the tungsten, the filler metal, and the base metal to be contaminated. When this happens, welding must be stopped and the contamination removed from everything or the weld will be unacceptable.

To become proficient with the GTAW process, a welder must learn and do all the things properly on the list given below.

Good body positioning and good eyesight are also very important. A welder has to be in a position to keep the gas cup from blocking the view of the weld puddle and also be able to judge the proper arc length.

Keys to proper GTAW welding are as follows:

- Setting the tungsten extension
- Starting and maintaining an arc
- Keeping the arc aligned on the weld joint
- Proper torch angles
- Proper filler metal angles
- Proper filler metal application
- Maintaining a weld puddle
- Stopping the arc

Setting the Tungsten Electrode Extension

The tungsten extension is determined primarily by the weld joint and the size of the gas cup. A general rule is to extend the tungsten a distance 1-1/2 to 2 tungsten diameters (Fig. 35). This distance will need to be increased for "T" joints (Fig. 36) that are welded with a medium to large-size gas cup and for making the root pass on most "V" groove welds. The tungsten can be set closer to the gas cup for a square butt or lap joint than it is for a "T" joint. The less tungsten there is extending past the gas cup, the less likely the tungsten will be contaminated. However, the tungsten should not be recessed inside the gas cup or set closer than one tungsten diameter to the gas cup. With either condition, the line of sight to the weld puddle is blocked and/or it will be impossible to get close enough to maintain the proper arc length. Fig. 35 and Fig. 36 illustrate various tungsten positionings.

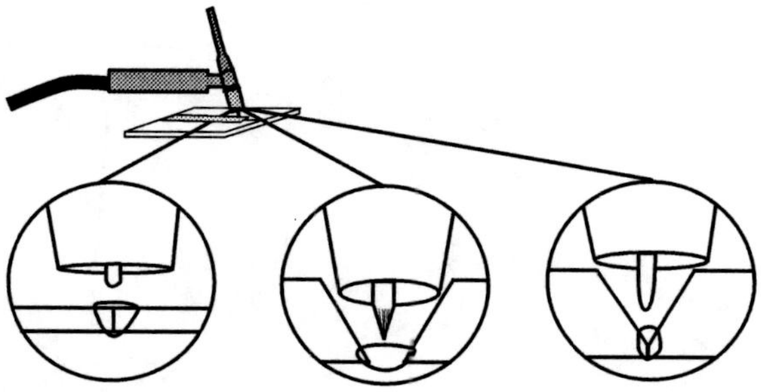

Fig. 35 — Tungsten Electrode Extensions for Groove Welds

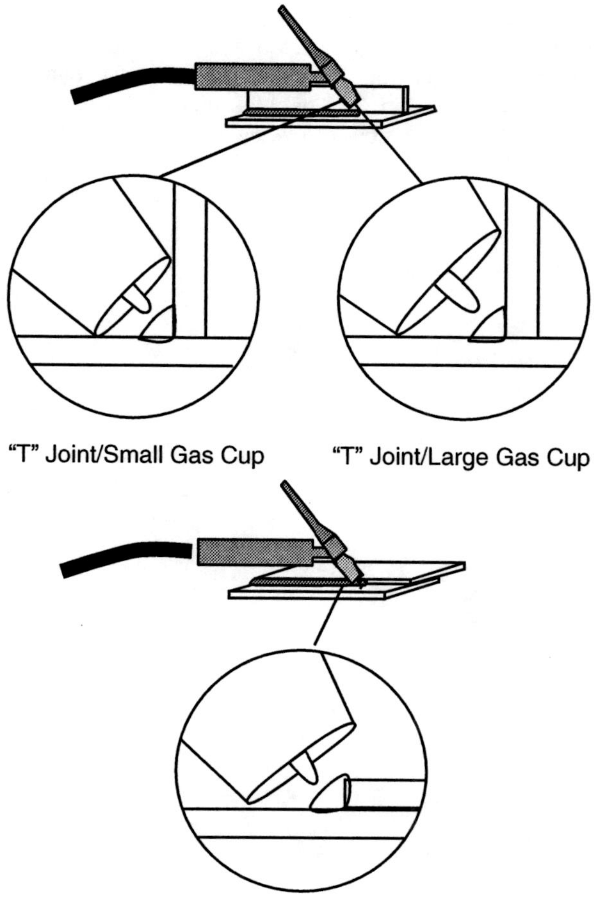

Fig. 36 — Tungsten Electrode Extensions for Fillet Welds

Starting and Maintaining an Arc

Starting an arc can be a challenge with the GTAW process. Scratch starting does not work well with a SMAW power source because of its electrical output characteristics (high open circuit voltage and high amperage when the electrode is short-circuited to the work). During a scratch start, it is easy to overheat the tungsten, melting it and contaminating both the electrode and base metal.

Since the tungsten is not compatible with any of the base metals being welded (unless you are welding tungsten), the arc must start without contaminating the electrode, the base metal, and/or the filler metal. The tungsten must be clean or the arc will be unstable (hard to keep in the weld joint). Also, trying to weld with contaminated base metal or filler metal will make the weld quality unacceptable. The use of high frequency (HF) and touch-and-lift start are the preferred methods of arc starting for most GTAW applications.

High-frequency arc starting is one of the most common methods of starting the arc. The tungsten is brought within approximately 1/8 inch (3.2 mm) from the work and the high-frequency current jumps the air gap. Once there is a path across the air gap, the welding current will follow and the arc is started. Fig. 37 illustrates resting the gas nozzle on the work to aid in high-frequency starting.

Touch-and-lift start is another method for starting an arc. This method is different and should not to be confused with the scratch start method, which is explained last. Touch-and-lift start takes a special power source circuitry to give a low voltage and low current start. Power sources equipped with this circuit will have a switch to turn it on and off. When in the ON position, the electrode is touched directly to the base metal for 2 to 3 seconds and then lifted straight up (Fig. 38). The tungsten is preheated while shorted to the work (hot tungsten starts arcs very easily). However, the voltage and amperage are so low, no melting of the electrode or base metal occurs, and no contamination takes place.

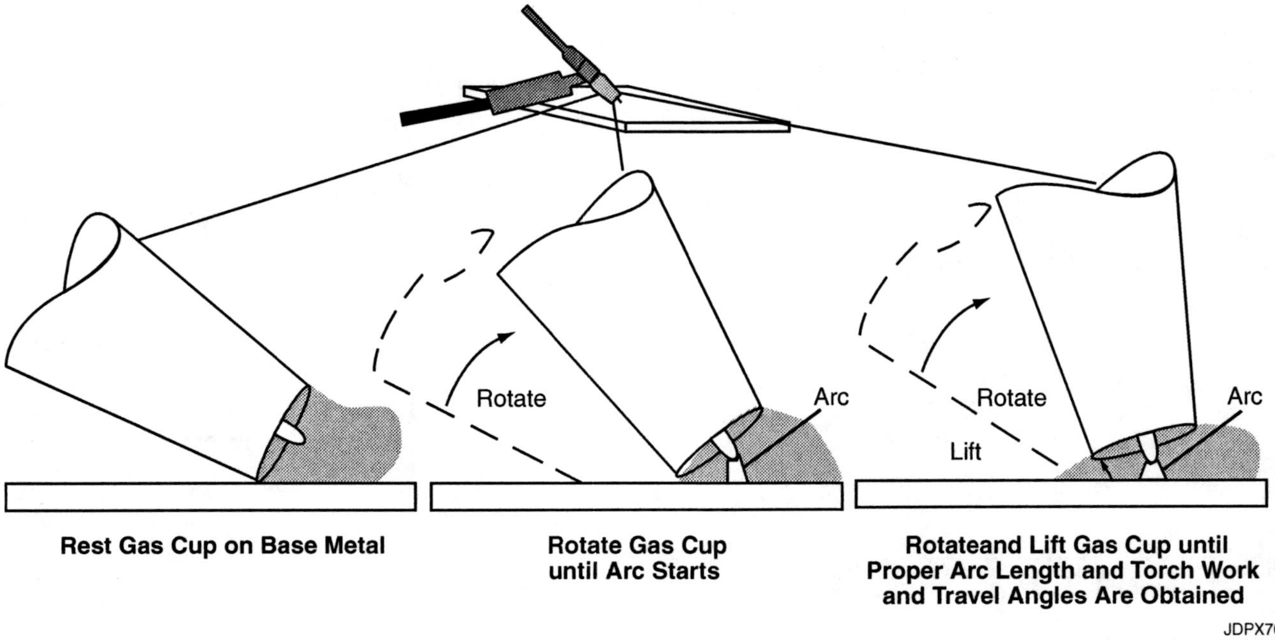

Fig. 37 — Gas Cup Movement for High-Frequency Arc Starting

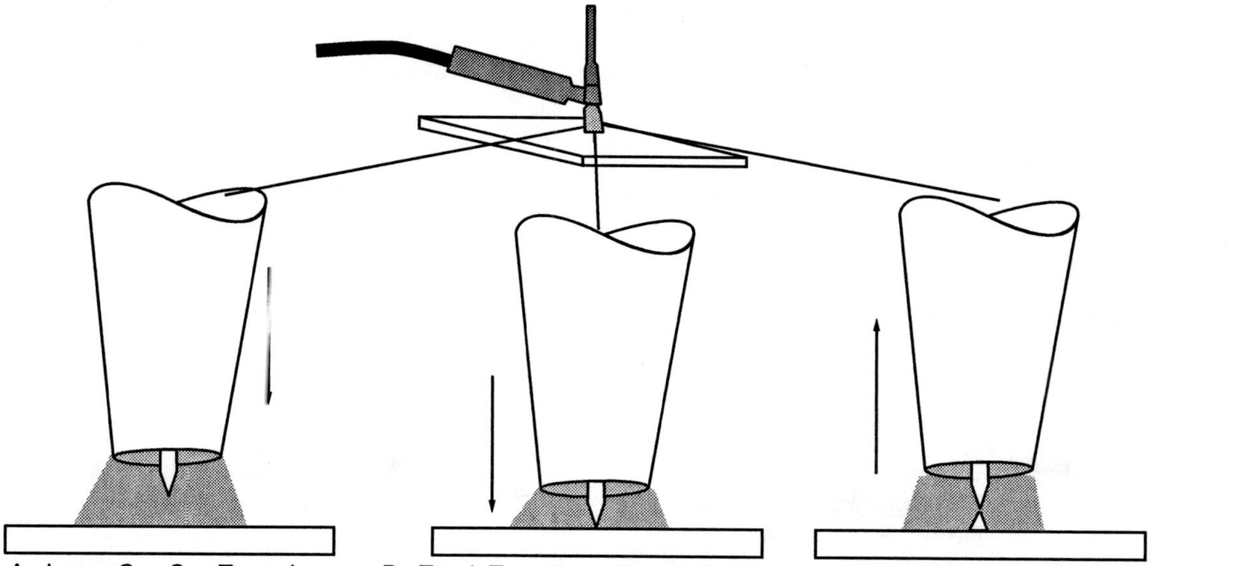

Fig. 38 — Gas Cup Movement for Lift Arc (Touch) Starting

Scratch starting is limited to applications that require DCEN. Starting an arc is similar to starting an arc with a SMAW electrode (like striking a match) except the operation must be done more carefully. A skillful hand is needed to get the arc started without contaminating the electrode and/or base metal and/or ruining the surface finish with arc strike marks. It is also very easy to break the point of the tungsten if the electrode happens to stick before the arc starts. Naturally, if this happens the broken piece of tungsten must be removed from the base metal before beginning to weld. Despite these limitations, scratch starting continues to be used in the field installation of stainless-steel piping and for making the root pass on carbon-steel pipe in power plants and chemical plants.

Electronic arc starts are used by some manufactures to control the polarity of the starts, so that whether AC or DCEN is selected, the arc starts with DCEP. The DCEP starts arcs very well and, before the heat can build up and damage the electrode, this low current disappears and the selected polarity and current type reappears.

Summary:

If there is difficulty in starting the arc, use the following list:

1. Use good-quality tungsten.
2. Use good weld cables that are as short as possible. Long cables should not be coiled up, and insulation must be not be cracked or missing. Otherwise, high frequency (HF) could be generated at several thousand volts and easily transferred into other objects.
3. Attach the work cable as close to the work as possible. Don't risk passing electricity through gears or bearings.
4. Remove any other cables attached to the power source, such as SMAW electrode leads.
5. Keep the torch power cable and connections from touching any metal parts that may be connected to the building ground.
6. Make sure all the connections are clean and tight; aluminum oxide is not a very good conductor.
7. Increase the high-frequency (HF) intensity knob, if available.
8. Check the spark gap adjustment of the high-frequency generator. This operation should be done only by a person qualified to perform maintenance on welding equipment.
9. Use pure argon. Helium makes arc starting more difficult.
10. Concrete can bleed off the HF through the weld cables. Insulate long runs of cable by placing them on boards.
11. Use the shortest torch possible. If a long distance from the power source is required, use one of the add-on HF units illustrated in Fig. 12.
12. Use the smallest diameter of tungsten for the amperage being used.

The arc length, which is actually the distance between the end of the tungsten and the weld puddle (Fig. 35, 36 and 39), should be approximately the same as the diameter of the tungsten. At this distance, the arc will be concentrated on a small area. This will provide the maximum rate of base metal heating and depth of penetration. At first it may be hard to see the arc and weld puddle, but holding an arc that close is necessary to make welds that are uniform and that meet inspection standards.

Learn to hold the proper arc length and avoid making the mistakes that are commonly made with the GTAW process. Holding an arc longer than the diameter of the electrode will increase the area heated and corresponding size of the weld. This translates into melting holes in thin metal (1/8 inch/3.2 mm and less) and making oversized welds. A long arc can also reduce gas coverage of the weld area that in turn will create weld defects. Holding an arc closer than the diameter of the electrode will increase the number of electrode contaminations with the base metal and filler rod. This translates into reduced welding time and increased time for cleaning the electrode and base metal.

Keeping the Arc Aligned with the Weld Joint

The size of the arc coming from the tungsten electrode is relatively small and must be kept aligned properly (Fig. 39-A and C) and with the intersection of the weld joint from the start to the finish of each weld. The concentration of the arc on one piece more than the other piece can cause holes to melt in the piece receiving the most arc energy. The piece receiving the least amount of arc energy (Fig. 39-B and D) may not reach melting temperature, which in turn will cause a lack of fusion and/or poor penetration.

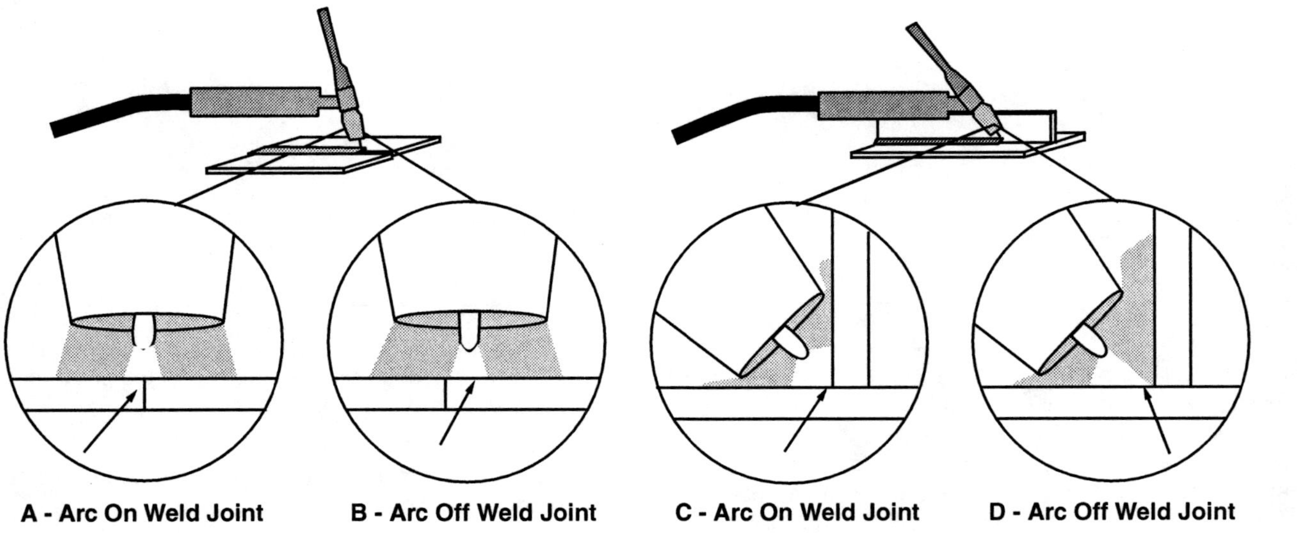

Fig. 39 — Arc Positioning on Butt and "T" Joints

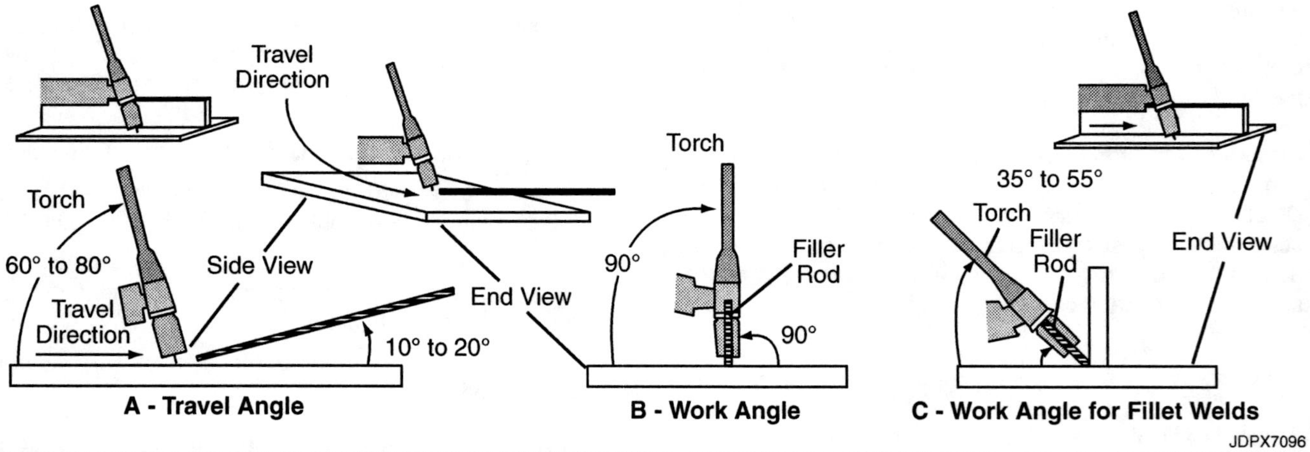

Fig. 40 — Work and Travel Angles for Making Welds in Flat Position

Torch Angles

Like the other welding processes covered in this text, the proper position to hold the GTAW torch is determined by the work and travel angle. As illustrated in Fig. 40-A, the torch is always pushed in relation to the direction of travel. As shown, the work angles will vary according to the type of weld joint and position of welding. Also, the work angle can be changed as needed to concentrate more heat on the thick piece and less heat on the thin piece when the base metal thickness is unequal.

Filler Metal Angles

As illustrated in Fig. 40, the filler metal should form a 10° to 20° angle with the work and be directly across from the GTAW torch. This angle is very important and is also the reason why people who have learned to oxyacetylene weld may have trouble learning the GTAW process. Experienced oxyacetylene welders have a tendency to hold the filler metal at a 45° angle rather than at the 10° to 20° degree angle necessary for GTAW. Holding the filler metal at the wrong angle makes it almost impossible to dip the filler metal into the weld puddle without a drop of metal from the filler rod first falling onto the base metal, causing a lack of fusion.

Filler metal contamination from the air is also a problem created by not holding the filler metal at the proper angle. The filler metal has to be kept in the gas envelope and out of the area of heat above the arc. If not, the end of the filler metal will contaminate the weld and cause weld defects. Also, the drops of filler metal will just be lying on the surface and the weld will not have adequate base metal fusion and penetration.

Filler Metal Application

The two methods of applying filler metal are called dipping the rod and laying the rod in the weld puddle. The dipping method is more common on aluminum and magnesium and the laying method is more common on stainless steel and carbon steel. If done properly, either method will give acceptable results in about any situation. However, the dipping method is the easiest way to achieve back side reinforcement (Fig. 41) when making a weld on aluminum or magnesium without having a special welding fixture. The bottom line is: which method is the easiest to learn and less tiring to the welder's hand and arm at the end of a workday?

Tips For Learning to Apply Filler Metal Properly:

1. Before starting to weld, get a pair of light-weight leather gloves made especially for GTAW. These gloves will make it easier to learn the feel of when the filler metal is melting properly with either method of filler metal application.
2. More heat will be required at the start of a weld, especially aluminum. As the end of the weld is reached, heat build-up will require that the amperage be reduced. A remote foot or hand control is very helpful in reducing this heat and filling the crater.
3. Welding on both edges (a butt joint or edge joint) will require less heat than a "T" and lap joint.
4. Adding the filler metal along the top edge of "T" and lap joints will keep it (the top edge) from melting away.
5. Adding the filler metal to the top leading edge of the weld puddle will help reduce undercut.
6. Make sure the weld puddle melts to the root of the joint or full penetration will not be possible.
7. To begin either method of filler metal application, center the electrode at the location the weld is to be started. Then start the arc and increase the amperage until a weld puddle forms that is 2 to 3 diameters of the filler metal (Fig. 41). When making "T" and lap joints, the edge of the top or vertical piece will want to melt easier than the flat surface on the bottom piece. Use the torch work angle and electrode alignment in the joint (Fig. 39) to obtain equal heating of both pieces.
8. Melt the filler metal in the weld puddle and not with the arc. There will be very little or no base metal penetration when the filler rod is melted with the arc. Even when the arc is held on the spot of liquid filler metal to make it flow in, there is no assurance of adequate base metal fusion.

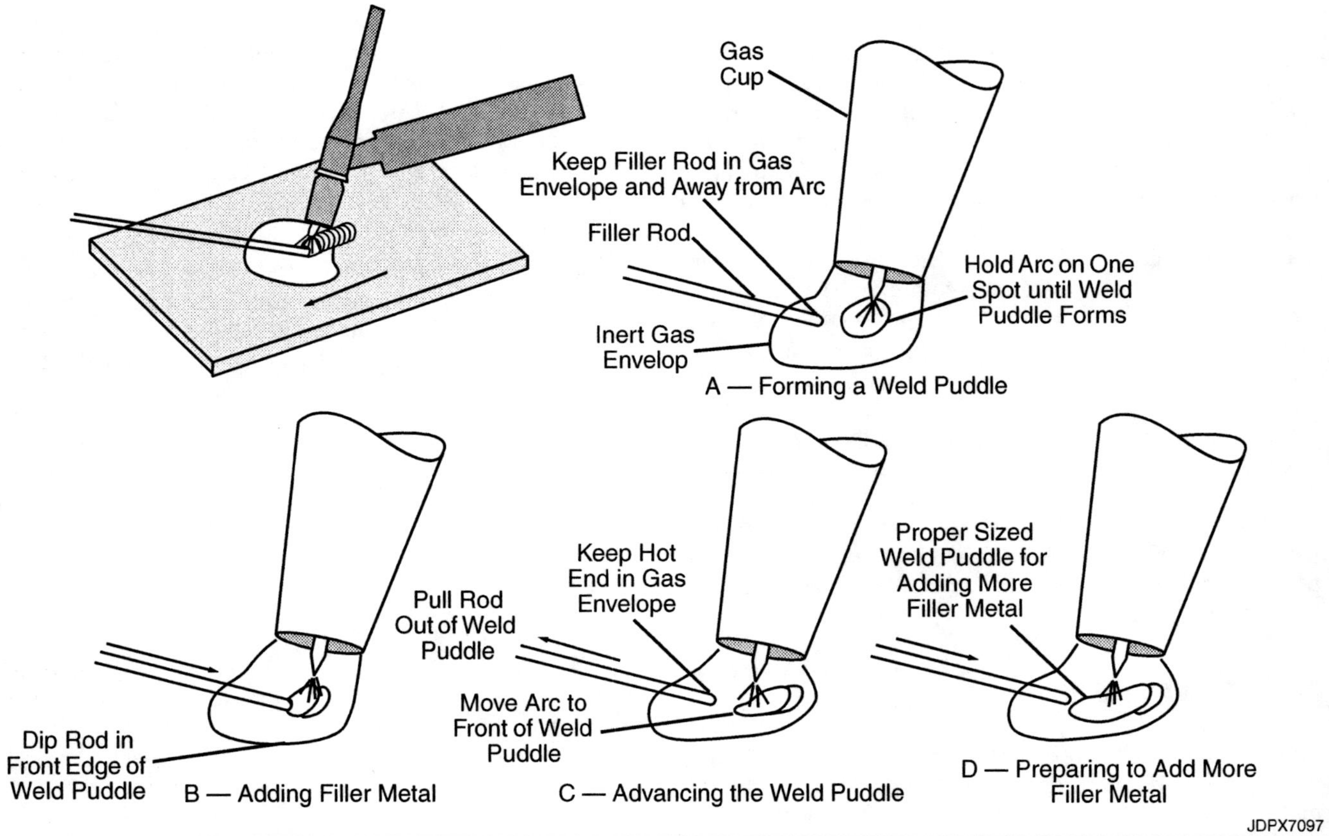

Fig. 41 — Adding Filler Metal Using the Dipping Technique

At a given amperage, travel speed and the addition of filler metal determine weld size and the corresponding depth of penetration.

Weld size is determined by travel speed, heat input (welding amperage) into the base metal, and the amount of filler metal added at any given time. For example, at a given travel speed, weld size can be increased by increasing amperage and increasing the addition of filler metal and vice versa. Weld size can also be increased by decreasing travel speed while maintaining the same rate for adding filler metal.

On material 1/8 inch (3.2 mm) or less in thickness it is possible to obtain satisfactory weld joints by just fusing the pieces together without adding filler metal.

CAUTION: When this procedure is followed, there should be adequate testing of the welds before the part is put into service to ensure there is adequate base metal penetration and adequate weld size, especially on fillet welds ("T" and lap joints) to provide adequate weld strength.

The dipping method (Fig. 41) is started by quickly dipping the filler rod in and out of the weld puddle without taking it (the filler metal) out of the gas envelope. Advance the tungsten to the front of the weld puddle. When the weld puddle has increased to the desired size, dip the filler rod again and keep repeating the process until the weld has been completed. To fill the crater, add filler metal without advancing the arc until the crater is of the proper height.

The laying method (Fig. 42) is started by laying the filler rod in the front on the weld puddle. As the filler metal is melted, it is fed or pushed into the leading edge of the weld puddle. The arc is advanced along the weld joint at a speed necessary to obtain the desired weld size and corresponding depth of penetration. No torch motion is necessary for making welds that are 3/16 inch (4.7 mm) or under. For welds that are larger than 3/16 inch (4.7 mm), a slight side-to-side motion is necessary to flatten and make the weld wash in (wet into the base metal), especially on stainless steel and the nickel alloys.

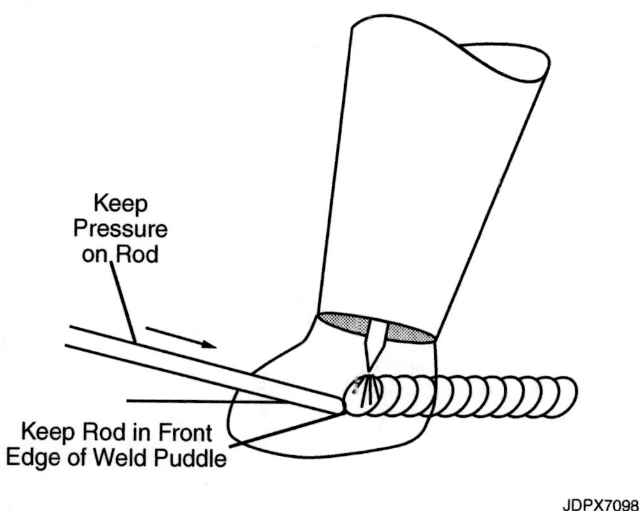

Fig. 42 — Adding Filler Metal Using the Laying Technique

A popular side-to-side motion used for making open groove welds on carbon steel and stainless steel is called "walking the cup" (Fig. 43). The tungsten is extended about 5/8 inch (15.9 mm), and the gas cup is rested on the sides of the groove. The travel angle is increased to 45°, and the torch is twisted back and forth with a wrist motion to obtain a weld 1/4 inch to 5/16 inch (6.4 mm to 7.9 mm) wide. The filler metal is pushed into the front of the weld puddle. Pressure on the filler rod and travel speed determine the amount of weld reinforcement and penetration. Gas flow is set from 35 CFH to 40 CFH (1 CMH to 1.1 CMH).

It takes several hours for an entry level welder to learn just the right amount of pressure to keep against the filler metal to form the proper size weld.

IMPORTANT: It is easy to apply too much pressure, causing the filler rod to slide across the weld puddle and into the tungsten. Also, it is easy to melt the filler metal without obtaining adequate base metal melting and penetration.

Maintaining a Weld Puddle

Maintaining a weld puddle for the entire length of a weld joint is a matter of timing and synchronizing the actions learned earlier in this section. A welder must constantly "read" the weld puddle and recognize when and where to add filler metal, and at the same time know how and when to vary the heat input into the weld puddle.

Weld size and/or penetration can be increased by decreasing travel speed or increasing amperage. Weld size and/or penetration can be decreased by increasing travel speed or decreasing amperage.

Achieving the proper combination of travel speed and amperage to obtain a desired weld size with adequate base metal penetration requires many hours of practice and coaching from someone who is proficient with the GTAW process.

Stopping the Arc

Stopping the arc is done with a power source equipped with a remote control by moving the control to the OFF position. As the control is moved to the OFF position, sufficient filler metal should be added to the weld puddle to fill the weld crater. The torch should continue to be held at the end of the weld after the arc has stopped and until the gas stops flowing. This hesitation is necessary to provide shielding to the molten metal and base metal until each has cooled below a temperature that can cause contamination by the air.

Stopping a DC arc on a weld being made with a power source that is not equipped with a remote contactor control is done by increasing travel speed until the weld puddle has been "out run" and then lifting the torch until the arc goes out. Crater filling is sometimes marginal, and there can be arc marks on unwanted areas of the part with this technique.

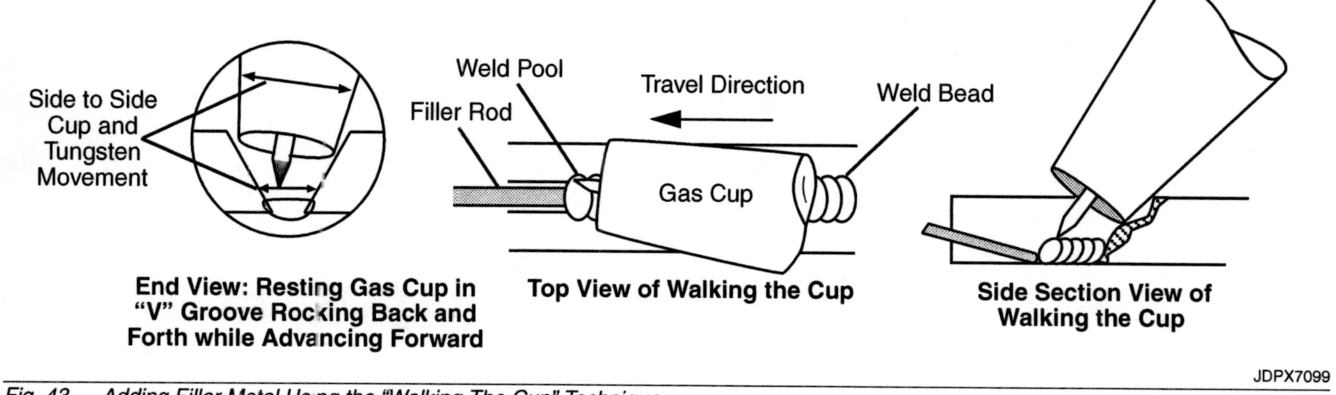

Fig. 43 — Adding Filler Metal Using the "Walking The Cup" Technique

Stopping an AC arc on a weld being made with a power source that is not equipped with a remote control is done by turning off the high frequency. A small switch similar to a fingertip control is attached to the torch.

The torch should continue to be held at the end of the weld until there has been adequate cooling. The gas flow is stopped by closing the manual gas valve on the torch body (Fig. 20).

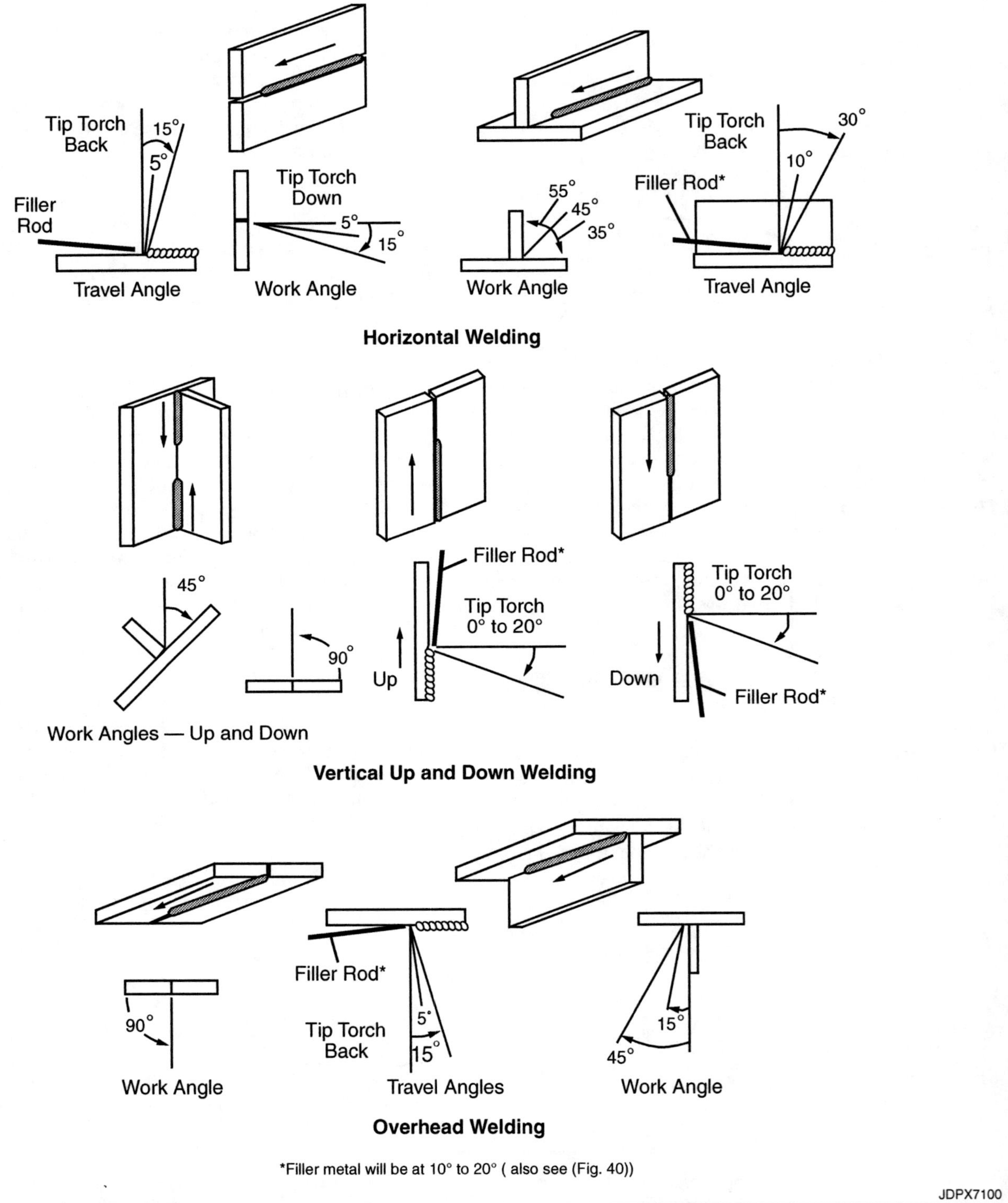

Fig. 44 — Torch Work and Travel Angles for a Right-Handed Person During Out-of-Position Welding

Out-of-Position Welding Techniques

Out-of-position welding techniques are not much different than for the flat position. The torch and filler metal angles are about the same as for the flat position (Fig. 44) . There is no molten metal being transferred through the arc. Thus, gravity has much less effect on the GTAW processes than it does on the processes that are transferring molten metal through the arc.

As long as the weld puddle remains small, surface tension will hold it in place. To keep the weld puddle small, decrease the amperage specifications in the data charts by 10 percent. Increasing the travel speed can also reduce the size of the weld puddle.

When welding must be done in the vertical position, traveling vertical up is generally preferred, but traveling down is acceptable. To begin a vertical up weld, start the weld puddle and then add the filler metal to create a shelf. This shelf allows the next dip of filler metal to be placed on top. The weld is actually a number of stair steps (made by each dip of the filler metal into the weld puddle) to the top of the joint.

Welding Suggestions

Planning and preparation are important for success. Creating optimum welding conditions will allow the best opportunity for making welds that have an acceptable appearance and that meet weld quality standards.

- Know and follow all of the safety precautions for the GTAW process. This includes everything from personal protective gear to a safe work area.
- Know how to operate and set the equipment. Follow the setup and operating instructions in the owner's manual supplied by the equipment manufacturer.
- Use and follow a written welding procedure. This includes metal preparation, filler metal selection, a plan for inspecting welds, and standards for determining weld acceptance and rejection.
- Know and follow the proper techniques for making welds. Torch positioning, weld puddle formation, and filler metal application are very important actions that must be done properly or weld quality will be poor.
- Don't try to weld with a contaminated tungsten or when the base metal has been contaminated. When the tungsten becomes contaminated, stop and clean it or remove the contaminated portion of the tungsten. Remove contamination from the base metal with the appropriate means, such as a wire brush or grinder. When necessary, remove oil, grease, and paint with an approved cleaner (one that does not produce flammable vapors or hazardous fumes when exposed to a welding arc) before grinding. Otherwise, grinding may embed the oil, grease or paint in the base metal and make cleaning even more difficult.

Steps for Shutting Down the Equipment and Stopping Work

1. Turn off cylinder valve or valve to flowmeter if on a manifold system.
2. Position the torch so accidental arcing cannot take place.
3. Activate the foot or hand control to open the gas solenoid valve and drain the gas supply system.
4. Back out the adjustment on the flowmeter valve.
5. Turn off the power source on-off switch and the primary fuse disconnect switch if easily accessible.
6. Wrap up the torch and work cable. Be careful, as the gas cup can be easily broken, and it may be hot enough to cause burn injuries or heat damage to nearby items.
7. Mark all hot metal HOT.
8. Properly dispose of filler metal trimmings, other residue, and scrap.

Troubleshooting of the Gas Tungsten Arc Welding (GTAW) Process

Troubleshooting a GTAW problem can best be done by breaking the process down into process variables, consumables, and equipment.

The following list contains information that can be used to troubleshoot typical problems that arise during welding. It is assumed proper welding conditions were obtained at the start of the welding job and a problem developed because something changed while welding was in progress. For example, the first 1 inch (25 mm) of welding was normal, or welding on the first part was normal, and something happened to create one of the problems described in this Chapter.

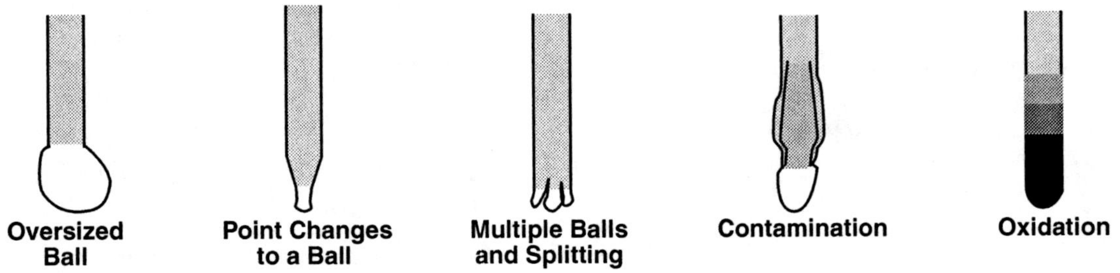

Fig. 45 — Examples of Common Tungsten Electrode Problems

Troubleshooting Gas Tungsten Arc Welding (GTAW) Problems

Problem	Probable Causes	Suggested Cures
1. Tungsten is being consumed and/or a large ball forms on the end (Fig. 45)		
A.	Wrong polarity or balance setting	Use DCEN or move balance toward maximum penetration
B.	Wrong tungsten type or diameter	Check tungsten alloy or use a larger diameter tungsten that will carry the necessary amperage
C.	Inadequate gas flow rate	Increase gas flow
D.	Torch body too hot	Make sure connection is tight; use water cooling
E.	Electrode oxidizes during cooling	Increase post-flow time; don't move torch rapidly until the electrode has cooled
F.	Improper shielding gas	Replace gas with a new cylinder of argon or helium
2. Tungsten will not stay pointed on DCEN (Fig. 45)		
A.	Wrong tungsten type	Use tungsten designed for DCEN
B.	Tungsten diameter too small	Use next larger diameter of tungsten that will carry the necessary amperage
3. Tungsten will not maintain a ball on ACHF (Fig. 45)		
A.	Wrong tungsten type	Don't use thoriated tungstens; use tungsten designed for ACHF
B.	Tungsten is contaminated	Clean or replace tungsten
4. Tungsten is rough and oversized near the end (Fig. 45)		
A.	Tungsten is contaminated	Clean or replace tungsten
5. Tungsten is dark gray on the end and discolored away from the end (Fig. 45)		
A.	Oxidation from the air	Increase post-flow time; don't move torch rapidly until the electrode has cooled properly
6. Base metal contaminated with tungsten (Fig. 46)		
A.	Scratch starting the arc	Use HF, touch and lift start, or piece of copper for starting
B.	Electrode melts into weld puddle	Check tungsten alloy; use a larger diameter tungsten electrode that will carry the necessary amperage
C.	Contacting electrode in weld puddle	Use proper arc length; decrease tungsten extension

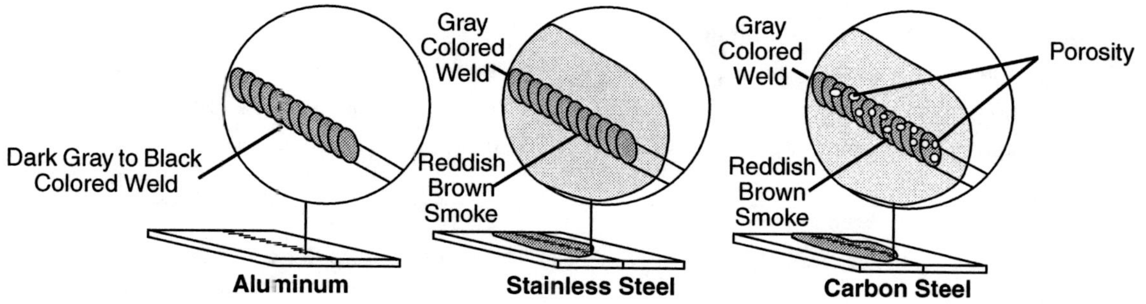

Fig. 46 — Recognizing Surface Oxidation on Gas Tungsten Arc Welds (GTAW)

Problem	Probable Causes	Suggested Cures
7.	**Porosity and/or poor weld color (Fig. 46)**	
	A. Inadequate gas flow rate	Increase gas flow
	B. Contaminated base metal	Clean/remove contaminated area
	C. Filler metal contaminated	Remove contamination
	D. Wrong filler metal	Use proper type of filler material
	E. Impurities in gas hose or line	Clean lines and/or purge lines before starting to weld
	F. Air is being drawin into system	Check for leaks: piping, connections, and hoses
	G. Air is getting into the torch	Torch assembled improperly, torch cap cracked or broken, bad torch O-ring, poor gas-cup-torch seal
8.	**Yellow powder or smoke on gas cup and work (Fig. 46)**	
	A. Inadequate gas flow rate	Increase gas flow
	B. Improper shielding gas	Replace gas with a new cylinder of argon or helium
	C. Inadequate post-flow time	Check post-flow time: increase to 1 second per 10 amps
	D. Contaminated tungsten	Keep tungsten away from sources of contamination
	E. Gas cup too small	Match gas cup to tungsten size and welding amperage
9.	**Unstable arc when using direct current**	
	A. Wrong polarity	Check polarity switch and/or cable connections
	B. Contaminated tungsten	Check for surface irregularities, remove contamination, reshape end to point or ball
	C. Long arc	Hold closer arc; extend tungsten when necessary
	D. Contaminated base metal	Clean metal properly before welding and/or remove contamination created during welding
	E. Joint too tight	Increase root opening, shorten arc, increase arc
	F. Tungsten improperly ground	Grind so grinding marks are parallel with tungsten
10.	**Unstable arc using alternating current**	
	A. Contamination in weld zone	Remove contamination from tungsten, filler rod and base metal
	B. Incorrect arc length	Use proper arc length
	C. Improper shielding gas	Use pure argon, decrease % of helium in mixture
	D. Incorrect balance	Adjust balance toward maximum penetration; set frequency in the 90 to 180 cycles per second range
	E. Rectification in weld puddle	Increase travel speed, set balance control to maximum penetration, don't dwell in weld puddle

Problem	Probable Causes	Suggested Cures
11. High frequency present but no welding arc		
A.	Welding circuit not complete	Check work connection and cables to power source
B.	No shielding gas	Check for adequate flow at cylinder or manifold
C.	Power source switches set wrong	Follow proper setup instructions
12. Arc wanders when using direct current		
A.	Long arc	Hold closer arc; extend tungsten when necessary
B.	Contaminated tungsten	Check for surface irregularities, remove contamination; reshape tungsten to a point
C.	Tungsten too large	Use smaller diameter tungsten; reshape tungsten to a point
D.	Extra resistance in welding circuit	Check cable connections and cables for heat build-up, tighten connections and cables for heat build-up, tighten connections or use larger diameter cables
E.	Turbulence in gas flow	Check flow rate, stop drafts/air movement, use gas lens, foreign material in gas cup
F.	Tungsten improperly ground	Grind so that grinding marks are parallel with tungsten
13. Arc wanders when using alternating current		
A.	Incorrect tungsten preparation	Use balled tungsten for sine wave and square wave power sources; use pointed tungsten for asymmetric or expanded balance control power sources
B.	Contamination	Remove contamination from tungsten, filler rod, and base metal
C.	Improper balance control setting	Move balance control toward maximum penetration
D.	Incorrect electrode diameter and type	Get smaller diameter of tungsten; get a tungsten that works well on alternating current
E.	Rectification in weld puddle	Increase travel speed, set balance control to maximum penetration, don't dwell in weld puddle
F.	Improper shielding gas flow	Check flow rate
G.	Ball not centered on tungsten	Reshape ball
14. Arc will not start or is difficult to start		
A.	No shielding gas	Check for adequate flow at cylinder or manifold
B.	Power source switches set wrong	Follow proper setup instructions
C.	Improper tungsten electrode	Use proper tungsten alloy for current and equipment
D.	Incorrect welding cable connections	Connect electrode cable to electrode terminal and work cable to work terminal
E.	Incomplete or poor welding circuit	Work clamp not attached or poor connection to work
F.	Tungsten too large	Use smaller diameter tungsten
G.	Loss of high frequency	Check/increase intensity control setting; torch cable fuse block contacting grounding surface; check torch weld cable for cracked insulation; check for poor connections in welding circuit; inspect spark gap points and adjust as needed

Troubleshooting Gas Tungsten Arc Weld (GTAW) Defects

Fillet Weld Examples | **Groove Weld Examples** | **Problem(s)** | **Corrective Action(s)**

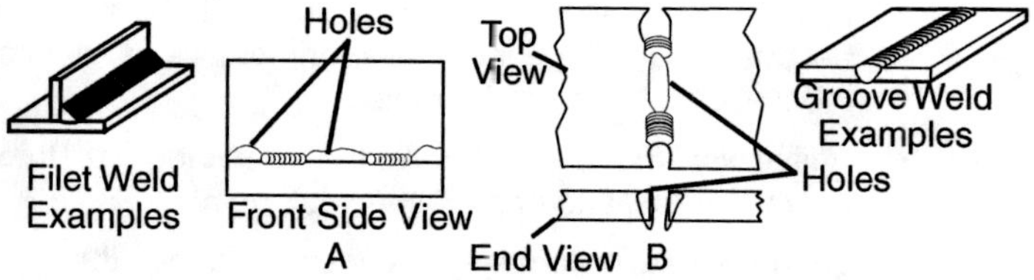

A, B Intermittent Holes Instead of a Complete Weld

1. Close up space between pieces if trying to fusion weld
2. Increase travel speed and/or decrease amperage
3. Add filler metal if fusion welding is not successful
4. Add more filler metal if already using filler metal
5. Use a backing bar to absorb heat when welding thin metals

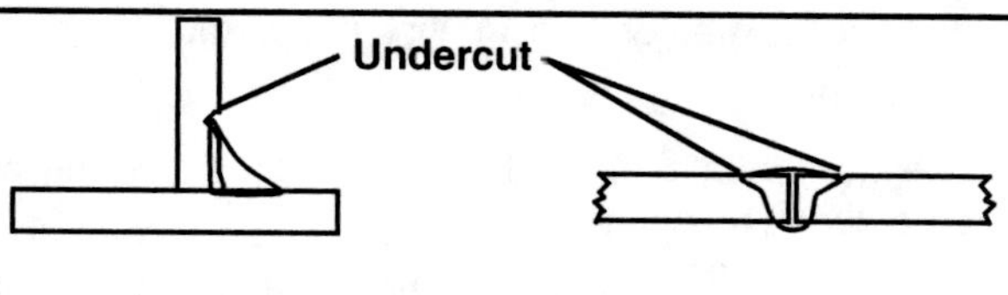

C Undercut Welds

1. Decrease weld puddle size by increasing travel speed or decreasing amperage
2. Add filler metal if fusion welding is not successful
3. Add more filler metal, if using filler metal

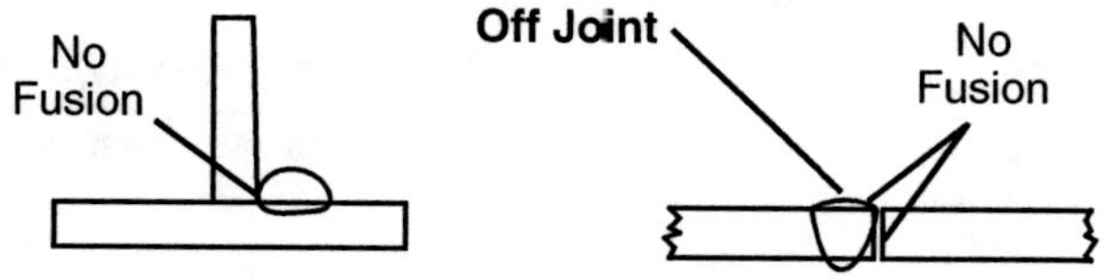

D Off-Joint Welds

1. Use proper work angle
2. Keep arc aligned with edges of weld joint so both pieces are melted by the arc
3. Correct vision problem, clean or replace cover glass, use lighter shade lens or get eyes checked

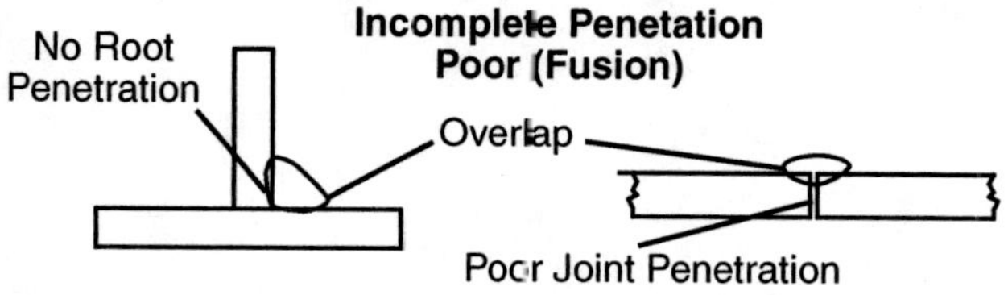

E Incomplete Penetration (Poor Fusion)

1. Don't add filler rod until metal forms a weld puddle
2. Don't melt rod with the arc. Melt rod in weld puddle
3. Increase amperage or decrease travel speed
4. For butt joints, increase root opening (space) between pieces and/or bevel both pieces

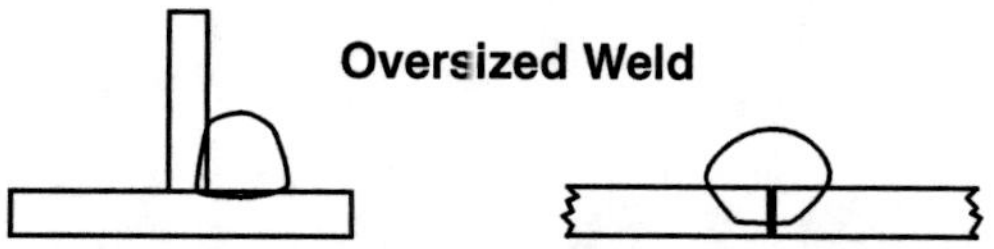

F Oversized Weld, Excess Reinforcement and Penetration

1. Increase travel speed and/or amperage
2. Do not use as much filler metal

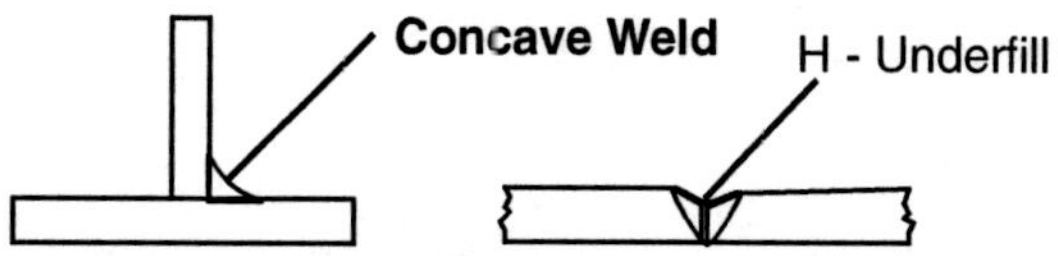

G, H Concave Weld or Underfilled Weld

1. Decrease amperage and/or increase weld travel speed if fusion welding
2. Use comparable filler material instead of trying to fusion weld

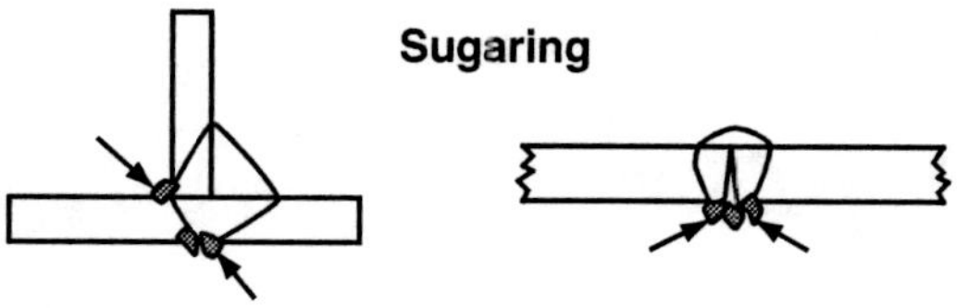

I Sugaring on Stainless Steel and Other Corrosion-Resistant Metals

1. Don't melt rod with arc and then try to melt it in with the arc
2. Fill the groove by making a second pass

*Examples are for single pass welds on 1/16" to 1/8" (1.6 mm to 3.2 mm) metal.

JDPX7105

Fig. 47 — Troubleshooting Gas Tungsten Arc Weld (GTAW) Defects

Acceptable Gas Tungsten Arc Welding (GTAW) Welds

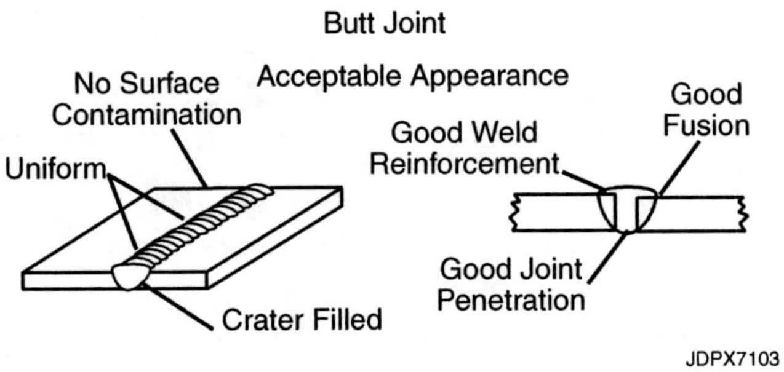

Fig. 48 — Acceptable Groove Weld on 1/8 inch (3.2 mm) Metal

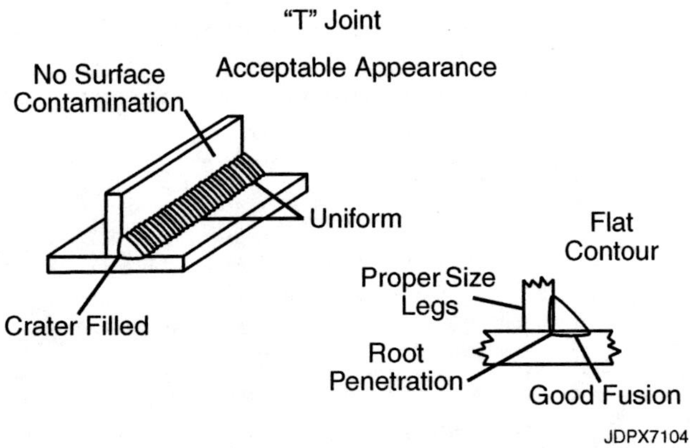

Fig. 49 — Acceptable Fillet Weld on 1/8 inch (3.2 mm) Metal

Acceptable welds of the groove or fillet type must meet visual inspection (Fig. 48 and 49). Welds must also meet structural strength requirements and show evidence of good workmanship, including uniformity. Because of the high manipulative skill required with GTAW, an acceptable-appearing weld would also have other properties that are acceptable.

Visual inspection includes: measuring weld size, penetration of the weld joint, reinforcement above the surface, depth of undercut, porosity or other uniformity problems on the weld face, and excessive smoke or discoloration on surrounding surfaces. Bending the piece in the weld area checks weld strength and fusion. These requirements are specified by the welding code, which applies to the function of the finished part.

Unacceptable Gas Tungsten Arc Welding (GTAW) Welds

Unacceptable welds are those with any of the following conditions. Also, check the welding code or other applicable standard for acceptance requirements as to what is excessive or unacceptable.

- A groove weld that doesn't completely fill the weld joint because of a lack of penetration and/or lack of reinforcement above the top of the metal.
- A fillet weld that does not have the proper leg size, throat measurement, and/or penetration at the root.
- Weld penetration and/or reinforcement above the top of the metal is greater than 1/8 inch (3.2 mm).
- Undercut is greater than 1/32 inch (0.83 mm).
- The weld breaks or the base metal breaks in the weld area during a bend test.
- The weld has not fused to the metal and/or contains tungsten inclusions.
- The weld face contains excess gas holes (porosity).
- Sugaring on the back side of welds on stainless steel and other corrosion resistant alloys.

The chart on the preceding page illustrates many of these weld defects.

Test Yourself

Questions

1. Gas tungsten arc welding (GTAW) is the AWS correct terminology for the process. List two other common names that describe it.
2. If the GTAW process operates at a temperature of 11,000°F (6000°C) and the tungsten electrode melts at 6170°F (3410°C), how can it be considered a non-consumable electrode process?
3. Which kind of welding power source is used for GTAW?
4. List the three types of alternating current used for GTAW of aluminum.
5. (Fill in blank.) DC _____ is a very common setting to use with GTAW.
6. (Fill in blanks.) The arc balance controls the amount of time during electrode positive which does the _______ and electrode negative which does the _______ in the AC cycle.
7. (Fill in blank.) Torches rated above _____ amps should be water-cooled.
8. What kind of a wire brush should be used for cleaning aluminum?
9. (Fill in blank.) The ________ is the torch part that must be of the proper size to hold the tungsten electrode.
10. What is the AWS classification, the alloy type, and percent for an electrode with an orange color code?
11. (Fill in blank.) A _____-shaped tungsten is used with DCEN, asymmetric wave, and extended balance-controlled power sources to concentrate the arc.
12. (Fill in blank.) A 0.060-in. dia. EWTh-2 operating on an asymmetric wave power source would have a maximum current-carrying capacity of _____ amps.
13. (Fill in blank.) An ___–_____ gas is used with GTAW.
14. Which type of inert gas creates the hottest arc, helium or argon?
15. Which gas makes it easier to start an arc, helium or argon?
16. (Fill in blank.) The flow rate for helium must be _____ times greater than argon because it is so light if floats as it leaves the gas nozzle.
17. (Fill in blanks.) Excess gas flow rates can cause ______ and ______.
18. When using a filler metal, the end not being melted presents what safety hazard?
19. When welding carbon steel, name one kind of filler metal that should not be used.
20. Argon and the cylinders it is supplied in present what two safety hazards?
21. (Fill in blanks.) A good system to follow when setting up the equipment is to start at _____ and work back to the _______.
22. (Fill in blank.) The aluminum oxide melts at a temperature ___ times greater than the aluminum base metal.
23. (Fill in blank.) The surface of a good GTA weld on stainless steel should be an ____ color.
24. (Fill in blank.) _______ steels require additional considerations to prevent porosity.
25. (Fill in blank.) According to Table 5, welding a 3/16-inch-thick aluminum fillet weld requires ____ maximum amperage.
26. (Fill in blank.) According to Table 6, welding a 3/16-inch-thick stainless steel fillet weld requires ____ maximum amperage.
27. (Fill in blank.) The arc length should be ___ times the electrode diameter.
28. (Fill in blank.) The proper filler metal angle is ____ degrees from the base metal.
29. (Fill in blank.) If the electrode becomes contaminated, _____ before continuing to weld.
30. Give one reason why it is easier to out-of-position weld with the GTAW process than with the SMAW process.
31. When shutting down the equipment, position the torch so that what safety hazard can be avoided?
32. On a fillet weld, list four things that must be correct for it to be acceptable.
33. What can happen to the tungsten electrode when too high an amperage is used?

Cutting and Gouging Metals

6

JDPX7018

Fig. 1 — Cutting Stainless Steel Pipe

Introduction

Cutting and gouging operations are necessary to shape pieces of metal so they will fit together into functional assemblies that can be joined together by welding.

The most common processes that use heat for cutting metals are oxyfuel, plasma, and laser beam, while the most common processes for gouging are oxyfuel, arc air (air carbon arc), and plasma.

An SMAW electrode and power source can also be used for cutting and gouging. However, even when special electrodes are used, the SMAW process is slow compared to these other processes and requires considerably more grinding, chipping, and cleaning before the part that has been cut or gouged is ready for welding. For these reasons, the other processes will be explained in more detail than SMAW cutting and gouging.

Cutting and Gouging Processes

Cutting processes can be broken down into either an "oxidation" process or a "thermal" process. With the oxidation process, heat from an oxyfuel flame is used to raise the surface temperature of carbon steel to its kindling point. Then a stream of pure oxygen is concentrated on the "hot spot," causing the iron in the steel to burn (oxidize) in the same manner wood or paper burns. Heat from the burning causes more steel (which is actually the fuel) to reach the kindling point and the cutting to continue as long as there is sufficient oxygen to support the combustion. The iron oxide or slag (as it is more commonly known) formed is removed from the cut by the force from the stream of oxygen.

Thermal cutting processes are used for cutting all types of metals, including those that cannot be cut with oxyfuel. With any of the thermal cutting processes, very intense heat melts an area. Then the force from the plasma stream or laser beam removes the molten metal. With the arc air process, compressed air is responsible for most of the metal removal (the force of the electric arc does have a slight affect on metal removal). For a more complete explanation of each process, read the sections in Chapter 1 that explain cutting, or the next chapter in this text on the laser processes.

Gouging processes can also be broken down into an oxidation process or a thermal process, with the same features and limitations.

The main difference between gouging and cutting is that more metal is removed during the gouging operation. Gouging is used to prepare a groove for welding or for removing a weld or weld defect. Gouging produces a groove that does not go the full thickness of the metal, and cutting produces a kerf from the top to the bottom of the metal. Plasma gouging and arc air gouging are more widely used than oxyfuel gouging.

Oxyfuel Cutting (OFC)

The equipment needed for oxyfuel cutting is basically the same as that for oxyfuel heating, brazing, and welding, with the exception of the torch and tip. The gas cylinders, regulators, hose, and sometimes the torch body are the same. The most common fuel gases are listed in Table 1. The type of gas selected depends on the cost of gas, flame heat intensity, and corresponding cutting speeds. Safety is also a factor.

Fuel	Maximum Flame Temperature in Oxygen		Maximum Burning Velocity in Oxygen	
	(°F)	(°C)	Ft./Sec	m/Sec
Acetylene	5589	3087	9.8	3
MAPP® Gas	5301	2927	7.9	2.4
Propane	4579	2526	5.5	1.7
Propylene	5250	2900		
Natural Gas	4600	2528		

Table 1 — Fuel Gases Used for Oxyfuel Cutting

Cutting Torches and Cutting Tips

Always use a cutting tip that is compatible with the cutting torch and the fuel gas which will be used. At first glance, all cutting torches and cutting tips look the same. However, a more careful examination will indicate there are differences between equipment manufacturers and variations between models of equipment from the same manufacturer. Trying to use a tip on a torch it was not designed for can lead to a serious accident. Also, never use fuel gas cutting tips with acetylene.

A cutting torch attaches to the torch body (Fig. 2-A) or directly to the fuel and oxygen hoses (Fig. 2-B). It contains valves for adjusting the gas flow necessary to obtain the proper preheat flames and a valve for starting and stopping the flow of oxygen needed for cutting.

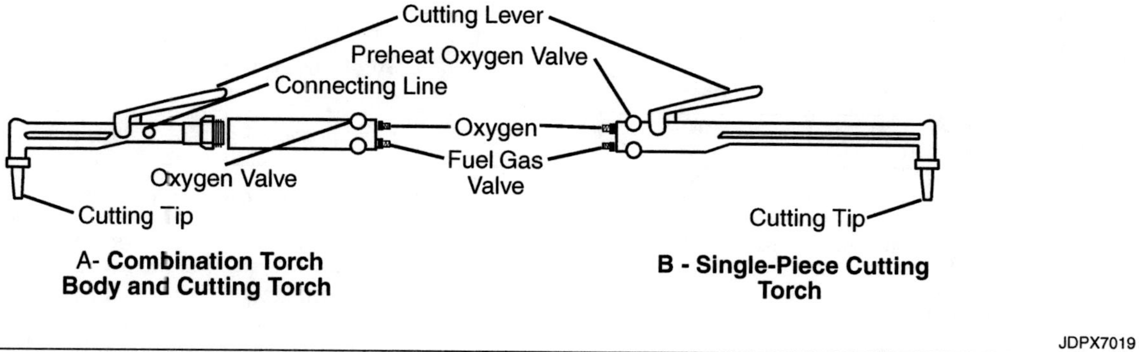

Fig. 2 — Two Basic Types of Cutting Torches for Oxyfuel Cutting

A cutting tip attaches to the end of the cutting torch, which has a threaded opening (Fig. 4-C). One end of the tip fits into the torch, and the other end is where the flame forms. The openings (on the flame end of the tip) around the outside are called preheat orifices and are for the mixed gases to exit and form the preheat flame. The center orifice is for the cutting oxygen to exit.

The type of fuel gas determines the style of cutting tip. Acetylene tips are of one-piece construction and have four to six preheat orifices (Fig. 3-B). Most of the tips for other types of fuel gases (Fig. 3-C) are of the two-piece construction type and have small rectangular slits for preheat orifices.

NOTE: Some equipment manufactures make MAPP® and propylene tips that are of one-piece construction. These tips have eight preheat orifices.

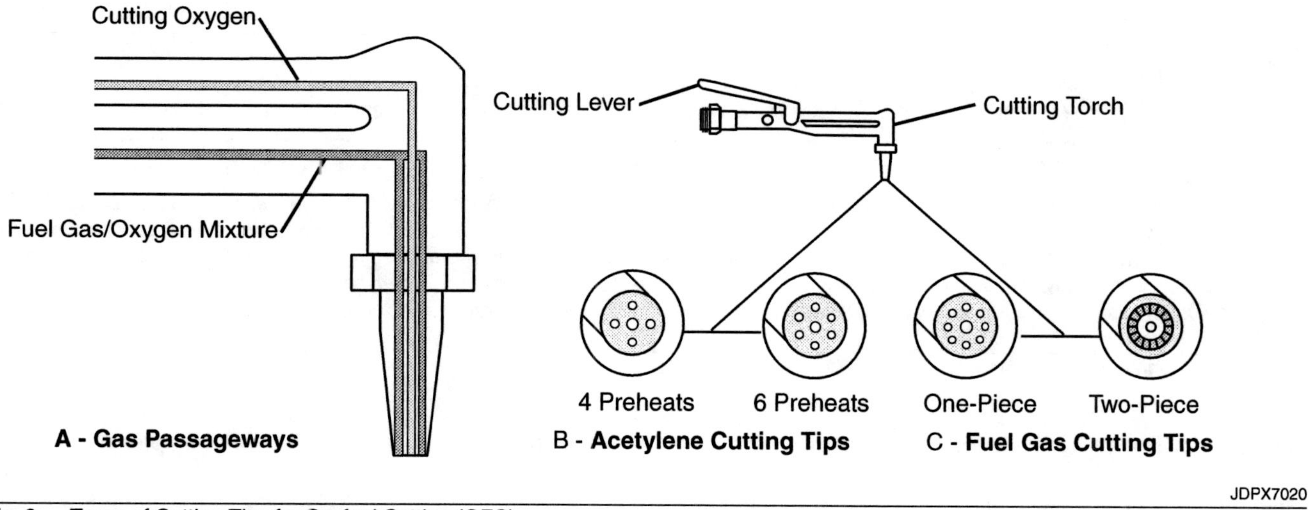

Fig. 3 — Types of Cutting Tips for Oxyfuel Cutting (OFC)

The metal thickness determines the size of the preheat orifices for each type of fuel gas. Thick pieces of metal require larger orifices (compared to thin pieces) because more fuel is needed for heating the metal to its kindling temperature.

The size of oxygen (cutting) orifice is also determined by the metal thickness, but doesn't vary much between the different fuel gases. The proper size is determined by volume of oxygen needed to cut a particular thickness at a given speed. Thick metals require larger oxygen orifices than thin metals.

Table 2 provides general guidelines for estimating tip size and gas flow. For best results as well as for safety issues, always use equipment manufacturer's information for determining the proper tip size and corresponding oxygen and fuel gas pressures.

Thickness of Steel in. (mm)		Diameter of Cutting Tip Orifice, in. (mm)		Cutting Speed in./min. (mm/mm)		Gas Consumptions, Ft^3/ h (m^3/h)							
						Cutting Oxygen		Acetylene		Natural Gas		Propane	
1/8	3.2	0.020-0.040	(0.5-1)	16-32	(400-800)	15-45	(0.4-1.3)	3-9	(0.08-0.25)	9-25	(0.25-0.7)	3-10	(0.08-0.3)
1/4	6.4	0.030-0.060	(0.8-1.5)	16-26	(400-660)	30-55	(0.8-1.6)	3-9	(0.08-0.25)	9-25	(0.25-0.7)	5-12	(0.14-0.34)
3/8	9.5	0.030-0.060	(0.8-1.5)	15-24	(380-610)	40-70	(1.1-2.0)	6-12	(0.17-0.34)	10-25	(0.3-0.7)	5-15	(0.14-0.4)
1/2	12.7	0.040-0.060	(1-1.5)	12-23	(305-585)	55-85	(1.6-2.4)	6-12	(0.17-0.34)	15-30	(0.4-0.8)	5-15	(0.14-0.4)
3/4	19	0.045-0.060	(1.1-1.5)	12-21	(305-535)	100-150	(2.8-4.2)	7-14	(0.20-0.40)	15-30	(0.4-0.8)	6-18	(0.17-0.5)
1	25.4	0.045-0.060	(1.1-1.5)	9-18	(230-455)	110-160	(3.1-4.5)	7-14	(0.20-0.40)	18-35	(0.5-1)	6-18	(0.17-0.5)
11/2	38	0.060-0.080	(1.5-2)	6-14	(150-355)	110-175	(3.1-5.0)	8-16	(0.2-0.45)	18-35	(0.5-1)	8-20	(0.2-0.6)
2	50.8	0.060-0.080	(1.5-2)	6-13	(150-330)	139-190	(3.6-5.4)	8-16	(0.2-0.45)	20-40	(0.6-1.1)	8-20	(0.2-0.6)
3	76.2	0.065-0.085	(1.7-2.2)	4-11	(100-280)	190-300	(5.4-8.5)	9-20	(0.25-0.6)	20-40	(0.6-1.1)	9-22	(0.25-0.61)
4	102	0.080-0.090	(2-2.3)	4-10	(100-255)	240-360	(6.8-10.2)	9-20	(0.25-0.6)	20-40	(0.6-1.1)	9-24	(0.25-0.7)
5	127	0.080-0.095	(2-2.4)	4-8	(100-200)	270-360	(7.6-10.2)	10-24	(0.3-0.7)	25-50	(0.7-1.4)	10-25	(0.3-0.7)
6	152	0.095-0.105	(2.4-2.7)	3-7	(75-180)	260-500	(7.4-14.2)	10-24	(0.3-0.7)	25-50	(0.7-1.4)	10-30	(0.3-0.8)
8	203	0.095-0.110	(2.4-2.8)	3-5	(75-120)	460-620	(13.9-17.6)	15-30	(0.4-0.8)	30-55	(0.8-1.5)	15-32	(0.4-0.9)
10	254	0.095-0.110	(2.4-2.8)	2-4	(50-100)	580-700	(16.4-19.8)	15-35	(0.4-1)	35-70	(1-2)	15-35	(0.4-1)
12	305	0.110-0.130	(2.8-3.3)	2-4	(50-100)	720-850	(20.4-24.1)	20-40	(0.6-1.1)	45-95	(1.3-2.7)	20-45	(0.6-1.3)

Preheat oxygen consumptions: Preheat oxygen for acetylene = 1.1 to 1.25 x acetylene flow (ft.^{3}/h or m^3/h); preheat oxygen for natural gas = 1.5 to 2.5 x natural gas flow (ft.^{3}/h or m^3/h); preheat oxygen for propane = 3.5 to 5 x propane flow (ft.^{3}/h or m^3/h).

Operating notes: Higher gas flows and lower speeds are generally associated with manual cutting, whereas lower gas flows and higher speeds apply to machine cutting. When cutting heavily scaled or rusted plate, use high gas flows and low speeds. Maximum indicated speeds apply to straight line cutting; for intricate shape cutting and best quality, lower speeds will be required.

JDPX7021

Table 2 — Cutting Chart for Estimating Oxyfuel Tip Size and Gas Flow

Oxyfuel Cutting (OFC) Safety Precautions

Most of the safety precautions for using oxyfuel cutting equipment are the same as the precautions for oxyfuel heating and welding. This includes knowing and following the proper operating instructions for opening cylinder valves and adjusting oxygen and fuel gas regulators.

Extra precautions should be taken to reduce the danger of fires and explosions from the sparks and hot metal created during cutting. This includes having the appropriate fire extinguisher on hand, having a fire watcher, and checking the area for fires for at least a half hour after cutting has been stopped.

Protect cement surfaces from the cutting flame and from hot slag and metal. Do not cut directly on cement surfaces such as floors and driveways. The hot area will explode, causing damage to the surface and burn injuries to anyone hit by the flying projectiles.

For more details on cutting safety, see Chapter 13.

Setting Up a Cutting Torch

1. Attach the cutting torch to the existing torch body (Fig. 4-A) or attach the oxygen and fuel gas hose to the cutting torch (Fig. 4-B).
2. Attach and secure the proper cutting tip to the cutting torch (Fig. 4-C).

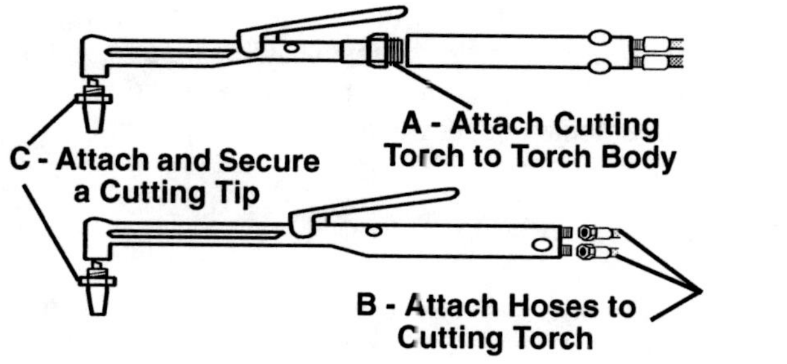

JDPX7022

Fig. 4 — Setting Up an Oxyfuel Cutting (OFC) Torch

3. Turn on the oxygen and fuel gas at the cylinders or manifold system.
4. Set the oxygen and fuel gas regulator pressure according to the equipment manufacturer's recommendations for the metal thickness that is to be cut.
5. Check all connections in the system for leaks.

Operating a Cutting Torch

The following procedure is for acetylene only. Check with the gas supplier or equipment manufacturer for instructions when using other fuel gases.

1. Open the oxygen valve on the torch or torch body completely by turning it in a counterclockwise direction (Fig. 2).
2. Purge the cutting oxygen passage by momentarily depressing the cutting oxygen lever (Fig. 2).
3. Open the fuel valve (Fig. 2) by turning it in a counterclockwise direction between 1/8 and 1/4 turn and ignite the gas by using a friction or spark type igniter. Make sure the torch is pointed away from other persons (as well as yourself), and from the cylinders and any flammable materials.
4. Keep opening the torch fuel valve until there is no soot (Fig. 5-C) and the flame has stopped excessive smoking (Fig. 5-A). If the flame leaves the end of the tip (Fig. 5-B), close the fuel valve by turning it in a clockwise direction until the flame returns to, or "touches," the end of the tip (Fig. 5-C). If adjustment of the torch valve does not make the flame stop smoking, the pressure is not correct for the tip or there is a significant pressure drop from the regulator to the torch. To correct the problem, increase working pressure or obtain a hose with a larger diameter that will provide adequate flow and pressure to the torch.

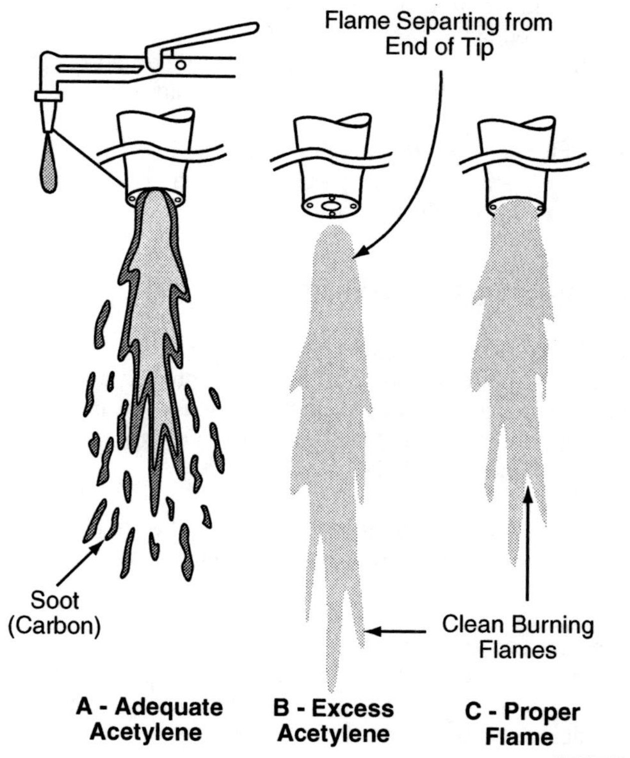

JDPX7023

Fig. 5 — Adjusting the Cutting Flame Size

5. Next, adjust the preheat flame to "neutral." Open the preheat oxygen valve (Fig. 2) by turning it counterclockwise until the bright inner cones with acetylene feathers appear in the flame (Fig. 6-A). Continue adding oxygen until the acetylene feathers merge with (or join) the inner cones to form the neutral flame illustrated in Fig. 6-B.

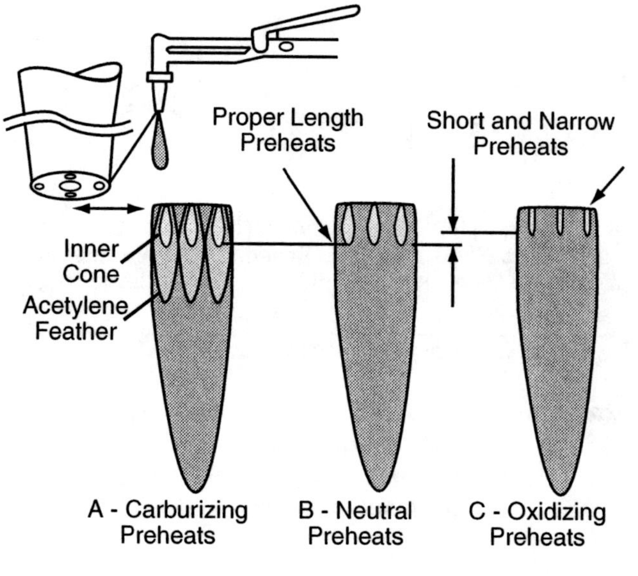

JDPX7024

Fig. 6 — Adjusting the Preheat Flame

6. Check the adjustment for a neutral flame by depressing the cutting oxygen lever. If the preheat flames change to a carburizing flame (Fig. 7-A), increase the preheat oxygen adjustment (while continuing to keep the cutting oxygen lever depressed) until a neutral flame is obtained.

Examples of carburizing (Fig. 6-A) and oxidizing flames (Fig. 6-C) are given as reference for the wrong types of flames to use for cutting.

Recognizing and Cleaning a Dirty Tip

A dirty tip will reduce cut quality and create slag that must be removed from the finished cuts. The absence of two straight bright lines in the center of the flame when the cutting oxygen is depressed indicates the cutting oxygen orifice is dirty or damaged (Fig. 7-C).

Expect cuts with rough edges unless two straight bright lines are present in the flame (Fig. 7-B) when the oxygen cutting lever is depressed.

Irregular preheat flames indicate that preheat orifices need cleaning or are damaged (Fig. 7-C, top part of flame). Cutting travel speed can be decreased if the preheat orifices are not kept clean.

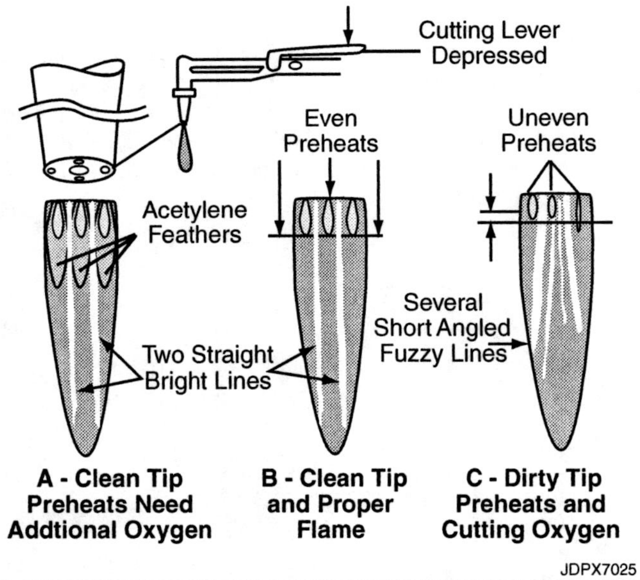

Fig. 7 — Checking the Flame and Condition of the Tip

IMPORTANT: Be very careful when cleaning the preheat orifices—the openings are small. It is easy to lodge or break a tip cleaner in the preheat orifices and ruin the whole tip.

To clean a cutting tip, use standard tip cleaners and follow the same procedures that were explained in Chapter 2 for cleaning a welding tip.

The tip should be replaced if cleaning does not restore the two straight bright lines to the center of the cutting flame when the oxygen lever is depressed. The lines need not be as clear and as distinct as those from a new tip to produce cuts of acceptable quality, but they must be present.

Making Cuts in Metal

Torch positioning and travel speed are the key factors in successful cutting once the proper tip has been chosen and the flame has been set properly.

A protractor should be used for judging and measuring the proper work angle and travel angle. When cutting bevels, the work angle becomes the bevel angle (see Fig. 8-B and Fig. 10-A). A straight edge such as a piece of angle iron or bar stock is useful for guiding the cutting torch in a straight line for making both square and bevel cuts.

Steps for Cutting and Piercing Metals

Begin by preheating the starting location until the metal turns a red color. As illustrated in Fig. 9, the preheat flames should be held above the surface of the metal 1/16 inch to 1/8 inch (1.6 mm to 3.2 mm).

Making a Square Cut

1. Position the torch so the work and travel angles are both at 90° (Fig. 8-A and Fig. 8-B).
2. With the cutting oxygen orifice positioned about 3/4 inch off the edge of the metal (Fig. 9-A), slowly depress the oxygen cutting lever and start moving forward into the metal. Fig. 9-B, Fig. 9-B, and Fig. 9-C illustrate an option for preheating and cutting metal thicker than 5/8 inch (16 mm).
3. Travel at a speed that leaves a smooth edge and a minimum amount of slag on the bottom edges. Traveling too fast will cause the cut to be lost—the cut won't be all the way to the bottom of the metal. Traveling too slow will cause a wide kerf and excess slag on the bottom surface. A very slow travel speed may produce so much slag the pieces will be "welded" back together with slag.

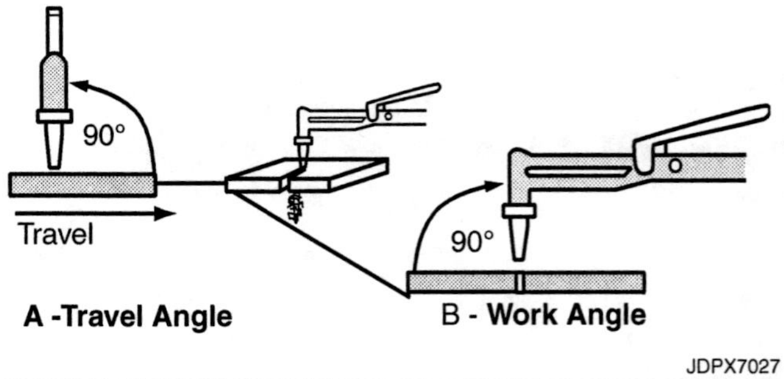

Fig. 8 — Making Square Cuts

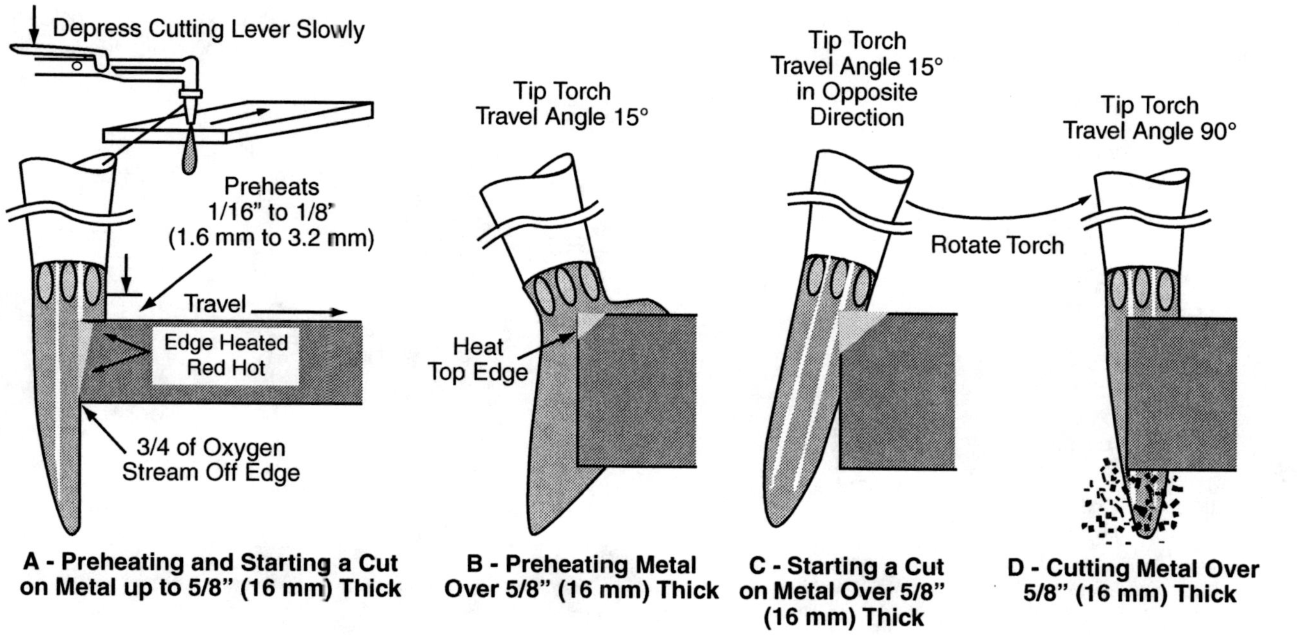

Fig. 9 — Hand Cutting Procedures—Preheating and Starting a Cut

Making a Bevel Cut

NOTE: Bevel angle refers to the amount of metal removed during cutting (Fig. 10) and not the angle of the edge of the piece that is left after cutting.

1. The tip size and oxygen pressure should be based on the depth of the bevel (Fig. 10-C) and not on the metal thickness. The depth of cut will be greater than the metal thickness, and the oxygen pressure will need to be increased accordingly.
2. Position the torch so the work angle is the same as the bevel angle.
3. With the cutting oxygen orifice positioned about 3/4 inch off the edge of the metal, slowly depress the oxygen cutting lever and start moving forward.

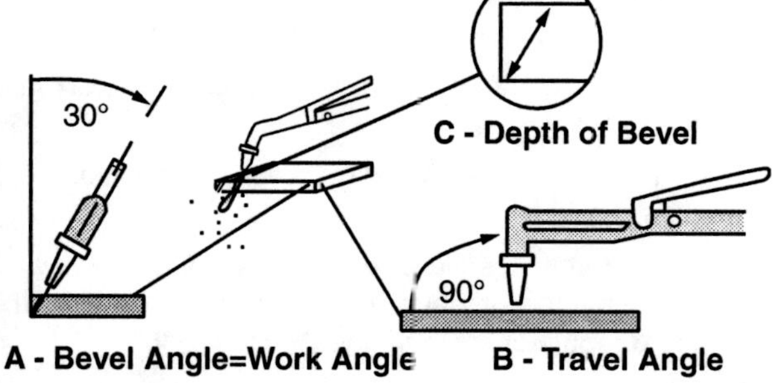

Fig. 10 — Making Bevel Cuts

4. Travel at a speed that leaves a smooth edge and a minimum amount of slag on the bottom edges. Traveling too fast will cause intermittent gouges in the edge, or the cut will be lost—the cut won't be all the way to the bottom of the metal. Traveling too slow will cause a wide kerf and excess slag on the bottom surface. A very slow travel speed may produce so much slag that the pieces will be "welded" back together with slag.

NOTE: It is normal to have large amounts of slag formed on the scrap piece when bevel cutting.

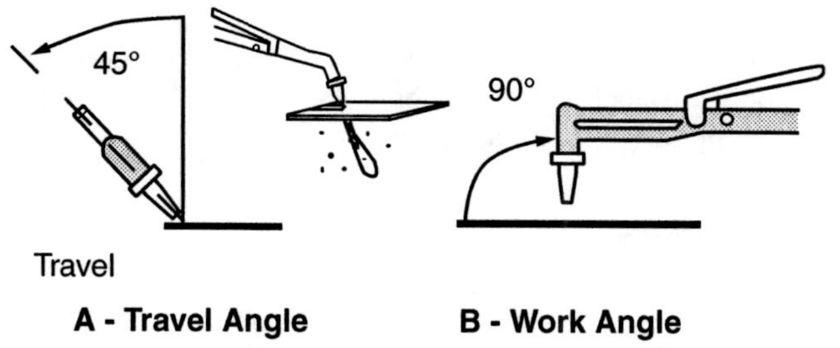

Fig. 11 — Hand Cutting of Thin Metal

Piercing of Metals

1. Position the torch so the work and travel angles are both 90° (Fig. 12).
2. Very slowly partially depress the oxygen cutting lever, and raise the tip enough to keep the preheats out of the slag as it builds up on the top of the surface. Keep ahead of the slag by moving in a 1/2 inch (12.7 mm) circle.

   An alternative method to moving in a circle is to hold the torch at a 10° to 20° angle. On metal 1/2 inch (12.7 mm) thick or less, it won't make much difference which method is used. However, on thicker metal, holding at much of an angle increases the depth of piercing and makes piercing more difficult. Use the circular method instead.

3. When the flame and slag reach the bottom side of the piece (Fig. 12-D), depress the oxygen cutting lever all the way and begin cutting by moving along the line of cut.

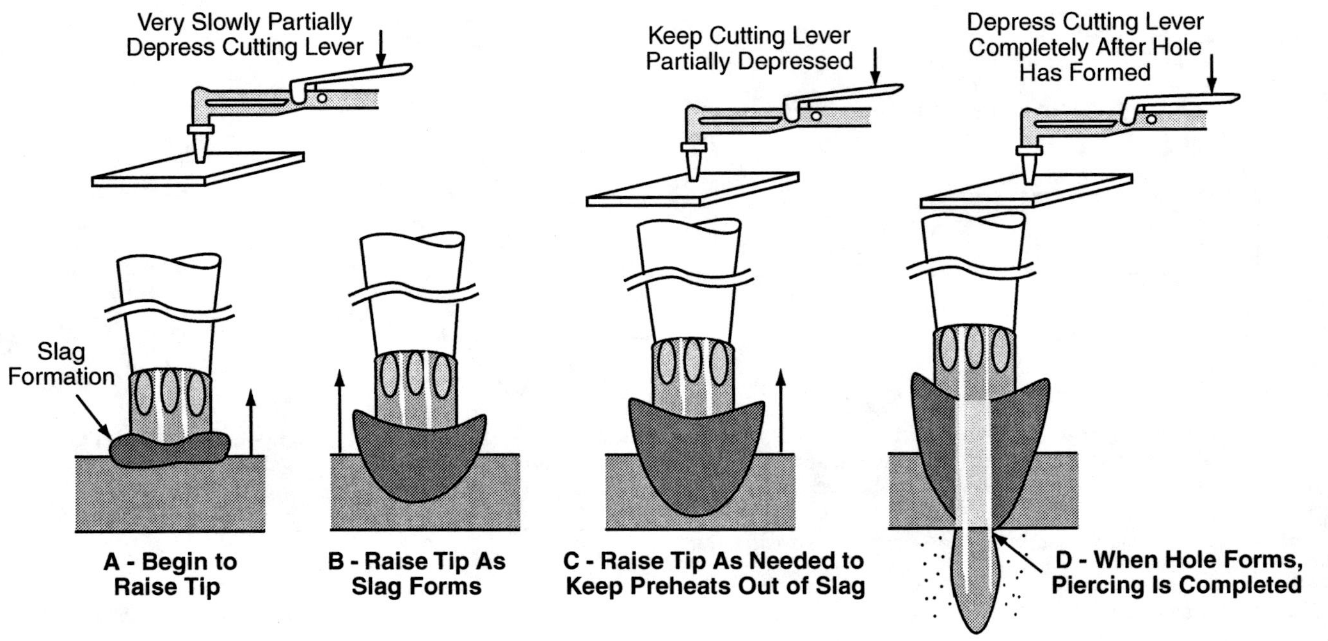

Fig. 12 — Piercing a Hole in Metal

Tips for Cutting Metals

1. It is easier to see the line of cut and watch it if body positioning is at the side of the cut rather than directly in front or back of the line of cut.
2. To reduce slag formation when cutting thin metal (less than 3/16 inch (4.8 mm), the travel angle (Fig. 11-A) can be tipped up to 45° in the opposite direction of travel.

Stopping Work

When cutting is to be stopped, extinguish the flame by turning the oxygen valve in a clockwise direction to close it, and then turning the acetylene valve in a clockwise direction to close it. With this procedure, there is not a “pop” when the flame goes out and you get a quick system check.

If the flame continues to burn with a blue color after the oxygen valve has been closed, it indicates there is an oxygen leak.

If the flame continues to burn with a yellow color when the acetylene valve is closed, it indicates the acetylene valve is leaking.

If the flame continues to burn with a blue color with both valves closed, both valves are leaking.

If any of these three conditions exist, the equipment should be taken out of use and sent in for repair.

When stopping work for more than a few minutes (as during lunch or overnight), close the cylinder valves and release all the pressure from the regulators and hoses by following these steps:

1. Close the oxygen cylinder by turning it in a clockwise direction.
2. Open the torch oxygen valve in a counterclockwise direction until the pressure has been released from the hose and regulator (both gauges* will read “0”).
3. Release the pressure on the adjusting screw by turning it in a counterclockwise direction until it turns freely.
4. Close the oxygen torch valve by turning it in a clockwise direction.
5. Close the acetylene cylinder by turning it in a clockwise direction.
6. Open the torch acetylene valve in a counterclockwise direction until the pressure has been released from the hose and regulator (both gauges* will read “0”).
7. Release the pressure on the adjusting screw by turning it in a counterclockwise direction until it turns freely.
8. Close the acetylene torch valve by turning it in a clockwise direction.

IMPORTANT: *Gauges that do not return to "0" should be replaced.

CAUTION: The acetylene and oxygen pressures should not be released simultaneously. Take the necessary precautions to prevent the release of oxygen and acetylene when they could create a fire or explosion hazard.

When acetylene is supplied from a manifold system that has a hydraulic flashback arrester, close the cylinder valves and release the pressure on the piping system using the release valve on the acetylene manifold. Close all work station valves and back out all work station and manifold regulator adjusting screws.

Safety caps should be placed on all cylinders if the regulators are removed for storage. When regulators are to be out of service for several weeks or longer, it is a good practice to turn in the pressure adjusting screw just enough to relieve the spring pressure on the valve seat. This aids in lengthening the life of the valve seat, especially when the seat is made of relatively soft material. Before putting the regulator into service again, the pressure-adjusting screw should be turned in a counterclockwise direction until the tension is released.

Safety When Stopping Work

Before leaving the work area for even a few minutes, follow these safety practices:

1. Mark all hot metal "HOT."
2. Check the surrounding work area for fires or smoldering combustibles that may have been overlooked when work was started.
3. Return to the area and check for smoldering combustibles (or fire) until it has been positively determined to be free of the danger from fire.

Acceptable Cuts

To be acceptable, cuts should require little or no cleaning before welding or the next manufacturing operation. The edges should not contain enough bevel from top to bottom to affect fit-up. Other dimensions should also be within manufacturing tolerances. See chart at the end of this chapter for acceptable cuts illustration.

A poor quality cut will need grinding to smooth the edges and/or have slag that must be removed before being acceptable for welding or other manufacturing operations. The cost of cleaning is less than the cost of cutting a new part.

Unacceptable Cuts

Cuts will be unacceptable when the cost of removing slag and of smoothing rough edges is greater than the cost of cutting a replacement piece. See Fig. 33 and Fig. 34 for illustrations of unacceptable cuts.

Other Flame Cutting Operations

Flame cutting is also used for stack cutting, lancing, gouging, and powder cutting.

Stack Cutting

Stack cutting is the cutting of several plates simultaneously while stacked together (Fig. 13). The advantage of stack cutting is increased productivity at lower unit cost. The resulting pieces will usually have less stressed edges with squarer surfaces, fewer burrs and less slag.

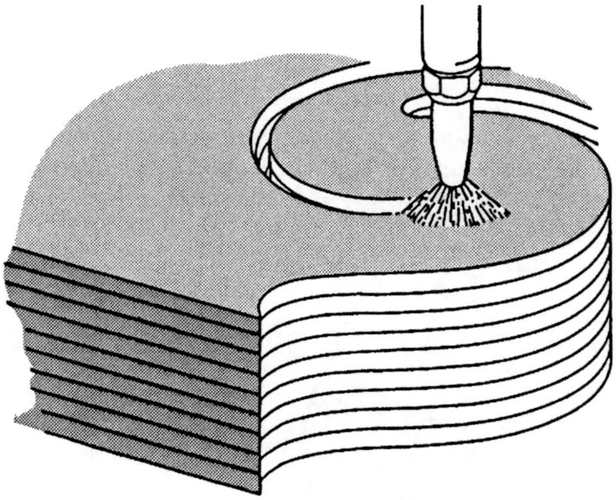

JDPX7031

Fig. 13 — Stack Cutting of Several Plates

In stack cutting, the pieces must form a solid slab. The surfaces must obviously be free of all foreign matter and be absolutely flat. Sufficient clamping devices are often required to eliminate all air gaps between the stacked plates.

Lancing

Lancing is used to cut a long or deep hole in a thick metal body. The lance is similar to a piece of steel pipe (3 or 19 mm) that is equipped with a valve to control the flow of oxygen (Fig. 14).

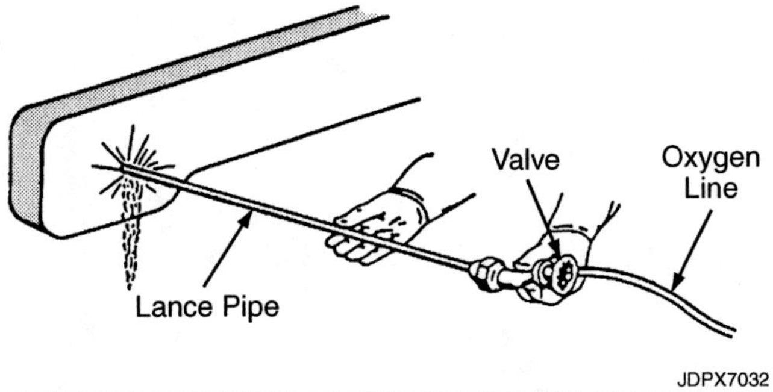

Fig. 14 — Lancing Is Used to Cut Deep Holes in Thick Metal Sections

Preheating is done with a welding or cutting torch or any other means for supplying heat. As soon as the preheating is completed, the lance pipe is brought over the area where the hole is to be pierced and the oxygen turned on. The lance pipe is rotated slightly to produce a hole larger than the diameter of the lance, permitting the slag to be blown out the hole.

Gouging

Gouging is another form of scarifying used to produce a fully controlled groove. The torch is held stationary until the correct preheat is reached, and the nozzle is then lowered and moved forward (Fig. 15).

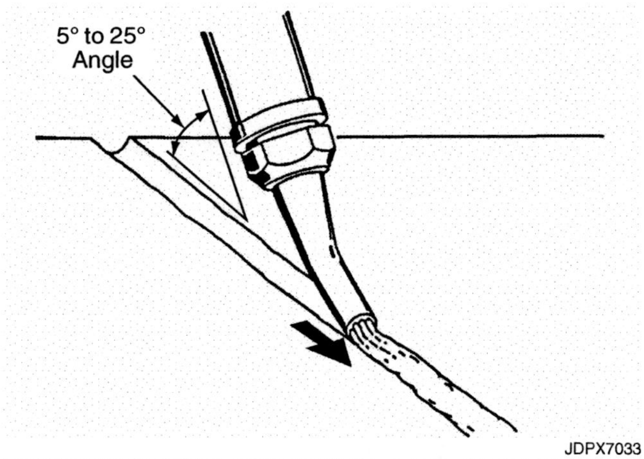

Fig. 15 — Gouging Is Used to Produce an Even Groove

Powder Cutting

The regular oxyfuel cutting process is ineffective for cutting metals that form refractory oxides such as aluminum, bronzes, stainless steel, cast iron, and high-nickel alloys.

One process of cutting these metals employs a special powder-cutting torch that is connected to a source of oxygen and fuel gas such as acetylene. An iron powder (some applications require the addition of aluminum powder) from the torch is directed on the line of cut before the cutting oxygen valve is opened. Plasma cutting is replacing this process in many applications (see later in this chapter).

Arc Cutting

Special cutting electrodes are available that will allow cutting metal with an arc welder. Coated electrodes such as E-6010 (used on DCEP) may also be used. If coated mild steel electrodes are used, soaking them in water 30 minutes before use will extend their life.

The diameter of the electrode will depend on the thickness of the metal to be cut and the amperage capacity of the power source. Table 3 provides guidelines for arc cutting.

CAUTION: Observe the duty cycle rating of the welding power source when doing arc cutting. This information is on the nameplate of the welder and in the owner's manual. Unless the power source is equipped with overload protection, failure to observe the duty cycle will ruin the power source by breaking down the insulation in the windings.

Cutting with the arc is possible because of the high temperature produced by the arc. Heat of the arc ranges from about 6500° to 10,000°F (3590° to 5540°C), while 2600°F (1430°C) will melt steel.

The limitation of the arc process for cutting is that it leaves a very rough, ragged edge. However, it is effective for cutting cast iron and steel for salvage purposes and in areas that are hard to reach.

Table 3 gives the approximate ampere setting for cutting.

Suggested Amperage Setting for Arc Cutting				
Metal Thickness		Electrode Diameter		Ampere Range
(Inch)	(mm)	(Inch)	(mm)	
Up to 1/8	(3.2)	3/32	(2.4)	75—100
Up to 1/8	(3.2)	1/8	(3.2)	125—140
Over 1/4	(6.4)	5/32	(4.0)	140—180

Table 3 — Amperage for Arc Cutting

In a cutting operation, the metal is placed in a flat position and the cut started at the bottom edge of the plate.

When the diameter of the electrode is larger than the thickness of the plate being cut, the electrode is simply moved in a straight line (Fig. 16).

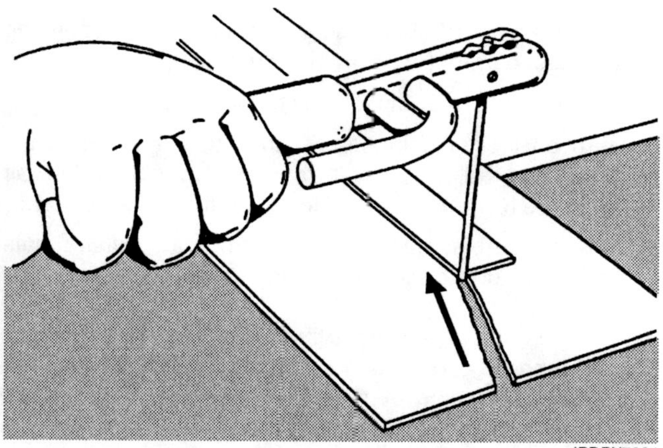

Fig. 16 — Cutting Thinner Plate With Shielded Metal Arc Welding (SMAW)

When the material is heavier than the electrode, a weaving motion is used to make the cut. The electrode is moved upward with a quick motion and then pushed downward as shown in Fig. 17. Flat stock over 1/8-inch (3.2 mm) thick is often easier to cut if placed in a vertical position. The cut is made from top to bottom as illustrated in Fig. 18.

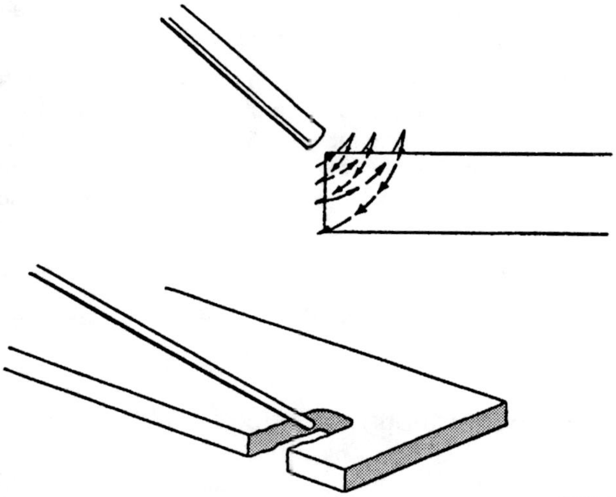

Fig. 17 — Cutting Heavier Plate With Shielded Metal Arc Welding (SMAW)

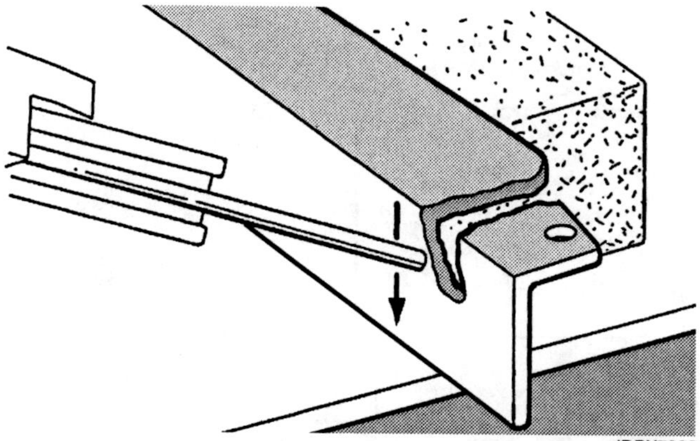

Fig. 18 — Arc Cutting Heavier Stock in a Vertical Position

Holes can be pierced by keeping the arc over the spot until the surface begins to melt. The arc is then brought down into the molten pool of metal and the electrode moved in a circular motion (Fig. 21)

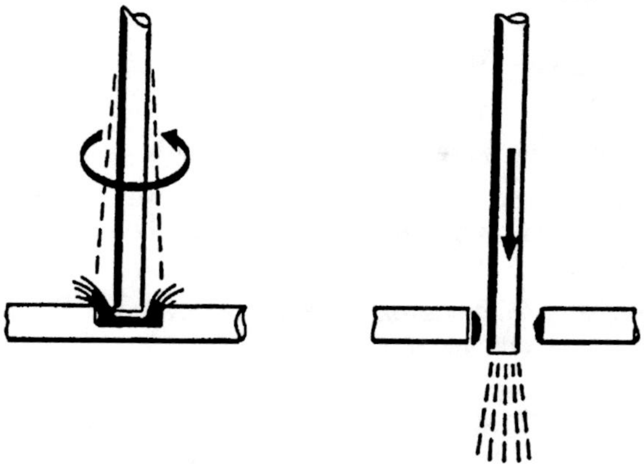

Fig. 19 — Piercing Holes With Arc

Piercing holes in metal over 1/4 inch (6.4 mm) thick is best done with the plate in vertical position. This permits the metal to run out of the hole.

Air Carbon Arc and Plasma Arc Cutting and Gouging

Two other types of arc cutting are used today:

- Air Carbon Arc
- Plasma Arc

An air carbon arc torch and a plasma torch are useful tools for preparation of a weld joint, backgouging, repairing a crack, removing defective welds, removing welds when dismantling a welded assembly, and for cutting metals that cannot be readily cut with another process (Fig. 20).

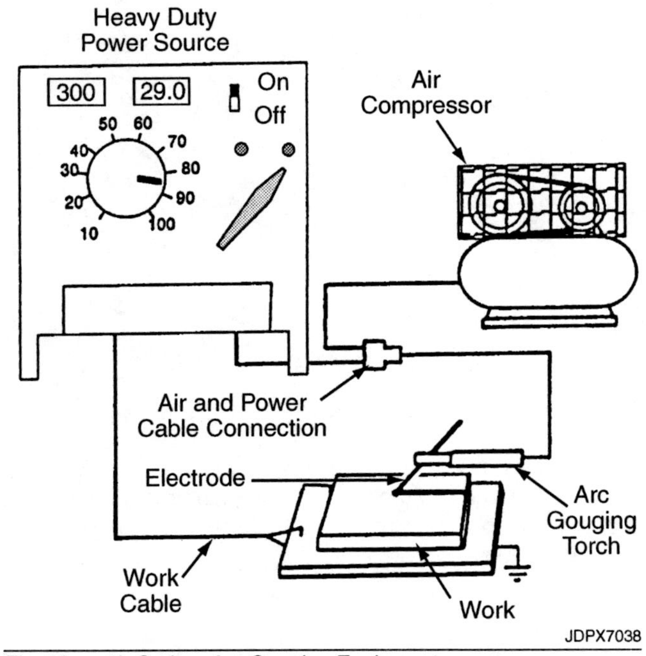

Fig. 20 — Air Carbon Arc Gouging Equipment

Setting Up and Using Air Carbon Arc Gouging Equipment

1. Turn off the power to the electrical outlet.
2. Turn off the on-off switch on the power source. (On engine-driven power sources, turn off the engine if it is running.)
3. Bolt the electrode cable from the power source to the connector on the arc torch.

IMPORTANT: Attaching the jaws of an electrode holder to the arc torch connection is a safety hazard and does not provide an acceptable electrical connection.

NOTE: *It will be necessary to replace the electrode holder (or make a separate cable) with a bolt-on lug to attach to the arc torch connector.*

4. Connect an air hose from the air sources to the arc torch air connection, and cover both connections with the boot cover on the arc torch.

IMPORTANT: If the air and electrode cable connections are not completely covered (insulated), they will arc against anything they touch that is part of the welding circuit. This will cause damage to the connections and to objects that make contact with the connection.

5. Attach the work clamp and cable plugs (or lugs) to the welding cable according to the equipment manufacturer's instructions.
6. Attach the cables to the designated locations on the power source to obtain the proper polarity for operating on direct current (or alternating current).
7. Attach the work clamp directly to the workpiece.
8. Turn on the air compressor, open the air valve at the tank, and adjust the air regulator to obtain the air pressure recommended by the arc torch manufacturer.
9. Select a carbon electrode that will operate within a safe amperage duty cycle range of the power source.

NOTE: *The carbon electrode manufacturer will provide an amperage operating range for each electrode. The power source manufacturer will provide duty cycle information at different amperage outputs. Match the amperage requirement of the carbon electrode to the duty cycle of the power source. For example: a 3/16-inch (4.8 mm) carbon will require from 200 to 250 amps to operate properly. This will require the power source to have at least a 50 percent duty cycle at 200 amps when actually arc gouging 5 minutes out of 10 minutes. More gouging time per 10-minute period will require a higher duty cycle rating of the power source.*

10. Before starting work, make sure all safety precautions have been taken—those given earlier for arc welding and the additional ones given for arc gouging.
11. Insert a carbon electrode in the electrode holder so the air blows under the end of the electrode (Fig. 21). Adjust the electrode so it extends 4 to 6 inches (102 to 152 mm) past the end of the electrode holder. Reposition the electrode when it has been consumed to about 2 inches (51 mm).
12. Turn on the power source.
13. Turn on the air by using the on-off valve on the torch and begin gouging.

Using the Air Carbon Arc Torch

The torch is held so the electrode slopes back from the direction of travel forming a 45° angle to the workpiece. The torch should not be held more than 4 inches (102 mm) from the work, and the orifices on the nozzle should be on the bottom side of the carbon so the air blasts are behind the electrode (Fig. 21). The arc should be kept as short as possible without touching the work and moved fast enough to keep up with the metal removal. The depth of the groove is controlled by the electrode angle and travel speed. For a narrow deep groove, a steep electrode angle and slow speed is used. For shallow grooves, use a flat electrode angle and a fast speed. It is very important that the air valve be turned on before striking an arc.

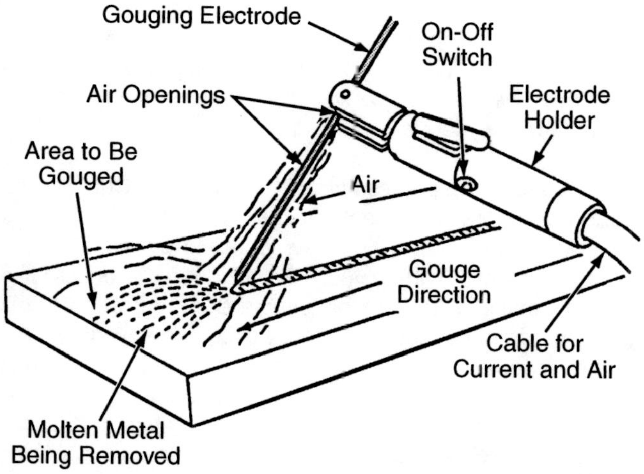

Fig. 21 — The Air Carbon Arc Processes

A fillet weld has been completely removed when the line of intersection of the two plates becomes visible (Fig. 22). When repairing a part that is cracked or when backgouging the back side of a groove weld to obtain 100 percent penetration, the proper depth has been reached when the bottom of the groove becomes smooth, indicating that there is no line of separation.

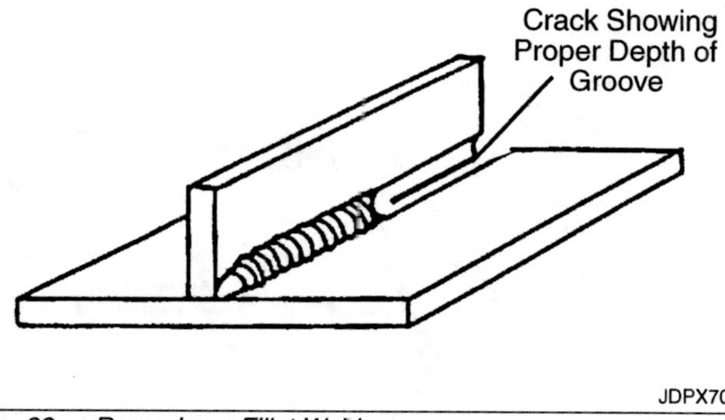

Fig. 22 — Removing a Fillet Weld

When the proper technique has been learned, the air carbon arc process produces little, if any, slag on the finished cut (Fig. 23). However, the slag that remains is rich in carbon and must be removed by either chipping or grinding before any welding is done. If the slag is not completely removed before welding, the resulting weld will be of very low quality.

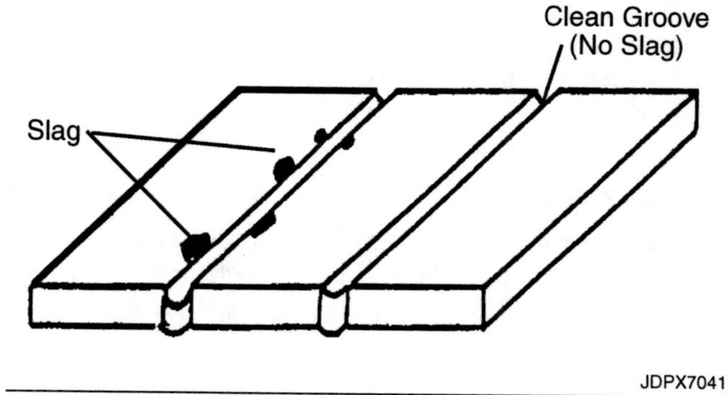

Fig. 23 — Back Gouging

A grinder should be used to check for slag that is in or attached to the sides of the completed groove. When grinding an area that was cut with the air carbon arc process, an increase in the volume of sparks being produced indicates an area of slag, while no change in the spark volume indicates that the area does not contain slag deposits.

Electrodes for manually operated torches are available in 12-inch (305 mm) lengths for either AC or DCEP operation. Either plain or copperclad electrodes can be used. Copperclad electrodes last longer, carry higher current, and produce more uniform cuts.

An adequate air compressor should be available. Air pressure requirements range from 60 to 100 psi (415 to 690 kPa) and air flow rates range from 8 to 30 cfm (0.23 to 0.85 m^3 per min). This means that usage will have to be intermittent if less than a 3 hp 2.2 kW compressor is available.

Before using this process, check the duty cycle for the power supply, the current requirements for the electrodes to be used, and the delivery capabilities of the air compressor that will be used. A heavy-duty AC or DC welding machine would be used for a power source. The carbon-graphite electrodes require high amperages that can ruin a light-duty power source in a short time if it is operated above its rating. Unsatisfactory results will be obtained by operating at amperages below those recommended by the manufacturer.

Stopping Work and Shutting Down Air Carbon Arc Gouging Equipment

1. Turn on the air by using the on-off valve on the torch (that is, after the carbon electrode has cooled to a safe handling temperature). Don't lay down the electrode holder—hold on to it until the power source has been turned off.
2. Turn off the power source and air valve at the compressor.
3. Remove the carbon electrode from the electrode holder and set it in a safe place on a heat-resistant surface.

4. Disconnect the air and electrode cable connections from the arc torch and disconnect the work cable from the work.
5. Wrap up the air hose and welding cables. Secure the arc torch, welding cables, and hose in a safe place as necessary in a cabinet or tool chest.
6. Mark all hot metal "HOT."
7. Return all reusable electrodes to the storage container (after they have cooled to room temperature).
8. Dispose of all slag, used electrodes, and scrap metal.
9. Turn off primary power to the power source and air compressor.
10. Check the area for fires or smoldering combustibles for at least one hour after stopping work.

Arc Gouging Safety

The same rules apply to arc gouging that apply to any electrical welding or cutting process. Additional rules include the following:

- Remove combustible materials from the work area and take adequate precautions necessary to prevent a fire or explosion.
- Use appropriate ear protective devices to prevent sparks or hot metal from getting into the ears and to reduce noise levels.
- Use adequate ventilation or a breathing device that provides protection from the fumes.
- Be sure the welding lens is the proper shade and offers adequate protection. The lens selected should be rated for the maximum amperage drawn by the carbon used.
- Clothing should provide adequate protection from the intense arc rays, the heat, and the hot sparks.
- To prevent arc damage to the work, be sure the electrical connection between the torch and power source are covered by the insulating boot.

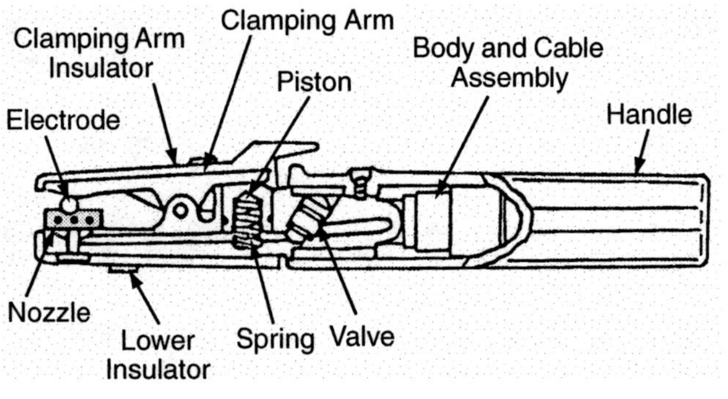

Fig. 24 — Typical Air Arc Gouging Torch

Plasma Arc Cutting (PAC)

Plasma arc cutting is widely used in factories for both manual and automated cutting applications. It is one of the most effective processes available for cutting both ferrous and nonferrous metals.

Advantages of the process include no preheat. Cuts can be started instantly. The heat-affected zone from cutting is minimal. High travel speeds are possible because it is a thermal process rather than an oxidation process like oxygen cutting. PAC can be used for cutting almost any material that will conduct electricity and, in many situations, at faster speeds than any other process. A plasma torch has the ability to cut metals such as aluminum, copper, brass, cast iron, carbon steel, nickel, stainless steel, refractory metal, and other ferrous and nonferrous metals.

Plasma Arc Cutting (PAC) Processes

The basic equipment needed for plasma cutting is illustrated in Fig. 25 and includes a special direct-current power source, a torch, a compressed air or gas supply, and corresponding connecting cables and hoses. With power source output and torch capacity ranging from 20 amps up to several hundred amps, it is possible to cut metals that range from 26 gauge (0.46 mm) up to 6 inches (152 mm) in thickness.

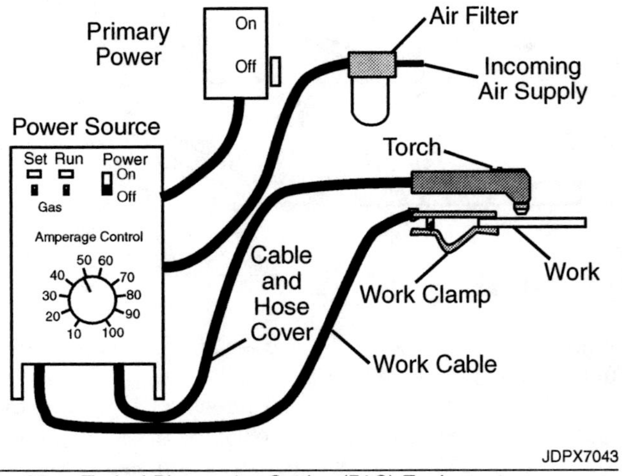

Fig. 25 — Typical Plasma Arc Cutting (PAC) Equipment

The plasma cutting process (Fig. 26) generates a high temperature, high velocity jet of electrically charged gas particles that can be used to cut both ferrous and nonferrous metals. The temperature of the jet or plasma stream can reach up to 50,000°F (27,800°C) and achieve a velocity close to supersonic speeds.

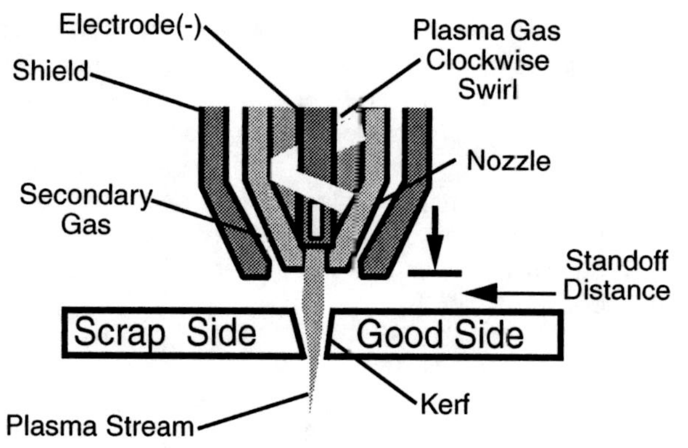

Fig. 26 — Plasma Arc Cutting (PAC) Process

Cutting speeds with the plasma process are much faster compared to the oxyfuel processes on carbon steel when the thickness is less than 3/4 inch (19 mm). These higher cutting speeds reduce or sometimes even eliminate the distortion (warping) that is caused by cutting carbon steel in an oxyfuel process. The high intensity of the plasma stream melts the metal and makes a kerf so fast that the area around the cutting doesn't have a chance to absorb the heat that causes the distortion associated with the oxyfuel processes.

The edges of plasma cut metals are not perfectly square. As the plasma stream swirls from the torch and strikes the metal surface, some of the intensity is lost as it travels to the bottom of the kerf. This causes the top of the kerf to be wider than the bottom, thus creating a slight bevel. Also, if the plasma gas is swirling in a clockwise direction the "bad" side of the cut will be on the left side of the kerf in relation to the torch travel direction (Fig. 26). The corresponding bevel can range from 1° to 3° on the "good" edge and 3° to 8° on the "bad" edge. This bevel must be taken into consideration when planning the tolerance and fit-up of the finished parts. As illustrated in (Fig. 26), when possible, the direction of swirl should be planned so that the "bad" edge becomes the scrap piece.

When air or pure nitrogen is used as the plasma gas, the edges of the cut may combine with the nitrogen to form nitrides. These nitrides can cause porosity during welding. If porosity is a problem when welding plasma cut metals, the edges will need to be ground to remove the thin layer of nitrides, and the dross (a slaglike material mostly on the bottom edges) will need to be removed from the edges.

There are five variations within the plasma arc cutting process. Within each process, cutting amperage, travel speed, and standoff are the main factors that affect the amount of bevel. For example, on a given thickness at given amperage, a low cutting speed with a close standoff (the distance from the nozzle to the work) will create less of a bevel than a high speed with a high standoff.

Let's review each of the five plasma arc cutting processes:

- Conventional plasma arc cutting (PAC) uses a single gas and has the most basic torch.
- Gas-shielded plasma arc cutting (PAC) can be of a single or multi-gas variety. The plasma gas does the cutting, and the shielding or secondary (the other gas) provides a shield around the arc. When a single gas such as air or nitrogen is used, the gas flow is divided so part of the flow becomes the plasma and part becomes the shield. The secondary gas helps blow metal out of the kerf, reduces metal spatter on the torch, cools the torch, and improves cut quality.
- Water-shielded plasma arc cutting (PAC) is actually a variation within the gas shielded process which substitutes water for the secondary gas. Additional benefits from the water shielding include a reduction in top edge rounding, smoke, and fumes, and a "cut edge" surface that is clean and shiny. Plasma cutting with mechanized equipment is the only time a system is set up with water shielding.
- Water-injected plasma arc cutting (PAC) channels water around the plasma stream before it leaves the torch. This more "focused" arc improves cut quality and improves cutting speed. Water-injected plasma arc cutting is also limited to use with mechanized equipment.
- High-tolerance plasma arc cutting (PAC) uses a smaller nozzle orifice and a higher flow of spinning gas to obtain cut quality on metals ranging from 20 gauge (0.91 mm) to 3/8 inch (9.5 mm) that are nearly as good as cuts obtained from laser cutting and at a much lower cost. High-tolerance plasma arc cutting is also limited to use with mechanized equipment.

Plasma Arc Cutting (PAC) Equipment

The power source has a high-voltage (90 to over 200 volts) direct current output with "amperage" being the adjustment controlling the thickness of the metal that can be cut. The power source is considered to be of the constant-current type and may have an output amperage control that is variably adjustable or fixed at a few settings such as "low" (20 amps)," "medium" (35 amps), and "high" (50 amps).

A general rule for figuring cutting capacity is that it takes 10 amps to cut 0.100 inch (2.5 mm). Piercing capacity should be no more than 3/4 the cutting capacity. For example, a 50-amp power source would be capable of cutting 1/2 inch (12.5 mm) metal and piercing up to but not over 3/8 inch (9.5 mm) metal.

As illustrated in (Fig. 27), the physical size cannot be used to judge the cutting capacity of a power source. New technology has resulted in the manufacture of some compact high-cutting capacity power sources. It is possible to obtain plasma arc cutting equipment having a cutting capacity of 3/8 inch (9.6 mm) that can easily be hand carried to a job site.

JDPX7045

Fig. 27 — Portable Plasma Arc Cutting (PAC) Power

The torch is where high voltage and high amperage turn a gas into the plasma stream that does the cutting. The options include torches that are hand held (usually up to 400 amps) or attached to mechanized equipment, air-cooled or water-cooled. In each case, a typical torch (Fig. 28) consists of a torch body, where the incoming power and gas connect, and a group of parts that are responsible for the formation of the plasma stream and providing passageways for cooling on water-cooled torches.

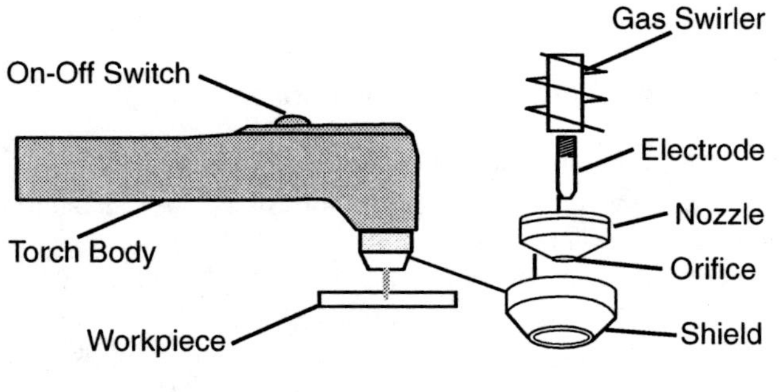

JDPX7046

Fig. 28 — Plasma Arc Cutting (PAC) Torch and Parts

The torch is designed with special insulation to prevent the high-voltage electrical current from reaching the person using the equipment. A protective cover encloses the power cable, control wires, and hoses that connect the power source to the torch. The equipment should not be abused and should always be checked for damage before starting work.

As illustrated in (Fig. 28), the torch parts that are necessary to form the plasma stream include an electrode, nozzle (sometimes called the tip), gas swirler, and shield.

Electrode—the electrode is negative and carries the current that forms the arc. A tungsten electrode is used when the plasma gas is nitrogen or argon/hydrogen. The tungsten can be reshaped and reused when it no longer functions properly. A copper electrode with a special metal insert called hafnium is used when the plasma gas is air or oxygen. Hafnium is used because it lasts much longer than tungsten when exposed to air or oxygen. When an electrode with an insert shows 1/16 inch (1.6 mm) wear (Fig. 29), it should be replaced.

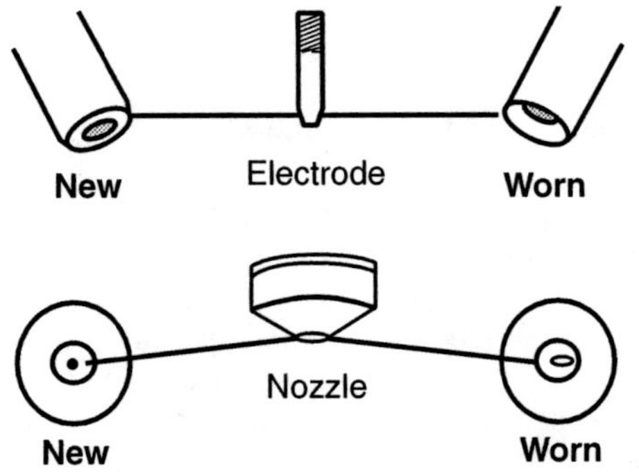

JDPX7047

Fig. 29 — New and Worn Plasma Arc Torch Parts

Nozzle—the nozzle (or tip as it is sometimes called) is a copper part (or copper part that has been plated) that contains a small hole or orifice which focuses the plasma stream as it leaves the end of the torch. The diameter of the opening (the orifice size) in the nozzle determines the amperage and the corresponding cutting capacity for the nozzle. The larger the opening in the orifice, the more amperage the nozzle is capable of handling and the thicker the material that can be cut. The orifice can range in size from 0.040 inch to 0.250 inch (1 to 6.4 mm) and should produce a kerf that is 1.5 to 2 times the diameter of the orifice. Operating at amperages slightly (about 10 percent) under the capacity of the nozzle will reduce gouging (Fig. 29) of the nozzle and increase part life. The direction of the swirl can be in a clockwise or counterclockwise direction.

Gas Swirler—the gas swirler makes the gas spin in a circle like a whirlwind or tornado. This spinning action is necessary to center the arc on the electrode and to focus the plasma stream as it passes through the orifice.

Plasma Arc Cutting (PAC) Safety

Follow the manufacturer's recommendations for setting and operating plasma cutting equipment. All of the hazards present when electric welding and oxyfuel cutting exist when plasma cutting. Before beginning to plasma cut, make sure everyone in the area has the proper eye protection (this includes a welding helmet with a lens dark enough for the cutting amperage) and protective clothing. Be sure the area is free of combustible materials, and be sure there is adequate ventilation to clear the fumes and gases produced during cutting. Take the necessary precautions to prevent sparks and hot metal from damaging the cover, cables, and hose going to the torch.

Setting Up and Using Plasma Arc Cutting (PAC) Equipment

NOTE: *Follow the manufacturer's recommendations for setting up and operating plasma cutting equipment with regard to the following:*

1. Power source connection to the primary power
2. Torch, air (or plasma gas), and work clamp connections
3. Air (or plasma gas) requirements
4. Torch assembly and setup
5. Power source amperage and gas setting
6. Torch operations

Plasma Arc Cutting Procedures

The type and thickness of metal will determine the amperage setting and corresponding nozzle and electrode size, and air (and/or gas) pressures. This information is in the operating instructions with the plasma cutting equipment.

Torch positioning and travel speed are the key variables to control for successful cutting, once everything is set up properly. (Setup includes attaching the work lead to the workpiece, turning on the power source, and making the necessary connections to an air supply.) A protractor should be used for judging and measuring the proper work angle and travel angle. When cutting bevels, the work angle becomes the bevel angle (Fig. 31-A and Fig. 31-B).

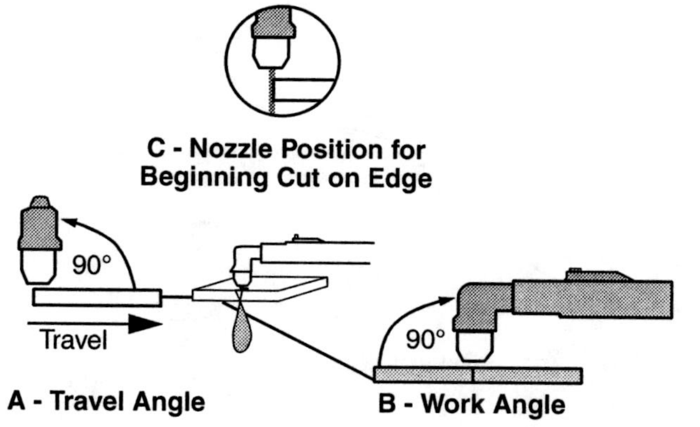

Fig. 30 — Making Square Cuts With Plasma Arc Cutting (PAC)

When possible, plan to use a straightedge (such as a piece of angle iron or bar stock) or pattern to guide the torch for making plasma cuts. The closeness of the torch to the work makes it more difficult to see and to follow the line of cut when plasma cutting. Also, the higher travel speeds possible when plasma cutting make it difficult to keep on the line when freehand cutting.

Steps for Cutting and Piercing

Making a Square Cut

1. The torch should be positioned so the work and travel angles are both 90° (Fig. 30).
2. The nozzle should be positioned so that most but not all of the plasma stream (Fig. 33) is over the piece to be cut. The standoff distance should be 1/16 inch (1.6 mm) for metal 1/8 inch (3.2 mm) thick or less, and 1/8 inch (3.2 mm) for metal over 1/8 inch (3.2 mm) through 1/2 inch (6.4 mm) thick. Holding the nozzle too high or too far from the edge "stretches" the arc. This makes it harder to start a cut and greatly reduces the life of the nozzle and electrode.
3. Begin cutting by depressing the on-off switch on the torch and beginning to move. Move at a speed that produces the least amount of dross. The travel speed will typically be much faster than when oxyfuel cutting. Traveling too fast will cause the cut not to reach the entire depth of the piece.

Making a Bevel Cut

NOTE: *Bevel angle refers to the amount of metal removed during cutting (Fig. 31) and not the angle of the edge of the piece that is left after cutting.*

1. Torch setup should be based on the depth of the bevel (Fig. 31) and not on the metal thickness. The depth of cut will be greater than the metal thickness, and the torch setup will need to be increased accordingly.

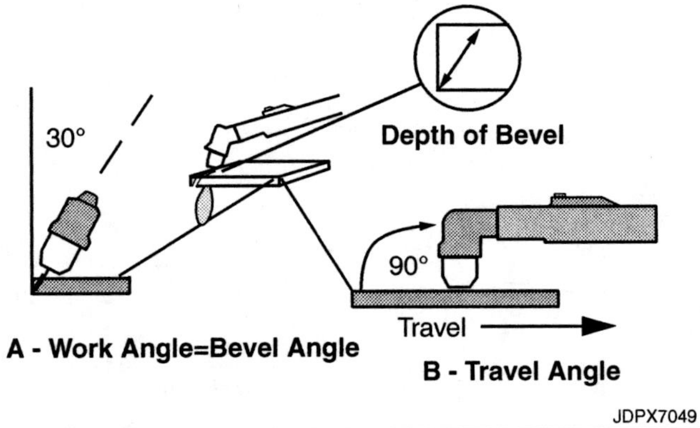

Fig. 31 — Making Bevel Cuts

2. Position the torch so the work angle is the same as the bevel angle.
3. The nozzle should be positioned so that most but not all of the orifice is over the piece to be cut. The standoff should be as close as possible because when the torch is at an angle it may not be possible to get as close as recommended for square cutting.
4. Begin cutting by depressing the on-off switch on the torch and beginning to move. Move at a speed that produces the least amount of dross on the "good" piece. It is normal for there to be large amounts of dross formed on the "scrap" piece when bevel cutting. Traveling too fast will cause intermittent gouges in the edge or the cut will be lost—the cut won't be all the way to the bottom of the metal. Traveling too slowly will cause a wide kerf and excess dross on the bottom surface. A very slow travel speed may produce so much dross the pieces will be "welded" back together with dross.

Piercing

The piercing capacity for a power source is normally no more than 3/4 the cutting capacity. So, before trying to pierce cut a piece of metal, read the operating instructions to determine the thickness that can be pierced with a given nozzle and amperage setting. Not following the proper procedure or trying to pierce metal thicknesses greater than the equipment manufacturer recommends can cause immediate damage to the nozzle and tip.

To pierce cut, follow these steps:

1. Position the torch so the work angle is about 45° and travel angle is 90° (Fig. 32-A).

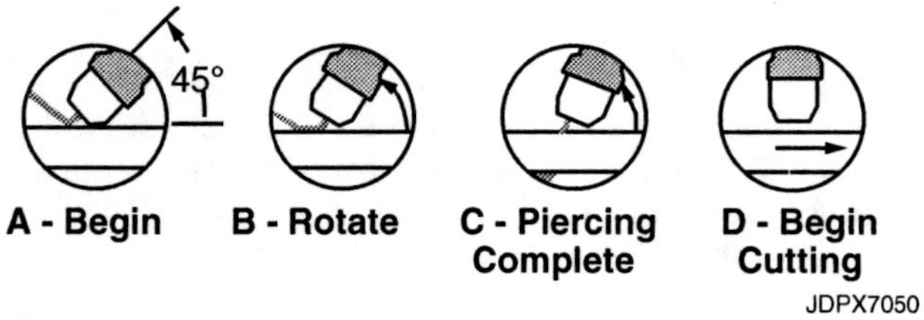

Fig. 32 — Piercing Metal

2. Depress the on-off switch and increase the work angle by rotating the torch (Fig. 32-B and C), and increase to 90° when the cut goes all the way through the piece (Fig. 32-D). Once a hole is pierced, move to the line of cut.

Stopping Work With Plasma Arc Cutting (PAC)

When the cutting project has been completed or the work period has ended, follow these steps for stopping work:

1. Turn off the air supply and release the pressure from the hose and torch.
2. Next, turn off the power source.
3. Secure the torch and connecting cable assembly at a location that protects them from damage from other activities in the work area.
4. Mark all hot metal "HOT."
5. Check the surrounding work area for fires or smoldering combustibles that may have been overlooked when work was started.
6. Return to the area and check for smoldering combustibles (or fire) until you are sure the area is free from the danger from fire.

Acceptable Plasma Arc Cuts (PAC)

To be acceptable, plasma cuts should require little or no cleaning before welding or before the next manufacturing operation. The edges should not contain enough bevel from top to bottom to affect fit-up. Other dimensions should be within manufacturing tolerances. A poor-quality cut will need grinding to smooth the edges and/or have dross removed before being acceptable for welding or other operations. The cost of cleaning is less than the cost of cutting a new part.

Unacceptable Plasma Arc Cuts (PAC)

Cuts will be unacceptable when the cost of removing dross and smoothing rough edges is greater than the cost of cutting a replacement piece.

Fig. 33 and Fig. 34 show examples of poor-quality cuts and gives corrective actions.

Recognizing Acceptable Oxyfuel Cuts (OFC) and Troubleshooting of Poor-Quality Cuts

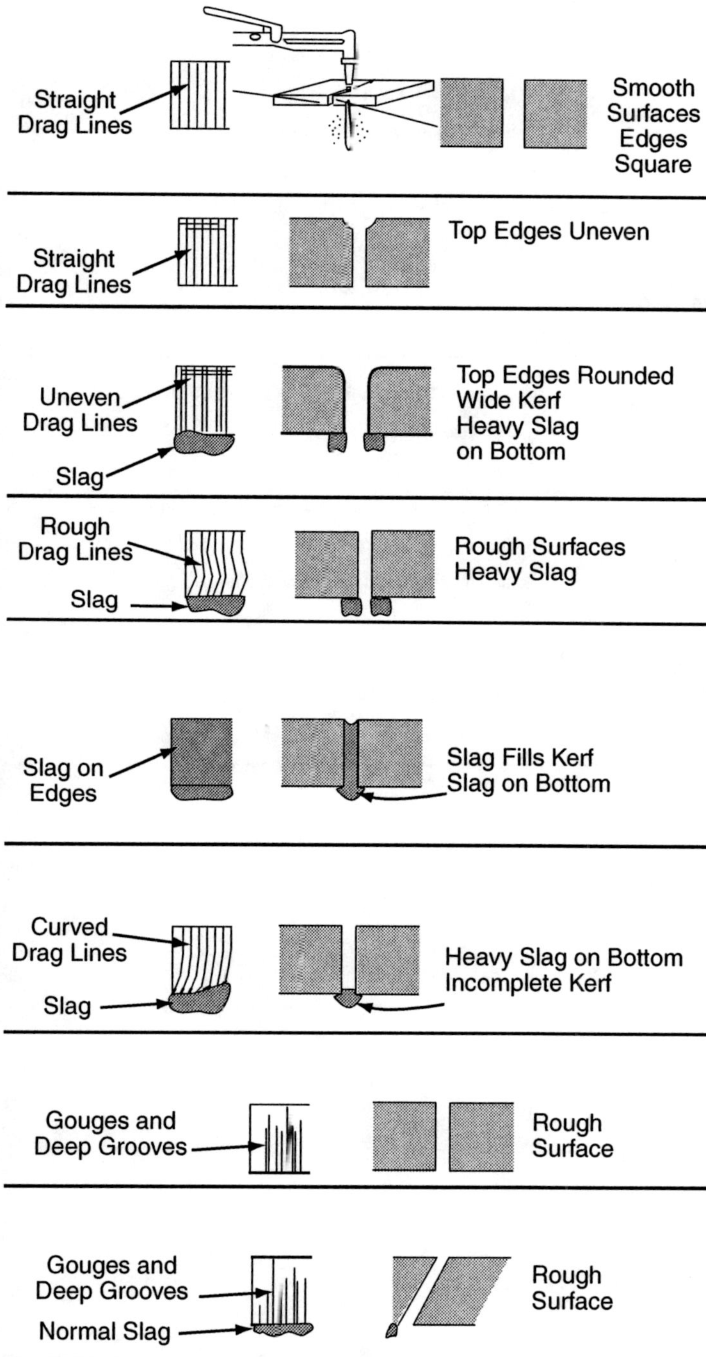

Problem(s) — **Corrective Action(s)**

A- An Acceptable Cut
1. Edges square
2. Smooth surfaces (No Grinding Required)
3. No slag
4. Kerf 3/32" to 1/8" (2.4 mm to 3.2 mm) wide
5. Drag lines straight

B- Top Edges Uneven
1. Preheats touching top edges — raise torch
2. Slow travel speed — increase travel speed

C- Top Edges Rounded and Slag on Bottom
1. Dirty or bad tip — clean or replace tip
2. Slow travel speed — increase travel speed
3. Flame too hot — use smaller tip
4. Flame too far away — hold proper preheat distance

D- Heavy Slag on Bottom
1. Check cutting orifice — clean or replace tip
2. Slow or uneven speed — travel at proper speed

E- Heavy Slag Fills Kerf
1. Slow travel speed — travel at proper speed
2. Flame too far away — hold proper preheat distance
3. Oxygen pressure low — increase oxygen pressure
4. Check cutting orifice — clean or replace tip
5. Cutting lever not fully depressed — press lever completely

F- Incomplete Kerf
1. Fast travel speed — travel at proper speed
2. Oxygen pressure low — increase oxygen pressure
3. Check cutting orifice — clean or replace tip
4. Cutting lever not fully depressed — press lever completely

G- Rough Edges with Gouges and Grooves (Square Cut)
1. Inadequate preheat flame — use larger tip
2. Inadequate preheating — decrease travel speed
3. Uneven travel speed — travel at proper speed

H- Rough Edges With Gouges and Grooves on (Bevel Cut)
1. Inadequate preheat flame — use larger tip
2. Inadequate preheating — decrease travel speed
3. Uneven travel speed — travel at proper speed

JDPX7051

Fig. 33 — Recognizing Acceptable Oxyfuel Cuts (OFC) and Troubleshooting of Poor-Quality Cuts

Recognizing Acceptable Plasma Arc Cuts (PAC) and Troubleshooting of Poor-Quality Cuts

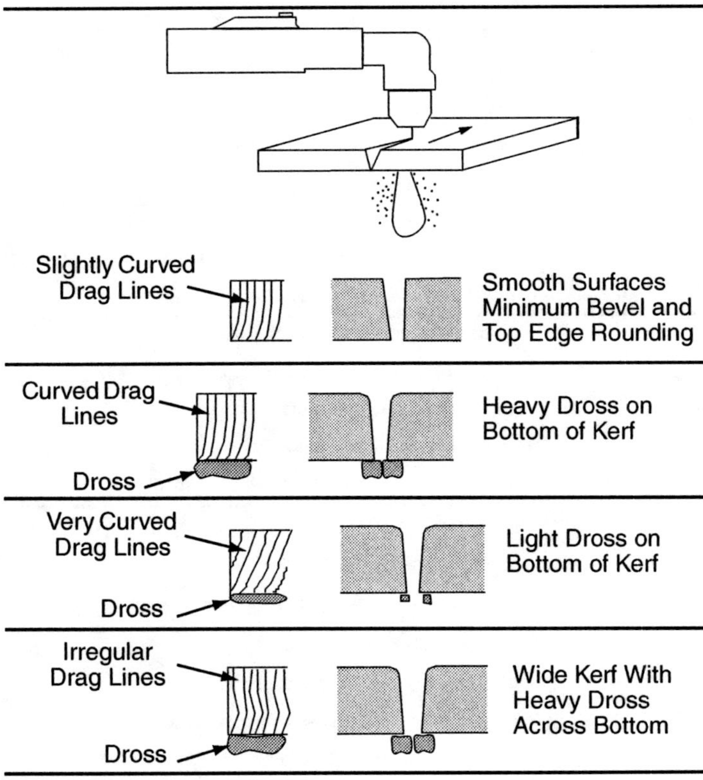

Problem(s) — **Corrective Action(s)**

A- An Acceptable Cut

1. Minimum beveling of edges
2. Minimum top edge rounding
3. Smooth surfaces (no grinding required)
4. No dross
5. Kerf 1-1/2 to 2 times orifice diameter

B- Low-Speed Dross on Bottom of Kerf (Easy to Remove)

1. Increase travel speed
2. Use smaller nozzle and a lower amperage

C- High-Speed Dross on Bottom of Kerf (Hard to Remove)

1. Decrease travel speed
2. Use a larger nozzle and a higher amperage

D- Wide Kerf and Heavy Dross on Bottom

1. Check nozzle orifice — Clean or replace
2. Check standoff — Too far away

JDPX7052

Fig. 34 — Recognizing Acceptable Plasma Arc Cuts (PAC) and Troubleshooting of Poor-Quality Cuts

Test Yourself

Questions

1. What are three processes that use heat for cutting metals?
2. How does cutting with the SMAW process compare to the processes in Question 1?
3. Break these cutting processes into two groups and explain the difference between them.
4. What is the difference between cutting and gouging?
5. How can you tell the difference between a one-piece acetylene cutting tip and a one-piece fuel gas cutting tip?
6. Give the function of the orifices at the "outlet" end of a cutting tip.
7. What determines the size of the orifice needed for a cutting project?
8. What can happen if a cutting flame or hot metal comes in contact with a cement surface like a floor?
9. List the three types of oxyacetylene flames and tell how to recognize the difference between them. Also tell which one should be used for cutting.
10. How can a clean cutting tip be identified without making a cut?
11. What is the proper distance to hold the cutting flame from the workpiece?
12. How should the cutting lever be operated when piercing a hole?
13. How can the proper depth of the groove needed for removing a fillet weld be recognized when using an arc air torch?
14. How can the proper depth of the groove needed for back gouging a groove weld be recognized when using an arc air torch?
15. What is the basic equipment needed for plasma arc cutting?
16. Give two differences between plasma arc cutting and oxyfuel cutting metal under 3/4 inch (19 mm).
17. How can the nitrides produced on plasma cut metal affect welding?
18. What is the rule for figuring the cutting capacity of a plasma power source?
19. List two types of electrodes used for plasma arc cutting and give the type of gas each should be used with.
20. Once the power source and torch have been set properly, what are the two key variables that must be controlled properly for successful cutting?
21. What is dross?

Laser Welding, Cutting, and Drilling

7

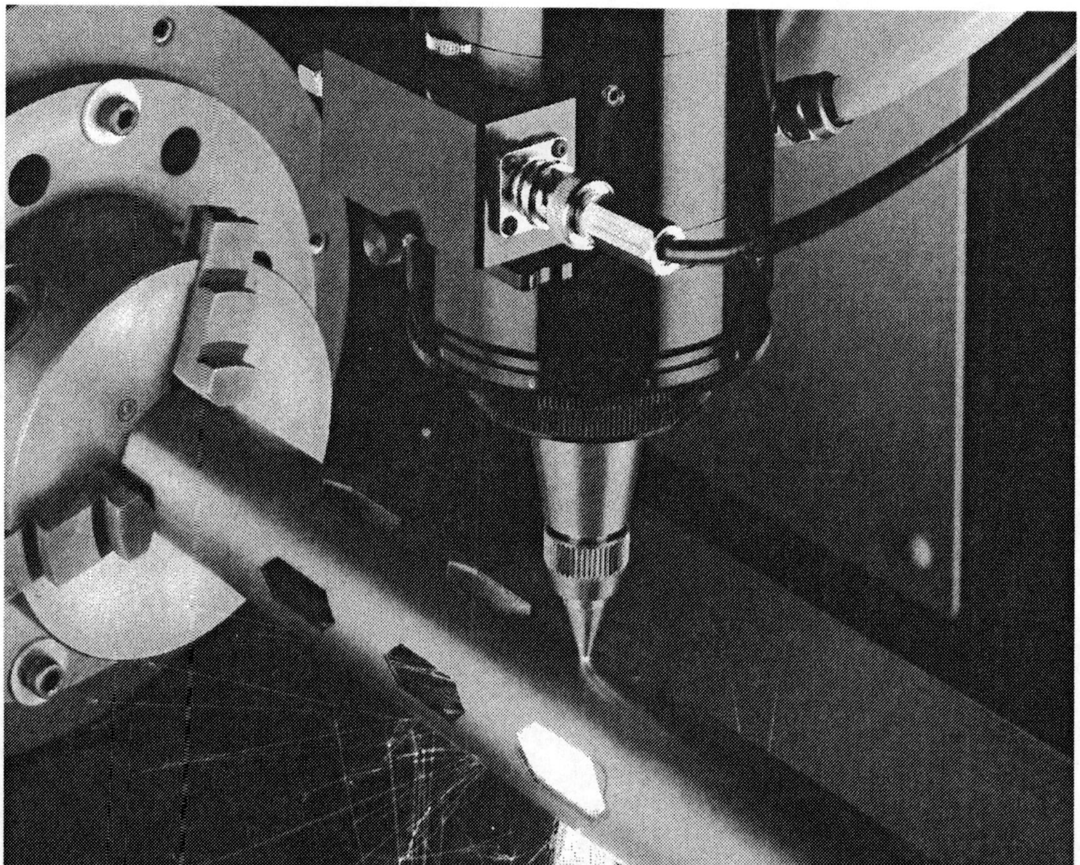

JDPX7013

Fig. 1 — Laser Cutting of a Metal Tube

Introduction

A laser is a device that generates a very concentrated beam of light. Applications that use the laser beam touch our life on a daily basis and range from a laser printer for a computer to a laser scalpel used in a hospital surgery room.

Welding and cutting are included in this range of laser applications. A tubing-cutting operation is illustrated in Fig. 1.

LASER is an acronym for Light Amplification by Stimulated Emission of Radiation. The first laser beam was produced in 1960 using a ruby crystal rod.

The laser is a very helpful tool and it is more common than one might think. A high-speed bar code reader at a checkout counters is but one example. A laser light show at a music concert or other attraction is another use. Public speakers use small pocket lasers for enhancing their presentations. On construction sites, lasers are used to measure distances or to aid in alignment of such things as drop ceilings. Lasers can also be used for marking material (scribing) or for sealing of nonmetallic materials. In automation, they can direct the robot welding arm along a weld joint (laser beam used for tracking). Medical and military uses of lasers are in the news daily—from bloodless surgery to laser-guided smart bombs. Lasers are very versatile and can be used to process metal, wood, plastics, and composites.

For the welding industry, lasers are used for welding (LBW), cutting (LBC), and drilling (LBD) operations. These are the three applications that will be covered in this chapter. Lasers can be easily adapted to computer control for doing complex contour work.

The laser is a very concentrated beam of light. It is coherent or focused light in that the light waves are synchronized and travel parallel to each other. It is also monochromatic; meaning the light has one frequency, one color. Fig. 2 shows an example of the laser light's ability to stay in a tight column, unlike most other light.

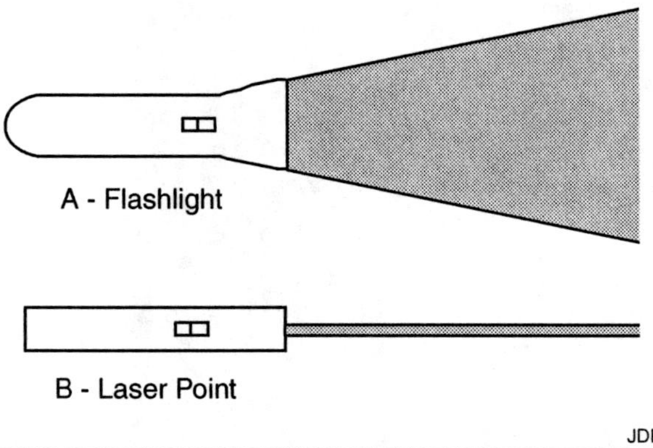

Fig. 2 — Flashlight Beam Compared to Laser Light Beam

Laser Processes

Because the laser beam can be used for welding, cutting, and drilling, these will be covered together. A laser light is produced when intense light or electric current excites certain materials.

The two different laser types are covered in Table 1. Lasers can be operated continuously or pulsed. The ruby or Nd-glass laser can only pulse at low frequencies (1-50 PPS), while the Nd-YAG or CO_2 can pulse at rates up to 2000 PPS. Pulsing reduces the heat buildup in the laser equipment as well as being effective for piercing and drilling applications.

Type	Solid State	Gas
Laser Material	Ruby Rod, Neodymium-Doped (Nd), Yttrium Aluminum Garnet (YAG) Neodymium-Glass (Nd-Glass)	Carbon Dioxide (CO_2) Nitrogen Helium or mixtures of these
Uses	Most used for drilling	Mostly used for cutting and welding

Table 1 — Types of Lasers for Welding, Cutting, or Drilling

Since the laser's light is used to heat the surface it is being focused on, it is considered a noncontact process. There is no physical contact of equipment with the part other than the beam being focused on it. The material being worked does not have to be a conductor, as in all the arc welding processes.

The concentrated heat of this light source has been measured as high as 65,000,000 watts per square inch (100,000 watts per square mm). Compare this to holding your hand a safe distance above a 100-watt light bulb to feel the heat. Then multiply this by 650,000 to get an idea of the amount of energy that is being released.

The laser is a melting and burning process. This heat source can weld 1/2 inch (13 mm) material at 60 inches per minute (25 mm per sec) using a 10 kW rated laser. Or with a 6 kW unit on thin material measuring 0.008" (0.2 mm) a laser can weld at 3000 inches per minute (1270 mm per sec), which is very fast travel speed. Units are commercially available with power levels up to 25 kW and can do full penetration single pass welds up to 1-1/4 inch (31.2 mm) thick.

For cutting, 1/2 inch (13 mm) is generally considered the limit because other processes are faster than laser on thicknesses greater than 1/2 inch (13 mm). Laser can cut materials up to 2 inches (50 mm) thick if speed is not a concern. Drilling can be done up to 1 inch (25 mm) in depth and at diameters in a range from 0.0001 to 0.060 inch (0.025 to 1.5 mm). However, holes smaller than 0.020 inch (0.5 mm) use the laser to its best potential.

Material thickness to be processed and travel speed is determined by:

1. Part geometry
2. Reflectivity of the material's surface
3. Heat conductivity
4. Vaporization point of the alloy being used
5. Type of alloy
6. Surface tension of the molten material

Laser Beam Welding (LBW) Process

LBW is generally done with an inert shielding gas to shield the weld pool. Since the laser beam is the heat source, it can be used either without filler metal (autogenous) or with filler metal.

Fig. 3 illustrates the keyhole method of laser welding. Because of the highly concentrated heat source and fast travel speed, there is little distortion created.

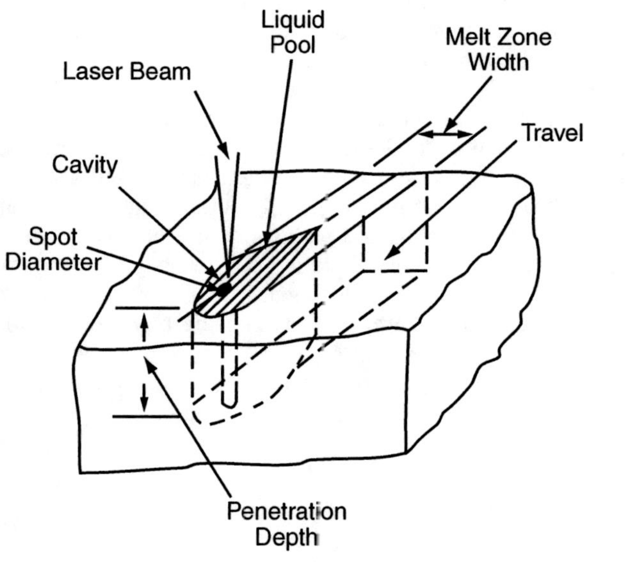

Fig. 3 — Laser Beam Welding (LBW)—Full Penetration Keyhole Method

Laser Beam Cutting (LBC) Processes

LBC is a very common use of the laser beam. This is done by reducing the spot size of the beam from 0.011 inch to 0.004 inch (0.252 mm to 0.085 mm), which turns the laser beam into a sharp cutting tool. It can be used to cut expensive alloys or traditional metals like stainless steel and copper. It can also cut nonmetals like plastic, wood, or cloth. Cut width or kerf can be very small—0.010 inch (0.25 mm). Very small heat affected zones (HAZ) are produced. Speed drops off dramatically when material approaches 1/2 inch (12 mm), where other cutting processes become more efficient and cost-effective. As thickness increases, blowouts may occur. Blowouts are unwanted molten metal flying out of the cut, interrupting the cut path.

When doing cutting, an assist gas is used. This helps improve combustion and physically blows metal from the kerf. Table 2 shows assist gases, materials they are used on, and various comments. High pressure up to 160 psi has been used on titanium to reduce cracks in the recast layer.

Assist Gas	Material	Comments
Air	Aluminum	Good results up to 0.060 in. (1.5 mm)
	Plastic	
	Wood	
	Composites	
	Alumina	All gases react similarly; air is the least expensive
	Glass	
	Quartz	
Oxygen	Carbon Steel	
	Stainless Steel	
	Copper	
Nitrogen	Stainless Steel	Clean, oxide-free edges to 1/8 in. (3 mm)
	Aluminum	
	Nickel Alloys	
Argon	Titanium	Inert assist gas required to produce good cutting of various materials

Table 2 — Types of Assist Gases for Laser Beam Cutting (LBC)

Pulsing the laser power can help overcome thermal problems. Fig. 4 shows a laser cutting operation.

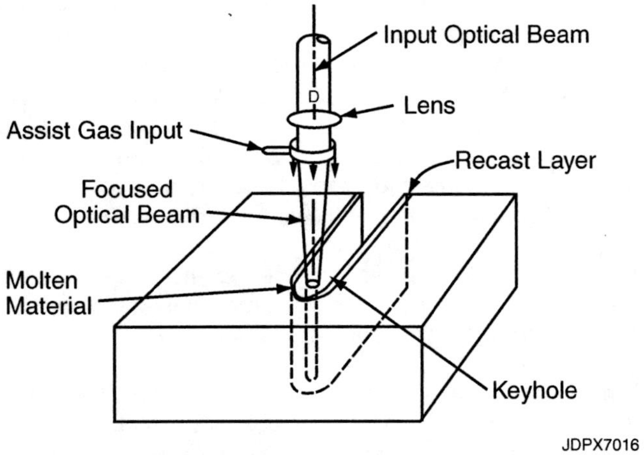

Fig. 4 — Laser Beam Cutting (LBC)

The surface of the material must be free of scale, coatings, dirt, and any impurities. On steel, it should be cleaned, pickled, and oil-free. The mill scale on hot-rolled steel creates problems, so cold-rolled steel has a much better surface condition for laser cutting. Cutting parameter will be greatly affected by the surface finish, tolerances, HAZ requirements, and flatness.

Laser Beam Drilling (LBD) Process

Laser beam drilling (LBD) can be done on very hard materials, like synthetic diamonds, tungsten carbide, quartz, glass, or ceramics. Holes as small as 0.0001 inch (0.025 mm) can be drilled. LBD is very fast but the possibility of blowouts can present safety concerns. A blowout is the throwing of molten metal outside the desired hole area. Blowouts occur when the hole diameter is small in relation to the thickness of the material being drilled. Continuous wave lasers have a much higher possibility of blowout when piercing metal or drilling. The pulse wave laser is typically used for this work. Lowering gas pressure to a minimum to just allow combustion, and checking focal point, can reduce blowouts. Also, the surface must be clean of any oil or dirt.

Equipment for the Laser Process

Fig. 5 shows a laser setup using the solid state Neodymium-doped, Yttrium Aluminum Garnet (YAG) material as a lasing source. The power supply, which operates at high voltage, supplies power to the flash lamps. A cryogenic cooler is used to cool the power supply and the laser rod and flash lamps. The rear mirror is fully reflective and bounces the coherent monochromatic beam back through the laser rod to the partially reflective front mirror. This helps in producing the coherent-synchronized wave light beam. Once the beam exits, it can be treated as light and be reflected, deflected, and focused. An energy monitor is placed in the beam to measure and compare the beam's power level. The beam dump and shutter control the beam. Various optics are placed in the beam path to expand or focus the beam. A 45° angle is used to deflect the beam's path 90°. A viewing head with safety filter is used to monitor the beam's operation and the workpiece being processed. This could be an LBW, LBC, or LBD application, depending on power level, spot size, focus point, constant or pulse power, use of assist gases, and their amount and type.

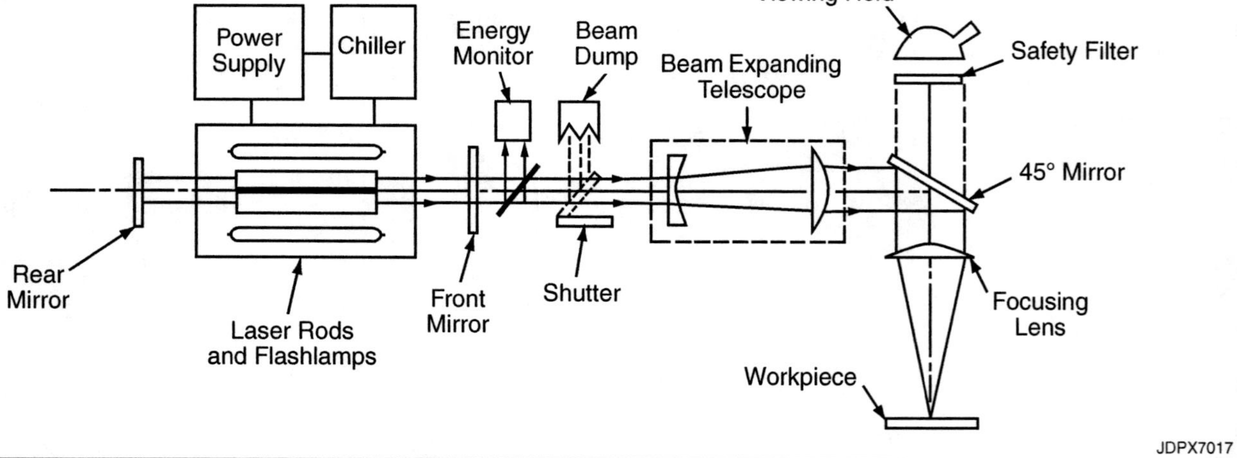

Fig. 5 — Components of a Typical Laser Beam Drilling (LBD) System

Laser Safety Rules

Working in an area where lasers are being used requires some special precautions. Some laser beams are not visible to the naked eye and in these cases, special care must be taken.

1. Eye protection, as even a small 5-milliwatt laser can damage the retina.
2. Safety goggles, which filter the specific laser wavelength.
3. To prevent burns, protect from contact with the beam by use of appropriate enclosures. Burns are deep and difficult to heal, and some beams are not visible.
4. Electric shock-high voltage is used throughout the equipment as well as capacitor storage devices, so keep enclosures in place and use proper electrical grounding.
5. Fumes and gases are produced with the variety of materials that can be processed. Highly toxic vapors can be produced, as well as metal fumes and ozone from the plasma, so use adequate ventilation and exhaust.
6. Chemical hazards can exist, so reference the safety data sheets on the equipment and cleaning materials.
7. Avoid direct touching to prevent "freeze" burns from the extremely cold nature of the cryogenic coolants.

Laser Setup

The proper setup of these fairly complex systems is very time consuming. The use of accurate records (procedures) greatly speeds the setup operation. The following list covers some of the issues:

1. Nozzle size
2. Power level (constant-pulsed)
3. Focal length
4. Assist gas (cutting or drilling)
5. Gas pressure
6. Speed

If these areas are not adjusted properly, the system won't work. If they are set correctly, the laser beam can easily work several times faster than other welding or cutting processes.

Acceptable Laser Results

Visual inspection is the first step in comparing current work with what has been done in the past on the same machine.

The color should match. When cutting, work should be free of dross and angle cut marks. Thermo-plastics will have a fire polished edge, while thermoset-plastics should have minimum charring or discoloration. Ceramics should be free of cracks.

Dimensional accuracy can be held to ±0.001 inch (0.025 mm).

Use destructive testing to determine strength and effects of thermal changes. The HAZ can be measured and is typically in a range of 0.001 to 0.010 inch (0.025 to 0.25 mm). Inspect for cracks in the base metal, HAZ, and weld metal, or for cutting in the recast layer.

Troubleshooting of Laser Applications

Refer to Table 3 as a guide to analyzing any problems in laser work.

Troubleshooting of Laser Applications		
Problem	**Cause**	**Cures**
Loss of speed	Loss of laser power	Check power meter reading
	Loss of intensity along beam path	Check for dirty reflecting mirror
	Loss of power density	Check focus spot size
	Change in chemical composition of base metal	Check chemical consistency of base metal
	Distorted laser output coupler	Check for source of distortion
	Light absorbing vapors in beam path	Use flow of gas to keep beam path clear
Direction Sensitivity	Laser beam is polarized	Get the beam in circular polarization, insert phase shift or circular polarizing optical devices
Accumulation of Oxide Slag/Wide Kerf on Cuts	Improper assist gas	Use oxygen for most metals and inert gas for flammable materials
	Cutting nozzle too high	Get nozzle closer to work
	Lens changed	With equipment change, make sure focus is on workpiece surface
	Lens defective and/or dirty	Replace or clean as needed
	Incorrect alignment of beam in head	Realign beam in center of nozzle
	Nozzle damage	Metal oxide blown in nozzle. Remove or replace nozzle. Adjust power level, travel speed, and/or height.

Table 3 — Laser Problems and Troubleshooting Guidelines

Test Yourself

Questions

1. (Fill in blanks.) LASER is an acronym and stands for _________ __________ by __________ __________ of __________.
2. Besides metal, lasers can be used to process other materials. List three of them.
3. (Fill in blank.) Laser light is coherent or synchronized so that the light waves travel _______ to each other.
4. (Fill in blanks.) Laser light is _______ and is made up of _______ wavelengths (colors) of light.
5. (Fill in blank.) Since laser is considered a __________ process, marring and deforming of the base material is not likely.
6. (Fill in blank.) The largest commercial units available are in the ______ kW range.
7. (Fill in blank.) Holes drilled in diameters of _______ use the laser to its best potential.
8. (Fill in blank.) Materials up to _______ inches have been laser welded to full penetration in one pass.
9. (Fill in blank.) When doing laser cutting of carbon steel, _________ is the assist gas used.
10. (Fill in blank.) On a laser cutting, the __________ layer is between the kerf and the base material.
11. (Fill in blank.) Eye protection is required when laser power is as low as _____ milliwatts.
12. (Fill in blank.) _______ lasers are visible to the naked eye.
13. (Fill in blank.) Dimensional accuracy of lasers can be + or – _____ inches.
14. List three problems associated with the laser processes.

Robotic Welding

8

Introduction

Welding robots are mechanical devices that can be programmed to repeat a sequence of tasks requiring movement between several consecutive starting and stopping points. They provide an alternative to situations that place a hardship on a human worker and to long-term repetitive jobs that are both tiring and boring to human workers.

Types of Robots

Robots are classified by the number of axis in which they rotate and by their source of power. A five-axis electrically powered robot is illustrated in Fig. 1. It is welding with twin GMAW (Gas Metal Arc Welding) torches in tandem.

JDPX6947

Fig. 1 — Five-Axis Electrically Powered Robot Welding With Twin Gas Metal Arc Welding (GMAW) Torches in Tandem

Electrically powered robots offer more consistent repeatability. Hydraulically powered robots can lift heavier loads, but require a warm-up period to reach their operating temperature before performing actual work.

There are at least two weld stations per robot. The robot may be at a fixed location (Fig. 2) or mounted on a track (Fig. 3). This arrangement allows one part to be welded while the other fixture is being reloaded with new parts.

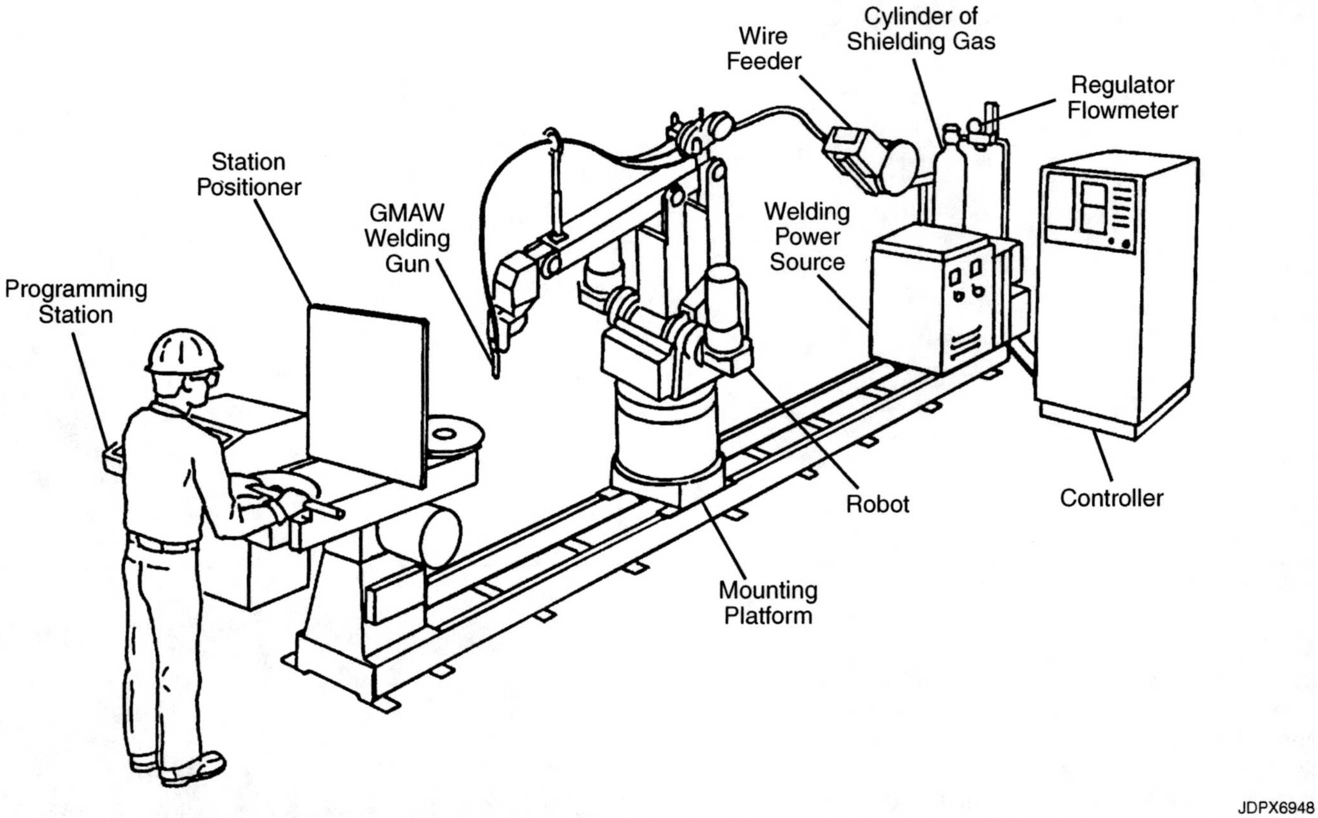

Fig. 2 — Robotic Welding System on a Track With Two Work Stations

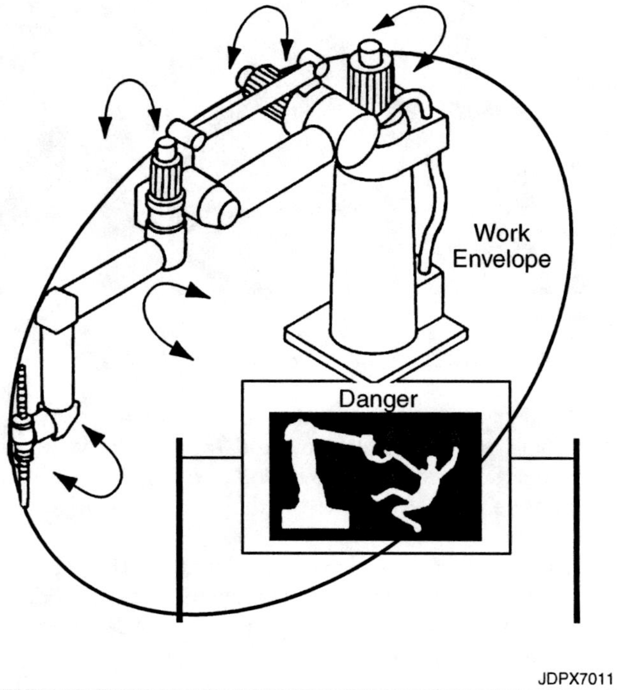

Fig. 3 — Robotic Envelope

Robotic Welding Applications

Manufacturers who join products with either the resistance or GMAW process use robots for making welds. Sophisticated controls must be used to coordinate movements of the robot with the operation of welding equipment. In most instances, computers provide instructions to controllers which drive DC electric motors and/or the hydraulic systems which power the robot.

From strictly an economic standpoint, to justify the expense of purchasing and maintaining a robotic system, the robot should be working 60 percent of the time. This means for 60 percent of a given work period, the robot must be supplied with materials, the completed parts must be removed from the work station, and the parts must be needed in another aspect of production or in sales.

In most applications, a robot's actual welding speed will not be much faster than that of a person. However, a robot can move from one weld to the next weld faster than a person. This feature makes a robot suitable for applications having numerous short start-stop type welds.

Tolerances for parts and positioning equipment must be more accurate than those sometimes used for manual welding. A robot can repeat movements to a location and stay on a work path within tolerances between 0.003 inch and 0.005 inch (0.076 mill to 0.127 mm). Shop experience indicates the total variation in part size, fit-up, and fixturing can be no more than plus or minus the wire diameter. Adaptive controls must be added to robotic systems subject to variations that exceed this limit.

Safety

Everyone responsible for operating and maintaining a robotic system should read and understand the safety instructions provided in the owner's manual for each piece of equipment in the system. Each person should be required to pass a written test before being permitted around the equipment.

Safety instructions for robotic welding should include the following:

1. Protecting operators and nearby workers from the arc's rays, welding sparks, hot metal, and welding fumes.
2. Keeping the work area free of combustibles, including trash containers.
3. Minimizing the danger from electrical shock by proper grounding procedures and by keeping the work area dry.
4. Protective devices such as guards and motion sensors should be in placed in locations that minimize the likelihood of unauthorized people getting into the work envelope and for operators coming in contact with the robot while it is in motion.
5. All power necessary to operate the robot and welding equipment should be turned off when making adjustments and when performing routine maintenance.
6. Common sense good housekeeping in the work area.

Test Yourself

Questions

1. What two situations are suited for robotic welding?
2. What are the two types of robots?
3. Why are there at least two work stations per robot?
4. What types of welding can be done by a robot?
5. What is a work envelope?
6. What are two of the most serious hazards associated with robotic welding?

Soldering, Brazing, and Braze Welding

9

JDPX6998

Fig. 1 — Brazing With a Torch

Introduction

Other methods besides fusion welding are often used to join metals.

The two most common processes are:

- Brazing
- Soldering

These methods are widely used in joining most metals, especially where it is not practical or economical to use regular fusion welding.

Both brazing and soldering join metals. A filler metal is melted between the two closely fitted metal surfaces. The major difference is that in brazing, the filler metal melts at a higher temperature and fuses by capillary action, giving a stronger joint.

The metal work to be joined is not melted in either process, but is only heated enough to cause the filler metal to fuse. This is what makes it different from fusion welding.

Brazing

Brazing is a process where metals are joined by heating to suitable temperatures above 800°F (425°C) and by using a nonferrous filler metal having a melting point below that of the base metals. The filler metal is distributed between the closely fitted surfaces of the joint by capillary action.

Most commercial metals can be brazed. Although a brazed joint has a relatively high tensile strength, it is not recommended when the full strength of a joint is required.

An important factor in brazing is that there is less danger of destroying the mechanical properties of the base metal since lower bonding temperatures are used than normally required for fusion welding.

Brazing can also be used to join dissimilar metals.

Requirements for Brazing

The success of any brazing operation depends on joints having relatively small clearances, and surfaces that are free of oxide and other contaminants. Cleaning is done by coating the surfaces with a special flux, which when heated is capable of dissolving all foreign matter. Once the surfaces are properly fluxed, a brazing filler metal is melted at some point along the seam. Capillary action then draws the molten brazing metal between the surfaces of the joint. Upon cooling, the solidified brazing metal forms a solid bond at the joint. During heating and cooling, precautions are taken to prevent any movement of the surfaces. For most production work, a fixture is used to hold parts in alignment during brazing.

Types of Joints for Brazing

The two basic joints for brazing are the lap and butt (Fig. 2). (Tee and corner joints are considered as butt joints.)

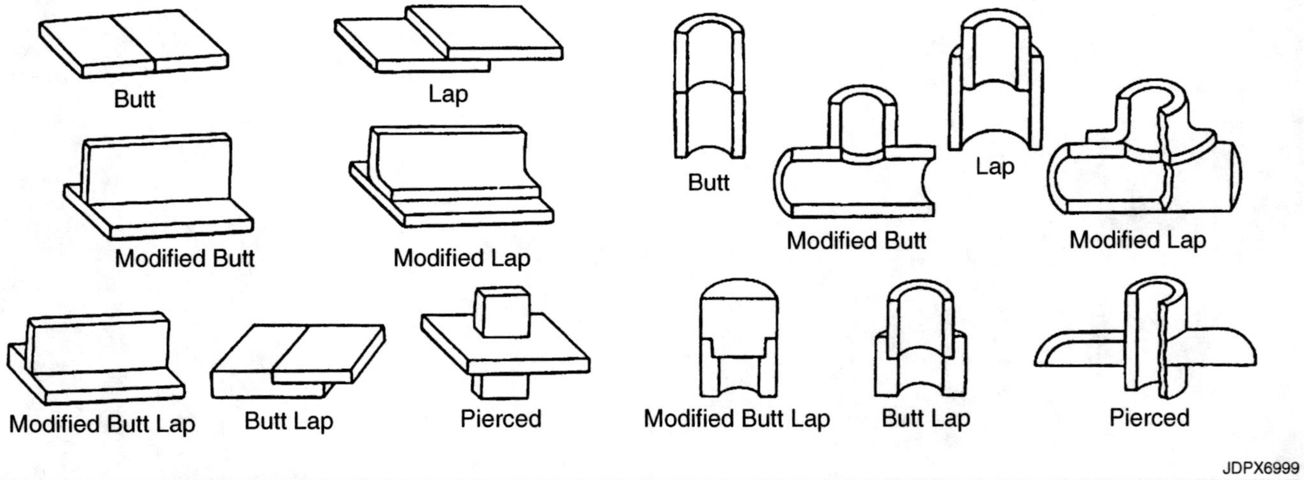

Fig. 2 — Types of Joints for Brazing and Soldering

The lap joint offers the maximum strength. The overlap should be at least three times the thickness of the thinnest section.

The strength of a butt joint can also be improved by increasing the cross-sectional area as shown or by using a sleeve over the joint.

Joint clearance, which is the distance between the interface of the pieces, is an important factor in the performance of a brazed joint, whether the joint is subjected to loadings of fatigue, impact, or static. Too tight a joint may hinder the plastic flow of the filler metal, while too much clearance may prevent the full effects of the capillary action, leaving voids and uneven distribution of filler metal.

In general, joint clearance should be between 0.001 and 0.010 inch (0.025 and 0.250 mm).

Filler Metals for Brazing

A brazing filler metal should meet these requirements:

1. Enough fluidity so the metal will flow evenly by capillary action.
2. Good melting to form a sound metal bond.
3. Melting point consistent with the type of metal to be joined.

Brazing filler metals fall into seven groups: silver, aluminum-silicon, copper-phosphorus, gold, copper and copper-zinc, magnesium, and nickel.

Table 1 lists the principal metals that are joined with these filler metals.

AWS Brazing	Metal Combinations for Which Various Fluxes Are Suitable		Effective Temperature Range of Flux, °F (°C)	Major Materials of Flux	Physical Form	Methods of Application
Flux Type No.	Base Metals	Filler Metals				
1	Aluminum and aluminum alloys	BAlSi	700—1190 (370—645)	Fluorides Chlorides	Powder	1, 2, 3, 4
2	Magnesium alloys	BMg	900—1200 (480—650)	Fluorides Chlorides	Powder	3, 4
3A	Copper and copper-base alloys (except those with aluminum); iron-based alloys; cast iron; carbon and alloy steel; nickel and nickel-based alloys; stainless steels; precious metals	BCuP BAg	1050—1600 (565—870)	Boric Acid Borates Fluorides Fluoborate Wetting Agent	Powder Paste Liquid	1, 2, 3
3B	Copper and copper-base alloys (except those with aluminum); iron-based alloys; cast iron; carbon and alloy steel; nickel and nickel-based alloys; stainless steels; precious metals	BCu BCuP BAg BAu R B CuZun BNi	1350—2100 (730—1150)	Boric Acid Borates Fluorides Fluoborate Wetting Agent	Powder Paste Liquid	1, 2, 3
4	Aluminum - bronze; aluminum - brass	BAg BCuZn, BCuP	1050—1600 (565—870)	Borates Fluorides Chlorides	Powder Paste	1, 2, 3
5	Copper and copper-base alloys (except those with aluminum); nickel and nickel-based alloys; stainless steels; carbon and alloy steel; cast iron based alloys; precious metals (except gold and silver)	BCu, BCuP BAg-(8-19) BAu, BCuZn Bni	1400—2200 (760—1205)	Borax Boric Acid Borates	Powder Paste Liquid	1, 2, 3

For proper method of application, see following pages, "Heating Methods for Brazing," methods 1–4.

Table 1 — Metals and Fluxes for Brazing

Fluxes

Any form of oxide on the surface of a metal will prevent a uniform flow of the brazing metals. For this reason, a flux of some kind is necessary to eliminate the oxide. The common commercial fluxes are in paste, liquid, or powder form. Fluxes have as their main ingredients: borates, fused borax, boric acid, fluorides, chlorides, and fluoborates. There is no single flux that can be used for all brazing operations. See Table 1 above. All traces of flux residues must be removed after brazing to prevent corrosion.

For some mass-production jobs, applying fluxes is a time-consuming task. So controlled atmospheres are used to remove oxide and prevent the formation of oxide during brazing. This method is often used with induction brazing of titanium, zirconium, and other refractory metals.

In a controlled atmosphere, a gas is continuously supplied to a furnace and circulated within it at slightly higher than atmospheric pressure. Gas may consist of high-purity hydrogen, carbon dioxide, carbon monoxide, nitrogen, argon, ammonia, or some form of combusted fuel gas.

Heating Methods for Brazing

Applying heat for brazing is done by a variety of methods depending on the kind of material to be brazed, quantity of production, and sizes of parts to be joined.

The following techniques are used:

1. **Torch heating**
2. **Furnace heating**
3. **Induction heating**
4. **Dip brazing**

Torch Heating

Torch heating is probably the most common for brazing. The gas mixture may be oxyacetylene, air-gas, gas-oxygen, or oxy-hydrogen. The type of gas mixture depends on the thermal conductivity, type, and thickness of the material to be joined.

Oxyacetylene is often more versatile for torch brazing because of its wide range of heat control. A slightly reducing (carburizing) flame is required, and care must be taken to prevent the core of the flame from comlng in contact with the metal. Close flame contact may cause the base metal to melt and restrict the flow of the brazing metal.

The air-gas torch provides the lowest heat and is far more adaptable for brazing thin sections. The air-gas mixture may consist of air at atmospheric pressure and city gas or air and acetylene.

The gas-oxygen process uses oxygen with natural gas, bottled gas, propane, or butane. This mixture produces a high flame temperature and is used where greater brazing heat is required.

The oxy-hydrogen torch is very useful for brazing aluminum and other nonferrous metals because of its low heat. This prevents possible overheating of the metal. Hydrogen also provides extra cleaning action and shielding during the brazing process.

Furnace Heating

Furnace heating is a production process for brazing parts that can be assembled and positioned on trays. The trays are loaded in a furnace either manually or automatically. Some automatic loading consists of a conveyor belt on which the parts are placed. The conveyor belt then moves into the furnace at regulated speeds (Fig. 3). The brazing filler metal, which can be in the form of wire, foil, powder or paste, is placed in the right position near the joint and the heat of the furnace melts the filler metal. Fluxing is used except when the heating is done in a controlled atmosphere.

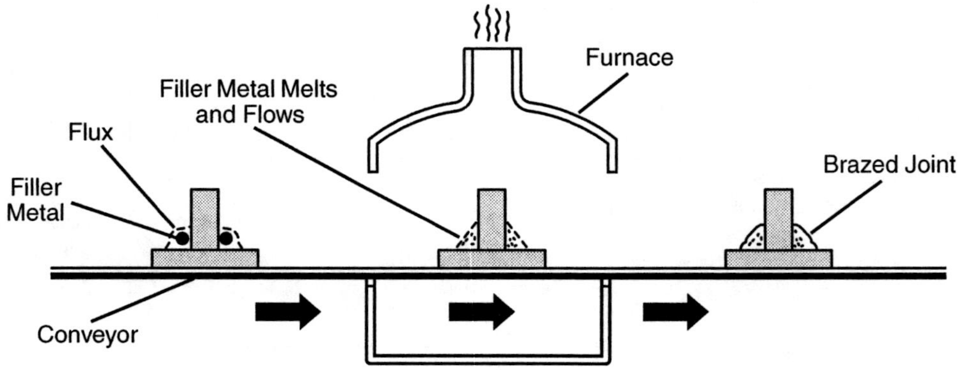

JDPX7000

Fig. 3 — Furnace Heating for Production Brazing

Most high-production brazing in a controlled atmospheric furnace is done with hydrogen and either endothermic or combusted gas. Inert gases such as helium or argon are also used for controlled atmospheric furnace brazing.

Some furnace brazing is done in a vacuum where continuous pumping prevents oxidation and removes volatile materials that are liberated during the brazing process. Vacuum brazing has wider applications in aerospace and nuclear industries where reactive materials are joined and any form of trapped fluxes must be avoided.

Induction Heating

Induction heating is a process where heat is generated by an inductor coil that is not in contact with the parts being brazed (Fig. 4).

A power source unit converts regular line 60-cycle current into a high-frequency low voltage current.

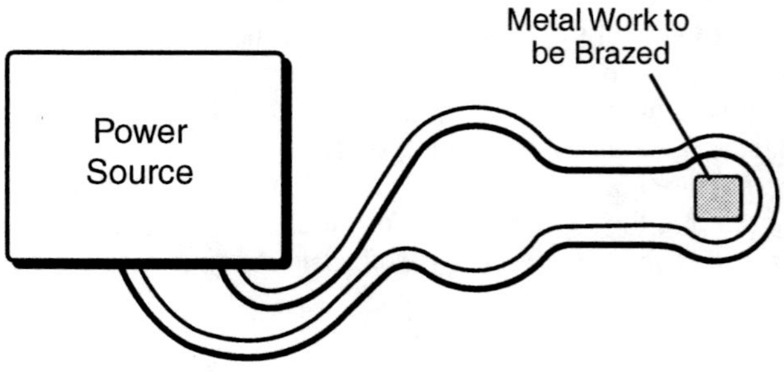

JDPX7001

Fig. 4 — Induction Heating Coils for Brazing

As the current flows through the inductor coil, which surrounds the object to be brazed, a magnetic field is created. When the conducting object is placed in a magnetic field, an electro-motive force is induced in the conductive material. This sets up a current, and the resistance of the object to the flow of current causes instant heating to occur. The heat is relatively near the surface of the metal, and any interior heating results from thermal conduction from the hot surface.

The size of the power source depends on the parts to be brazed and the production rate. Low-frequency power units produce deeper heating and are designed for brazing heavy sections, while high-frequency units produce light or shallow heating.

Induction heating is used in brazing parts that can be aligned in a fixture and require rapid heating. It is very economical for quantity production when parts can be adapted to the induction furnace.

Dip Brazing

Dip brazing is broken down into two methods:

1. Molten metal bath
2. Molten flux bath

Molten metal bath is a technique that consists of immersing parts in a bath of molten brazing metal. The brazing material is melted in a crucible with a cover of flux maintained over the molten filler metal. The parts to be brazed are first cleaned and fluxed and then dipped into the bath. The process is limited to brazing small assemblies such as wire connections or metal strips when they can be easily held in fixtures.

Molten flux bath dip brazing is often used to fabricate radiators or other cooling units.

Flux in the form of a chemical salt is melted in a container or flux pot by a gas flame or electrical resistance. The most common method is passing an electric current through the bath. With resistance heating the initial charge of flux is first melted by some other heating source and the molten salt poured into the fluxing container. Once the flux is in a molten state, the resistance of the bath to the electrical current provides sufficient heat to keep the flux at the proper temperature. A thermocouple immersed in the molten bath and a temperature control unit keeps the molten flux at the required brazing heat.

The parts to be brazed are cleaned and assembled in a suitable fixture. Brazing filler metal in the form of rings, washers, slugs, or paste mixture is placed on the base metal. In production brazing with molten salts, the parts and holding fixtures are usually preheated in a furnace to a temperature near the melting point of the molten flux. The parts are then dipped in the molten flux where the heat is sufficient to melt the brazing material.

Braze Welding

Braze welding is slightly different from regular brazing. While brazing is the joining of two surfaces by a thin bond of brazing metal, braze welding is much like fusion welding except that the base metal is not melted. The base metal is brought up to its tinning temperature and then a brazing alloy is deposited over the seam with a filler rod.

Although the base metal is never actually melted, the bond formed by the brazing metals are often comparable to those secured through fusion welding.

Bronze welding is a typical braze welding operation. This technique is frequently used for joining or repairing such metals as cast iron, malleable castings, copper, brass, and various dissimilar metals (Fig. 5). The bronze filler rod that is applied to the seam consists of copper and zinc with small quantities of tin, iron, manganese, and silicon.

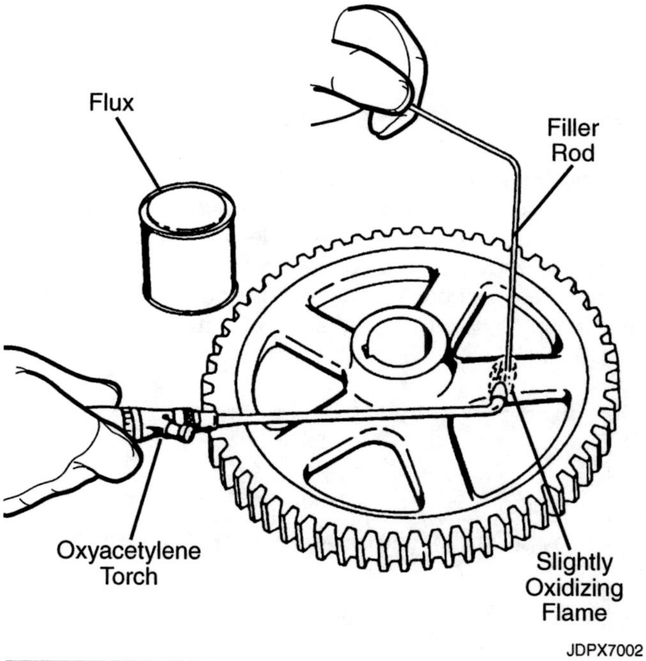

Fig. 5 — Braze Welding

The welding procedure involves cleaning the surface and then applying a coating of flux to diffuse the oxide. The flux is applied by dipping the heated rod into the powdered flux. The flux can also be mixed with water and spread over the seam.

Any type of joint is suitable for bronze welding. On thick sections, the edges should be beveled to form a 90° V-groove.

Generally, bronze welding is performed with an oxyacetylene torch with a flame that is slightly oxidizing. The flame is concentrated on the starting end until the metal begins to turn a dull red color. A small amount of bronze is melted on the surface and allowed to spread along the entire seam. The flow of this thin film of bronze is known as the tinning operation. Unless the surfaces are tinned properly, the remaining bronzing cannot be done successfully.

If the surface of the metal is heated properly, the bronze should spread out evenly over the metal. A surface that is too hot will cause the bronze to bubble or run around like drops of water on a warm stove. When the bronze forms into balls that tend to roll off, the base metal is not hot enough.

Once the base metal is well tinned, a weld bead is deposited over the seam using a slight circular torch motion. The rod must be constantly dipped into the flux as the weld progresses.

Methods for Brazing

Brazing With Carbon Electrodes

Brazing can be performed with a carbon arc. The process is very similar to the metal electrode arc except that the electrode does not melt or provide filler metal. Here the carbon electrode serves only as the medium of generating the arc. The electrode is either of pure graphite or in the form of a copper-coated carbon and held in a special holder (Fig. 6).

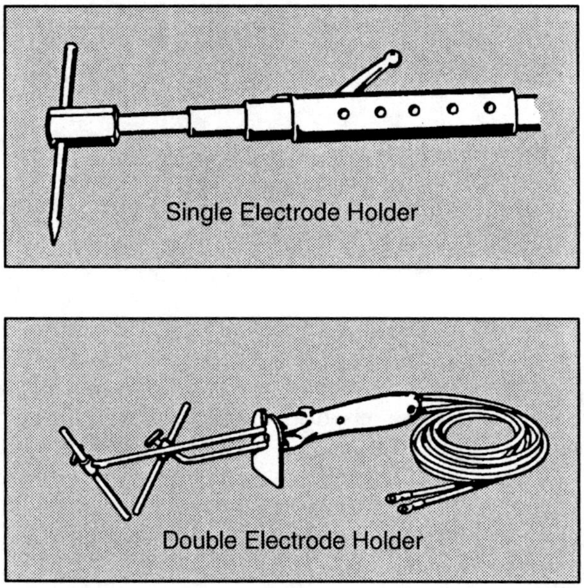

Fig. 6 — Carbon Arc Electrode Holder

The electrode for brazing using a DC source is shaped by grinding it to a long tapering point (Fig. 7).

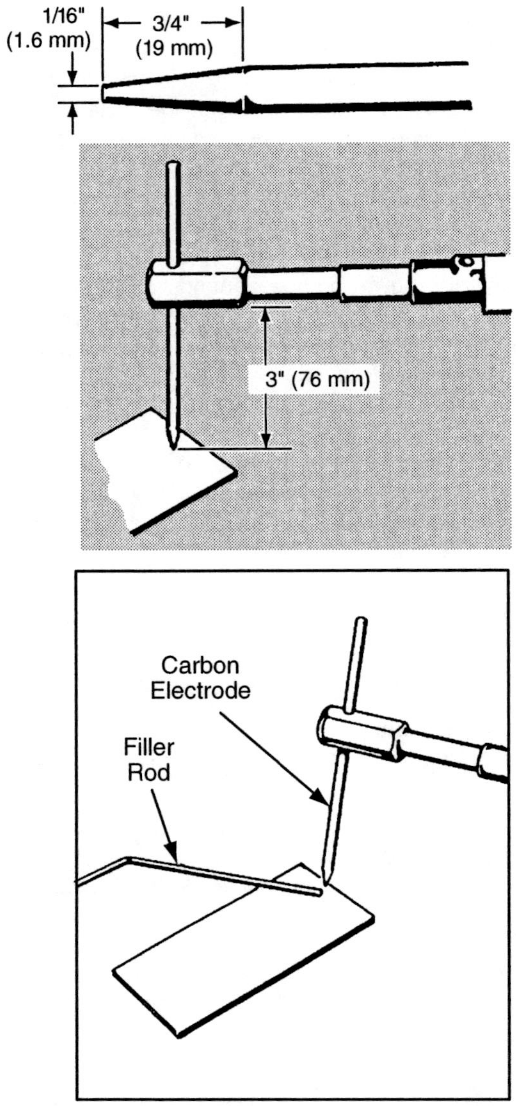

Fig. 7 — Brazing With the Carbon Arc

Either a DC or AC power source can be used for brazing. With an AC unit, the arc is formed between two carbon electrodes held in a holder as shown in (Fig. 6). In the single carbon electrode with DC current, the machine must be set for straight polarity. Reverse polarity produces too much of an unstable arc and causes greater quantities of vaporized carbon to enter the molten brazing material.

Brazing is done by striking an arc and feeding a brazing rod into the arc. Carbon arc brazing is done with a bare silicon bronze rod or a flux-coated brazing rod. The filler rod is held as illustrated in (Fig. 7) and moved along the surface with the arc over the rod.

With the AC holder, the formed arc between the two electrodes remains intact even though the electrodes are withdrawing from the work. The arc is established by moving the electrodes with a push button located on the holder until the electrodes touch. Pressure on the button is then released to permit the points of the electrode to part and establish the arc. When the distance between the two electrodes is correct, there will be a quiet, soft flame. Striking the arc with a DC current is done in the same way as in SMAW.

Brazing With a Liquid Flux

Liquid flux eliminates the need of applying flux to metal surfaces before beginning the actual brazing operation.

A liquid fluxing material is placed in a separate tank that is connected to the regular gas line. The flow of gas into the tank vaporizes the flux and induces it into the gas stream, where it passes through the welding torch and into the flame. As the flame contacts the metal, the flux in the flame cleans and fluxes the surface for brazing. No pre-cleaning or post-cleaning of the parts is necessary.

As a rule, less heat is required with liquid flux than when a powder or wet flux is used. The weld metal is not oxidized and the strength of the brazed joint is greater.

Except for the fluxing tank, no other special equipment is needed as a standard welding torch is utilized.

Liquifluxer can be used for brazing brass, bronze, copper, cast aluminum, cast and malleable iron, nickel, and steel.

Soldering

Soldering is the joining of two metal pieces by using a nonferrous filler metal that melts below 800°F (427°C) or below the temperature of the base metal. The filler metal is called solder and is distributed between the surfaces by capillary action. Soldering with a torch is shown in (Fig. 8)

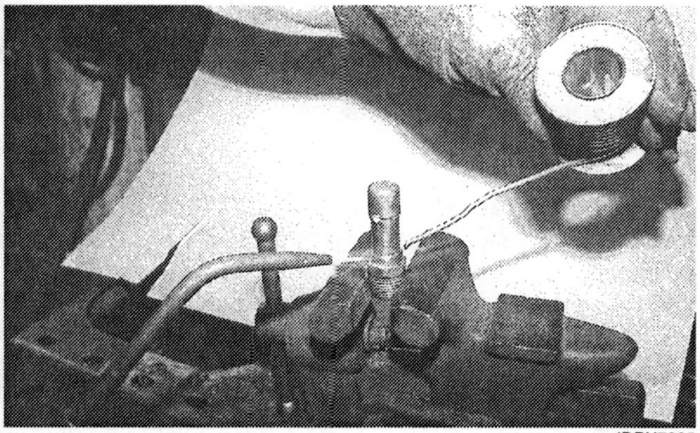

JDPX7005

Fig. 8 — Soldering With a Torch

CAUTION: Do not breathe fumes from soldering. Some solders contain lead, cadmium, and materials that are health hazards. See American National Standard Z49.1 Safety in Welding and Cutting and OSHA STANDARDS 29 CFR 1910 for specific hazards and precautions for soldering.

Soldering is used to join surfaces not under high-strength forces, since solder has a relatively low tensile strength. Soldered joints are also not suitable where temperatures approach the melting point of solder.

Soldering is a simple operation, providing five basic requirements are observed:

1. The right type and amount of solder and flux are used for the base metal being joined.
2. The pieces being soldered fit tightly together.
3. The pieces are absolutely clean.
4. The pieces are held together until the solder solidifies.
5. The correct amount of heat is applied to the seam.

See Table 2 for metals that can be soldered.

Metals and Soldering				
Base Metal, Alloy, or Applied Finish	Flux Requirements			Soldering Not Recommended
	Non-Corrosive	Corrosive	Special Flux and/or Solder	
Aluminum			X	
Aluminum-Bronze			X	
Beryllium				X
Beryllium-Copper		X		
Brass	X	X		
Cadmium	X	X		
Cast Iron			X	
Chromium				X
Copper	X	X		
Copper-Chromium		X		
Copper-Nickel		X		
Copper-Silicon		X		
Gold	X			
Inconel			X	
Lead	X	X		
Magnesium			X	
Manganese-Bronze (High Tensile)				X
Monel		X		
Nickel		X		
Nichrome			X	
Palladium	X			
Platinum	X			
Rhodium		X		
Silver	X	X		
Stainless Steel			X	
Tin	X	X		
Tin-Bronze	X	X		
Tin-Lead	X	X		
Tin-nickel	X	X		
Tin-Zinc	X	X		
Titanium				X
Zinc		X		
Zinc Die Casting			X	

Table 2 — Metals That Can Be Soldered

Most soft solders are alloys of tin and lead. The percentage of each metal determines its melting point and other characteristics. The melting range is from about 370°F (188°C) for a mixture of 70 percent tin and 30 percent lead to about 590°F (310°C) for 5 percent tin and 95 percent lead.

The most common general-purpose solder is known as half-and-half or 50-50 solder. It contains 50 percent tin and 50 percent lead and melts at about 471°F (244°C).

A variety of other alloys are used for soldering purposes. See Table 3 for composition of solders.

Solders					
ASTM Number	Composition %				Melting Range
	Tin	Lead	Antimony	Copper	
5A	5	bal.	0.12 max.	0.08	572–596
5B	5	bal.	0.50 max.	0.08	572–596
10B	10	bal.	0.50 max.	0.08	514–573
20B	20	bal.	0.50 max.	0.08	364–535
20C	20	bal.	1.0–1.2	0.08	364–535
30A	30	bal.	0.25	0.08	361–477
30B	30	bal.	0.50	0.08	361–477
40A	40	bal.	0.12	0.08	361–455
50A	50	bal.	0.12	0.08	361–421
60A	60	bal.	0.12	0.08	361–374
63A	63	37	0.12	0.08	361–361

Table 3 — Composition of Solders

In general, alloys with a low tin content have higher melting points and do not flow as readily as the high-tin alloys. The low-tin alloys are less expensive and are used where large volumes of soldering are done. Solders with a high amount of tin have better wetting properties and produce less cracking. High-tin solders are used considerably in electrical work.

Solders with a tin content of 60 percent or more are classified as fine solders and are used in soldering instruments where temperatures are critical.

Many commercial types of metals can be soldered. Table 2 lists the various metals that are recommended or not recommended for soldering.

Special solders are also available. A tin-antimony solder is designed to solder food-handling vessels where lead contamination must be avoided. Tin-zinc solders are intended primarily for joining aluminum. Lead-silver solders are used where strength at high temperatures is required.

Solders are available in bar, cake, solid wire, flux-core wire, ribbon, and paste forms. Flux-core wire solder has an acid or resin flux in the center of the wire. With these solders, no additional flux is needed.

Fluxes

Oxides and rust will form on surfaces of most metals when exposed to air. Solder will not adhere if such impurities are present. By applying a flux, the oxide is removed and the formation of new oxide during the soldering process is prevented. Fluxes also increase the wetting action, allowing the solder to flow more freely.

Fluxes come in paste, liquid, powder, and cake form. Some are general-purpose fluxes usable on almost all or most metals. Others are special fluxes such as those for aluminum soldering.

All fluxes are classified as corrosive or noncorrosive. Although the corrosive types are most effective, they must be washed from the metal after soldering. They should never be used for electrical or electronic work.

Resin is the most common noncorrosive flux. Zinc chloride is the most frequently used corrosive flux.

Soldering Equipment

In any soldering operation, both pieces of metal to be joined must be hot enough to melt the solder. A strong bond is achieved only if the molten solder spreads evenly over the surface. A number of devices are available for heating purposes. The type used depends on the size and shape of the assembly. The major types of soldering equipment are:

1. Soldering coppers
2. Electric soldering tools
3. Flame-heating devices

Soldering Coppers

A soldering copper consists of a forged piece of copper fastened to an iron rod with a wooden handle on one end. These coppers vary in size with heads forged in several shapes (Fig. 9).

Generally, a lightweight copper is used for soldering light-gauge metal and a heavyweight copper for soldering heavy-gauge metal. A lightweight copper on heavy metal does not hold enough heat to heat the metal or allow the solder to flow smoothly. Soldering coppers are heated in a furnace or with a torch.

Electric Soldering Tools

These devices are often more convenient than soldering coppers because they maintain uniform heat (Fig. 9). They vary in sizes from 25 to 550 watts.

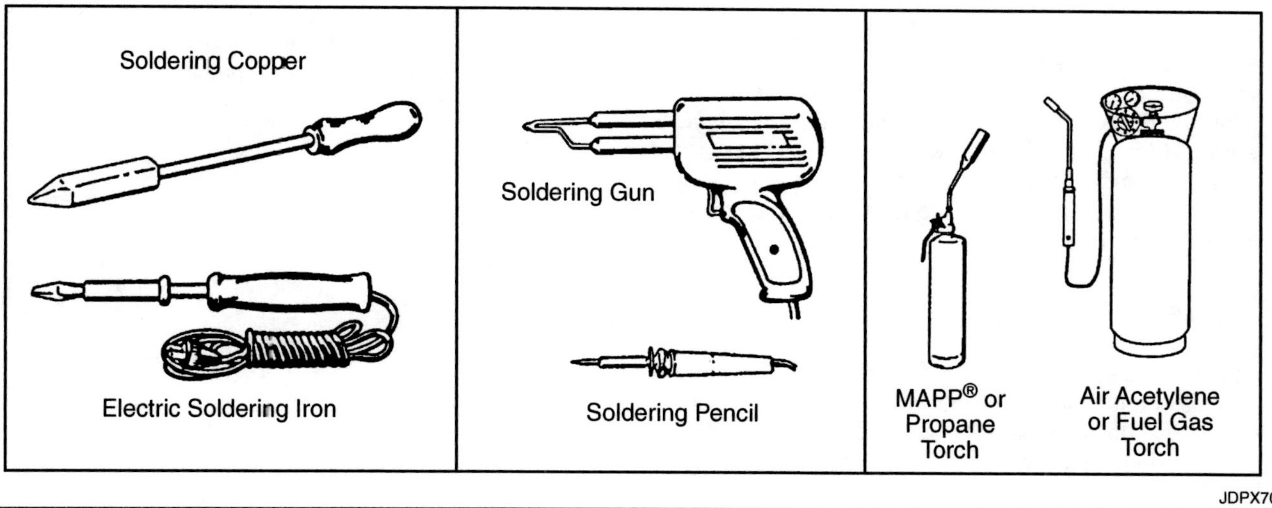

Fig. 9 — Soldering Equipment

Lightweight, low-voltage irons with replaceable heating elements and tips are called soldering pencils and are preferred for electrical and electronic work. Electric soldering guns produce instant heat at the tip of a long small point when the trigger is pulled. On most guns, the trigger also turns on a light that focuses at the point. For these reasons, the soldering gun is very popular for electronic work.

Torches

Some soldering operations are impossible or very difficult to perform with a soldering copper or iron. For these jobs, a flame is used as the source of heat (Fig. 9).

The flame is produced with either an acetylene or fuel gas air torch.

The most efficient, safe, and versatile gas torch is one that burns natural gas and air. These torches are often equipped with changeable tips, which can produce a wide range of flame sizes. The torch has one valve which controls the flow of gas.

To light the torch, the gas valve is opened slightly and ignited.

The length of the flame is controlled by the amount of gas allowed to flow to the tip.

Bottled-gas torches are also used for soldering, especially when the work is not at a fixed station where a gas-air torch is available. The bottled-gas torch must be operated with care; follow the manufacturer's instructions carefully.

Preparing to Solder

1. Parts to be soldered must fit perfectly so the solder can travel by capillary action between the two surfaces. Solder will cease to flow where there is a gap between the two work pieces.
2. Parts to be soldered must be absolutely clean because the solder will not stick to a dirty, oily, or oxide-coated surface. Dirt and grease can be removed with a cleaning solvent. Steel wool or other abrasive cloth is used to eliminate the oxide. Applying a flux completes the cleaning and keeps the metal free from oxide during heating and soldering.
3. Parts must be held together during soldering so there is no movement. Any movement during the heating will cause the pieces to be misaligned and the slightest disturbance of the solder as it solidifies will cause a weak joint or connection that leaks.
4. Parts to be soldered must have a suitable joint design to withstand the loads imposed on it. The strength of the joint cannot depend on the bond itself because a soldered joint, for example, will seldom develop shear strength greater than 250 psi (1700 kPa). If greater strength is needed, make some type of mechanical joint by lapping or folding before you solder.

Tinning a Soldering Copper

The point of a soldering copper must be covered with a thin coat of solder to operate properly. A tinning operation is performed by first smoothing the surfaces of the point with a file and then heating the copper until it is hot enough to melt solder. The point is next rubbed on a block of sal ammoniac and a small amount of solder melted during the rubbing action (Fig. 10).

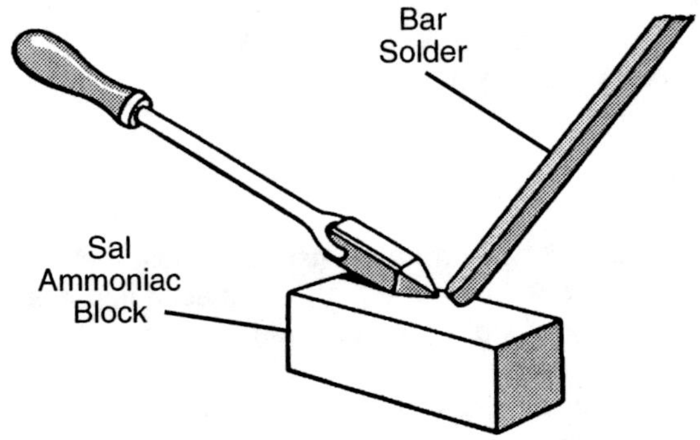

Fig. 10 — Tinning a Soldering Copper Over a Block of Sal Ammoniac

Tinning can also be done by dipping the point in a liquid or paste flux and applying solder.

Methods of Soldering

Sweat Soldering

In this process, two surfaces are soldered without the solder being visible. This is done by applying a coating of solder to each of the surfaces to be joined (Fig. 10). The coated surfaces are then placed together and a heated copper held over the seam. To avoid smearing the exposed surfaces of the metal with solder, the excess solder on the copper is removed by quickly wiping the joint with a damp cloth.

When the solder between the two surfaces begins to melt and show evidence of flowing out from the edges, pressure is applied on the metal. As the copper is drawn slowly along the seam, pressure must be maintained until the solder solidifies.

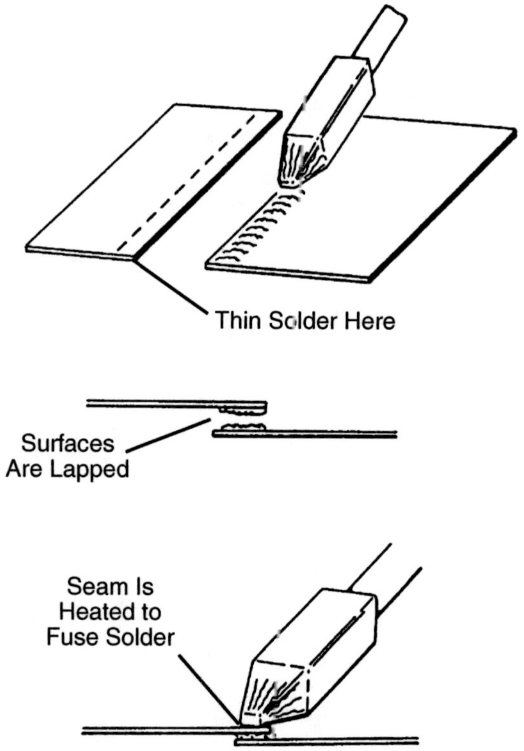

Fig. 11 — Sweat Soldering a Joint

Seam Soldering

Seam soldering is running a layer of solder along the outside edge of the joint (Fig. 12). As a rule, the surfaces of the joint are first tacked to hold them in position. Tacking Is done by depositing a small drop of solder at several places on the seam. The point of the copper is then moved along the edge of the seam as shown and solder applied directly in front of the point.

Fig. 12 — Seam Soldering a Joint

Flame Soldering

Many soldering jobs are difficult when using a soldering copper or electric soldering iron. In these cases a torch flame is often more practical, especially where fast soldering is required. The flame can be adjusted to the required size and directed to the exact spot where it is needed and moved to heat both pieces properly as well as to melt the solder.

Induction Soldering

This is used where large-scale production is involved and the pieces can be kept in alignment by a fixture. The equipment is similar to that used for induction brazing.

Resistance Soldering

Here the electrical resistance of the metal provides the necessary heat for soldering. The work to be soldered is placed between a ground and one or two movable electrodes. Solder is fed directly into the joint when the surfaces of the work pieces reach the proper temperature.

Dip Soldering

Dip soldering consists of a pot with molten solder. The joint to be soldered is simply dipped into the molten solder. The bath supplies both the heat and solder to complete the soldering operation.

This soldering technique is very economical since an assembly with several joints can be soldered at once.

Any dip soldering operation requires suitable jigs and fixtures to keep the unit in alignment until the solder has solidified. These fixtures must be well made to prevent movement of the parts during the operation when they are being dipped and handled.

Oven Soldering

Gas or electric ovens are used for soldering when the entire assembly can be heated without damaging any of its components. Since the assembly must be moved before and during the solidification of the solder, suitable jigs and fixtures are required to hold parts together. Without proper clamping, the joint will fail because solder is easily disturbed before it becomes solid.

Test Yourself

Questions

1. How does braze welding differ from regular fusion welding?
2. In some instances, why is braze welding preferred over fusion welding?
3. Why is a flux required for brazing?
4. What three types, of heating devices may be used for brazing?
5. In brazing with the carbon arc, how does the carbon electrode for DC current differ from the one used with AC power supply?
6. What is the difference between braze welding and regular brazing?
7. Give one example of when soldering not recommended for joining metals.
8. Most solders are alloys of _____ and _____.
9. What is the advantage of an electric soldering iron over the soldering copper?
10. What is meant by tinning a copper?
11. How does sweat soldering differ from seam soldering?

Metal Surfacing

10

Fig. 1 — Surfacing of Metal Parts

Introduction

Surfacing is the depositing of metal on tools and machine parts to increase their size, wear life, or other properties.

Examples of surfacing are the building up of worn shafts and gear teeth, hardening of cutting tools (Fig. 2), or replacing metal that has corroded away.

JDPX6991

Fig. 2 — Surfacing of Bucket Teeth

There are two major types of surfacing:

- Hard surfacing
- Metallizing

Hard surfacing is a process that gives a hard, tough overlay to resist corrosion, abrasion, and impact loads. (It is sometimes called "hardfacing.")

Metallizing is a spray coating process in which fine particles of metal are deposited on surfaces where metal has worn away or where contours must be built up.

Hard Surfacing

Hard surfacing is much like fusion welding; a special rod, wire, or electrode is mixed with the base metal by heating and melting it.

For good surfacing, the correct filler metal must be used to meet the service requirements of the pieces.

In general, parts that require surfacing get three types of wear:

- Abrasion
- Impact
- Corrosion

Abrasion happens with surfaces that are subjected to continuous grinding, rubbing, or gouging actions. For example, low-stress scratching abrasion or erosion is characterized by soil sliding past a plowshare, high-stress grinding action is typified by stone being crushed between metal faces, and gouging abrasion is exemplified by ripper teeth crushing into bedrock.

Impact is the absorption of kinetic energy and its release by elastic action. With impact, metal is lost or deformed as a result of chipping, upsetting, cracking, or crushing forces. In abrasion the wearing forces move parallel to the surface of a component, while impact forces are more or less perpendicular to the absorbing members.

Corrosion is the destruction of a surface from atmospheric contamination, chemicals, and oxidation or scaling at high temperatures.

Unless you understand the service condition of the parts to be repaired, it is impossible to select the correct wear-resistant material. Thus, surfacing a metal for corrosion wear may not be suitable for impact resistance.

In most cases, surfacing requires the use of a single metal product. However, there are instances when the use of two alloys provides the best results. For example, if a part is extremely worn, a less expensive material may be used, followed by the appropriate surface deposit. The buildup metal may be a carbon steel of sufficient strength to resist deformation and still support the surface. On the other hand, if the part is subject to impact loading or high temperatures, carbon steel will not be adequate for buildup and an alloy buildup material will have to be used.

Metals That Can Be Surfaced

Metals that can be surfaced are in two groups:

1. Metals whose physical properties are NOT changed nor are not likely to crack during surfacing. Includes low-carbon and medium-carbon steels, low-alloy steels, and stainless steels.
2. Metals whose physical properties ARE changed during surfacing. These metals have usually been hardened and any exposure to heat may undo their heat treatment and produce cracks. Includes high-carbon steels, cast irons, and other alloy steels.

Metals in Group 1 can be surfaced without any particular precautions since no harmful cracking will result.

With metals in Group 2, special care must be taken to reduce the sudden shock of localized heat. This is done by reducing the hardness by annealing or through gradual and uniform preheating and post cooling. Preheating from 300° to 500°F (150° to 260°C) will usually prevent weld hardening in medium and high-carbon steels. High-carbon alloy steels and wear-resistant alloy steels require preheating to the same temperature. After surfacing is completed, reheating to 800° to 1300°F (425° to 705°C) and slow cooling should follow.

Surfacing Materials

There are many different types of surfacing materials. Most of them have a base of iron, nickel, copper, or cobalt. Auxiliary elements may be carbon, chromium, molybdenum, tungsten, silicon, manganese, nitrogen, vanadium, and titanium.

The alloying elements form hard carbides that contribute to the matrix properties of the surfacing metals. Thus, a high percentage of tungsten or chromium with a high carbon content will form high carbide crystals that are harder than quartz. Materials having a high chromium content provide excellent resistance to oxidation and scaling. Nickel, cobalt, and chromium are particularly effective for corrosion resistance.

The matrix of surfacing metals is either martensitic, pearlitic, or austenitic. Martensite is the hardest and strongest. Pearlite is moderately tough and hard. Austenite is soft, tough, and has good impact qualities. (For details, see Chapter 12.)

Surfacing metals are available as rods for oxyacetylene welding, electrodes for shielded metal-arc welding, and hard wire for automatic welding (Fig. 3). Some surfacing materials come in tubular rods that contain a mixture of powder metal, powder ferralloys, and fluxing ingredients. The same material is available in powder form for surfacing with the carbon arc and the oxyacetylene powder spray torch.

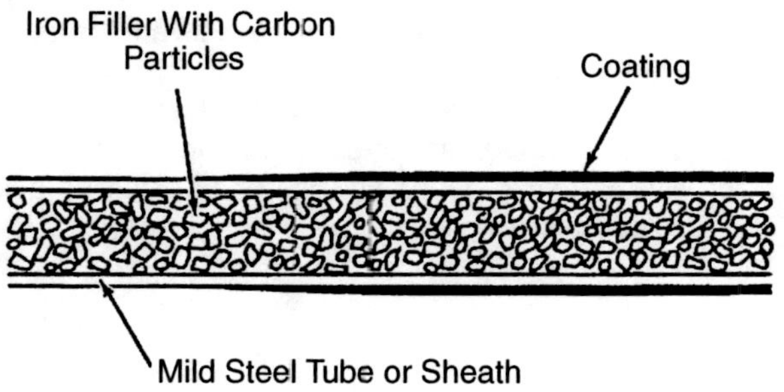

Fig. 3 — Metal Surfacing Electrode

Although many surfacing materials are designated by manufacturers' trade names, some also carry AWS-ASTM codes.

AWS-ASTM has classified surfacing materials into the following types: high speed steel (Fe5), austenitic manganese steel (FeMn), austenitic high chromium iron (FeCr), cobalt-base metals (CoCr-A and CoCr-C) copper-base alloy metals (CuZn, CuSi, and CuAl), and nickel chromium-boron metal (NiCr).

The coding system identifies the important elements of the surfacing material. The prefix R designates a welding rod and the prefix E an electrode. Suffixes are sometimes used to identify a classification within a basic group. Certain materials also carry a further subdivision in the form of digits after the suffix.

Surfacing With Oxyacetylene

The oxyacetylene surfacing process is very useful in depositing overlays on small parts such as engine valves, plowshares, tools, and other parts.

With the oxyacetylene flame, tiny areas can be surfaced and thin layers applied smoothly. Preheating and slow cooling are readily controlled, minimizing cracking even with brittle surfacing materials. The prime limitation of the oxyacetylene technique is its low depositing speed.

Metals for surfacing with oxyacetylene generally consist of low-melting, high-carbon filler rods such as high chromium or Cr-Co-W alloy. As a rule, a slightly carburizing flame is recommended as this will add carbon to the deposit.

The surfacing operation is started by preheating the surface to produce a "sweating" condition. During the preheating cycle, the tip of the surfacing rod is held on the fringe of the flame. The rod is then moved into the center of the flame and melted (Fig. 4). The actual depositing of the filler rod is done with a regular welding technique using a slightly weaving motion.

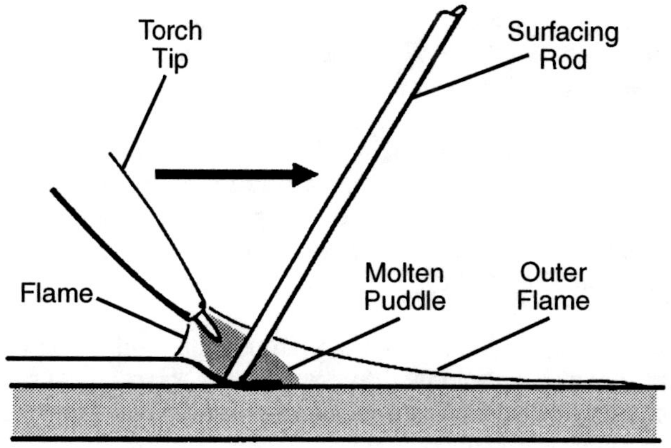

Fig. 4 — Metal Surfacing With Oxyacetylene Torch

Surfacing With Shielded Metal Arc Welding (SMAW)

Surfacing with the shielded metal arc is more widely used because of its fast deposition rate. It also has wide application where large areas have to be surfaced or for heavy parts that normally would require excessive time to heat with the oxyacetylene flame. This method is especially suitable for depositing overlays on manganese steel and other steel alloys where heat buildup must be restricted.

Either AC or DC current produces satisfactory welds. The electrodes may be coated solid wires or hollow tubes containing alloy powder and flux.

Types of Surfacing Electrodes

Surfacing electrodes are normally classified in five ways:

1. Resistance to severe impact
2. Resistance to very severe abrasion
3. Resistance to corrosion and abrasion at high temperatures
4. Resistance to severe abrasion with moderate impact
5. Resistance to abrasion with moderate to heavy impact

(Fig. 5) shows typical parts that require surfacing in each of the five groups.

Resistance to Severe Impact

These electrodes are intended for repair and buildup of manganese steel, of which most impact resistance parts are made. The electrodes are either solid drawn, that is, they contain all alloying elements in the wire itself, or composite in which all of the alloying elements are in the flux coating.

Deposits of these electrodes are not hard but are very tough. They are often referred to as selfhardening (also called work hardening) because the deposited surface hardens as it is pounded or "worked." While the outside surface is hard, the material underneath remains soft.

Typical equipment requiring impact protection includes: crusher parts, bucket fronts, hammermills, and rolling mill parts, jaws and teeth on dredges, ditchers and power shovels, and other similar heavy Industrial and construction equipment (Fig. 5, No. 1).

Stainless steel electrodes are often used for surfacing parts that must resist impact forces without cracking. These electrodes offer the least resistance to abrasion in the "as deposited" condition; however, they will work harden. Stainless steel and low hydrogen electrodes are frequently used as base layers for other surfacing electrodes.

Resistance to Very Severe Abrasion

Electrodes in this group are of the tungsten carbide and chromium carbide types. These electrodes deposit a very hard abrasive-resistant material. They are not suitable for impact wear, since the material they deposit chips and cracks when subjected to shock.

Tungsten electrodes have tiny crystals of tungsten carbide embedded in the steel alloy. When applied on a surface, the steel wears away, leaving toothlike particles of tungsten carbide exposed. Since tungsten carbide is very hard, the exposed particles make the edge of the part self-sharpening. The property is particularly desirable for earth digging equipment, scraping tools, plowshares, rotary digger blades, cultivator sweeps, and similar machinery (Fig. 5, No. 2).

Still another type of tungsten carbide electrode deposits fine particles of tungsten carbide which are so close that they form a smooth cutting edge. These electrodes are useful in repairing steel cutting edges such as lathe tool bits.

Chromium carbide electrodes are slightly less hard and less abrasion-resistant than the tungsten carbide type but are tougher. Most of them are not affected by heat treatment and are too hard to be machined. In addition to being hard, chromium carbide electrodes produce surfaces that provide better protection against corrosion.

Resistance to Corrosion and Abrasion at High Temperatures

A great deal of wear on metal surfaces occurs at high temperatures, particularly where metal-to-metal contact is involved. In cases where friction is the cause of high operating temperature, special cobalt base alloy electrodes are used for surfacing. These electrodes are capable of withstanding temperatures of 1200°F (650°C) and higher. Cobalt alloy electrodes are excellent for surfacing valves and valve seat facings, chain-saw guide bars, pump shafts, rocker arms, arbors, and conveyor screws (Fig. 5, No. 3).

Nickel-base alloys also are effective for surfacing metals that are subjected to corrosion and oxidation at high temperatures. For surfacing corrosion-resistant parts that are not used at elevated temperatures, copper-base alloys are sometimes recommended. Copper-base alloy filler metals have wide application in surfacing areas that come in contact with corrosion-generating materials such as acids, mild alkalies, and salt water.

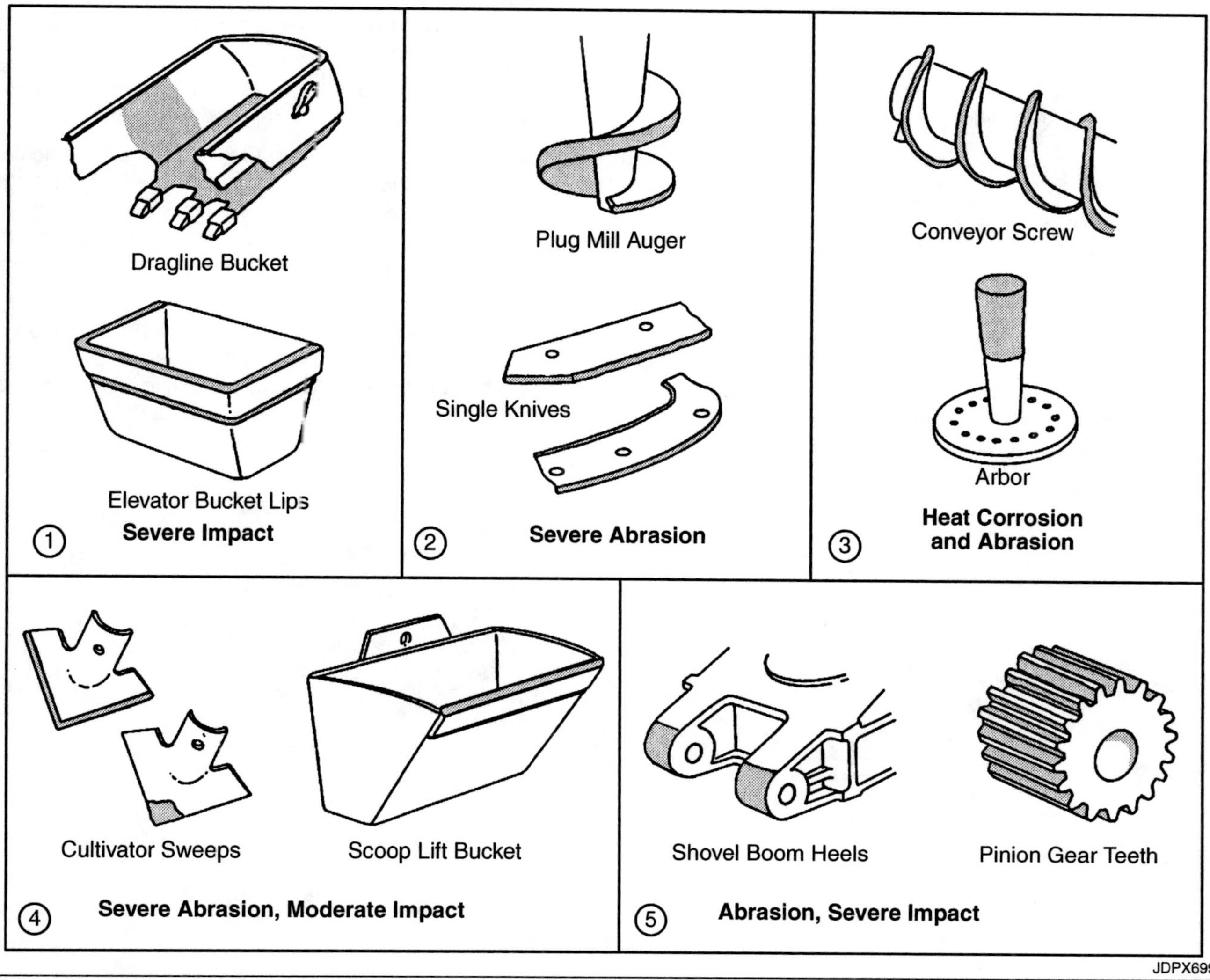

Fig. 5 — Typical Parts That Require Metal Surfacing

Resistance to Severe Abrasion With Moderate Impact

Industrial, transportation, and construction equipment is frequently subject to a combination of wear factors, abrasion, and impact. Alloys used for this are carbon-chromium types ranking in hardness from 54 to 65 Rockwell C. In abrasion resistance they are second only to the tungsten carbide alloys. These carbon-chromium alloys lay down a smooth-surfaced deposit that is excellent for moving, cutting, or handling of loose sliding materials such as earth, sand, gravel, and cement(Fig. 5, No. 4).

Resistance to Abrasion With Moderate to Heavy Impact

Many industrial machines such as cable tools, mud pumps, bucket fronts, and churn drills require surfacing alloys that are resistant to abrasion as well as to moderate or heavy impact. These alloys have an iron base and can be heat treated and sharpened to a knife edge (Fig. 5, No. 5)

How to Surface With an Arc

1. Thoroughly clean the metal to be surfaced. Remove all rust, scale, paint, etc.
2. Set the power source for only enough amperage to maintain an arc. This is important to minimize dilution of the deposit by the base metal.
3. For best results, place the work in a flat position when surfacing (Fig. 6).

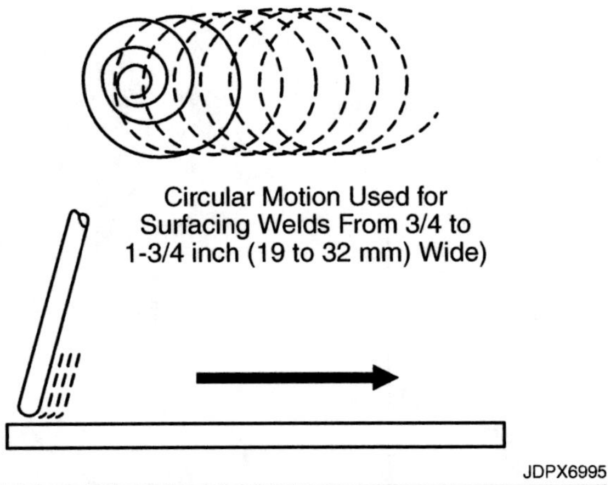

Fig. 6 — Surfacing With an Arc Welding Electrode

4. Use a medium-long arc and move with a straight or weaving motion-circular motion for wider surfacing welds as shown in Fig. 7.
5. If several layers are to be deposited, remove all slag before extra layers are made.
6. Overlap welds as shown in Fig. 7. Carefully manipulate the electrode as necessary for complete fusion into the adjoining welds. This can be done by holding the electrode over the deposited weld a moment to allow the heat to build up in the adjoining welds. This will also prevent undercutting.
7. Use a whipping action when surfacing an area along a thin edge. Hold the arc over the heavy portion and then whip it out to the thin edge. In this way a shallow deposit is made before the heat builds up enough in the base metal to burn through.

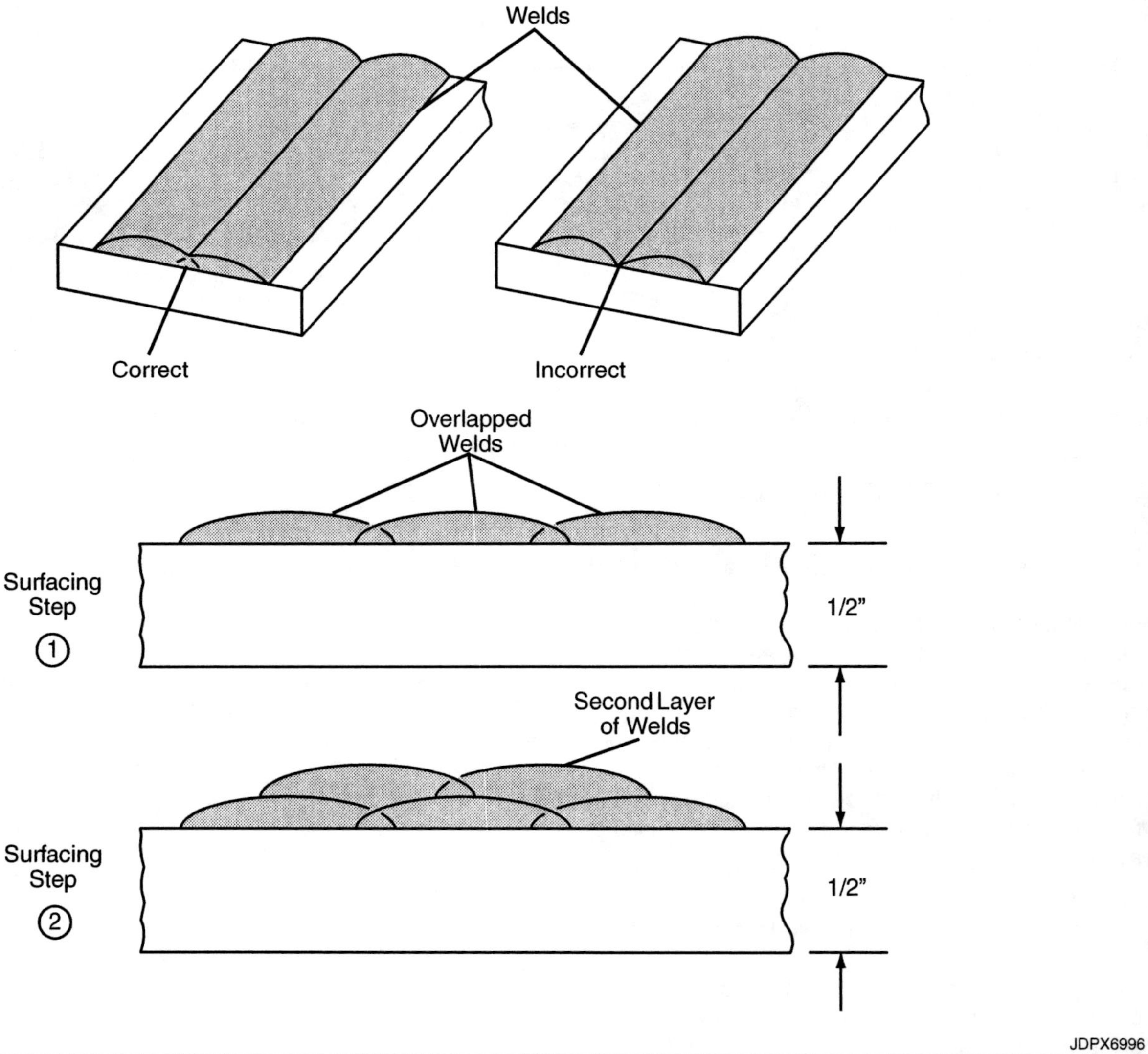

Fig. 7 — Proper Way to Overlap Surfacing Beads When Surfacing With an Arc

Metallizing

Metallizing is a process of depositing fine semi-molten particles or metal powder onto the surface of a metal to form an adherent coating. It is also referred to as "metal spraying" (Fig. 8).

The powder or metal particles are impelled through an intense heat and are impinged on the surface, where they form thin layers of metallic lamellates. The powder or wire rod is fed into an oxygen-fuel gas flame, and the small semi-molten droplets are driven onto the surface by a stream of high pressure air. The minute particles strike the surface at estimated speeds of 250 to 500 ft. (76 to 152 m) per second, depending on the gun design.

The metal spray coheres by the mechanical interlocking and fusion of the tiny metallic particles and by the bonding of the thin oxide film that forms on the particles while in motion.

Where Metallizing Is Used

Metallizing has wide use in the machine field where worn surfaces need to be restored to their original sizes but where low tensile strength and porosity are not objectionable. It is very useful for parts where welding or brazing heat is impractical or for applying deposits of dissimilar metals which otherwise are not possible.

The process is unique since there is no limit to the size of the object or structure that can be coated. An extra feature of metallizing is that no preheating or postheating is required.

How Metallizing Works

The success of any spraying process depends on having a clean surface that has been properly roughened. All traces of oil, dirt, scale, rust, etc., must be removed. The roughing of the surface provides mechanical anchorages for the sprayed metal particles.

Grit blasting is probably the most common method used for surface roughing. Blasting abrasives may be steel grit, hard sand, aluminum oxide, or silicon carbide.

Sometimes after a surface is prepared for spraying, a thin layer of molybdenum is sprayed on. This produces a fusion bond that gives greater adherence to the spraying coats. Another method is to go over the area, producing ragged threads.

The sprayed metal coatings are somewhat porous, but in machine elements this is an advantage. The porous coating absorbs oil, which provides more complete lubrication. Porosity becomes more critical in applications that are subject to severe attacks of corrosive materials.

Porosity can be partly controlled by the adjustment of gas and air as well as the distance of the gun from the workpiece. However, too much reduction of porosity will usually result in hard, brittle, and highly oxidized coatings.

Wire Spray Guns

Metallizing is done with a special spray gun (Fig. 8), which weighs from 3 to 6 pounds (1.4 to 2.7 kg) and handles 20 gauge to 3/16-inch (0.9 to 4.8 mm) diameter wire. Usually these guns will spray 4 to 12 pounds (1.8 to 5.5 kg) of metal per hour. Larger guns are often mounted on a fixture.

Oxidation normally occurs in the melting flame and during the flight of the metal particles to the surface. As a rule, little oxidation will take place as the metal is melted unless the gas-fuel mixture is oxidizing. The greatest causes of oxidation are overheating of the coating, excessive use of oxygen, and spraying too great a distance from the workpiece.

The gun has two major parts: the wire feeder and the gas head. The power unit feeds the wire into the nozzle of the gun. The gas head controls the flow of oxygen, fuel gas, and compressed air. The nozzle has a center orifice through which the wire is fed (Fig. 8). Around this orifice are a number of gas jets that provide the flame and high velocity air stream. As the wire comes through the orifice, it is melted and atomized by the flame. The fine molten particles are picked up by the air stream and projected against the work. The most common gas for the oxyfuel flame is acetylene, which produces a temperature exceeding 5600°F (3100°C), although hydrogen or propane is sometimes used for lower melting metals.

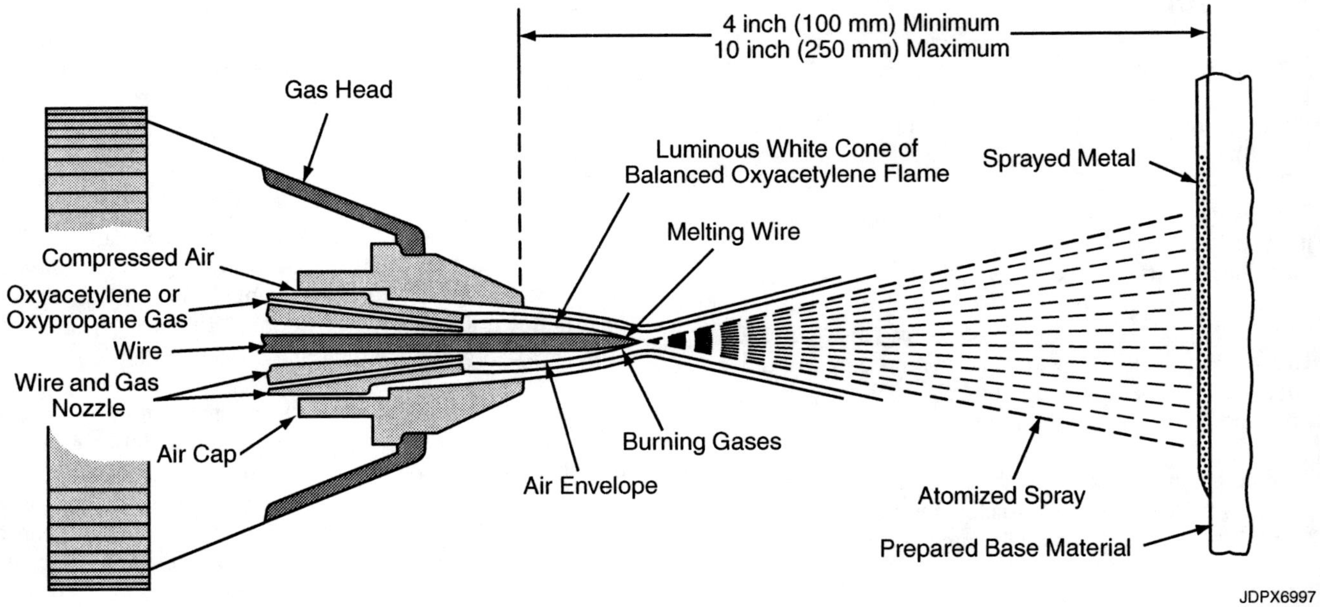

Fig. 8 — Metallizing Spray Gun

Powder-Type Guns

Guns used for spraying metal powder are somewhat similar to wire guns except that they do not have a power unit. The powder is carried from a reservoir by compressed air to the gun where it is forced through the central orifice of the gun. The gas flame then melts the tiny particles which are carried to the work by compressed air as in the wire-spray gun.

How to Metal Spray

1. Do any metal spraying operation in a well-ventilated area. This is necessary to remove dust particles and fumes, which are extremely hazardous to health. If positive ventilation is not possible, wear a respirator.
2. The wire speed, amount of spray, gas, and oxygen pressure must be regulated per recommendations for the equipment used and the type of metallizing. As a general rule, air pressure is set for 60 psi (415 kPa). The use of a flowmeter will ensure more accurate control of the gases. A slight increase in air pressure provides a finer coating, while a decrease of air pressure produces a coarser coating.
3. The tip of the melting wire should project beyond the end of the air cap. This length depends to a large extent on the material being used. A recommended practice is to speed up the wire until chunks are ejected. Then reduce the wire feed until the ejection of chunks stops.
4. Keep each coating as light as possible-about 0.003 to 0.005 inch (0.08 to 0.13 mm) in thickness. Too heavy a coat will produce an irregular and stratified surface. The actual movement of the gun is very similar to paint spraying. Keep the nozzle 4 to 10 inches (100 to 250 mm) away from the surface and move with a uniform motion. If the gun is held too close to the work, minute cracks will form in the coating. Too great a distance will produce a soft spongy deposit. The rate of gun travel is also important. When the travel is too rapid, the coating develops a high oxide content.
5. When spraying a flat surface, move the gun back and forth to allow a full uniform deposit. Spraying should begin beyond the edge of the area to be covered and continued beyond the end of the area. After the first layer, the work or gun is often rotated 90 degrees. This technique is repeated for each subsequent coating until the required thickness is built up. On cylindrical pieces, the work is generally fastened in a lathe with the gun mounted on the traveling carriage.

Torch Spraying

Surfacing and buildup of small objects is simplified by using a spray torch. This is a hopper and spray control that fits on any regular oxyacetylene torch body.

Metal powder is simply placed in the hopper, the torch is lighted, and the tip directed over the area to receive an overlay. The powder is pulled into the oxyacetylene stream and becomes fluid when it hits the flame. The surface of the metal must be at "sweat" temperature (liquescent).

Fig. 1 illustrates a spray torch being used to build up a turbocharger shaft as the shaft rotates in a lathe.

Test Yourself

Questions

1. What is the difference between surfacing and metallizing?
2. Parts that are surfaced are subject to what three basic types of wear?
3. In surfacing metals that are very susceptible to cracking, what can be done to minimize the sudden shock of localized heat?
4. Why is the shielded metal arc used more than oxyacetylene in surfacing?
5. When surfacing with an arc, should amperage be as high or as low as possible?
6. When surfacing with an arc, should the arc be medium short or medium long?
7. In metallizing, how do the tiny particles of metal become bonded to a surface?
8. What are the two procedures in preparing a surface for metallizing?

Properties of Metals

11

JDPX6978

Fig. 1 — Magnified Views of Two Carbon Steels

Introduction

To weld, you must have a basic knowledge of metals. All of the problems of metal casting are present in fusion welding — expansion, contraction, shrinkage, cracks, hardening, and brittleness.

Welding affects any metal. Sometimes the effects are slight, but they can be harmful if certain rules are not followed.

Metal Numbering Systems

Several standardizing organizations have numbering systems for the designation of metals. Basic information is given about these systems so that the proper standardizing organization can be identified from the number designation for a piece of metal. These organizations are listed below and should be contacted if more information is needed on a particular system or about a specific metal.

- Society of Automotive Engineers (SAE)
- American Society for Testing and Materials (ASTM)
- American Iron and Steel Institute (AISI)
- American National Standards Institute (ANSI)
- American Society of Mechanical Engineers (ASME)
- Aluminum Association (AA)

Unified Numbering System (UNS)

The SAE and ASTM developed the Unified Numbering System (UNS) as a means for correlating the different numbers for some of the common metals and alloys. In many instances, the UNS numbers contain the same number sequences given by the original organization. For example, an AISI 1018 steel would be a G10180 in the Unified Numbering System. A UNS designation begins with a letter followed with a 5-digit number. The numbers within each grouping begin with 00001 and end with 99999. Table 1 gives the essentials of the system.

Nonferrous Metals and Alloys	
Letter	**Metal**
A	Aluminum and aluminum alloys
C	Copper and copper alloys
E	Rare earth and rare earth-like metals and alloys
L	Low melting metals and alloys
M	Miscellaneous nonferrous metals and alloys
P	Precious metals and alloys
R	Reactive and refractor metals and alloys
Z	Zinc and zinc alloys
Ferrous Metals and Alloys	
Letter	**Metal**
D	Specified mechanical property steels
F	Cast irons
G	AISI and SAE carbon and alloys steels*
H	AISI H-steels
J	Cast steels*
K	Miscellaneous steels and ferrous alloys
S	Heat and corrosion resistant (stainless) steels
T	Tool steels
** Except tool steels*	

Table 1 — Letter Designation of the Unified Numbering System

AISI-SAE System for Carbon and Alloy Steels

This system uses mostly a 4-digit number (sometimes 5 digits or 3 digits) for designating different steels. (See Table 2) The first digit gives the kind of steel, the second digit gives (usually) the approximate percentage of the major alloy represented by the first digit, and the last two digits give the average percentage of carbon the steel contains in points or hundredths of 1%. (One percent, 1.00%). For example, the explanation for each of the numbers in the steel designated 1050 would be:

1 = carbon steel, 0= no major alloys, 50 = 50 points (0.5% or 0.005) carbon. Table 2 lists the essentials of the system.

AISI Number System for Stainless Steel

The stainless steels are designated by a 3-digit number. Also, stainless steel is sometimes grouped according to its resistance to corrosion and its carbon content. Stainless steels that have a much lower carbon content compared to a regular stainless steel will contain the letter "L" after the number designation. For example, a 304 stainless would have 0.08% (max.) carbon and a 304L would have 0.03% (max.) carbon. Table 3 lists the essentials of the system.

AISI-SAE Number System for Tool Steels

Seven categories for commonly used tool steels are given in Table 4. Each group type is indicated by a letter and in most instances followed by a number. Examples of individual tool steels would be D2, H13, M2, 01, and S7.

Type of Steel	Number
Carbon	10XX
Carbon	11XX
Carbon	12XX
Carbon	15XX
Manganese	13XX
Nickel	23XX
Nickel	25XX
Nickel-chromium	31XX
Nickel-chromium	32XX
Nickel-chromium	33XX
Nickel-chromium	34XX
Molybdenum	40XX
Molybdenum	44XX
Chromium-molybdenum	41XX
Nickel-chromium-molybdenum	43XX
Nickel-chromium-molybdenum	47XX
Nickel-chromium-molybdenum	81XX
Nickel-chromium-molybdenum	86XX
Nickel-chromium-molybdenum	87XX
Nickel-chromium-molybdenum	88XX
Nickel-chromium-molybdenum	93XX
Nickel-chromium-molybdenum	94XX
Nickel-chromium-molybdenum	97XX
Nickel-chromium-molybdenum	98XX
Nickel-molybdenum	46XX
Nickel-molybdenum	48XX
Chromium	50XX
Chromium	51XX
Chromium-carbon	50XXX
Chromium-carbon	51XXX
Chromium-carbon	52XXX
Chromium-vanadium	61XX
Tungsten-chromium-high carbon	72XX
Silicon-manganese	92XX
High-strength low-alloy	9XX

Table 2 — The AISI-SAE Numbering System for Carbon Steel

AISI System for Stainless Steel	
Number (Series)	Major Alloying Elements
2XX	Chromium-nickel-manganese
3XX	Chromium-nickel
4XX	Chromium (low—under 4%)
5XX	Chromium (high—under 4 to 6%)

Table 3 — AISI System for Stainless Steel

AISI-SAE System for Stainless Tool Steels		
Letter	Group Type	Category
M	Molybdenum	High Speed
T	Tungsten	High Speed
H1-19	Chromium	Hot Work
H20-39	Tungsten	Hot Work
H40-59	Molybdenum	Hot Work
D	High Carbon & Chromium	Cold Work
A	Medium Alloy, Air Hardening	Cold Work
O	Oil Hardening	Cold Work
S		Shock Resisting
P		Mold Steels
L	Low Alloy	Special Purpose
F	Carbon Tungsten	Special Purpose
W		Water Hardening

Table 4 — AISI-SAE System for Stainless Tool Steels

ASTM System for Carbon Steel

The ASTM uses a letter and 2- or 3-digit number for designating steel. Each designation may also have a grade or class. Designations for two common mild steels are A36 and A53 Grade B.

AA System for Aluminum

The Aluminum Association uses a 4-digit numbering system for designating wrought (rolled or extruded shapes) and cast aluminum alloys. Examples of these numbers would be XXXX for wrought alloys and XXX.X for cast alloys of aluminum. Table 5 illustrates the AA number system.

The alloy numbers may be followed with a letter suffix and sometimes a number which further indicates if the aluminum has been annealed, hardened, heat-treated, or tempered. These letters are "F," "0," "H," "W," and "T." The number for a popular structural aluminum is designated as a 6061 T-6. The "T-6" means the aluminum has been solution heat-treated and then artificially aged to obtain the desired hardness (temper).

AA Numbering System		
Number Wrought	Cast	Major Alloy
1XXX	1XX.X	Aluminum 99.0% and greater
2XXX	2XX.X	Copper
3XXX		Manganese
	3XX.X	Silicon w/ Copper and/or Magnesium
4XXX	4XX.X	Silicon
5XXX	5XX.X	Magnesium
6XXX		Magnesium and Silicon
	6XX.X	Unused series
7XXX	7XX.X	Zinc
8XXX		Other elements
	8XX.X	Tin
9XXX		Unused series
	9XX.X	Other elements

Table 5 — Aluminum Association's Numbering System

Ferrous and Nonferrous Metals

Metals are divided into two major groups:

- Ferrous metals
- Nonferrous metals

FERROUS metals contain iron and its alloys such as steel (Fig. 2).

NONFERROUS metals are those with little or no iron (Fig. 2). They include copper, brass, zinc, bronze, lead, and aluminum.

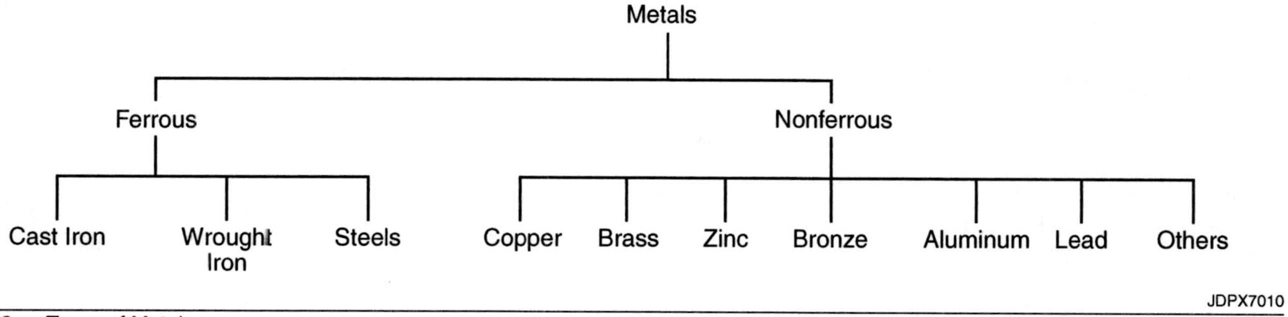

Fig. 2 — Types of Metals

Most welding is done on ferrous metals, but recent developments have made welding possible on most nonferrous metals as well.

Melting Points of Metals

The melting points of various metals and alloys are vital to good welding. The chart in Fig. 4 gives this data.

Iron and Steel

Iron is produced by reducing iron oxide, commonly called iron ore, to pig iron by means of a blast furnace.

Carbon steel is an alloy of iron and controlled amounts of carbon. Alloy steel is a combination of carbon steel and controlled amounts of other desirable metal elements.

The percentage of carbon content determines the type of carbon steel. For example, wrought iron has 0.003 percent carbon, meaning three thousandths of one percent. Low carbon steel contains less than 0.30 percent carbon. Medium carbon steel varies between 0.30 and 0.45 percent carbon content. High carbon steel contains approximately 0.45 to 0.75 percent carbon, and very high carbon steel contains between 0.75 and 1.50 percent carbon.

Cast iron contains 2.25 to 4 percent carbon. Some is in the noncombined form and some is in the form of graphite in gray cast iron.

The carbon generally combines with iron to form iron carbide, a very hard, brittle substance. This action means that as the carbon content of the steel increases, the hardness also tends to increase.

Various heat treatments are used to enable steel to retain the strength of its higher carbon content, and yet not have the extreme brittleness usually found with high carbon steels. Also, certain other substances such as nickel, chromium, manganese, vanadium, and other alloying metals may be added to steel to improve its physical properties.

A welder must also have an understanding of the impurities occasionally found in metals and their effect upon the weldability of the metal.

Two of the harmful impurities sometimes found in steels are phosphorus and sulphur. Their presence in the steel is due to their presence in the ore, or to the method of manufacture. Both of these impurities are detrimental to the welding qualities of steel; therefore, during manufacture, extreme care is always taken to keep the impurities at a minimum (0.50 percent or less).

Sulphur improves the machining qualities of steel, but it is harmful to its hot-forming properties.

During welding, sulphur or phosphorus tend to form gas in the molten metal, resulting in gas pockets in the welds. Brittleness is also increased because of the sulphites and phosphates formed.

Another impurity is dirt or slag (iron oxide) imbedded in the metal. These impurities may produce porosity in the weld and reduce the physical properties of the metal.

Physical Properties of Metals

A physical property is a characteristic of a metal which may be observed or measured.

The physical properties of steel are affected by the following:

- Carbon content
- Impurities
- Addition of various alloying metals
- Heat treatment

Testing machines are being used in welding shops to allow operators to identify metals and to check on the physical properties of welds.

The vital physical properties of steels are:

- Tensile strength
- Ductility
- Hardness
- Brittleness
- Compression strength
- Elongation
- Malleability
- Toughness
- Grain size

Fig. 3 illustrates the types of stresses imposed on metal structures.

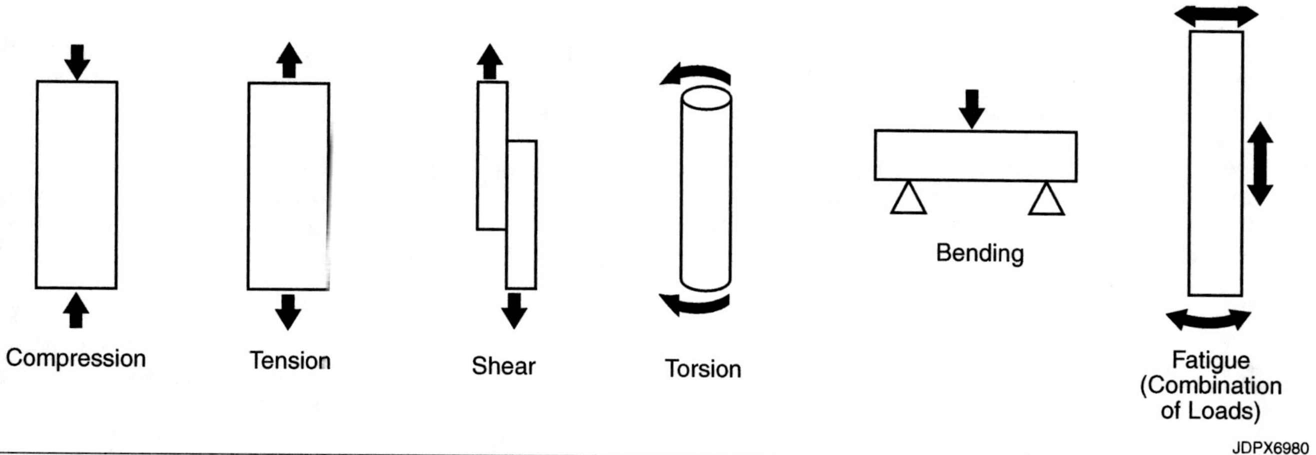

Fig. 3 — Types of Stresses or Loads on Metal Structures

Tensile strength is the ability of a metal to resist being pulled apart. This can be measured on a tensile testing machine which puts a stretching load on the metal.

Ductility is the ability of a metal to be stretched. A very ductile metal such as copper or aluminum may be pulled through dies to form wire. The ductility of iron and steel may be measured during a tensile test.

Hardness is the quality which allows a metal to resist penetration. The Rockwell, Brinell, or Scleroscope methods may be used to test hardness in iron or steel.

Brittleness in a metal means it will fracture easily if bent sharply or struck. Cast iron, especially in thin sections, is very brittle.

Compression strength of a metal is a measure of how much squeezing force it can withstand before it fails. The metal to be tested is mounted in a tensile tester, but instead of pulling on the metal, it is squeezed or compressed.

Yield strength is the force in psi or kgSmm at which a weld specimen starts to pull apart or assumes a specified limiting permanent set.

Elongation is measured by how much a metal will stretch before it breaks. The percent of elongation is measured during a tensile test. Two marks exactly 2 inches apart are placed on the tensile sample and the distance between the marks is again measured after the sample breaks. The original distance between the marks is compared to the increase in length to determine the percent of elongation. Cast iron is brittle, so its percent of elongation is very low.

Malleability is the tendency of a metal to be hammered into shape. Low carbon steel is malleable, while cast iron is not.

A metal is said to be tough if it can withstand repeated bending or twisting.

The examination of the grain size of a known metal under a microscope will give a good idea of the brittleness of a metal, its tensile strength, the heat treatment, and its ductility.

Before these properties can be studied in detail, you must have an understanding of the effect of carbon on the properties of steel, and a good knowledge of alloys in general.

Alloy Metals

An alloy metal is an intimate mixture of two or more metals.

Any ferrous or nonferrous metal may be alloyed to form an alloy metal with new and desirable characteristics.

Steel is an alloy of iron and controlled amounts of carbon. Alloy steels are created by adding other metals to plain carbon steel.

Some metals which are alloyed with carbon steel and their qualities are:

Chromium — Increases resistance to corrosion; improves the responsiveness to heat treatment.

Manganese — Increases strength and responsiveness to heat treatment.

Molybdenum — Increases toughness and improves the strength of steel at higher temperatures.

Nickel — Increases strength and toughness; increases impact resistance.

Tungsten — Produces dense, fine grain; helps steel to retain its hardness; helps steel to retain strength under high heat.

Vanadium — Retards grain growth and improves impact resistance.

The melting point of a metal is changed somewhat whenever an alloying metal is added.

Note in Fig. 4 the various melting points of steels as compared to cast iron. For example, while cast iron melts at about 2300°F (1260°C), a high carbon steel melts at about 2550°F (1400°C).

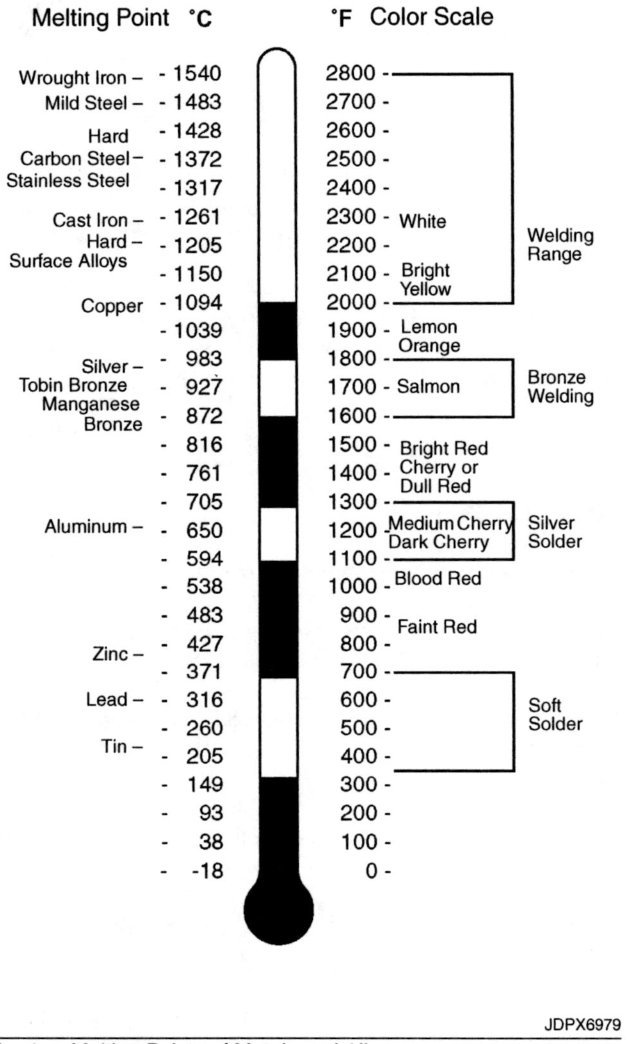

Fig. 4 — Melting Points of Metals and Alloys

Another example of a simple alloy is the combination of lead and tin. The melting temperature of lead is 635°F (335°C), while tin has a melting temperature of 432°F (222°C). However, as the two metals are mixed, any combination of the two results in a lower melting temperature than 635°F (335°C). At a certain proportion of the metals, the lowest melting temperature is reached, which is lower in a simple alloy than the melting temperature of either of the two pure metals separately.

Cooling Curves

The temperature of a metal (not an alloy) under atmospheric pressure remains constant during the change from the liquid to the solid state.

As most metals are heated to their melting temperature, a thermometer will show a constant rise in temperature per unit of time, with a constant heat input, until the metal reaches its melting temperature.

However, as the metal melts, the temperature will remain constant for a length of time. During this time, the molecular structure of the metal absorbs heat energy, which is required to change the position of the metal atoms in the structure of the molecule as the metal changes from a solid to a liquid state. After the metal is melted, the temperature will again rise as the metal is heated.

As a metal cools from the liquid state, the temperature drops until the metal solidifies. At this point, the temperature will remain constant while the atoms in the metal molecules change back to the solid state molecular structure. During this time, while the atoms are changing position in the molecule, the molecule releases heat. After the metal has solidified, the temperature will again drop until it reaches room temperature.

In these alloys, the absorption of energy during heating, or energy release during cooling, causes a delay in the temperature change while internal energy changes are taking place, even though the alloy stays in the solid condition.

These points signify a crystalline structure change in the solid metal. They cause the metal to expand (absorbing heat), or to contract while cooling as heat energy is suddenly released. These points are called critical points.

The heat treatment of a metal through this crystalline structure change is of great importance.

Heating Metals During Welding

The carbon content of steel has a vital effect on how various temperatures change its properties.

The heating of steel during fusion welding will ruin the heat treatment or any other properties added during manufacture. To renew these properties, the metal must again be heat treated or mechanically reworked.

See later in this chapter under "Heat Treatment of metals" for details.

Common Crystal Patterns of Metals

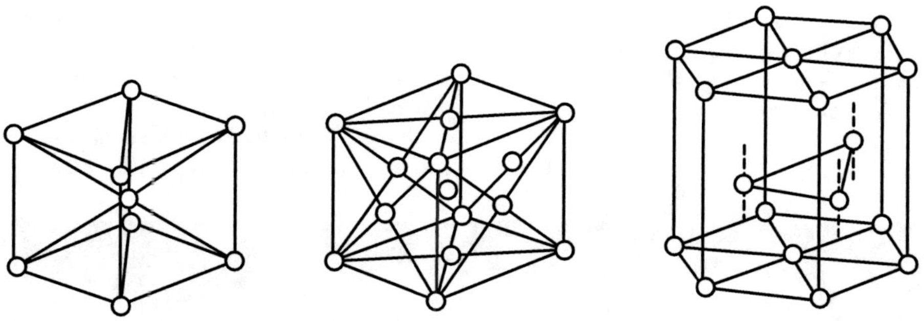

Fig. 5 — Common Crystal Patterns of Metals

Heating is related to the structure of the crystals in metals. Note in Fig. 5 the schematic shape of crystals for different metals. When iron is heated, for example, it has a different structure than when cooled.

When various amounts of carbon are added to iron, the structure of the metal is changed (Fig. 6). The ferrite structure is pure iron at lower temperatures, while the pearlite is iron with carbon added. The martensite is iron with carbon which is rapidly cooled and is hard and brittle.

Various heat treatments are used to get different iron and steel properties — hard or soft, ductile or brittle, etc.

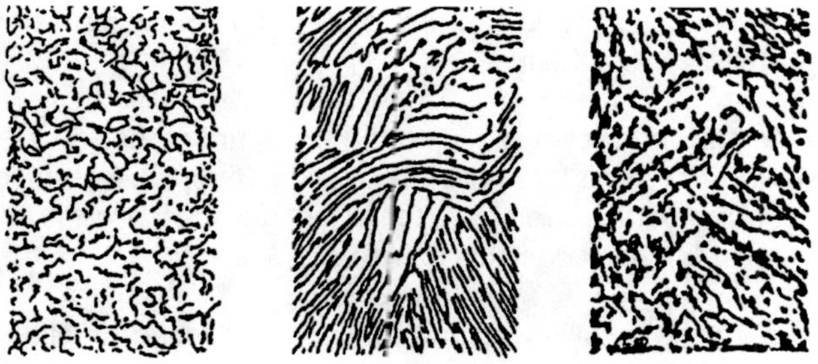

Fig. 6 — Different Iron Structures (Magnified)

How to Identify Iron and Steel

Because of the effect on the properties of steel caused by three variables, carbon content, temperature, and time, a welder must be able to determine accurately the nature of the steel being handled.

The manufacturer's specifications for the particular steel are most desirable. Whenever possible, the welder should obtain these specifications and keep them on file. At the same time, the welder should mark the metal to correspond with filed information.

Where manufacturer's specifications are not available, other methods may be used to determine the nature of the metal.

The most common tests for shop use are:

1. Spark test
2. File test
3. Fracture test
4. Color test
5. Density or specific gravity test
6. Ring or sound of the metal
7. Magnetic test
8. Chip test

Of these tests, number 2 through 8 should take place almost subconsciously as the welder works on the metal.

The spark test must be done under carefully prepared conditions.

Together or in the proper combinations, these tests will tell, with remarkable accuracy, the properties and identity of the metal.

Spark Test

A spark test can be used to separate ferrous from nonferrous metals and to identify different ferrous metals from each other. A first step in learning to identify a metal that has been plated or painted is with a spark test. Nonferrous metals do not produce sparks, while ferrous metals do.

In addition to using the spark patterns illustrated in Fig. 7, compare the spark patterns from known examples of metals to the spark patterns of the unknown metal. Pay particular attention for the presence of the different parts of the spark patterns of the unknown metal illustrated in Fig. 8. Some sparks will only have a few of these parts and others may have all of the parts. The presence or absence of one of the parts of a spark makes it less difficult to learn the identity of an unknown metal when compared to a known example.

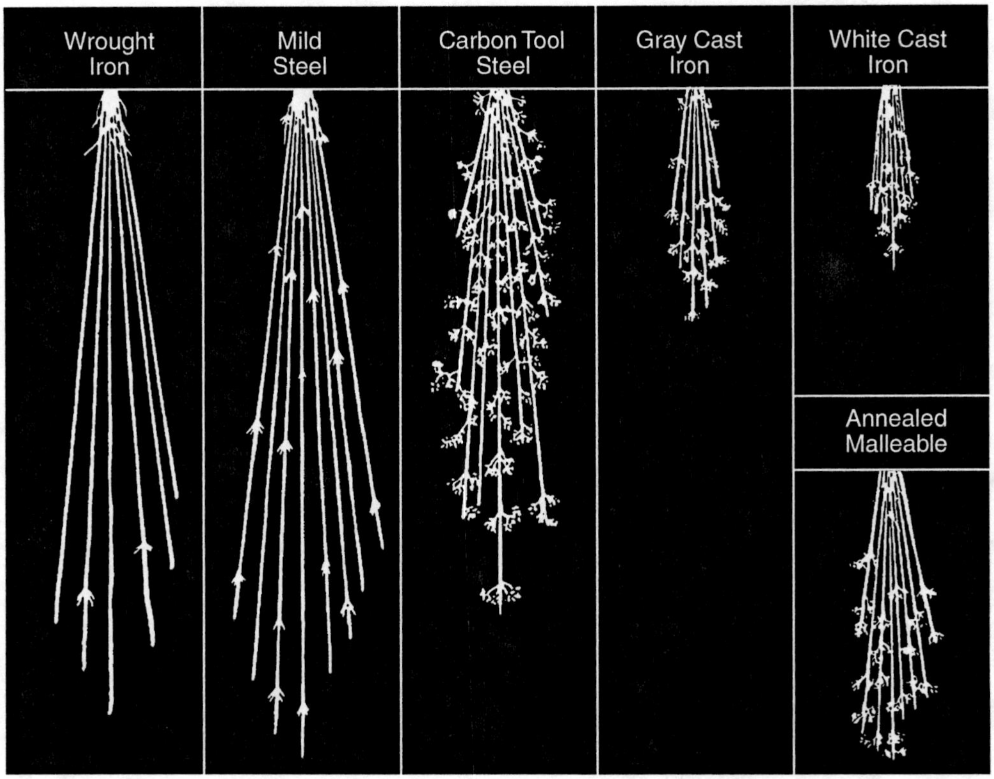

Fig. 7 — Spark Test for Common Cast Irons and Steels

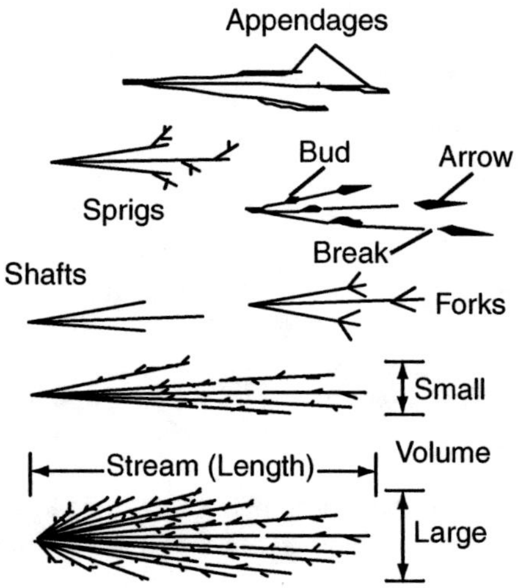

Fig. 8 — Parts of a Spark from Metal

A power grinder is used as the spark test equipment.

When grinding, you must always wear goggles. The grinder must be inspected to see that it is in good condition before proceeding with the test.

When testing a sample, if you touch the rim of the revolving wheel lightly with the metal, the friction of the wheel surface heats the particles removed to the incandescent and burning temperature as it contacts the metal.

The sparks resulting from the contact are found to differ in character for different steels.

The lighter the contact, the better. Use a black background to make it easier to identify the spark.

The theory of the spark test is that when a metal is heated, the different parts of the metal oxidize at different rates and the oxidation colors are different.

Relatively pure iron, when heated by the grinding wheel, does not oxidize quickly. Therefore, the sparks are long and fade out on cooling. As the carbon content of steel or cast iron increases, the compounds of carbon and iron have different ignition temperatures, and therefore the characteristics of the sparks differ.

Four characteristics of the spark generally tell the nature and condition of the steel. They are:

- Color of spark
- Length of spark
- Number of explosions (spurts) along the stream of the individual sparks
- Shape of the explosions (forking or repeating)

For example, a mild steel containing 20 points of carbon (0.20 percent) will show a long white spark that will jump approximately 70 inches (1780 mm). from the power grinder (using a 12-inch (305-mm) wheel). Some of these sparks will suddenly explode, shooting off smaller sparks at approximately 45° angles to the direction of travel of the original spark.

A 30-point carbon steel, will have sparks almost identical to the 20-point carbon steel, except that more streamlined sparks will explode while the total length of the sparks will decrease slightly. This shows that as the carbon content of steel increases, the explosions of the sparks become more frequent. Also, as the carbon content increases, the length of the spark is decreased, because of the interruption during their flight of the individual, explosions, or extremely rapid oxidation of some of the materials in the incandescent particle.

In Fig. 7, for example, a (high) carbon tool steel containing 80-points carbon has a very short spark with explosions occurring so rapidly that it is hard to identify the longer streamlined iron shapes. The sparks dissipate very quickly.

When higher carbon content metals are tested, you may become confused between extremely high carbon steel and cast iron because the spark is somewhat the same. Heat treatments have the same confusing effect on the nature of the spark.

The shape of cast iron and high carbon steel sparks are quite different.

Also, the cast iron spark, when leaving the point of contact, is a dull red and the spark jumps only 20 to 25 inches (500 to 635 mm). from the wheel; however, the spark from a high carbon steel of approximately 130-point carbon is white, is faster in disappearing, and the length of the spark is usually a little longer. Both of these metals show a small spark explosion at the end of the spark flight. The amount of the spark is considerably less than one would expect from such a high carbon content.

File Test

A comparison of the filings from members of the "white metals" can be used to separate zinc die cast (also called "pot metal"), magnesium, and aluminum from each other. The filings from magnesium will almost "float like a feather" and the fillings from zinc die cast will "fall like a rock." Aluminum will be in between. Known examples of these metals to use for the comparison will make the identification process much easier. Also, clean the file or have a separate file to use with each piece. Otherwise the filings will become mixed and it will be impossible to see a difference between the known and unknown examples.

A file can also be used to check the relative hardness of a piece of ferrous metal or to see if the piece is heat treatable. Strike an arc with a GTAW torch on an edge (or someplace that will not affect the functioning of the piece) and then see if a file will cut the area melted by the arc. If a file will not cut the arc strike area, the metal is heat treatable and will require a more exact identification and special welding procedures. If the area is not noticeably affected (hardened) by the arc strike, normal welding procedures and filler metals can be used.

Other Metal Identification Tests

As an extra precaution, other tests are occasionally used by the welder:

Fracture Test

The fracture test is widely used and is the breaking of a portion of the metal in two. If it is a repair job, the fractured surface is ready for inspection. The appearance of the surface where the metal is cracked shows the grain structure of the steel. If the grains are large, the metal is brittle and weak; if the grains are small, the metal is usually hard and brittle, or soft and ductile. The fracture shows the color of the metal, which is a good way to tell one metal from another. The test also indicates the type of metal by the ease with which it can be fractured.

Color Test

The two main divisions of metals include the irons and steels, which are indicated by their typical gray-white color, and the nonferrous metals, which come in two general color classifications of yellow and white. Copper may be rather easily identified by a welder and the same applies to brass and bronze. Aluminum, white metal, aluminum alloys, zinc, and the like, are all of somewhat the same silver-gray color, although they may vary in shade.

Density or Specific Gravity Test

The metals may also be differentiated by means of the weight or density test of the specimen. A good example of identification by density or specific gravity is identifying aluminum and lead. Roughly speaking, their colors are somewhat similar, but anyone can readily distinguish between the two metals because of their respective weights.

Ring or Sound Test

The ring or sound test uses the sound of the metal as an easy means of identifying certain metals, after one has had some experience. It is used extensively for identifying heat-treated steels from annealing steels. It is also used to detect alloys from the base metal. An example of the latter is aluminum and duralumin. The pure aluminum sheet has a duller sound, or ring, than the duralumin, which is somewhat harder and has a more distinct ringing sound.

Magnetic Test

The magnetic test is a basic test used to identify iron and steel metals from the nonferrous metals. Generally, all steels are affected by magnetism while the nonferrous metals are not. However, some stainless steels are not magnetic.

Chip Test

Another test which requires some experience is the chipping test of a metal. Here the cutting action of the chisel indicates its structure and heat treatment.

Cast iron, for example, when chipped, breaks off in small particles, but with mild steel the chip tends to curl and cling to the original piece (Fig. 9).

The higher-carbon, heat-treated steels cannot be tested in this way because of their hardness.

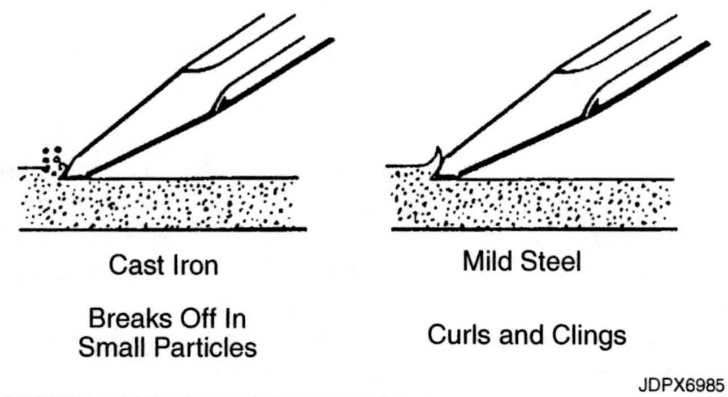

Fig. 9 — Chip Test for Iron and Steel

How to Identify Iron and Steel Alloys

Today's iron has been alloyed with carbon to produce a wide variety of steels for modern uses.

The two main fields are stainless and low-carbon alloy steels. These metals are very difficult to identify, so other tests are needed.

Spark Test for Alloy Steels

The appearance of the power grinder spark for alloy steels varies considerably, depending on the alloys included in the metal.

The basic test, the iron-carbon spark, is identical with the straight, iron-carbon steels. The number of the spark explosions increases as the length of the spark flight decreases, and the color of the spark becomes brighter and brighter as the carbon content increases.

The major spark appearance variations the alloy elements contribute to a steel are the multitude of right-angle, branch-off sparks that result from the different incandescent colors of the alloys in the metal (the change of color of the spark).

For example, a manganese steel has only a slight hint of the spark explosion, but the manganese causes the sparks to shoot out at right angles to the flight of the original spark. These sparks tend further to explode (repeating sparks), producing the appearance of a leafless tree branch (Fig. 10).

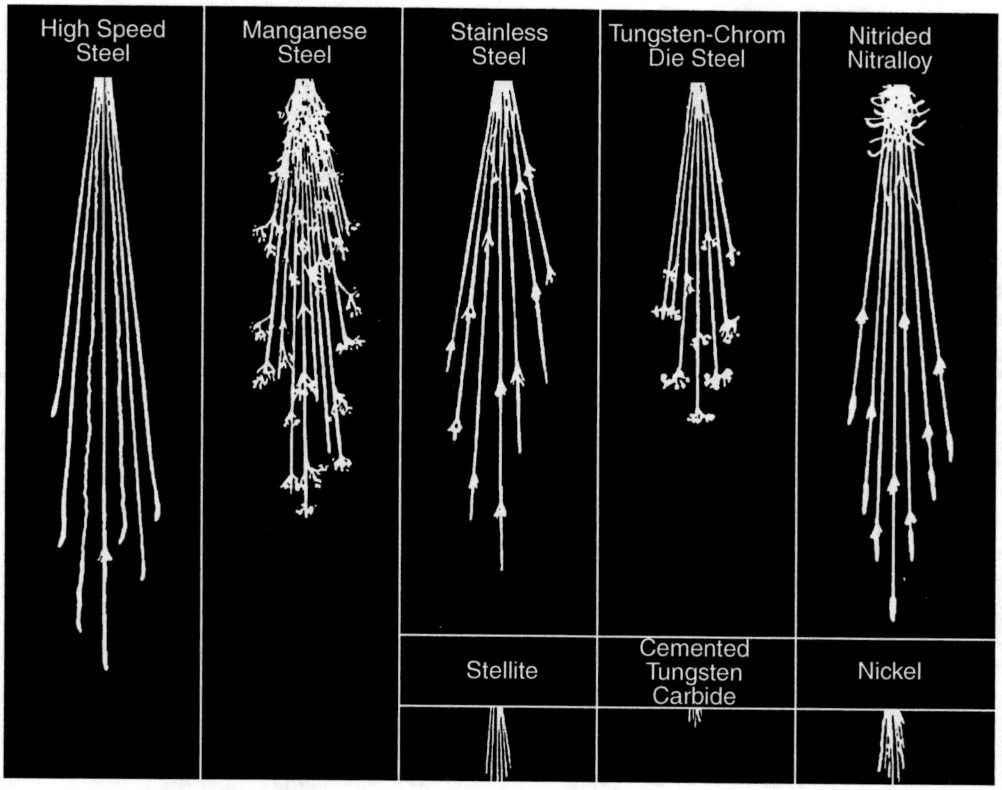

Fig. 10 — Spark Test For Alloy Steels and Other Alloy Metals

Tungsten-chromium steels, which are used for high-speed work, show the typical, high-speed steel spark, except that the spark turns to a chromium yellow (straw color) at the end of its flight. The chromium and the tungsten reduce the carbon spark, causing it to be very fine or thin with repeating spurts.

Torch Test for Alloy Steels

Generally, the higher the alloy content, the more difficult it is to weld the metal. Under the torch flame, these metals have a tendency to boil, because of the action of the alloys in them. Special fluxes are often used when welding them.

Stainless steels are difficult to flame weld, and a flux is needed to produce good welds. Without the flux, the metal boils and a porous puddle is produced.

Manganese steel is often used for surfaces which are exposed to abrasive wear, such as power shovel buckets. The usual application of welding in this case is to build up the worn surfaces (Fig. 8). The steel melts readily under the torch flame and presents few difficulties in welding.

Nickel steels can be identified under the torch flame by the boiling action of the metal.

Nickel-chrome steels behave in much the same way.

Other Tests for Alloy Steels

The color test, ring test, magnetic test, fracture test, and the chip test are all applicable to alloy steels, but perhaps the easiest one is the color test.

The different alloys tend to change the color of the metal, which in some compositions is easy to identify.

Stainless steels, for example, have a distinct silvery color that sets them apart from the other steel alloys.

Some of the metals are magnetic, while others are not Identification can be done to a certain extent by using this test.

The chip test is not widely used because of the similarity in the appearance of practically all the stainless steel alloys. However, some of these alloys are considerably harder than others, and the chip test will bring this out quite clearly.

The ring or metallic sound of the metal is another test which can be applied to steel alloys with considerable accuracy. However, this test should be used in addition to the other tests.

Heat Treatment of Metals

The welder should be particularly interested in heat treatment of metals, because he or she must know what welding does to the heat treatment of the metal and its effect on the physical properties.

It is necessary to know if a metal must be preheated for welding, and if the metal should be heated during the welding operation. In addition, you must know what heat treating procedure to use to bring back, as nearly as possible, the original properties of the metal. (This is the postheat treatment.)

The most common application of heat treating in a welding shop is annealing. Annealing relieves the metal of internal stresses and strains caused by the expansion and contraction during welding and improves the properties of the metal in the weld.

Most structural welding in shops involves only the knowledge of how a metal can be annealed.

In maintenance welding — involving the repair of broken parts — the welding operator needs an accurate knowledge of all phases of heat treating.

Fig. 11 — Heat Treating to Harden Gear Teeth

Purposes of Heat Treatment

All metals can be heat treated. Some metals are affected very little, but some, particularly most steels, are greatly affected.

Heat treating may serve the following purposes:

- Develop ductility
- Improve machining qualities
- Relieve stresses
- Change grain size
- Increase hardness or tensile strength
- Change chemical composition of metal surface, as in case hardening
- Alter magnetic properties
- Modify electrical conduction
- Induce toughness
- Recrystallize metal which has been cold worked

When heat treating, there are three factors of prime importance:

- The temperature to which the metal is heated
- The length of time that the metal is held at that temperature and speed of cooling
- The material surrounding the metal when it is heated, as in case hardening

When a weld is made, the metal in and around the weld joint is heated to a variety of temperatures as the distance from the weld joint increases. This drop in the metal temperature from the weld joint outward is called a temperature gradient.

In steels, some of the metal in the area of the weld is heated through only one critical point, some through two points, and some may be heated to only 500°F (260°C) as shown in Fig. 12

Because of the uneven heating, the strength, ductility, grain size, and other metal properties may vary greatly, and this may affect the strength of the metal in the weld area.

The welding operator may use preheating, concurrent (continuous) heating, and/or postheating to avoid temperature gradients in the weld area.

Heating the metal before welding helps to prevent internal stresses and strains which may cause weld failure. The welder may use continuous heating for the same purpose, and may use postheating to relieve stresses and bring all the metal in the area of the weld to the same heat treated condition.

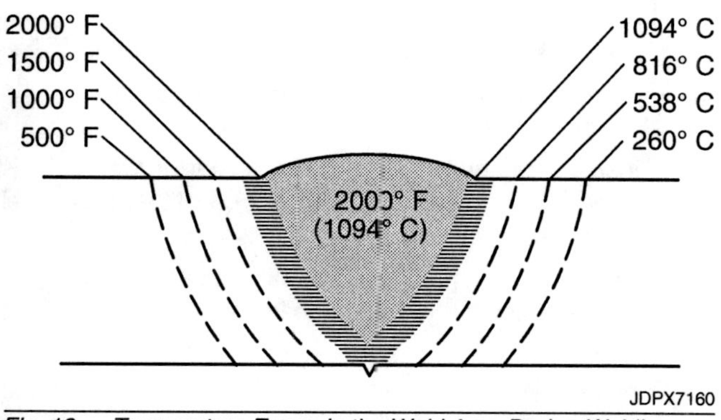

Fig. 12 — Temperature Zones in the Weld Area During Welding

How to Heat Metals

Heating metals is a complex operation. All metals expand when heated, and rather drastic changes in volume take place as metals are heated above their critical temperatures. The same is true when they are cooled below their critical temperatures (where a crystalline structure change takes place).

Unequal heating or cooling may cause cracks, or at least a warping of the structure. However, this action is not always a disadvantage, as warped parts can also be straightened using this technique.

Local heating can be done by using either of these flames:

- Air-fuel gas
- Oxy-fuel gas

Or it can be done by:

- Electrical resistance heating
- Induction heating
- Localized furnaces

ELECTRICAL RESISTANCE HEATING is done by either putting electrical resistance units against the parts to be heated, or by using electric resistance units to heat a furnace.

INDUCTION HEATING is a way of heating using high-frequency alternating current to create eddy currents in the metal. This method heats a metal throughout its thickness.

LOCALIZED FURNACES are normally used for general heating of metals. The furnace can be heated with direct heat or indirect heat, using air-fuel gas, electrical induction, or electrical resistance units.

How Metals Are Cooled

The speed and evenness of cooling a metal object is a major factor In the physical properties of the metal.

Welds performed on steels at room temperature cool quickly, and because the cooling takes place from the molten condition, the crystals are large. Cooling may be done in various ways. The heat loss is a combination of:

- Convection
- Conduction
- Radiation

The problem in heat treating is to cool the inside of the metal at the same speed as the surface.

Air is a common gas-cooling medium. Others are thermal convection cooling (still air), forced convection cooling (fans), and water cooling.

Properties of Steel

Steel can be obtained with various carbon contents and alloying metals. The following chart shows the carbon content of some common items made of steel (most are fractions of one percent).

The family of steels starts with low-carbon steel (almost wrought iron). As the carbon content increases, the steel becomes harder, stronger, and more brittle until approximately 1.75 percent of combined carbon is reached. This is the maximum percent of carbon that will combine with iron to form steel. Table 6

Metal Part (of Steel)	Carbon Content %
Boiler plate	0.12
Castings, Low-carbon steel	0.20 (max.)
Case hardened steel	0.12
Cold chisels	0.75
Files	1.25
Forgings	0.30
Gears	0.35
Hammers	0.65
Lathe tools	1.10
Machinery steel	0.35
Metal tools	0.95
Nails	0.10
Pipe, steel	0.10
Rails	0.60
Rivets	0.05
Set screws	0.65
Shafts	0.50
Springs	1.00
Steel for stamping	0.90
Tubing	0.08

Table 6 — Carbon Content of Common Items Made of Steel

Crystalline Structure of Steel

Practically all heat treatments deal with the crystalline structure of steel, or the grain size. They are also concerned with combinations of iron and carbon, and with the distribution of iron carbide in the metal.

In 10-point carbon steel, which contains 1/10 of one percent carbon, all of the carbon is in chemical combination with iron, forming iron carbide, or Fe_3C. There is considerable free iron remaining, which in metallurgy is called ferrite.

Steel containing a large amount of ferrite cannot be hardened. The only heat treatment possible is to alter the grain structure and size.

Heating a steel to its upper range (critical temperature) does not greatly affect the steel because it is the iron carbide which determines the hardness.

Iron carbide, mixed in various proportions with ferrite, gives several combinations of hardness and brittleness, depending on various heat treatments. The hardness levels have been given names.

In the order of their hardness, these steels are:

1. Austenite, soft, ductile, tough, and wear resistant.
2. Martensite (Fig. 13), is the first stage of austenite decomposition and a brittle constituent.
3. Troosite, not quite as hard as martensite, but is considerably tougher. This is the constituent of steel used in files.
4. Sorbite, not as hard as troostite, but it is tougher. Because of its high yield point it is the structure of spring steels.
5. Pearlite (Fig. 13), the toughest structure and has the highest breaking strength.

How Various Heat Treatments of Steels Are Obtained

The various forms of steel are obtained by the following steps:

1. Heating
2. Quenching
3. Drawing
4. Tempering

Eighty-five point carbon steel has the correct amount of carbon to form steel that has the lowest critical temperature. When properly heat treated, the whole structure of this steel becomes pearlite. Pearlite is a steel which, under the microscope, has alternate layers or rows of iron carbide and ferrite (Fig. 13). This grain structure, when magnified, looks similar to a fingerprint printed with black ink on white paper. This is the only steel that has just one critical temperature.

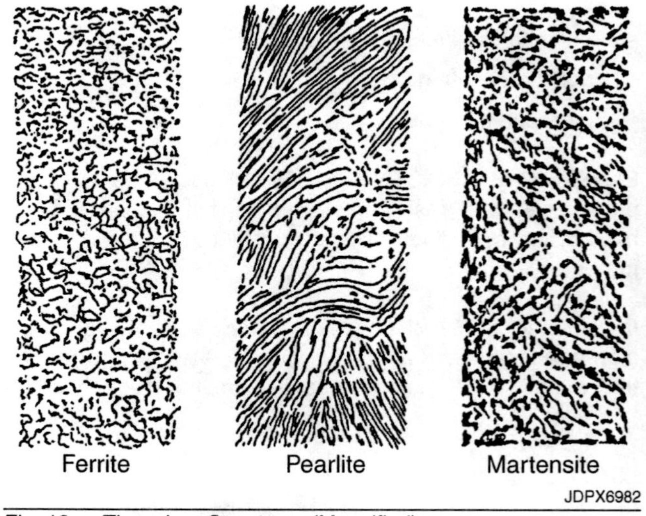

Fig. 13 — Three Iron Structures (Magnified)

The hardness of a steel which has been heated above the critical temperature, as in welding, may be affected by controlling the rate of cooling.

If the eighty-five point carbon steel is heated to 1250°F (675°C) and is then quenched in cold brine, austenite is formed.

If the steel is heated to 1250°F (675°C) and is quenched in cold water, martensite is formed

By heating the steel to its critical temperature and quenching it in oil, troostite is formed.

If it is heated to 1250°F (675°C), and quenched in a molten metal bath, sorbite and pearlite are formed.

The speed of cooling affects the distribution and clustering of the iron carbide. Quenching, or the rate of cooling, has much to do with the type of steel obtained:

- Heat the eighty-five point carbon steel to 1250°F (675°C).
- If an austenite is desired, quench in cold brine.
- If you want martensitic steel, reheat the sample to 400°F (205°C) and cool again.
- If troostite is desired, reheat the sample to 600°F (315°C) and cool.
- If sorbite is desired, reheat to 800°F (425°C) and cool.
- When the metal is reheated to 1000°F (540°C) and cooled, pearlite is formed.

The temperatures to which these various forms of steel are heated overlap, so that in many cases, one may have a steel which is part sorbite and part troostite, or part sorbite and part pearlite.

These temperatures must be held constant (25°F/14°C) and the part must be left in the furnace at this temperature until it is fully heated throughout its thickness.

Annealing Steel

Annealing includes several types of heat treating operations.

Generally, annealing is considered a heat treatment which refines the grain and leaves the metal in a soft condition.

An annealing operation may be performed for several reasons:

- So the metal may be worked cold.
- To make the steel machinable.
- To relieve the internal stresses (also called "stress relieving") produced while the metal was being shaped or welded.

When performing field welds, practical ways to preheat and anneal metal are with an oxyacetylene torch, or by induction heating.

In induction heating, electrical coils are wrapped around the part and high-frequency AC electricity is passed through.

To fully anneal most steels, do the following:

1. Heat the steel slightly above its critical temperature (50°F/10°C).
2. Cool the steel as slowly as possible.

If the grain size is of no importance you do not have to be careful of the exact temperature. As the grain size increases, the elastic limit and ductility of the metal decrease.

When steel is heated to the critical temperature, it changes its crystalline structure and reaches its most refined grain condition at its lowest critical temperature.

However, if the steel is heated above this temperature, the grain size starts to grow and becomes larger as the temperature rises. Upon being cooled, the steel will retain the size of the maximum crystalline structure reached. The crystals will continue to grow if the piece is kept at the maximum temperature or if kept above its critical temperature until a maximum size is reached. Therefore, the speed of welding and the speed of cooling affects crystal size.

This is particularly important in welding, because both the weld metal and the metal adjacent to the weld are heated considerably above the critical temperature.

When cooled, regardless of the speed of cooling, the crystalline structure in the weld and in the steel adjacent to the weld will be large and the strength and hardness of the metal will be reduced.

Two methods may be used to strengthen and harden the metal:

1. Pound the metal with a hammer as it is cooling (peening). The peening will refine the grain structure.
2. Allow the metal to cool to room temperature. Reheat to just slightly above the critical temperature of the metal and then allow it to cool slowly.

Peening affects the grain size only in the area worked, usually at the surface. Therefore, method number 2 is preferred.

Industries with the facilities put the whole steel structure that has been welded in an annealing furnace. As a result, the whole structure is annealed instead of just the weld and the parts adjacent to it.

Commercial annealing (partial annealing) may be done at a temperature between 1100° and 1200°F (595° and 650°C). This process is used mainly to relieve internal stresses.

Hardening Steel

Steels of less than 50 points of carbon content cannot be successfully hardened by heat treatment. Steels above 170-points of carbon content are also difficult to harden by heat treatment. Within these extremes it is possible to obtain almost any degree of hardness by using the proper carbon content and the proper heat treatment.

It is also common practice to obtain a refinement of the grain structure along with the hardness. Generally, to harden a steel specimen in this range of carbon content, the steel is heated just above its critical temperature and then cooled rapidly. The critical temperatures for steel with various carbon contents are shown in the chart below.

Percent of Carbon in Steel	Critical Temperature for Hardening and Full Annealing °F (°C)
0.10	1675–1760 (915–960)
0.20	1625–1700 (885–925)
0.30	1560–1650 (850–900)
0.40	1500–1600 (815–870)
0.50	1450–1560 (790–850)
0.60	1440–1520 (785–825)
0.70	1400–1490 (760–810)
0.80	1370–1450 (745–790)
0.90	1350–1440 (730–780)
1.00	1350–1440 (730–780)
1.10	1350–1440 (730–780)
1.30	1350–1440 (730–780)
1.50	1350–1440 (730–780)
1.70	1350–1440 (730–780)
1.90	1350–1440 (730–780)
2.00	1350–1440 (730–780)
3.00	1350–1440 (730–780)
4.00	1350–1400 (730–780)

Table 7 — Percent of Carbon In Steel and Critical Temperature for Hardening and Full Annealing

As we mentioned earlier, the rapidity of cooling determines to a great extent the hardness and brittleness of the metal. Cold brines, cold water, air, and molten metal baths, are used as mediums for cooling metal rapidly.

The water cooling and the brine cooling results in extreme brittleness along with hardness. Air cooling and molten-metal-bath cooling tends to relieve much of the brittleness, but at a sacrifice of some hardness.

The hardness of the metal seems to depend on the distribution and structure of the iron carbide throughout the steel. When cooled rapidly, there is little opportunity for the iron carbide to separate or segregate, but as the rate of cooling is slowed, the carbides have time to separate and the metal becomes progressively softer.

Tempering Steel (Drawing)

Quenching a steel sample in a molten-metal bath will produce a hard metal without too much brittleness. However, a more effective method is called “drawing the hardness” or tempering.

By reheating the steel to a temperature between 400° and 1000°F (205° and 538°C) after hardening, much of the brittleness in the steel can be relieved. Most of the internal stress is also relieved.

Reheating the metal to 600°F (316°C) (blue color on the steel surface) is most common.

Flame Hardening

Oxyacetylene flame hardening is being widely used today in manufacturing.

In many parts such as gears, shafts, and connecting rods, it is advisable to produce as hard a wearing surface as possible, and yet keep the inside of the structure in a ductile and tough condition. The hard surface provides long wear, while the tough interior is tough enough to withstand shocks without breaking.

In flame hardening, the part is heat treated and then drawn to produce a tough, ductile structure throughout. The surfaces to be hardened are then put in a flame-hardening machine. A multiple-tipped oxyacetylene flame is passed over the surface to heat it quickly to a high temperature. Following the flame, a heavy stream of water is fed to the surface to cool it quickly.

The depth of the hardness can be accurately controlled by the temperature of the cooling water.

Case Hardening

Another way of producing a metal part with a tough interior and a hard-wearing surface is known as case hardening.

The process is to make the part out of low carbon steel, so any machining required may be done easily. The surfaces to be hardened are then exposed to carbon while they are at a high temperature. The heated iron will absorb some of the carbon, increasing the carbon content of the surface, which may then be hardened by the simple heat treating process.

This is called the spherodizing process. Typical examples of this method are piston pins and camshafts in engines.

Where it is not desirable to case harden certain portions of the part exposed to the carbon at high temperatures, these portions may first be copper plated.

Generally, the carbon will penetrate into the steel to a depth of 1/64 in. (0.4 mm) per hour. However, the depth will vary with the temperature, the kind of carbon used, and the steel to be case hardened.

To heat treat a case hardened part, do the following:

1. Heat the steel to its critical temperature and quench it. This produces a glass-hard exterior on parts that have been exposed to the carbon bath. This also produces a tough interior with a refined grain that is of low carbon content and which is therefore strong and ductile.
2. Then reheat to a temperature between 400° and 900°F (205° and 480°C) to draw (temper) the surface.

Heat Treating of Alloy Steels

The heat treatment of alloy steels is difficult because of the many alloy ingredients, and the hundreds of combinations of these alloy metals with the iron carbon steels.

Generally, alloy metals lower the critical temperature of the iron carbide steel.

Some alloy metals have a special influence on the characteristics of the heat treatment. For example, iron carbide steel, with manganese added to it as an alloying metal, will become a self-hardening steel.

If a sample of manganese steel is heated to its critical temperature (1350°F [732°C]) and cooled in air, the sample is as hard as an ordinary iron carbide steel sample quenched in cold water. Further, this alloy has the characteristic of maintaining its hardness at high temperatures. This characteristic makes it useful as a metal cutting tool.

Chromium-tungsten high-speed steel also has the property of maintaining hardness at a high temperature.

It is difficult to anneal these steels because of their self-hardening properties.

To heat treat alloy steels properly you should, if possible, obtain the exact percentage of the constituents in the steel and use the recommended process as specified by the manufacturer of the steel.

Heat Treating of Cast Iron

There are several forms of cast iron:

- White cast iron
- Gray cast iron
- Nodular iron and malleable iron

All are similar cast irons in their carbon content and method of manufacture, except for the nodular irons, which have another alloying metal in them.

They differ because of the heat treatments.

WHITE CAST IRON is formed by heating cast iron to the critical temperature (1650° to 2090°F [900° to 1145°C]) and cooling rapidly. This cast iron gets its name from the appearance of the fracture, which is white as a result of the carbon being in solution with the iron, forming cementite and pearlite.

If cast iron is heated to its critical temperature, and then cooled slowly (in sand or a furnace), the slow change through the temperature ranges permits enough time for some of the excessive carbon to separate from the iron to form microscopic flakes of graphite. When a sample of this type of cast iron is broken, the fracture has a grayish color because of the number of graphite flakes imbedded in the cast iron body.

White cast iron, because of the excessive amount of carbide, is too hard to machine, whereas gray cast iron, because of the presence of the graphite flakes and the reduction of the amount of carbide is machinable.

MALLEABLE IRON is cast iron heat treated to make it more ductile and more resistant to shock. This type of cast iron is formed by heating a casting to its critical temperature and cooling it rapidly, forming white cast iron. Then reheating to a temperature between 1450° and 1650°F (790° and 900°C), and keeping it at this temperature for 24 hours for each 1 inch (25 mm) of casting thickness. Being kept at this high temperature for this long causes much of the carbon to separate slowly from the iron in the casting, forming globular carbon (spheroids of carbon or small microscopic pools of carbon surrounded by low carbon iron). After heat treatment at this high temperature for the proper length of time, the casting is cooled slowly.

We can now see that if a malleable iron casting is welded, the casting will lose its malleable qualities in the weld and in the vicinity of the weld.

For this reason, never fusion weld on malleable iron castings. Instead, they should be brazed or braze welded to keep the temperature below the point where white cast iron is again formed.

Test Yourself

Questions

1. What are the two major groups of metals?
2. Which melts at a high temperature, cast iron or mild steel?
3. What is the difference between an iron and a steel?
4. Match the items on the left below with the correct item at the right.

a. Tensile strength	1. Ability to resist penetration
b. Ductility	2. Ability to resist pulling apart
c. Malleability	3. How easily a metal fractures
d. Hardness	4. Ability to be stretched
e. Brittleness	5. Tendency to be hammered into shape

5. If you chip a metal with a cold chisel and it curls, would it be a mild steel or a cast Iron?
6. Annealing is the ______________ of metals by heat treatment.
7. How does malleable iron differ from cast Iron?
8. What happens if malleable cast iron is welded?
9. List three organizations that have systems for numbering metals and give an example of a number from each system.
10. How many points of carbon are there in an AISI-SAE 1035 steel?
11. What would be the percent of carbon in an AISI-SAE 1035 steel?
12. What type of metal would have the numbers 6061 T-6? What would be the major alloys in this metal?
13. Give the numbers for a shock resistant tool steel.
14. Give the numbers for a hot work tool steel.
15. Give the numbers for a stainless steel.
16. What does the "L" after a stainless steel mean?
17. What is the first step in identifying a metal that has a plating or painted surface?
18. List two things that can be learned toward identifying a metal using a file test.
19. List three "white metals."
20. What is another name for "pot metal"?

Design of Welded Units

12

Fig. 1 — Large Welded Unit

Introduction

When designing any product, you must resolve whether metal parts are to be welded, cast, or pressed. Too often welding is avoided when it may actually have advantages over other techniques.

For example, it is often easier to build a welded structure by using rolled plates and structural shapes instead of a casting because of greater savings in weight and cost. Similarly, heavy framed units for large machine components will often be easier to construct by welding.

The vital point in any design is to achieve maximum savings in cost but at the same time develop a product that is pleasing in appearance, has sufficient strength, and is safe.

Advantages of Welded Construction

Welding is not the best solution to all design problems. Nevertheless, welding frequently has advantages. These advantages can be summarized as follows:

- Welding is more flexible and has many basic cost savings over casting and press forming operations. However, this is not necessarily true where high-quantity production runs are required and parts are small.
- Welding is often three or four times stronger than other fabricating processes and consequently can absorb higher loads without increasing the weight of the material.
- A weld is frequently two to three times stiffer than steel castings, yet is more ductile.
- There is less tendency for a welded product to crack due to porosity and shrinkage stresses that are often encountered in a steel casting.
- Investment in equipment is generally less for welding than for other manufacturing processes.
- Welded structures provide a greater freedom of design and still meet functional requirements.
- Welded parts are much more resistant to shock and impact forces than most castings and often provide more safety.

Welding Design Factors

Correct joint design is necessary if welded assemblies are to perform their service and safety requirements.

One of the first requisites of joint design is the function of the part.

Next, consideration must be given to the kind of loads imposed on the welded section. Thus, an analysis must be made whether the applied stress is one of tension, compression, shear, torsion, or bending.

Joint configuration is next determined on the basis of the design that provides the greatest resistance to the applied load. For example, some types of joints are best if the stress comes from one direction, while other joints are more suitable for stress that is of a more complex nature.

The problem of whether a joint is under a static or dynamic load is another factor that affects joint design.

Finally the efficiency of a joint, that is, its total strength as compared with the actual base metal, is an important factor.

The principal design factors that are to be covered here are joint design, joint location, weld size, and joint strength.

The Best Welded Design

To produce a part at the lowest overall cost, you must evaluate the design to make sure that best use is made of the metallurgical and physical properties of the materials. Consider these factors:

- The design should satisfy the strength and rigidity requirements of the product. Designing beyond these requirements means that more material and labor will be used, adding to the costs.
- The safety factor for the design should meet all criteria.
- The appearance of the product should be pleaslng, but where areas are hidden from view appearance should not be a critical factor. Less expensive materials can sometimes be used where appearance is not important. For example, hot-rolled steel will frequently prove as effective as cold-rolled steel.
- Since more weight means higher costs, analyze the members to see if stiffeners could be employed for rigidity as a means of reducing weight.
- Wherever possible, use cheaper low-carbon steel — except that a high-grade steel is best where the added strength reduces the cost.
- Standard sizes and shapes should be specified wherever it is practical.
- When a hard surface is required, plan for surfacing, heat treating, or plating.

Design of Joints

Joint design is determined by the cost of preparing the joint, the accessibility of the weld, its adaptability for the product, and the type of loading the weld is required to withstand.

The basic joints used in arc welding are:

- Butt joints
- Tee joints
- Lap joints
- Corner joints

Each group has several variations to provide for different needs.

Butt Joints

The SQUARE groove (Fig. 2-A) is primarily for materials that are 1/4 inch (6.4 mm) or lighter and require full and complete penetration for optimum strength. The joint is reasonably strong in static tension, but is not recommended when tension due to bending is concentrated at the root of the weld. It should never be used when it is subjected to fatigue or impact loads, especially at low temperatures. The preparation of the joint is relatively simple since it requires only matching the edges of the plates; consequently the cost is low.

The SINGLE "V" groove (Fig. 2-B) is used on plate 3/8 inch (9.6 mm) or greater in thickness. Preparation is more costly because a special beveling operation is required and more filler material is necessary. The joint is strong in static loading, but like the square groove it is not particularly suitable when tension due to bending is concentrated at the root of the weld.

The DOUBLE "V" groove (Fig. 2-C) is best for all load conditions. It is often specified for stock that is heavier than stock used for a single "V." For maximum strength the penetration must be complete on both sides. The cost of preparing the joint is higher than the single "V" but less filler material is required. To keep the joint symmetrical and reduce warpage the weld deposit must be alternated, welding first on one side and then the other.

The SINGLE "U" groove (Fig. 2-D) readily meets all ordinary load conditions and is used for work requiring high quality. It has the greatest applications for joining plates 1/2 inch to 3/4 inch (12.8 to 19.1 mm).

The DOUBLE "U" groove (Fig. 2-E) suitable for plate 3/4 inch (19.1 mm) or heavier where welding can readily be accomplished on both sides. Although the preparation cost is higher than the single "U" butt joint, less weld metal is needed. The joint meets all regular loading conditions.

Grooved Butt Joints

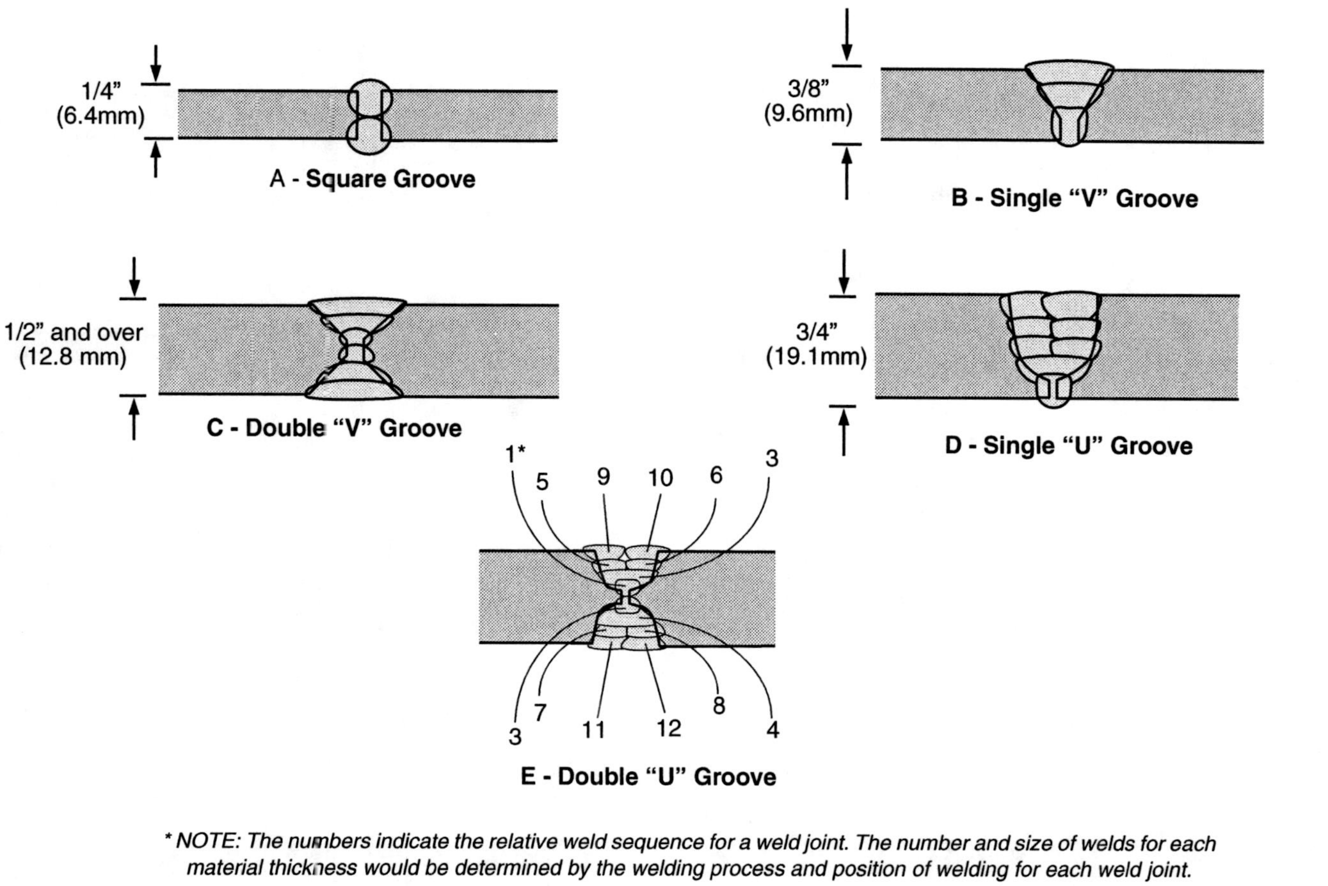

Fig. 2 — Types of Grooved Butt Joints

Tee Joints

The SQUARE groove (Fig. 3-A) requires a fillet weld which can be made on one or both sides. It can be used for light to fairly thick materials where loads subject the weld to longitudinal shear. Since its stress distribution is not uniform, care must be taken in specifying this joint where severe impact or heavy transverse loads are encountered. For maximum strength, considerable weld metal is required.

The SINGLE BEVEL groove (Fig. 3-B) will withstand more severe loadings than the square tee joint due to better distribution of stresses. It is generally confined to plates 1/2 inch (12.8 mm) or less in thickness where welding can be done from one side only.

The DOUBLE BEVEL groove (Fig. 3-C) is used where heavy loads are applied in both longitudinal and transverse shear and where welding can be done on both sides.

The SINGLE "J" groove (Fig. 3-D) is used on plates 1 inch (25 mm) or more in thickness where welding is limited to one side. It is especially suitable for severe loads.

The DOUBLE "J" groove (Fig. 3-E) is suitable in heavy plates 1-1/2 inch (38 mm) or more thick where unusually severe loads must be absorbed. Joint location should permit welding on both sides.

Tee ("T") Joints

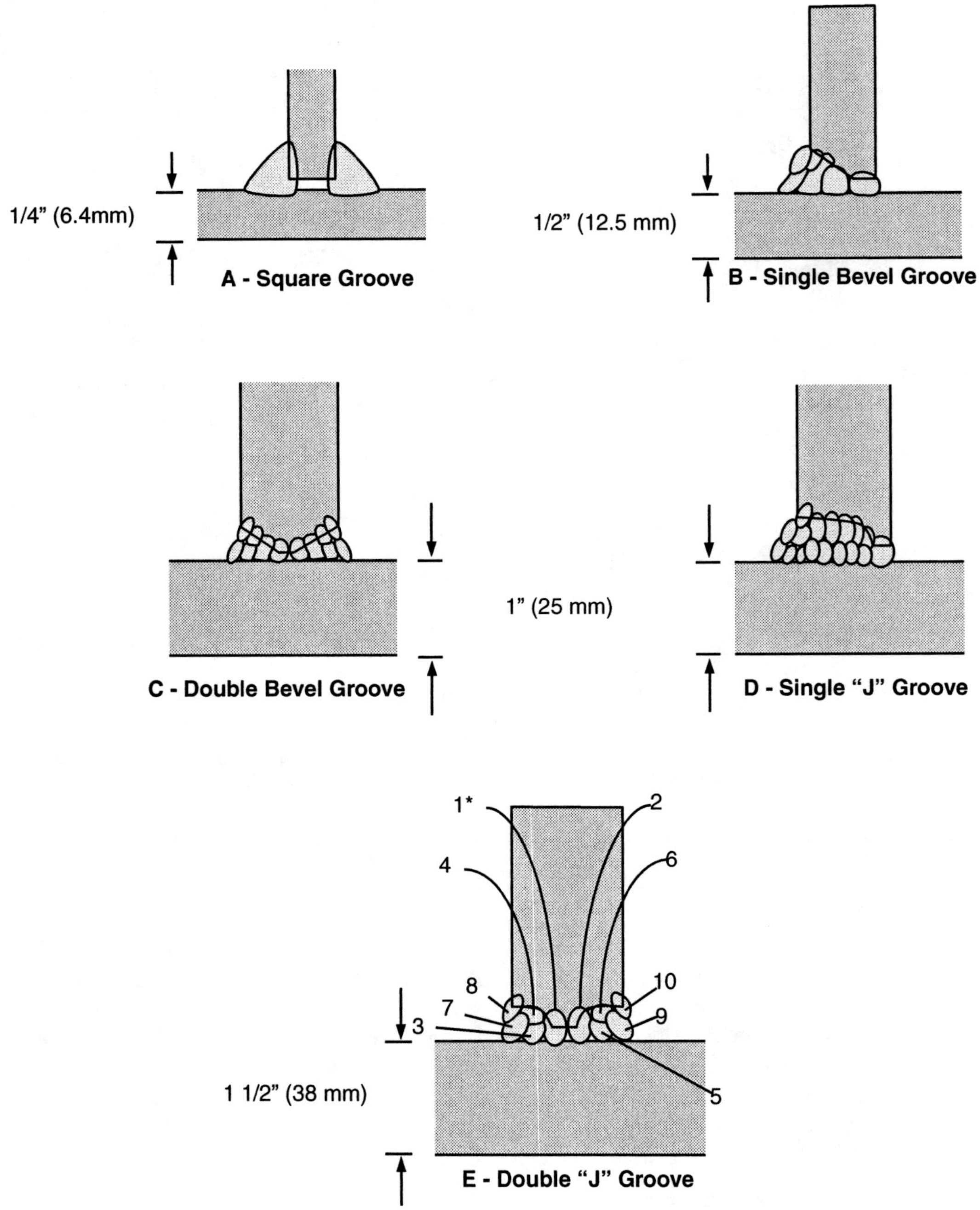

* *NOTE: The numbers indicate the relative weld sequence for a weld joint. The number and size of welds for each material thickness would be determined by the welding process and position of welding for each weld joint.*

JDPX6962

Fig. 3 — Types of Grooved Tee ("T") Joints

Lap Joints

The DOUBLE FILLET LAP JOINT (Fig. 4-B) can withstand more severe loads than the SINGLE FILLET LAP JOINT (Fig. 4-A). It is one of the more widely used joints in welding.

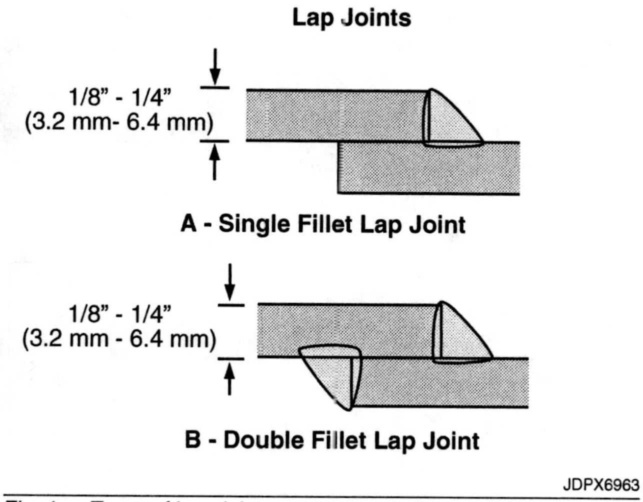

Fig. 4 — Types of Lap Joints

Corner Joints

The FLUSH CORNER JOINT (Fig. 5-A) is designed primarily for welding sheet that is 12 gauge and lighter. It is restricted to lighter materials because deep penetration is sometimes difficult and it supports only moderate loads.

The HALF-OPEN CORNER JOINT (Fig. 5-B) is usually more adaptable for materials heavier than 12 gauge. Penetration is better than in the flush corner joint.

The FULL-OPEN CORNER JOINT (Fig. 5-C) is used where welding can be done from both sides. Plates of all thickness can be welded. This joint is strong and capable of carrying heavy loads. It is also recommended for fatigue and impact applications because of good stress distribution.

The EDGE JOINT (Fig. 5-D) is suitable for plate 1/16 inch (1.6 mm) or less in thickness and can sustain only light loads.

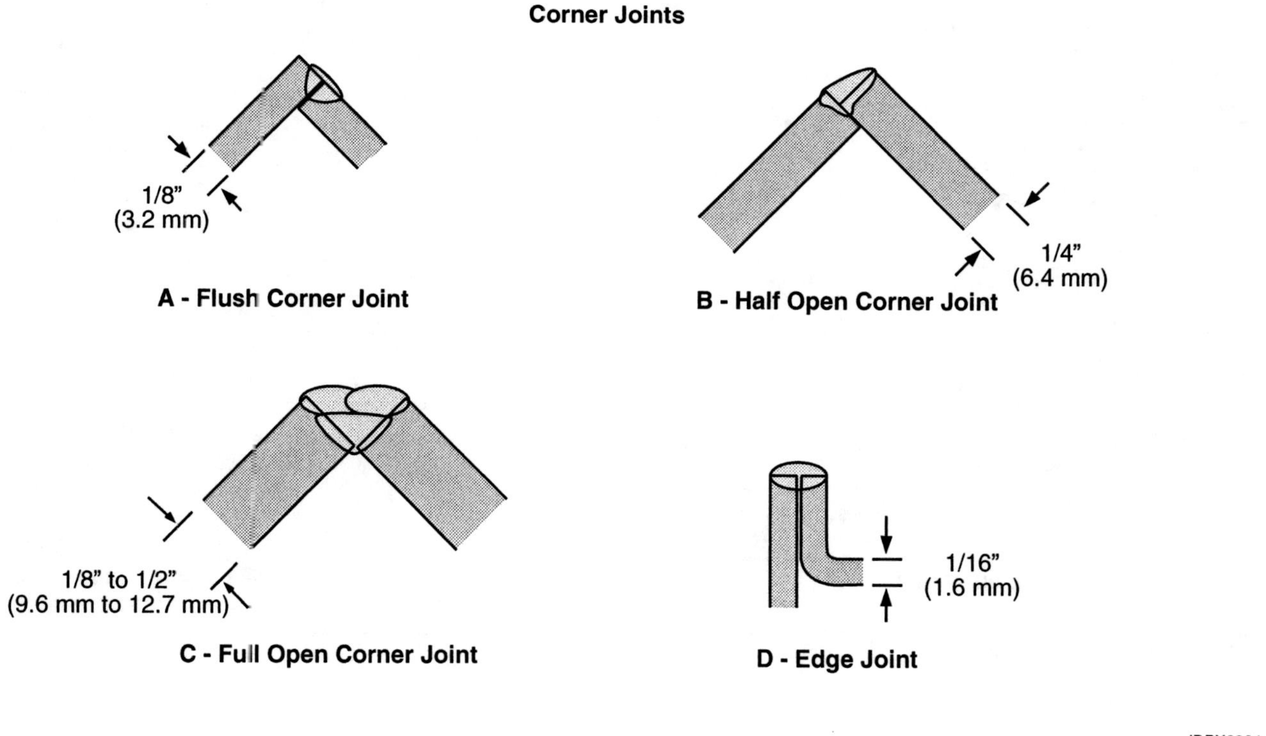

Fig. 5 — Types of Corner Joints

Selecting the Proper Joint

The square groove is the most economical joint, but is often good only for products that are to be made of light-gauge material.

Whenever metal is beveled, costs are higher. However, there is often no alternative, especially if heavy plates are to be welded. Without proper edge beveling, complete penetration is impossible. And without good penetration, the resulting weld joint will be weak.

For the greatest economy, select a root opening (Fig. 2) and groove angle that requires the least amount of weld metal.

The type of groove angle is frequently governed by the accessibility of the weld. For example, the position of the members may be such that only a single-bevel groove joint is practical. The accessibility of the weld also determines whether a weld is to be made on one side or on both sides.

The required strength of a joint must always be a major factor of joint configuration. Only by actually understanding the type and magnitude of various loadings is it possible to select the joint that will do the job.

Fig. 6 shows the recommended proportions for various joints to produce complete penetration.

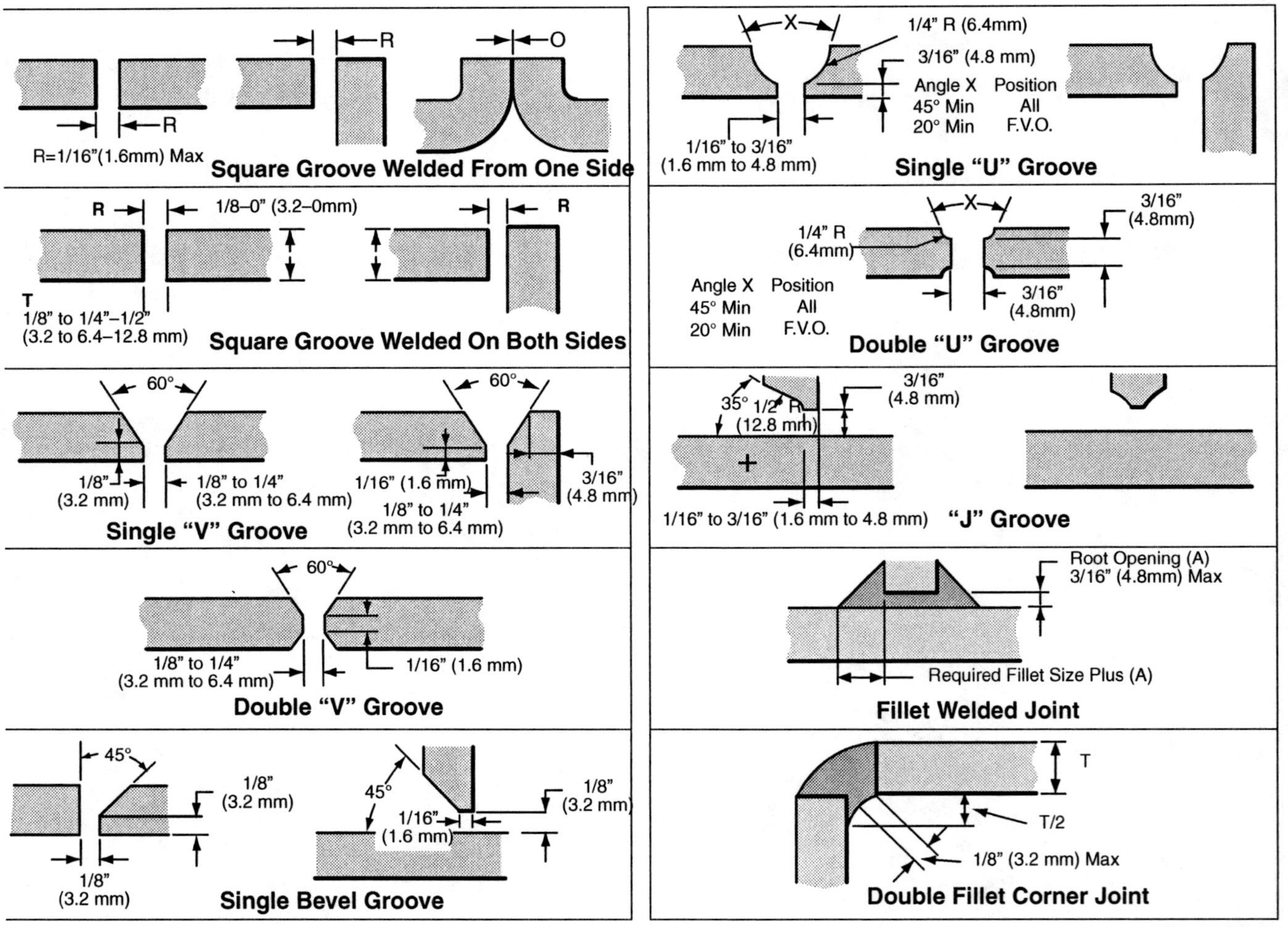

Fig. 6 — Recommended Proportions for Various Joints

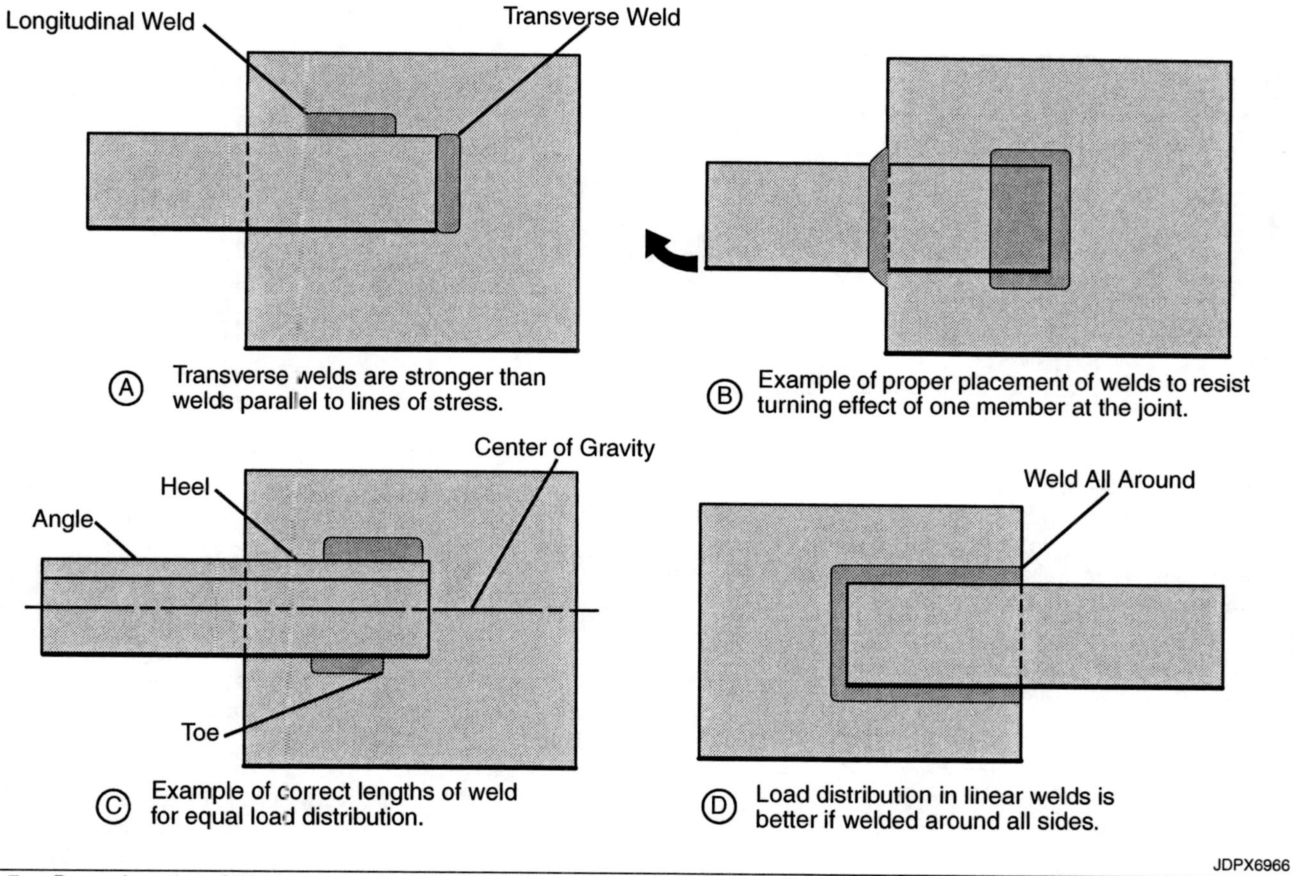

Fig. 7 — Proper Location of Welds

Locating the Welds

Welding is a low-cost method of joining parts, but unless careful thought is given to joint design and location of welds, much of the cost saving can be nullified.

There are many factors in proper weld location. Only a few can be illustrated here.

1. The location of the weld in its relationship to the members being joined has a decided effect on the strength of the joint. Normally welds made across the lines of stress are much stronger than welds which are parallel to stress lines (Fig. 7-A). However, transverse welds may cause the most stress in the base metal.
2. A member that is subjected to a movement (turning load) at a joint has greater resistance to turning motions if the welds are spaced apart instead of being close together (Fig. 7-B).
3. When joining shaped sections such as channels, angles, etc., consideration must be given to proper distribution of loads on the welds. Notice in (Fig. 7-C) that the length of the weld at the heel of the angle is greater than at the toe. This eliminates the tendency for the angle to turn and generate highly eccentric loads at the joint.
4. Wherever possible, locate the welded members to avoid bending or shearing actions. Greater strength is achieved if joints are kept symmetrical. Load distribution is more uniform on symmetrical than non-symmetrical joints.
5. To get uniform distribution of loads in linear welds, it is better to weld all around the joint than simply on two sides (Fig. 7-D). (The welded piece shown can also be cut at an angle before welding for better strength.)
6. Abrupt surface changes in a weld joint should be avoided since greater local stresses will be concentrated at the point of surface change. There is a more uniform transfer of stress through a weld if the surface flow is gradual (Fig. 8).

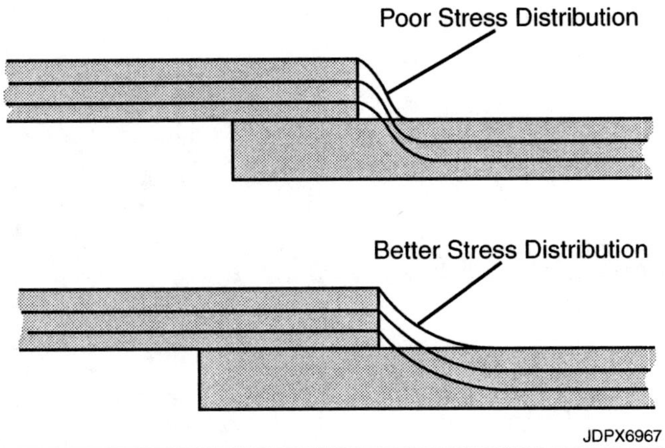

Fig. 8 — Avoid Abrupt Surface Changes in a Joint

7. Weld joints should be located so they are readily accessible and welding can be done with comparative ease. Notice in (Fig. 9) that the weld at (A) is almost impossible to make, whereas the welds at (B) are easily accessible.

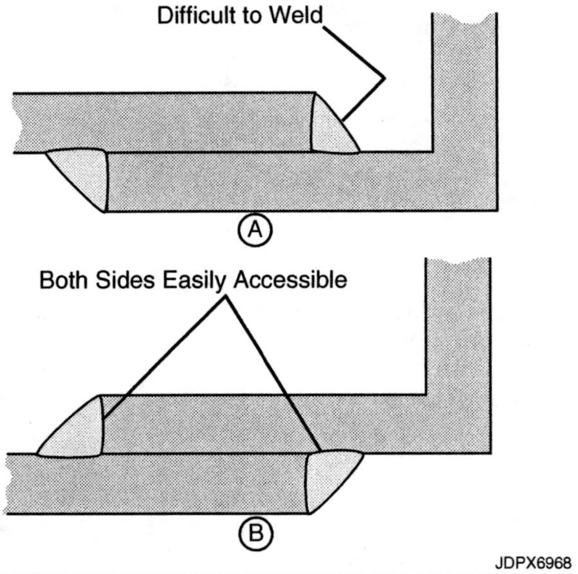

Fig. 9 — Be Sure that Weld Joints Are Easy to Reach

8. When an angle is used as a stiffener joined to a plate, use method (A) in (Fig. 10) rather than (B). The weld location at (B) becomes more expensive, since both beveling and machining the weld are required.

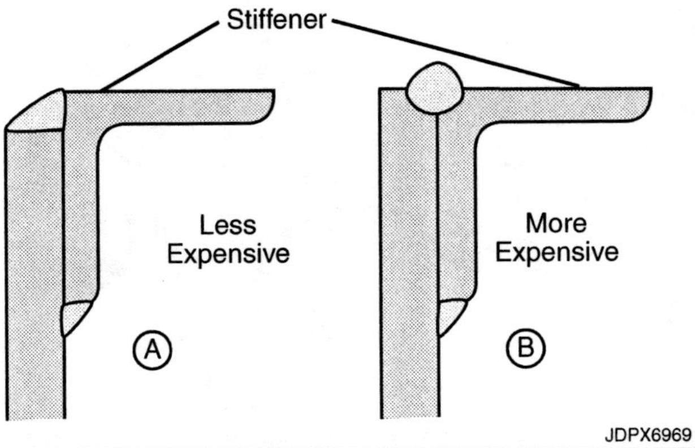

Fig. 10 — Location of Stiffener for Economy in Welding

Weld Size

Weld size affects cost as well as strength and distortion. It is a common practice to overweld just to be on the "safe side." The volume of metal deposited increases as the square of the length of a leg. In addition to increasing cost, the excess weld metal may generate an uneven load distribution that may result in a concentration of stress and lead to premature failure.

In a butt joint type groove, where full strength of the joint is required, the weld must have the same thickness as the connecting plates. As a rule the unit stress and weld size are not calculated because a butt joint has strength equal or greater than the base metal.

The strength of a fillet weld is based on the effective throat thickness, which is the shortest distance from the root to the face of the weld (Fig. 11) . For an equal leg (45°) fillet weld, the throat is 0.707 (sine of 45°) times the leg size of the weld.

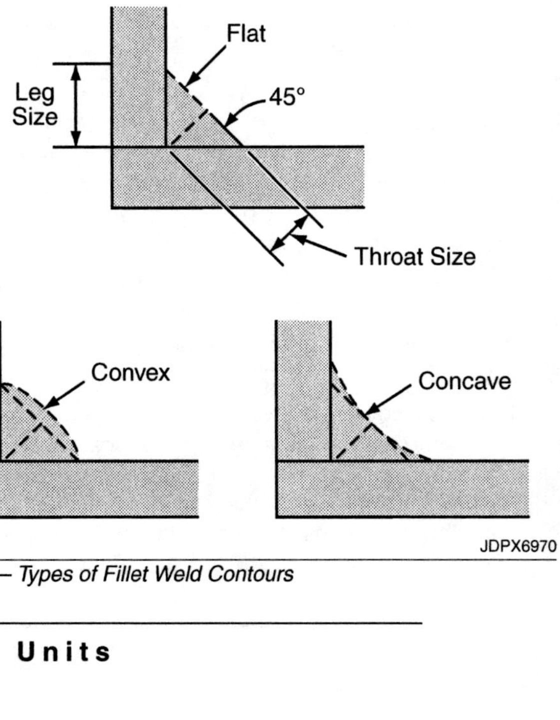

Fig. 11 — Types of Fillet Weld Contours

To check the size of the fillet welds, it is necessary to make gauges to fit the required contour of the weld deposit. Weld contours may be concave, flat, or convex with equal or unequal legs. The correct fillet size is determined by finding the leg length of the largest isosceles triangle that can be inscribed within the cross section of the required weld metal. See Fig. 10, Fig. 15, and Fig. 16 Chapter 1.

A flat fillet weld has a flat surface and is considered the most ideal weld. A concave fillet weld produces a smooth weld flow from one member to another but is often subject to shrinkage cracks. A convex fillet weld may result in excess weld metal but has less tendency to crack from shrinking.

To develop the full strength of the plates being connected, a fillet weld must have a parallel loading equal to the shear strength of the plate as computed on the basis of a minimum unit shear stress of 15,000 psi (103 mPa). It must also have a transverse loading equivalent to the tensile stress of the plate as computed at a unit tensile stress of 20,000 psi (138 mPa).

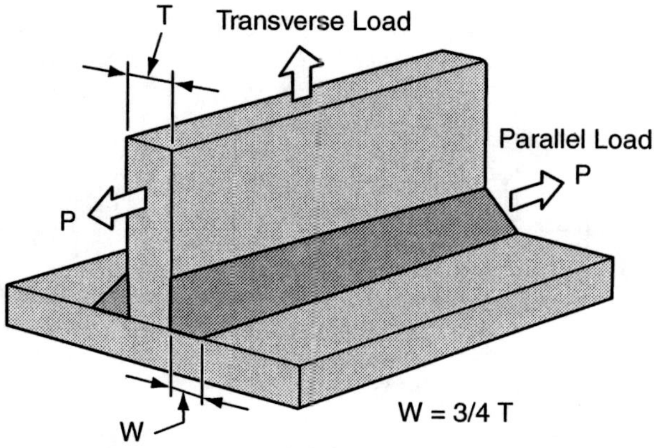

Fig. 12 — A Good Fillet Weld Must Have a Proper Ratio Between Parallel and Transverse Loads

In general, this relationship is achieved by maintaining a fillet width equal to three quarters of the plate thickness. See Fig. 12 and Table 1.

Recommended Fillet Weld Size For Best Plate Strength (Welded on Both Sides)			
Fillet Weld Size		Plate Thickness	
(Inch)	(mm)	(Inch)	(mm)
5/32	(4)	3/16	(4.8)
3/16	(4.8)	1/4	(6.4)
1/4	(6.4)	5/16	(7.9)
5/16	(7.9)	3/8	(9.5)
3/8	(9.5)	1/2	(12.7)
1/2	(12.7)	5/8	(15.9)
5/8	(15.9)	3/4	(19)
3/4	(19)	1	(25.4)

Table 1 — Recommended Fillet Weld Size for Best Plate Strength

Defects in Welds

When welding, care must be taken to prevent these defects which can weaken the weld:

- Grain growth
- Slag inclusions
- Segregation
- Porosity
- Residual stress

Grain Growth

When welding, a wide temperature differential exists between the molten metal of the actual weld and the edges of the heat-affected zone of the base metal. This temperature may range from a point far above the critical temperature down to an area unaffected by the heat. Thus the grain size can be expected to be large at the molten zone of the weld puddle and to gradually reduce in size until re-crystallization is reached.

Effective control of preheating and postheating will often minimize the formation of coarse grain structure. Where heavy welds require successive passes, use the heat of each succeeding pass to refine the grain of the previous pass. This can be done only if the metal is allowed to cool below the lower critical temperature between each pass.

High carbon and alloy steels are particularly vulnerable to coarse grain growth if cooled rapidly. The rapid cooling develops a martensitic condition, leaving a hard and brittle weld.

Slag Inclusions

Inclusions are impurities or foreign substances which are trapped in a molten puddle during the welding process. Any inclusion tends to weaken a weld because it has the same effects as a crack (Fig. 13).

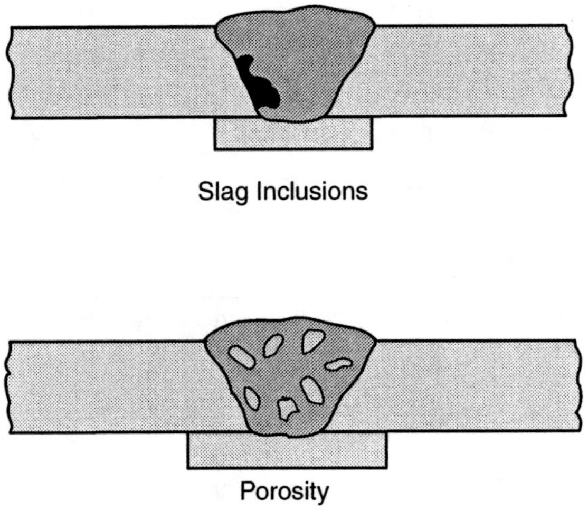

JDPX6972

Fig. 13 — Welding Defects

A typical example of an inclusion is slag that normally forms over a deposited weld. If the electrode is not manipulated correctly, the force of the arc causes some of the slag particles to be trapped into the molten pool and become lodged in the metal. Inclusions are more likely to occur in overhead welding since the tendency is to not keep the pool molten too long to prevent it from dripping off the seam. Improper cleaning between passes on a multiple pass weld also causes slag inclusions.

Segregation

In segregation, some regions of the metal are enriched with an alloy ingredient, while surrounding areas are actually impoverished. For example, when metal begins to solidify, tiny crystals form along grain boundaries. These crystals tend to exclude alloying elements. As other crystals form, they become progressively richer in alloying elements, leaving other regions without the benefits of the alloying ingredients. Segregation can be remedied by proper heat treating.

Porosity

Porosity (Fig. 13) is the formation of tiny holes generated by atmospheric contamination. Some metals have a high affinity for oxygen and nitrogen when in a molten state. Unless a good protective shield is provided over the molten metal, gas will enter the metal and weaken it.

Residual Stresses

Any rapid heating and cooling of weld metal will produce thermal stresses that may weaken the welded joint. Heating causes rapid expansion of the metal around the weld, while rapid cooling results in contraction of metal, which creates residual stress.

If large areas of unaffected metal surround the weld area or if the section is rigidly clamped, the solidifying weld metal readily develops internal or locked-up stresses. These stresses are not serious in welding ductile metal because the flow of metal will usually relieve the stresses. However, they may become very serious in metals of great rigidity unless provisions are made to compensate for expansion and contraction.

Residual stresses form cracks in the weld metal and in the base metal surrounding the weld (Fig. 14).

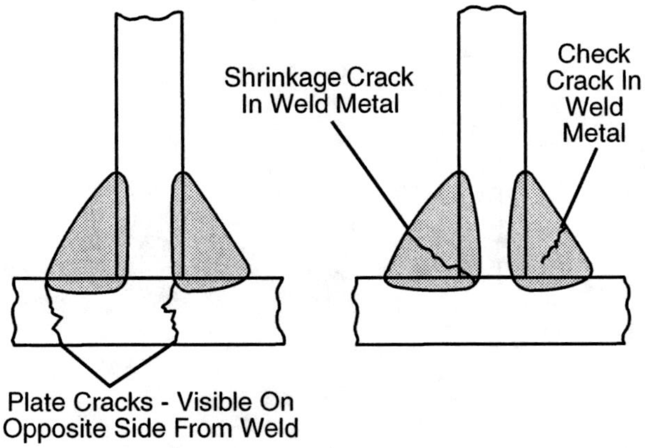

JDPX6973

Fig. 14 — Residual Stresses Cause Cracks in the Weld Metal and the Plates

Control of Residual Stresses

Control of residual stresses can be achieved by following these steps:

1. Properly Preparing and Assembling Edges:

Bevel edges of plates and allow sufficient space between them to prevent the movement of the component parts during welding. Weld the joints held most solidly in a fixture first. Thus a long longitudinal seam is welded before a short transverse seam.

2. Controlling the Heat Input:

The faster a weld can be made, the less heat is absorbed in the base metal.

A technique often used to reduce input is the intermittent or skip weld (Fig. 15). Instead of making one continuous weld, a short weld is made at the beginning of the joint (1); next a few inches are welded at the center of the seam (2); and then a short length is welded near the end of the joint (3). Finally, the welding is returned to where the first weld ended, and the cycle is repeated as shown at (4), (5), and (6).

In the back-step or step-back method, short sections of weld are deposited as shown in Fig. 15 instead of making one continuous weld.

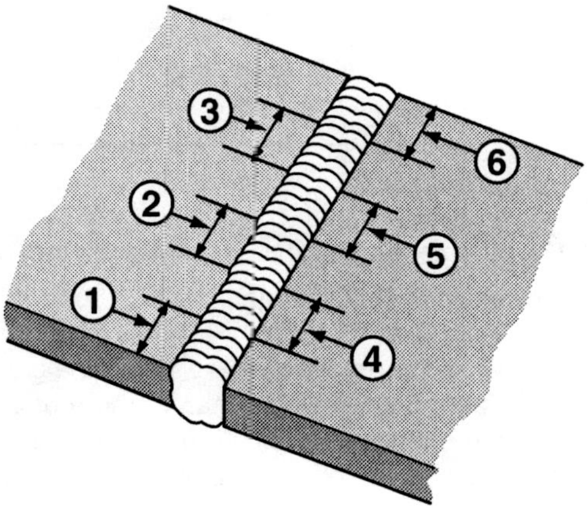

Intermittent or Skip Weld

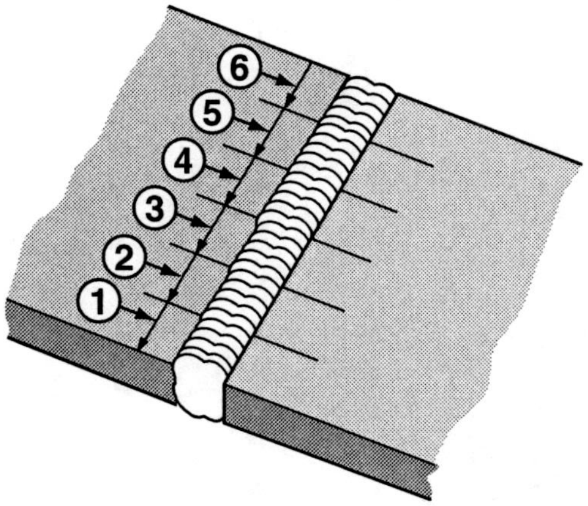

Back-Step or Step-Back Weld

JDPX6974

Fig. 15 — How to Control the Heat Input During Welding

3. Preheating the Metal:

On many pieces, particularly alloy steels and cast iron, expansion and contraction forces can be controlled by preheating the entire structure before the welding is started. To be effective, preheating must be kept uniform throughout the welding operation, and after the weld is completed the piece must be allowed to cool slowly.

4. Peening the Weld Metal as It Cools:

To help the weld metal stretch as it cools, a common practice is to peen it lightly with the round end of a ball peen hammer. However, peening should be done with care because too much hammering will stress the weld or cause it to work-harden and become brittle.

5. Heat Treating the Metal:

Heat treating is a common way to relieve stress in metals. The welded component is placed in a furnace capable of uniform heating and temperature control. The metal must be kept in a soaking temperature until it is heated throughout.

Correct temperatures are important to prevent damage to the metal being treated. For example, mild steels require temperatures of 1100° to 1200°F (594° to 650°C), while other alloy steels must be heated to temperatures of 1600°F (872°C) or more.

After the proper soaking period, the heat must be reduced gradually to atmospheric temperature.

6. Using Jigs and Fixtures:

The use of jigs and fixtures will help prevent distortion, since holding the metal in a fixed position prevents excessive movements. A jig or a fixture is any device that holds the metal rigidly in position during toe welding operation.

Fig. 16 shows a simple way to hold pieces firmly in a flat position. These heavy plates not only prevent distortion but they also serve as chill blocks to avoid excessive heat building up in the work.

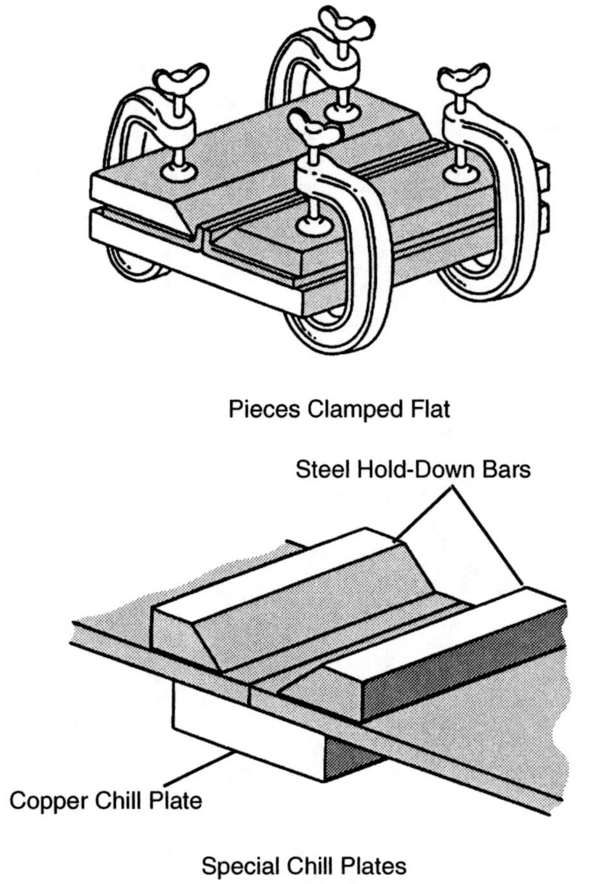

Fig. 16 — Use of Jigs and Fixtures

Special chill plates made of copper or other metal having good conductivity are particularly effective in dissipating heat away from the weld area (Fig. 16).

7. Using Fewest Weld Passes:

Distortion can be kept to a minimum by using as few passes as possible over the seam. Fig. 17 shows that six passes made with large electrodes are often better than many passes with smaller electrodes.

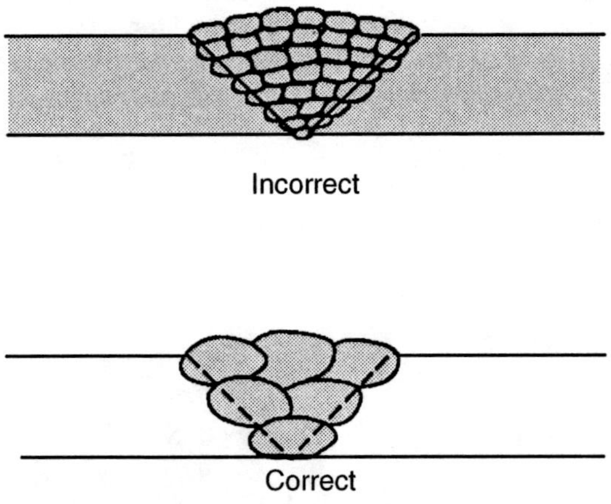

Fig. 17 — Use Fewest Weld Passes to Reduce Distortion

8. Allowing for Heat Distortion:

When a single "V" groove joint is welded, the greater amount of hot metal at the top rather than at the root of the "V" will cause more contraction across the top of the welded joint. The result is a distortion of the plate as shown in Fig. 18-A.

In a Tee-joint, the weld along the seam will bend both the upright and flat piece. See Fig. 18-B.

To reduce distortion, angle the pieces slightly in the opposite direction in which contraction is to take place. Upon cooling, the contraction forces will pull the pieces back into position.

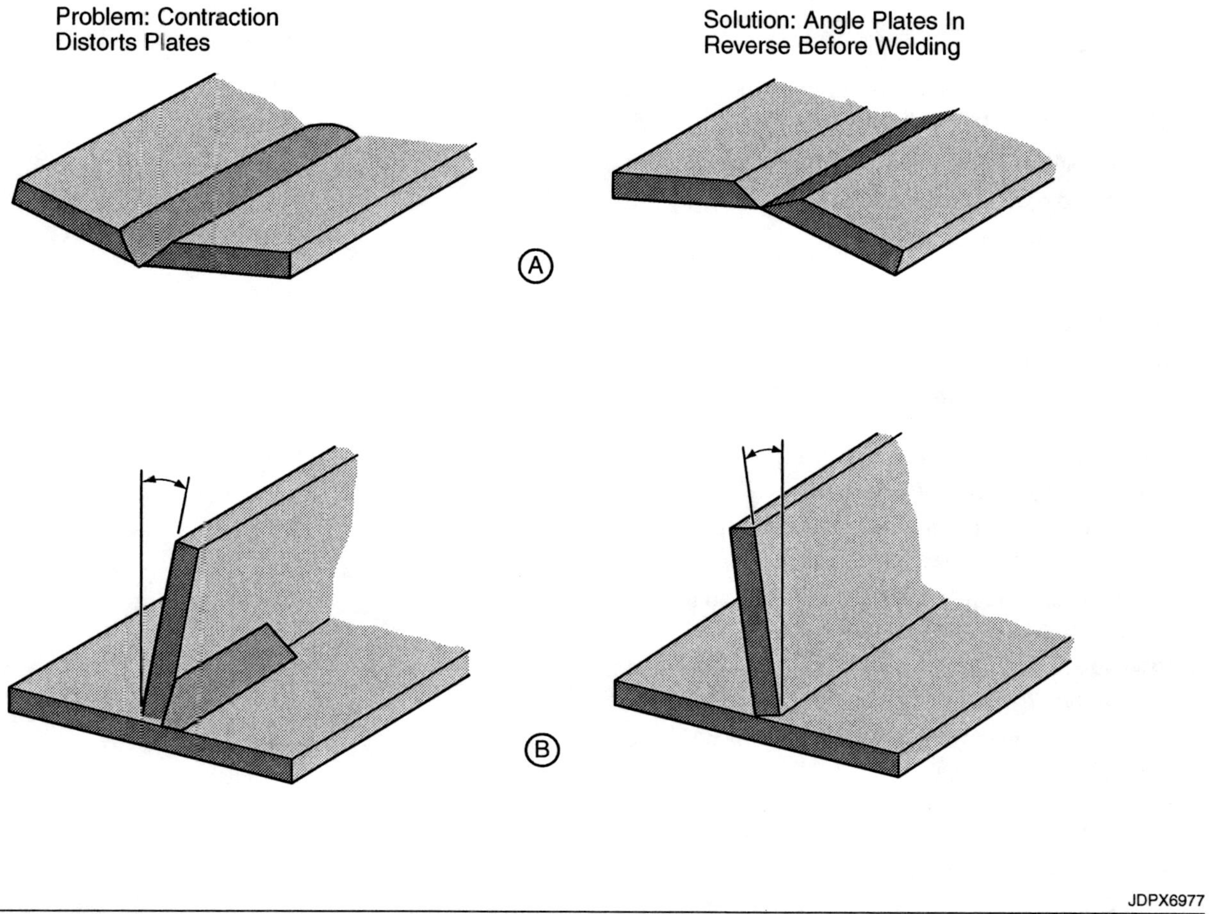

Fig. 18 — Distortion Is Reduced by Proper Positioning of Parts

Other Welding Problems

Welding problems related to wrong welding techniques — such as undercutting — are covered in earlier sections of this manual.

Review these areas for a supplement to the welding defects covered here.

Test Yourself

Questions

1. What advantages can a welded construction give over a metal casting or pressed unit? (List three)
2. Draw a simple sketch of these types of joints:
   a. Square groove butt joint
   b. Double "V" groove butt joint
   c. Tee joint
   d. Double bevel tee joint
   e. Double fillet lap joint
   Using a colored pencil, draw in good welds for each joint.
3. Which is recommended for heavier metal plates, a square butt joint or a "V" type butt joint?
4. Which are normally stronger, welds which are parallel or transverse to the stress lines of the joint?
5. When a joint receives turning loads, should the welds be spaced apart or close together?
6. (Select one). In general, fillet welds should be (a) 1/2, (b) 2/3, (c) 3/4 the thickness of the metal plate.
7. Match the items on the left below with the correct one on the right.

a. Porosity	1. Cracks in weld or base metal
b. Grain growth	2. Coarseness of metal near welds
c. Slag inclusions	3. Foreign substances in the weld
d. Residual stresses	4. Gas cavities In the weld

8. What is skip welding and why is it used?

Welding Safety

13

Introduction

We learn by experience. But learning safety through your own personal experience would be foolhardy—and unnecessary.

The safety rules for welding given here are based on someone else's experience. For your own good, learn these rules before you start welding. Someone learned them the hard way.

For more complete safety recommendations, refer to the joint American Welding Society AWS-ANSI Standard Z49.1.

When safety precautions refer to acetylene, they apply as well to other fuel gases such as natural gas, propane, and hydrogen.

Protective Equipment for Electric and Oxyfuel Processes

Wear ear protection at all times. Welding equipment, as well as the welding process, is noisy. Hot sparks also can get into unprotected ears (Fig. 1).

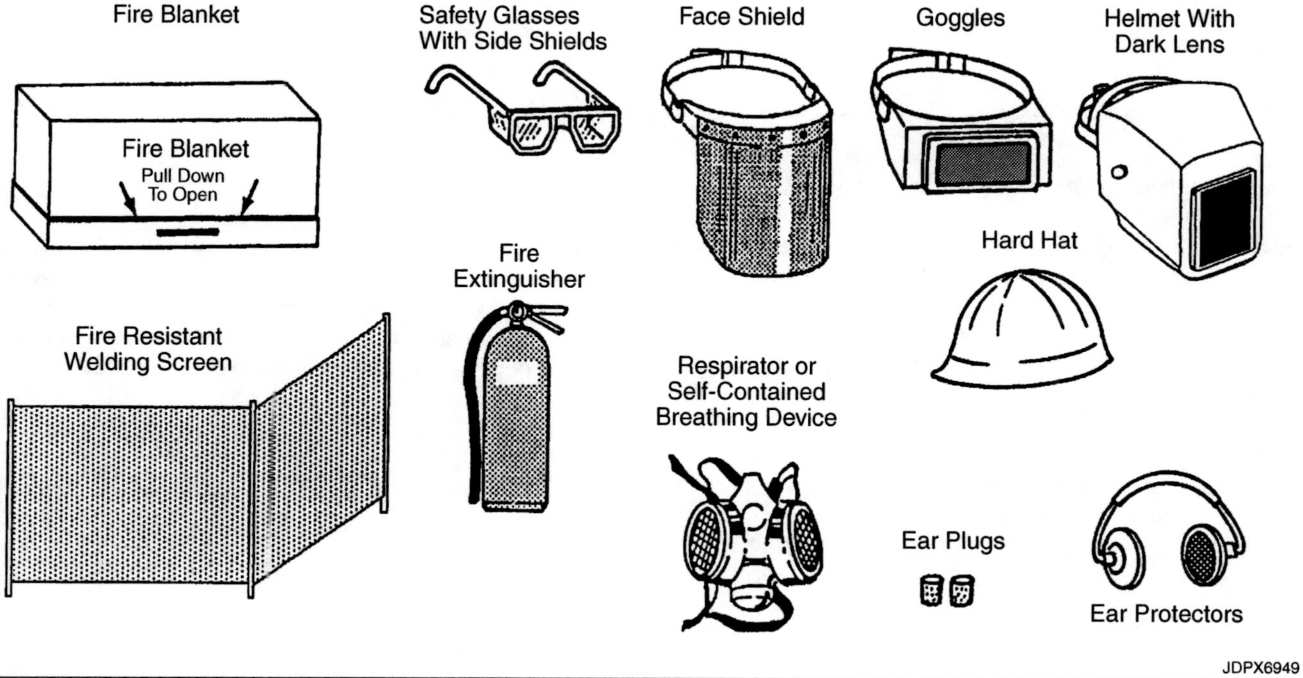

Fig. 1 — Protective Equipment for Welding

Always wear goggles with suitable filter lenses when using a torch. Also wear a head shield or helmet with suitable filter plates when arc welding. Wear flash goggles having side shields and lenses at all times—even when adjusting controls, etc. Goggles and helmets protect your eyes from sparks and flying slag, and also from the strong light and injurious rays of the flame or arc. They also help you to see your work better.

CAUTION: Avoid fabrics that are so thin (or loosely woven) that they do not block the arc rays from reaching your skin. Avoid fabrics that contain synthetics, which could melt to your skin if contacted by hot metal or from being close to the arc. Avoid fabrics that are highly flammable (nylon and polyester) which could start on fire or melt to your skin if contacted by hot metal or a spark.

Wear leather or treated cloth gloves and aprons, and wear suitable shoes and other protective clothing (Fig. 1) and (Fig. 2).

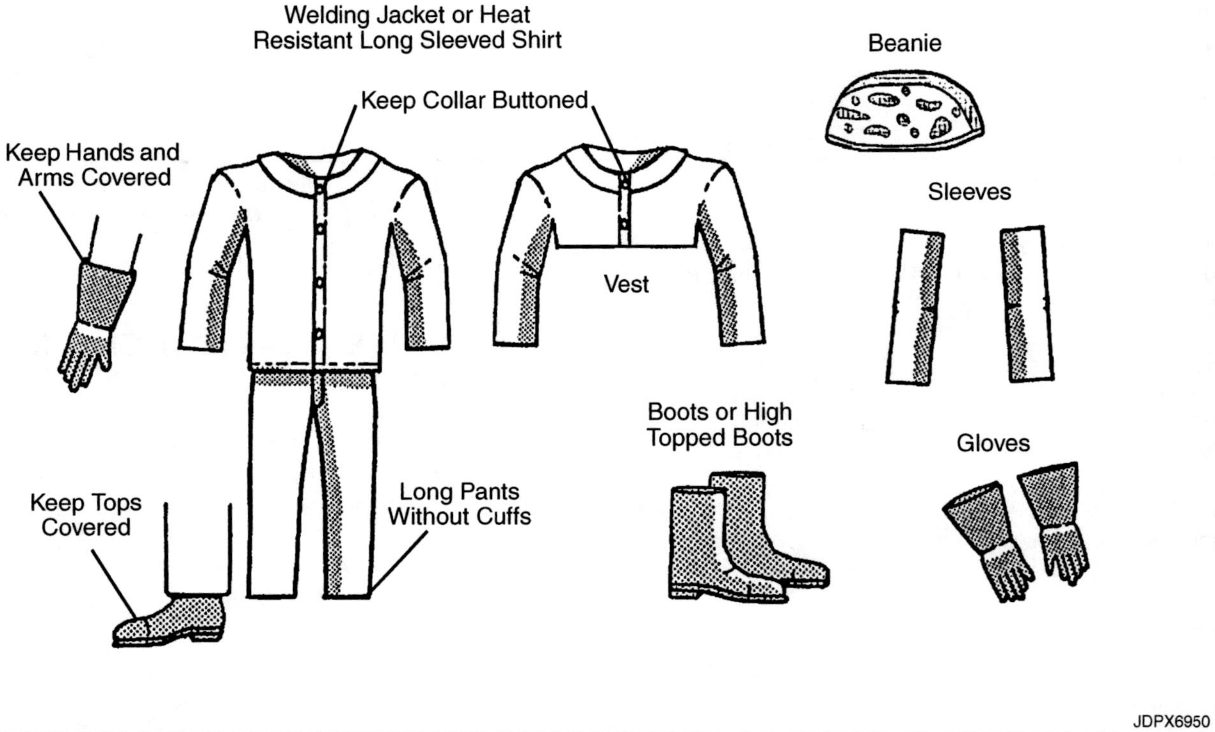

Fig. 2 — Protective Clothing for Welders

Keep protective equipment dry and free of oil.

Take care that your own clothing is not oily and that pockets and cuffs are not open and ready to receive sparks or hot slag.

Don't use rope to suspend staging for support of cutting or welding operations.

Do not weld metals coated with zinc or paint. Do not weld galvanized metals unless you have the proper breathing apparatus and your work place Is well ventilated.

Ventilation

Working areas must have adequate ventilation. When toxic fumes from lead, cadmium, or beryllium materials or any other substances are present in harmful concentrations, always use an air-supplied respirator. Make sure that oxygen concentrations above normal cannot develop in the work area (Fig. 3).

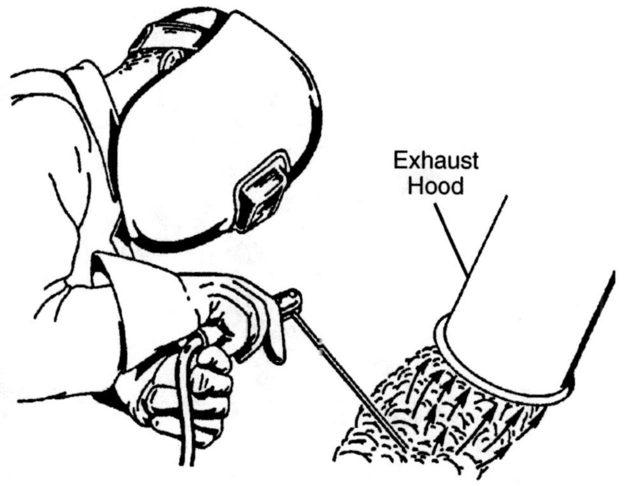

Fig. 3 — Welding Areas Must Have Ventilation

Never use oxygen for ventilation. That is, never try to replace oxygen in the atmosphere that has been consumed during a welding or cutting operation—ventilate with air.

Avoid leaks or open valves in small enclosures. Leave gas cylinders outside when working in such enclosures. Don't weld or cut for at least 15 minutes after contamination of clothing with oxygen.

Working on Tanks and Containers

Welding of containers should be done only by personnel familiar with the American Welding Society Publication F4.1 or the American Petroleum Institute Publication 2201.

Don't weld or cut containers such as drums, barrels, or tanks until you know there is no danger of fire or explosion.

Don't depend on your eyes or nose to decide if it is safe to weld or cut a closed container—find out what was in the container or use an explosimeter. A very small amount of residual flammable gas or liquid can cause a serious explosion.

Fire Prevention

Fires may result from gas or arc welding and cutting operations if combustible materials are allowed to come in contact with flying sparks, falling slag, hot metal, or the arc or flame itself.

Remember:

- Flying sparks can travel as much as 35 feet (11 m).
- Falling slag can pass through cracks out of sight of the goggled operator.
- Metal being cut or welded (or the arc or flame Itself) can be hot enough to ignite combustibles.

Fig. 4 — Fire Prevention While Welding

To prevent fires:

1. Keep flames and sparks away from cylinders and hose.
2. Move the job, if possible, to an area free of combustibles. Never work near explosive atmospheres. Avoid paint spray rooms, dip tanks, storage areas, ventilators, etc.
3. Move combustibles at least 30 to 40 feet (9.1 to 12 m) away from the job. Sweep the floor before lighting the torch. Wet down wooden floors.
4. Cover combustibles with fire-resistant shields or covers, if they cannot be moved. Covers should be large enough to completely shield the material and tight enough to prevent sparks from going under, over, or between the covers to the material. Use heat-resistant shields to protect nearby walls, ceilings, and floors.
5. Use standby watchers equipped with suitable fire extinguishes where the risk of serious fire damage is great.

6. Inspect the area on completion of the job to make sure that it is free of sparks, glowing embers, and flames. Check 30 minutes later.
7. Do not bring cylinders into confined spaces.
8. Avoid oxygen leaks in enclosed spaces. Do not leave oxy-fuel gas torches in enclosed spaces when not in use.
9. Never use oxygen to dust off clothing or work. A spark will cause clothing in an oxygen atmosphere to burst into flame immediately.
10. Never use oxygen in pneumatic tools or in oil preheating burners to start engines, to blow out pipelines, to create pressure in a container—or anywhere—as a substitute for compressed air or other gases.

Oxyacetylene Welding (OAW) Safety

Oxyacetylene welding safety will be covered under the following subjects:

- Cylinders
- Torches and Regulators
- Hoses
- Setting Up Equipment
- Adjusting Pressures
- Operating the Torch
- Backfires and Flashbacks
- Stopping Work

General Rules for Safe Handling of Oxyacetylene Equipment

- Keep oxyacetylene equipment clean, free of oil, and in good working condition.
- Avoid oxygen and acetylene leaks.
- Open oxygen valves slowly.
- Purge oxygen and acetylene lines before lighting torch.
- Keep heat, flame, and sparks away from combustibles.

Cylinders

- Use approved cylinders. They are safe as long as they are properly handled. Don't drop cylinders.

JDPX6958

Fig. 5 — Cylinders Can Explode if Damaged

CAUTION: Cylinders can explode if damaged. Welding cylinders contain gas under high pressure. If damaged, a cylinder can explode. Since gas cylinders are normally part of the welding process, be sure to treat them carefully.

- Always call acetylene "acetylene"—not "gas."
- Always call oxygen "oxygen"—not "air."
- The word "gas" is a general term and confusion here is dangerous. The use of the term "air" instead of the word "oxygen" may lead to serious accidents.
- Identify gas content by the name marked on the cylinder. If the cylinder is unmarked, do not use it! Return it to the supplier. Do not rely on color codes.
- Never use a cylinder or its contents for other than intended purposes.
- Keep oil and grease away from cylinders and cylinder valves.
- Separate oxygen cylinders from fuel gas cylinders in storage.
- Keep cylinders away from exposure to sparks, hot slag, open flame, and all possible sources of excessive heat.
- Be careful that cylinders are not placed so as to become a part of an electrical circuit. Avoid third rails, wires and electric welding circuits.
- Never strike an arc on a cylinder and never tap an electrode against a cylinder.
- Take care to place cylinders out of passageways for personnel and equipment.
- Always secure upright cylinders to a suitable support with a chain or other holder (Fig. 6).

IMPORTANT: Always secure cylinders in a vertical position on a cart or in a cylinder rack. When transporting, always secure cylinders to a suitable support with a chain or other holder.

Fig. 6 — Always Secure Upright Cylinders

- Always transport, store, and use acetylene cylinders in the vertical position to avoid "spitting" acetone.
- When transporting cylinders by crane, use a cradle, platform, or other suitable support. Never lift the cylinders by slings, by the caps, or by electric magnets.
- Never use cylinders as supports or rollers.
- Never try to mix any gases in a cylinder.
- Never try to refill a cylinder.
- Mark empty cylinders "Empty" or "MT."
- Send "Empties" back to the supplier promptly.
- Keep "Empties" and "Fulls" separate.
- Never tamper with or alter cylinder numbers or other markings. This is not only foolish but may be illegal.
- Do not tamper with or change fittings on cylinders.
- If valves cannot be opened by hand, do not use a hammer or wrench. Notify supplier.
- When acetylene cylinder valves are open, they may have leaks that cannot be stopped by closing the valve and tightening the packing gland. Leaks may also be found at acetylene fuse plugs or in the cylinder. If these types of leaks are found, (1) remove cylinder out of doors, away from possible sources of ignition; (2) tag cylinder properly to explain the trouble; and (3) notify the supplier as soon as possible.
- Protect cylinder valves from bumps, falls, falling objects, and weather. Keep them covered with cylinder caps when moving cylinders.
- Keep valves closed on empty cylinders.
- If an adapter is required between cylinder and regulator, always use a standard adapter. These may be obtained from your supplier. Where right-hand or left-hand threads are used on adapter, use two wrenches to ensure leak-proof connections.
- Never use other than approved wrenches for opening acetylene cylinder valves not equipped with handles. These should be obtained from the suppliers.
- Do not store cylinders in unventilated areas.

Torches and Regulators

Do not work with defective equipment. Have leaking or damaged equipment repaired by an authorized service person.

Give your oxyacetylene equipment the same good care and attention you would give any other dependable tool. Your work will be easier and safer.

Keep your equipment clean. Oil and grease in the presence of oxygen may burn with explosive violence if ignited. Do not use oil, grease, or other readily oxidizable substances on any torch or regulator. Do not handle equipment with oily rags or gloves.

Check systems for leaks under operating pressure.

Inspect union nuts, connections, and all seating surfaces on regulators and torches before use. Damaged connectors should be removed and faulty seats repaired at once—they are apt to cause fires, backfires, or flashbacks.

Never connect a regulator to a cylinder containing a gas other than that for which the regulator is designed.

If a regulator shows excessive creep (pressure building up when torch valves are closed), close cylinder valve and have the regulator repaired at once.

Have pressure gauges on the regulator tested periodically for accuracy.

If the pressure gauge needle does not return to the stop pin when pressure is released, do not merely adjust the pin. Have the gauge checked for accuracy.

Never test or calibrate regulator pressure gauges with oil.

Never use a torch as a hammer or to knock slag from work. Deformation of the torch or tips may cause trouble. Keep slag hammers and wire brushes handy.

Never use a hard, sharp tool for cleaning tips, except where specifically recommended by the tip manufacturer. Use appropriate tip cleaners or a copper or brass wire. When possible, clean orifices from the inner end. An oversize or bellmouth orifice may lead to trouble.

Make connections using wrenches designed for that purpose (Fig. 7). A standard wrench that fits your connections may be obtained from your supplier. Wrenches need not be used on connections using an O-ring seal, which can be made gas tight by hand.

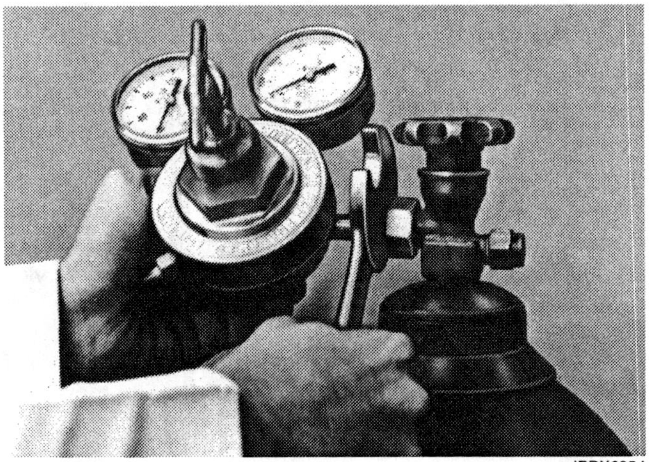

Fig. 7 — Use Proper Wrenches to Make Connections

If a leak develops around the torch valve stem, tighten the packing nut. If necessary, have valve repaired by a person qualified to repair oxy-fuel equipment.

The gases used on oxyacetylene jobs should never be drawn from cylinders except through properly attached pressure regulators approved for oxygen or acetylene service. Uncontrolled, pressure is dangerous in itself.

Hoses

All new hose may contain small particles of manufacturing residue—blow this out before using. Do not use compressed air hose. This hose may contain oil.

Use only hose designed for use with the gas source to which it is to be connected.

Hoses should be color-coded to avoid mixing, and a groove in the nut will generally indicate a left-hand thread. Oxygen hoses are usually green, while acetylene hoses are normally red—but be sure.

Standard hose connections are threaded, right-hand for oxygen and left-hand for acetylene, to prevent accidental interchange of oxygen and acetylene hoses. Do not interchange fittings on hoses or use other than standard hose fittings.

Never use ordinary wire as a binding in making up hose connections. Use binders or clamps designed for hose use.

Use only standard brass splices for splicing hose—never use copper tubing.

Avoid long lengths of hose. They tend to kink and to be mistreated. When long lengths are necessary, be sure all connections are tight.

Keep oil and grease away from hoses.

Protect hoses from sparks, hot slag, hot objects, sharp edges, and open flames.

If a hose is burned by a flashback, discard that length of hose. A flashback burns the inner walls and makes the hose unsafe.

Examine all hose periodically for leaks, worn places, and loose connections. Escaping acetylene may start a serious fire or cause severe burns.

Test for leaks by immersing the hose—under normal working pressure—in water. Repair leaks and worn places at once by cutting the hose and remaking the joint with standard fittings.

Do not try to repair hoses with tape.

Setting Up Oxyacetylene Welding (OAW) Equipment

See that your cylinders and equipment are clean, in a safe location, in good working order, and adequate for the job intended.

Be sure that your gloves are not oily and that your clothing is designed to protect you properly.

Before connecting regulators to cylinders, crack the cylinder valve carefully to blow out any foreign matter that might harm seats, clog orifices, or become ignited.

Never crack an acetylene valve near other welding work, sparks, open flames, or persons.

Never force connections.

Before opening the cylinder or manifold valve, make sure that the regulator is drained of gas by engaging regulator adjustment screw and allowing gas to flow out of the regulator. This will allow the high-pressure seat to open when pressurizing the regulator and will prevent excessive compression temperatures at the high-pressure seat.

After all the residual gas has been drained, completely release the adjusting screw. This will prevent the regulator from feeding gas into the line when the regulator is pressurized, until the adjusting screw is turned in.

Stand to one side of the regulator gauge when the cylinder valve is opened. Open all cylinder valves slowly. When pressurizing the regulator with the HP gauge, particularly the oxygen regulator, open valve slightly so that the HP gauge pointer moves slowly to the maximum reading.

Never open an acetylene cylinder valve more than 1-1/2 turns and preferably less than a single turn. But open oxygen cylinder valves completely. Opening acetylene cylinder valves wide is unnecessary, but opening oxygen valves wide seals the clearance around the stem and prevents leakage. Inert gas cylinders also must be opened completely when in use.

Leave the valve wrench in position whenever acetylene cylinder valves are open. This will allow quick closing in an emergency.

Standard hose is not always color-coded, although connections are thread-coded. Follow your connections through when setting up, and make sure oxygen and acetylene lines are not interconnected.

Be sure all connections are tight. Never try to locate a leak with a flame—use soapy water.

Coil excess hose neatly so it cannot kink or tangle. Place hose so it will not be stepped on or run over. Make sure it is clear of sparks and slag from your work.

Adjusting Pressures

Improper pressures are not only wasteful, they prevent you from doing your best work and they can be dangerous. Flashbacks can be caused by improper pressures.

Use manufacturer's charts as a guide to obtain the correct pressures for your job.

The use of acetylene at pressures in excess of 15 psig (103 kPa) or 30 psig (206 kPa) for welding, cutting, and heating (or in pressurized chambers) is not recommended.

In adjusting oxygen pressure, first be sure that the torch acetylene valve is closed. Open the torch oxygen valve. Adjust your oxygen regulator until low-pressure gauge indicates proper pressure, while torch valve is open and oxygen is flowing through the tip. Close torch oxygen valve.

In adjusting acetylene pressure, first be sure that the torch acetylene valve is closed. Open the torch oxygen valve. Adjust the oxygen regulator until low-pressure gauge indicates proper pressure, while torch valve is open and oxygen is flowing through the tip. Close torch oxygen valve.

In adjusting acetylene pressure, first be sure that the torch oxygen valve is closed. Then follow same procedure as for oxygen.

Operating the Torch

Purge each hose before lighting the torch.

Light torch with flint striker (Fig. 8).

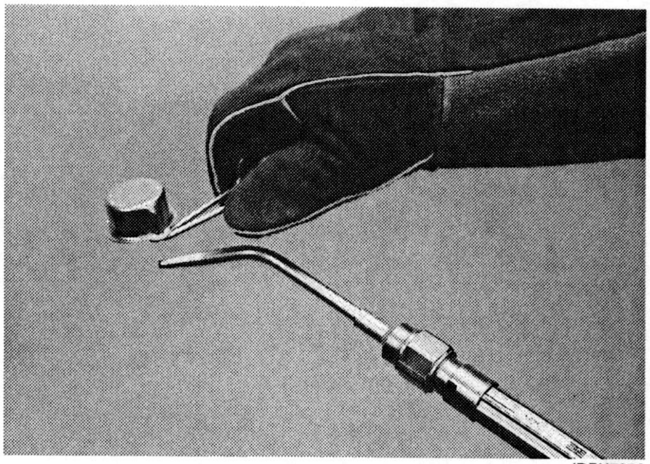

JDPX7053

Fig. 8 — Light Torch With Flint Striker—Don't Use Matches

Don't use matches.

Never direct gas stream toward yourself or others. When lighting or handling torch, be careful to keep flame and sparks directed away from personnel, flammables, and equipment. (See "Fire Prevention.")

Do not bring cylinders into confined spaces.

Take all precautions for ventilation, protective clothing, etc., covered in earlier chapters.

Recognizing and Troubleshooting Backfires and Flashbacks

A backfire is a loud noise associated with the momentary extinguishment or re-ignition of the flame at the torch tip. It may be caused by touching the tip against the work, by particles entering the tip and obstructing the gas flow, or by overheating the tip. Sometimes the trouble will clear itself immediately and, if work Is hot enough, the torch will relight automatically. If this does not happen, close oxygen torch valves immediately (cutting and then preheating valve). Close acetylene valve. Then relight, using standard lighting procedure, which includes purging the hose lines. Before relighting, check your pressures.

A flashback is the burning back of the flame into the tip or torch (or the hose if an explosive mixture is present in one of the lines). It is sometimes accompanied by a hissing or squealing sound and a smoky or sharp-pointed flame. This condition requires extinguishing the flame by closing first the torch oxygen valve and then the torch acetylene valve. Then wait a few moments to be sure the fire in the torch or hose has a chance to burn out. Flashbacks can be caused by failure to purge, improper pressures, distorted or loose tips or mixer seats, kinked hose, clogged tip or torch orifices, or overheated tip or torch.

Flashbacks mean that something is radically wrong with your equipment or your handling of it. Before attempting to relight your torch, check for trouble as follows:

First of all, did you purge your lines individually before lighting? Did you check your pressures? Are they near recommended pressures? Then remove hoses from torch and inspect for damage. If the flame has burned back into the hose, cut back to undamaged portion of the hose and remake connections (with standard clamp device). In setting up, be sure hose does not lie where it may be stepped on or run over. Avoid long lengths of hose. Use recommended pressures.

Stopping Work

Close oxygen torch valve first, then acetylene torch valve, making sure that flame left at the tip is extinguished and the acetylene torch valve is closed tight.

When shutting down for a brief interval, close torch valves only and leave the hose and torch in orderly fashion so they will not be damaged or disturbed.

Never hang a torch or hose on a regulator or cylinder valve unless the cylinder and torch valves are closed, and the hose is drained of gas.

Do not crimp hose to stop flow of gases temporarily as in changing torch or tip.

When work is completed or when leaving area for any reason, proceed after closing torch valves as follows:

1. Close cylinder valves.
2. Drain all oxygen from hose and regulator by opening torch oxygen valve. Then close torch oxygen valve.
3. Drain acetylene in the same manner.
4. Release regulator pressure adjusting screws.

When shutting down for an extended period, disconnect all apparatus and store it. In this way, unauthorized personnel will not be using the equipment, and equipment will not be damaged accidentally.

Make sure cylinder valve is tightly closed before removing regulator.

Safety in Electric Welding and Cutting

Fig. 9 — Safety in Electric Welding and Cutting

IMPORTANT: **Always follow local, state, and national regulations regarding the installation of electrical wiring and other devices necessary for the connection of welding equipment. This includes complying to the provisions in the latest National Electrical Code (NEC).**

Electric welding includes SMAW, GMAW, FCAW, and GTAW welding. Only general safety measures can be indicated for these areas, because electric welding equipment varies considerably in size and type. Equipment may range from a small portable shielded metal-arc welder to highly mechanized production gas-metal arc welders. In each case, follow the specific recommendations of the manufacturer.

Safety practices that are generally common to all types of arc welding operations are as follows:

1. Welding equipment should be installed according to provisions of the National Electric Code.
2. A welding power source should be equipped with a power disconnect switch that is conveniently located at or near the source so the power can be shut off quickly.
3. Repairs to welding equipment should never be made unless the power to the source is shut off. The high voltage used for arc welding can inflict severe and fatal injuries.
4. Welding power sources must be properly grounded. Stray current can cause severe shock when ungrounded parts are touched.

5. Never change the polarity switch while the power source is under a load. Wait until the source idles and the circuit is open. Otherwise, the contact surface of the switch may be burned and the resulting arcing could cause an injury.
6. Welding cables should not be overloaded nor a power source operated with poor connections. Operating with currents beyond the rated cable capacity causes overheating. Poor connections may cause the cable to arc when it touches metal grounded in the welding circuit.
7. Avoid damp areas, and keep hands and clothing dry at all times. Dampness on the body may cause an electric shock.
8. Never strike an arc on a compressed gas cylinder.
9. Do not strike an arc if someone is nearby without proper eye protection. Arc rays are harmful to the eyes and skin. Use a screen to protect others (Fig. 10).

JDPX6956

Fig. 10 — Use a Protective Screen to Protect Others From Arc Ray

10. Robotic welding power sources should be effectively guarded at all times.
11. Suitable spark shields must be used around resistance spot welding equipment.
12. Keep the uninsulated portion of the electrode holder from touching the welding ground when the current is on. This will cause arcing. The arcing can damage the equipment, and the flash of light can injure your eyes.
13. Keep welding cables dry and free of oil and grease.
14. Never carry welding cables coiled around the shoulders when they are carrying power.
15. Observe all the general precautions on protective clothing, ventilation, and fire prevention, which are outlined in the earlier part of this chapter. Be sure to wear adequate ear protection.
16. Put hot electrodes in a container and not on the floor or in the work area.
17. Remove electrodes from the holder when you stop working.
18. Check your welding helmet for cracks and other defects such as broken lenses by holding it in front of a direct light source before welding. Do not use a helmet that allows light to shine through cracks or separations in the material.
19. Avoid breathing fumes when welding galvanized metal, paint, or zinc.

Safety in Cutting

Fires often occur in cutting simply because precautions were not taken. Too often an operator forgets that sparks and falling slag can travel as much as 35 feet (10.5 m) and can pass through cracks out of sight of the goggled operator.

Observe these rules when cutting:

1. Never use a cutting torch where sparks will be a hazard, such as near rooms containing flammable materials, especially paint dipping or spraying rooms.
2. If you must do cutting over a wooden floor, sweep the floor clean and wet it down before starting. Use a bucket or pan containing water or sand to catch the dripping slag.
3. Keep a fire extinguisher nearby whenever any cutting is done.
4. Whenever possible, do cutting in wide open areas so sparks and slag will not become lodged in crevices or cracks.
5. If cutting is to be done near flammable materials and the materials cannot be moved, suitable fire-resistant guards must be securely installed.
6. In shops where there is a dirty or gassy atmosphere, take extra precautions to avoid explosions resulting from electric sparks or open fire during a cutting operation.
7. Never weld or cut on closed containers (Fig. 11).
8. Do not weld or cut directly on concrete surfaces.

JDPX6957

Fig. 11 — Never Weld or Cut on Closed Containers

Test Yourself

Questions

1. List protective equipment for welders.
2. List protective clothing for welders.
3. (Fill in blank.) Never use _____ for ventilation.
4. Personnel welding on containers should be familiar with what standard?
5. Why should containers that have held gas or liquid which will readily dissolve in water be thoroughly cleaned and before they are filled with an inert gas?
6. How far can flying sparks travel?
7. (Fill in blanks.) Thirty minutes after the completion of a welding job the work area should be inspected for _______, _________, and _________.
8. What can result from using oxygen to dust off clothing?
9. (Fill in blanks.) Keep _______ and _______ away from cylinders.
10. (Fill in blank.) Transport, store, and use acetylene cylinders in the _______ position to avoid "spitting" acetone.
11. What may oil and grease do in the presence of oxygen?
12. (Fill in blanks.) A notch or groove in the nut on a gas fitting indicates a _______-hand thread. Standard hose connections are threaded _______-hand for oxygen and _______-hand for acetylene or other fuel gases.
13. (Fill in blanks.) Before connecting regulators to cylinders, _______ the cylinder valve carefully to blow out any foreign matter that might harm seats, clog orifices, or become ignited.
14. (Fill in blanks.) Stand to _______ _______ of the regulator gauge when the cylinder valve is opened.
15. How many turns should an acetylene cylinder be opened?
16. (Fill in blank.) Open oxygen cylinder valves _______.
17. (Fill in blanks.) When arc welding, avoid _______ areas and keep hands and clothing _______ at all times. Dampness on the body may cause an _______.
18. (Fill in blank.) Avoid breathing _______ when welding galvanized metal, paint, or zinc.
19. (Fill in blank.) Robotic welding machines should be effectively _______ at all times.
20. (Fill in blanks.) Never strike an arc on a _______ gas _______.

Appendix

JDPX6991

Fig. 1 — Surfacing Bucket Teeth

Standard Welding Symbols[1]

Basic Welding Symbols and Their Location Significance

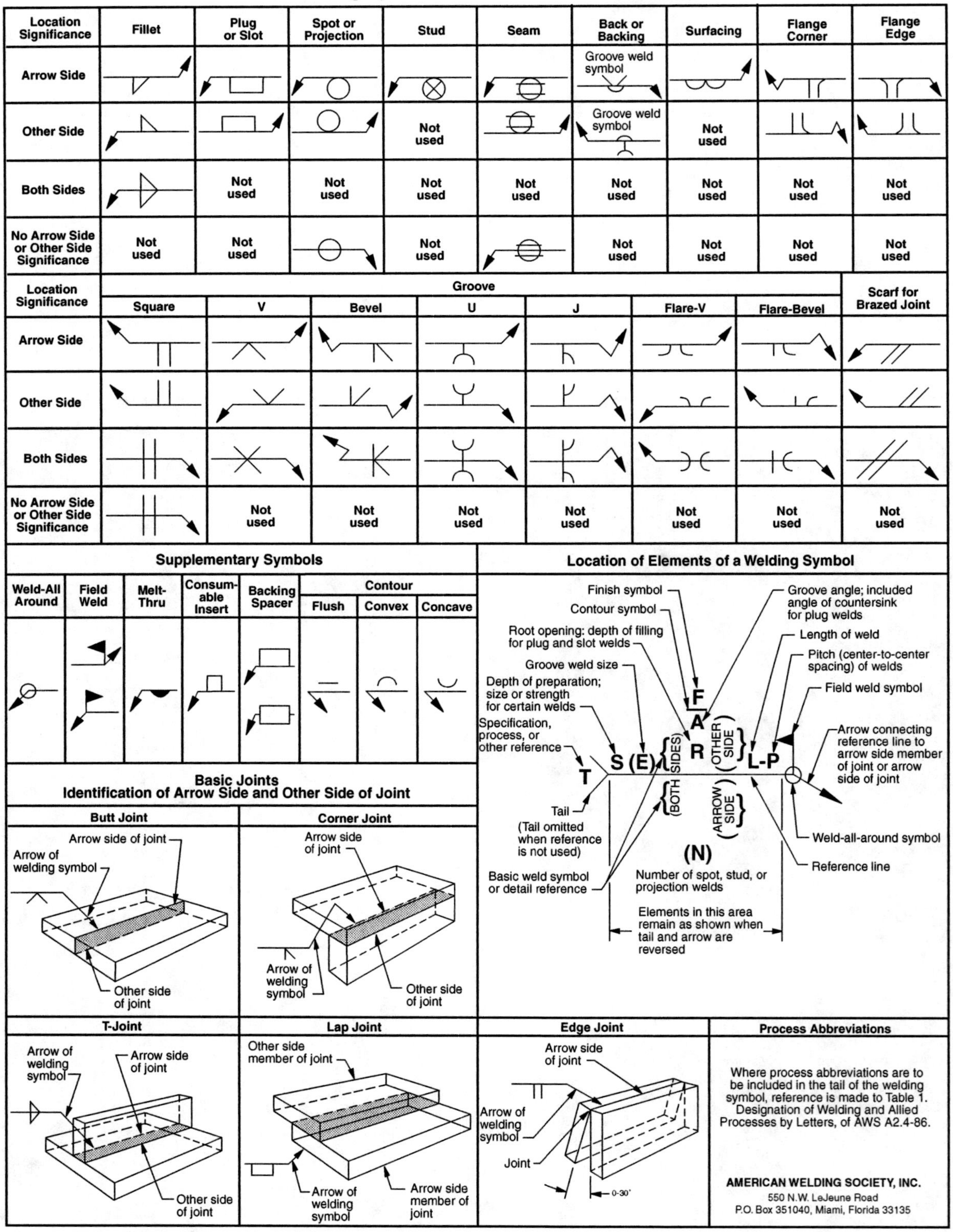

1 *It should be understood that this chart is intended only as a shop aid. The only complete and official presentation of the standard welding symbols is the American Welding Society, Inc. For more information, see the American Welding Society, Inc., www.aws.org.*

JDPX7279

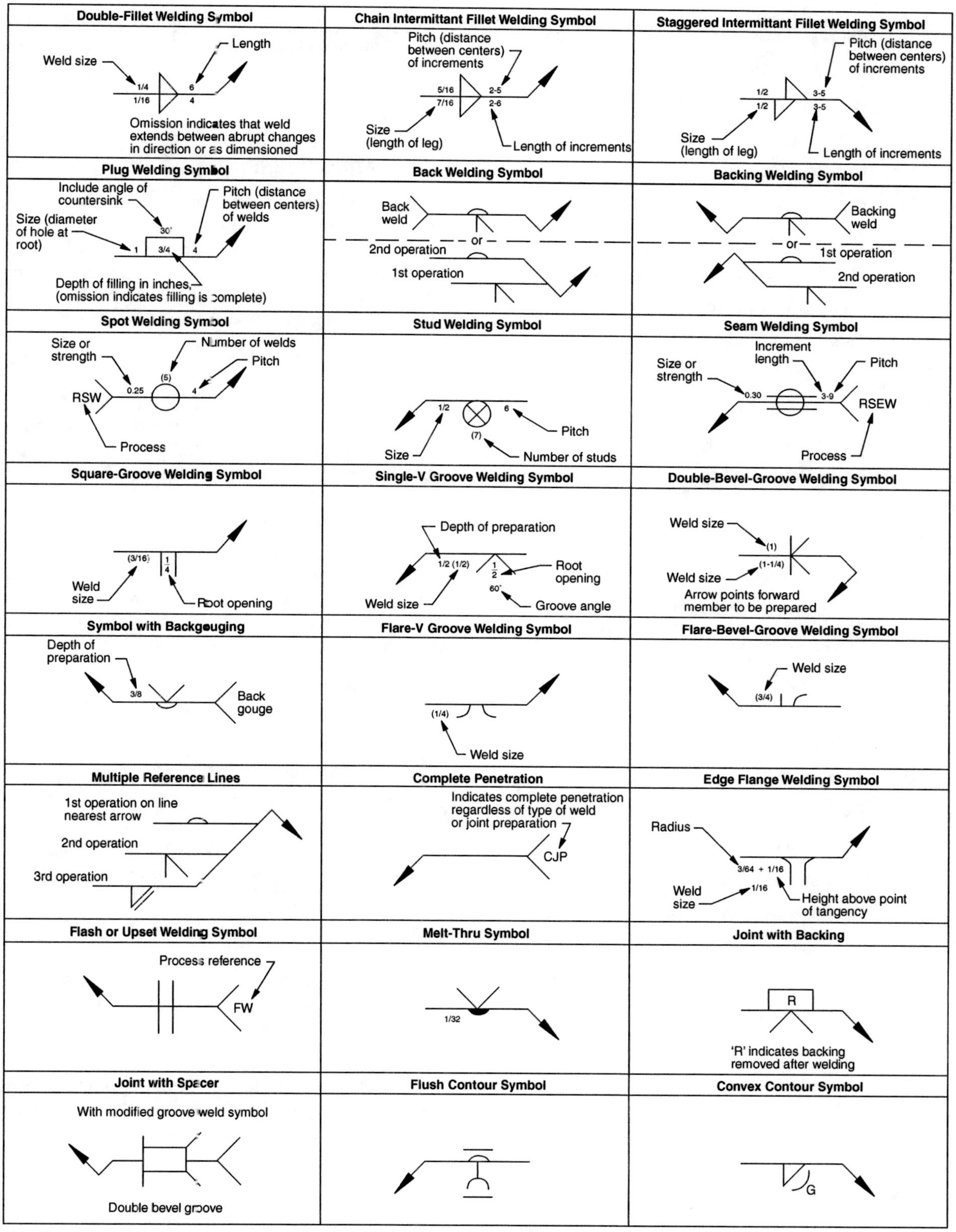

*It should be understood that these charts are intended only as a shop aids. The only complete and official presentation of the standard welding symbols is in A2.4.

JDPX7280

Abbreviations/Acronyms

AC	Alternating Current
API	American Petroleum Institute
ASME	American Society of Mechanical Engineers
AWS	American Welding Society
CC	Constant Current
CV	Constant Voltage
DC	Direct Current
DCEN	Direct Current Electrode Negative
DCSP	Direct Current Electrode Positive
FCAW	Flux-Cored Arc Welding
GMAW	Gas Metal Arc Welding
GTAW	Gas Tungsten Arc Welding
LBC	Laser Beam Cutting
LBD	Laser Beam Drilling
LBW	Laser Beam Welding
MCAW	Metal-Cored Arc Welding
MIG	Metal Inert Gas
NEC	National Electrical Code
OAC	Oxyacetylene Cutting
OAW	Oxyacetylene Welding
OFC	Oxyfuel Cutting
OFW	Oxyfuel Welding
PA	Plasma Arc
PAC	Plasma Arc Cutting
R	Resistance
SAW	Submerged Arc Welding
SMAW	Shielded Metal Arc Welding
TIG	Tungsten Inert Gas

Definitions of Terms and Symbols

This section explains the meaning of terms most used by welders. Technical engineering terms have been simplified.

A

ALTERNATING CURRENT

Electricity that reverses its direction of electron flow regularly and periodically. The period of alternating depends upon the cycle of the current. For 60-cycle current, the current goes in one direction and then in the other direction 60 times in the same second, so the current changes its direction 120 times in one second.

ACETYLENE

Gas composed of two parts of carbon and two parts of hydrogen. When burning in an atmosphere of oxygen, it produces one of the highest flame temperatures obtainable.

ACETYLENE CYLINDER

Specially built container manufactured according to I.C.C. Standards. Used to store and ship acetylene. (Occasionally called Tank or Bottle.)

ACETYLENE HOSE

See Hose.

ALLOY

A mixture of two or more metals.

ANNEALING

Softening metals by heat treatment. This most commonly consists of heating the metals up to a critical temperature and then cooling them slowly.

ANODE

Electrical term for Positive. It is that part of an electrical circuit to which electrons are flowing.

ARC

Term given to the flow of electricity through a gaseous space or air gap. In arc welding, this flow of electricity through the air produces high temperatures.

ARC BLOW

Tendency for an arc to wander or whip from its normal path because of a magnetic field building up in the workplace during arc welding. It occurs only when using direct current. It is most troublesome near the ends of a weld and on the first weld pass. Relocating the position of the work clamp (ground) sometimes helps. Switching to AC eliminates the problem.

ARC FLASH

The name given to a condition of the human eye caused by being exposed to the arc's rays without the proper eye protection. It may take several hours before an arc flash is felt. There will be a feeling like someone put hot sand in the person's eyes. It is caused by tiny water blisters forming on the surface of the eye that result from a "sunburn."

ARC VOLTAGE

Electrical potential (pressure or voltage) across the arc.

ARC WELDING

Fusing two metals together using an electric arc as the source of heat.

AUTOMATIC WELDING

Welding in which the work, the torch, and/or the arc is mechanically moved and controls are used to regulate the speed and/or the direction of travel.

AXIS OF A WELD

An imaginary line that passes through the center line of a weld and runs in a parallel direction with the weld.

B

BACKFIRE

A short pop of the torch flame followed by extinguishing of the flame or continued burning of the gases.

BACKHAND WELDING

Welding in the direction opposite the direction the gas flame is pointing.

BACKING

Some material placed on the root side of a weld to aid control of penetration.

BACK STEP WELDING

A method of sequencing welds used to control distortion. Short beads are run in one direction with each bead being started a short distance (1/2—12 inches/13—305 mm) from and running to the beginning point of the previous bead. Welding is actually progressing along the joint in a direction opposite the direction of each weld bead.

BASE METAL

Metal to be welded, cut, or brazed. Also called the work.

BEAD

The metal that is deposited in one pass. A narrow bead that is deposited in a straight direction without any movements is called a stringer bead. A wide bead that is deposited with a weaving motion is called a weave bead.

BEVEL

Removing the edge of a plate at an angle from top to bottom in preparation for welding. Bevel angles are measured by the amount of metal removed. For example, if two pieces that were beveled 30 degrees were butted together, the "V" groove formed would have an included angle of 60 degrees.

BIRD'S NEST

A condition which occurs when the wire wraps around the drive rolls on a GMAW wire feeder. Usually caused by the drive rolls being adjusted too tightly or the wire guides not being adjusted close enough to the drive rolls.

BLOWPIPE

Another term applied to the oxyacetylene torch.

BOND

Junction of the weld metal and the base metal.

BRAZING

Joining two metals together with an alloy that has a lower melting point than either of the metals being joined and without melting the base metal(s). Brazing temperatures are considered to be in a range of 800°F (425°C) and above.

BRONZE WELDING

Using a bronze alloy as filler metal when doing fusion welding. Aluminum bronze and silicon bronze are typical filler metals used in the TIG process.

BUILD-UP

Amount a weld face is extended above the surface of the metals being joined.

BURNED METAL

Term occasionally applied to the metal which has been combined with oxygen so that some of the carbon changed into carbon dioxide and some of the iron into iron oxide.

BURNING

Violent combination of oxygen with any substance which produces heat. This word is sometimes used instead of Flame Cutting.

BUTT JOINT

An assembly in which the two pieces joined are in the same plane, with the edge of one piece touching the edge of the other.

C

CABLE

See Lead Wire.

CAPILLARY ACTION

Property of a liquid to move into small spaces if it has the ability to "wet" these surfaces.

CARBON

An element that, when combined with iron, forms various kinds of steel. In steel, it is the changing carbon content that changes the physical properties of the steel. Carbon is also used in a solid form as an electrode for arc welding, as a mold to hold weld metal, and for motor brushes.

CARBURIZING

A carburizing flame is an oxygen-fuel gas flame with a slight excess of the fuel gas. Also called a reducing flame. It is named according to the relative length of the acetylene feather to the inner cone. For example, a 3X carburizing flame would have an acetylene feather 3 times the length of the inner cone.

CASE HARDENING

Adding carbon to the surface of a mild steel object and heat treating to produce a hard surface.

CASTINGS

Metallic forms which are produced by pouring molten metal into a shaped container (mold).

CATHODE

Electrical term for negative.

CELSIUS

Units of metric system temperature scale. Sometimes called centigrade. Symbol is °C (as in 22°C).

CERAMIC CUP

The cup which directs gas from the TIG torch body to the weld zone.

CHAMFERING

Removing the top edge of a plate at an angle in preparation for welding. A 1/16—1/8 inch (1.6—3.2 mm) square edge called the land (also shoulder or root face) is left on the bottom side of the piece.

COATED ELECTRODE

See Covered Electrode.

CONCAVE WELD FACE

A weld having the center of its face below the weld edges.

CONE

Inner visible flame shape of a neutral or near neutral flame. Also called inner cone.

CONTINUOUS WELD

A weld that extends from one end to the other end of two welded pieces.

CONVEX WELD

A weld with the face above the weld edges.

CORNER JOINT

A weld joint formed by the edges of two pieces of metal touching each other at an angle of approximately 90 degrees. If welding is done on the inside it is called an inside corner joint, and if the welding is done on the outside it is called an outside corner joint.

COVERED ELECTRODE

Metal rod used in shielded metal arc welding that has a covering of materials to aid in the arc welding process.

CRACKING

Action of opening a valve slightly and then closing the valve immediately. This is a method of clearing foreign material from the valve opening on a high-pressure cylinder before attaching a regulator or a regulator-flowmeter.

CRATER

The depression at the end of a weld created by not stopping the arc properly.

CROWN

Curved or convex surface of finished weld proper.

CUTTING FLAME

Cutting by a rapid oxidation process at a high temperature. It is produced by a gas flame accompanied by a jet action that blows the oxides away from the cut.

CYLINDER

A container used to hold gases used in welding. See Oxygen, Acetylene.

D

DIRECT CURRENT

Electric current which flows in only one direction.

DIRECT CURRENT ELECTRODE NEGATIVE

Sometimes referred to as straight polarity.

DIRECT CURRENT ELECTRODE POSITIVE

Sometimes referred to as reverse polarity.

DEOXIDIZERS

An element or elements added to welding filler metals that chemically remove oxygen from the weld puddle.

DISTORTION

The warping of a piece of metal by uncontrolled and unequal heating and cooling.

DOWNHAND WELDING

Welding in a vertical down direction.

DROSS

The name given the material which adheres to a piece of metal that has been cut with a plasma torch. It is similar to the slag produced from oxyfuel cutting.

DUAL SCHEDULE

Having two sets of welding parameters (or welding schedules) programmed into a continuous wire welding system that can be accessed by moving a switch on the welding gun. For example, by having this feature it is easy to weld parts of different thickness or fit-ups without having to physically go to the welding power source to change the setting.

E

EDGE JOINT

Joint formed when two pieces of metal are placed on top of each other with at least one side of the edges being even with one another.

ELECTRODE

The part of the welding circuit that is responsible for forming the arc when electricity is flowing from the power source. An example of a consumable electrode is when the electrode becomes part of the base metal as in shielded arc welding. An example of a non-consumable electrode is when the electrode does not become part of the base metal as with the GTAW process.

ELONGATION

Percentage increase in the length of a specimen when stressed to its yield strength.

EROSION

Reducing in size of an object because of a liquid or gas impact on the object.

EUTECTIC ALLOY

A mixture of metals that has a melting point lower than that of any of the metals in the mixture, or any other mixture of these metals.

F

FACE OF WELD

The exposed surface of the weld.

FAHRENHEIT

A temperature scale used in most English-speaking countries in the industrial and engineering areas. Symbol is °F (as in 72°F).

FILLER ROD

Metal wire that is melted and added to the welding puddle to produce the necessary increase in bead thickness. See Welding Rod.

FILLET WELD

The name given to a weld formed by two pieces of metal whose welded surfaces are approximately 90 degrees to each other. Lap joints and "T" joints are examples of fillet welds.

FLAME CUTTING

Cutting performed by an oxygen-fuel gas torch flame that has an oxygen jet.

FLASH

Impact of electric arc rays against the human eye. Also the fin of surplus metal formed at the seam of a resistance weld. Also see Arc Flash.

FLAT POSITION

See the American Welding Society welding position diagrams in Chapter 1.

FLOWMETER

A device attached to the supply of gas used for regulating the flow rate of a gas in cubic feet per hour (CFH) at a specified pressure.

FLUX

Chemical(s) used to clean impurities from the metal and to protect the metal from the surrounding air during soldering, brazing, and welding.

FOREHAND WELDING

Fusing metal in the same direction as the torch flame points.

FORGING

Metallic shapes being derived by either hammering or squeezing the original piece of metal into the desired shapes or thicknesses.

FUSION

Intimate mixing of molten metals.

G

GAS METAL ARC WELDING

(GMAW) An electric welding process that uses a continuously fed wire electrode that melts to form a weld bead. The weld must be protected from the surrounding air by a shielding gas. (Also referred to as MIG.)

GAS POCKETS

Cavities in weld metal caused by entrapped gas. (Also called porosity.)

GAS TUNGSTEN ARC WELDING

(GTAW) An electric welding process which uses a tungsten as an electrode for forming an arc to the work. An inert gas must be used for shielding. (Also referred to as TIG.)

GENERATOR

A mechanism which generates or produces some substance, i.e., electric generator, acetylene generator.

GLOBULAR TRANSFER

The range between short arc transfer and spray arc transfer. Welding conditions in this range are unacceptable when using solid wire because an erratic arc produces excessive quantities of large droplets of spatter and penetration is not consistent.

GOUGING

Cutting of a groove in the surface of a metal using the gas cutting torch of an arc-air cutting torch.

GROOVE WELD

A weld that is formed by filling the opening or groove between two plates. The edges of the plates may be prepared so they form several shapes. Common types of grooves include bevel grooves, "V" grooves, "J" grooves, "U" grooves, and combinations of these shapes. Also see the AWS weld symbol chart and the AWS weld position diagrams.

H

HAND SHIELD

See Shield.

HEAT

The use of an oxyfuel torch to preheat (warm) a piece of metal before welding or bending.

HEAT-AFFECTED ZONE

That part of the base metal that has been altered by the heat from the welding, brazing, or cutting operation.

HEAT CONDUCTIVITY

The rate or speed heat transfers through a substance.

HELMET

A protective shield that must be used by a person watching arc welding. The opening in the front must contain the proper shade lens to filter out invisible rays and to reduce the intensity of the visible light produced by an electric arc.

HORIZONTAL POSITION

See the American Welding Society welding position diagrams in Chapter 1.

HOSE

Small diameter tubing that carries welding gases. It is specially constructed of rubber and fabric to withstand welding gas pressures and yet remain flexible.

HYDROGEN

A gas formed of the single element, hydrogen. It is considered one of the most active gases. When combined with oxygen, it forms a very clean flame which, however, does not produce very much heat.

I

INERT GAS

A gas that does not chemically combine or react with another substance.

INFRARED RAYS

A type of invisible radiation given off by the electric arc. Short- and long-term injury can occur to a person's eyes from exposure to infrared radiation without the proper shade welding lens.

INSIDE CORNER WELD

See Corner Joint.

INTERMITTENT WELD

Joining two pieces and leaving equally spaced unwelded sections in the joint. Also see AWS Weld Symbols Chart.

J

JOINT

The intersection where two pieces are joined together by a welding process.

L

LAP JOINT

A joint in which the edges of the two metals to be joined overlap one another.

LAYER

A certain weld metal thickness made of one or more passes.

LEAD WIRE

Electric wire from the power source to the electrode holder or to the work (ground) clamp(s).

LEG OF FILLET WELD

Distance from the point where the base metals touch to the toe of the fillet weld.

LENS

A specially treated glass used in an arc welding helmet or in a pair of oxyfuel cutting goggles to filter out harmful radiation. Different numbers (shades) are required for each welding or cutting process — the larger the number, the darker the lens. Consult the owner's manual or the equipment manufacturer's manual for the proper shade lens.

M

MALLEABLE CASTINGS

Cast forms of metal which have been heat-treated to reduce their brittleness.

MAPP

A stabilized methyl acetylene-propadiene fuel gas. It is a BOC Gases product (it was patented originally by Dow Chemical Co.).

METAL INERT GAS

(MIG) A term used to describe gas metal arc welding (GMAW).

MIXING CHAMBER

A special area in a welding torch where the welding gases are intimately mixed prior to combustion.

N

NEUTRAL FLAME

The type of flame produced from the combustion of equal portions of oxygen and a fuel gas. This flame is used for many applications because there is no chemical activity in the flame envelope to affect the molten metal.

NOZZLE

The part of a GMAW gun that directs the shielding gas around the weld. It is also called a gas cup.

O

OPEN CIRCUIT VOLTAGE

The voltage measurement before the arc is struck. It is several volts higher than the arc voltage.

ORIFICE

Opening through which gases flow. It is usually the final opening, or any opening controlled by a valve.

OUTSIDE CORNER WELD

See Corner Joint.

OVERHEAD POSITION

See the American Welding Society welding position diagrams in Chapter 1.

OVERLAP

A weld defect caused by the weld extending beyond the toe of the weld without fusing into the base metal.

OXIDIZING

Combining oxygen with any other substance. For example, a metal is oxidized when the metal is burned (i.e., oxygen is combined with all the metal or parts of it).

OXIDIZING FLAME

The type of flame produced by an excess of oxygen in the oxygen-fuel mixture. The free oxygen will combine to form oxides or may actually burn the molten metal as in the oxyfuel cutting process.

OXYACETYLENE WELDING (OAW)

See Oxygen Acetylene Welding.

OXYGEN

A gas form of the element oxygen. It is stored as either a liquid in special containers or as a high-pressure gas in cylinders. When it very actively supports combustion it is called burning; when it slowly combines with a substance it is called oxidation.

OXYGEN ACETYLENE CUTTING

A cutting process used on steel. Acetylene is the fuel used for preheating the steel to its kindling temperature. When the proper temperature is reached, a jet of pure oxygen is directed onto the heated area and a kerf is burned through the steel.

OXYGEN ACETYLENE WELDING

A method of welding which uses a fuel combination of two gases, oxygen and acetylene, as a heat source.

OXYGEN CYLINDER

A specially built container manufactured according to I.C.C. standard, used to store and ship oxygen.

OXYGEN HOSE

See Hose.

OXYGEN REGULATOR

A device used to reduce oxygen cylinder pressure to a safe working pressure. They are never to be used as acetylene regulators. Also see Regulator.

P

PASS

Weld metal created by one progression along the weld joint.

PENETRATION

Depth of fusion into the base metal as measured from the surface of the base metal.

PLASMA

Temporary physical condition of a gas after it has been exposed to, and has reacted to, an electric arc. This principle is used in welding and cutting applications.

PLASMA GOUGING

A process of metal removal using a special power source and torch that provides a plasma stream capable of removing welds or cutting grooves on any metal that will conduct electricity. It is a much faster process than the air carbon arc process. The completed groove requires very little if any cleaning when finished.

PLUG WELD

See the AWS weld symbol chart.

POROSITY

Presence of gas pockets or voids in the metal.

POSTHEATING

Heating a piece of metal after welding. It is used to relieve the stresses created from welding or bending (forming).

PREHEATING

Heating a piece of metal before welding, cutting, or bending (forming). Heating thick metal sections and heating metals that are heat treatable, reduces and sometimes eliminates weld cracking.

PUDDLE

A small area of molten metal (material) that must be present in fusion welding. It occurs at the place where the heat is applied.

PULSED SPRAY TRANSFER (GMAW-P)

One of the continuous wire processes that uses a special power source for varying the welding current between a high and low setting instead of keeping it constant. Metal is transferred only at the high value. Some power sources also allow the pulse frequency to be varied.

R

RADIOGRAPH

A photograph obtained by passing X-rays or gamma rays through the object to be photographed. The variations in the object's density is recorded on a photographic film.

RAYS

See Infrared Rays and Ultraviolet Rays.

REDUCING FLAME

An oxygen-fuel gas flame with a slight excess of fuel gas. Also called a carburizing flame.

REGULATOR

A device used on a gas cylinder or on a manifold system to reduce high gas pressures to safe working pressures. Single stage regulators perform this reduction in one stage. Two stage regulators do this in two stages and provide better regulation when the source pressure fluctuates.

REGULATOR FLOWMETER

A device used on a gas cylinder or on a manifold system that first reduces high gas pressures to safe working pressures and then allows adjustment of the rate of gas flow in cubic feet per hour (CFH).

REINFORCEMENT OF WELD

Excess metal on the face of a weld.

RESISTANCE WELDING

Welding metals using the resistance of metals to the flow of electricity as the source of heat.

REVERSE POLARITY

See Direct Current Electrode Positive.

ROOT OF WELD

That part of a weld farthest from the application of weld heat and/or filler metal side.

ROPING

When the GMAW gun pushes away from the work and/or when the wire turns red and forms a coil of "rope" on the work rather than forming a weld.

ROSEBUD

The name given to a large welding tip or a special tip used for heating purposes.

S

SHIELD

An eye and face protector held in the hand. It enables a person to look directly at the electric arc through a special lens without being harmed.

SHIELDED METAL ARC WELDING (SMAW)

A form of electric welding in which a heavy flux-coated electrode is used.

SHORT ARC or SHORT CIRCUITING TRANSFER (GMAW-S)

A gas metal arc welding process which uses a low arc voltage, where the arc is continuously interrupted as the molten electrode metal bridges the arc gap.

SLAG HOLE

A visible hole on the surface of a weld formed by trapped slag.

SLAG INCLUSIONS

Slag that is trapped inside a weld. It is not visible by looking at the outside surface and can be located only with special weld testing equipment.

SOLDERING

Joining two metals together with an alloy that has a lower melting point than either of the metals being joined. Soldering temperatures are considered to be under 800°F (427°C). Solder joints will not withstand the working temperatures, pressures, and stresses that brazed joints will withstand.

SPRAY ARC or SPRAY TRANSFER (GMAW)

A gas metal arc welding process that has an arc voltage high enough to continuously transfer the electrode metal across the arc in small globules.

STRAIGHT POLARITY

See Direct Current Electrode Negative.

STRESS

Load imposed on an object.

STRESS RELIEVING

Even heating of a welded area to a temperature below the critical temperature followed by a slow, even cooling.

STUBBING

See Roping.

T

TACK WELD

Small weld used to hold together components of an assembly. It may be temporary or permanent.

TANK

See Cylinder.

"T" JOINT

Joint formed by placing one metal against another at an angle of 90 degrees to form an upside down "T." Also see fillet weld.

TENSILE STRENGTH

The amount of force (measured in psi or kg/mm^2) necessary to pull a weld test coupon apart.

THROAT OF A FILLET WELD (ACTUAL THROAT)

Distance from the weld root to the weld face.

TUNGSTEN INERT GAS

(TIG) Term used to describe gas tungsten arc welding (GTAW).

TINNING

Term applied to soldering, where the metals to be soldered together are first given a coating of soldering metal.

TIP

Part of the torch at the end where the gas burns, producing the high-temperature flame. In resistance welding, the electrode ends are sometimes called tips. Also the name given for the copper tube the wire exits from on a GMAW gun.

TOE OF WELD

Junction of the face of the weld and the base metal.

TORCH

The mechanism which the operator holds during gas welding and brazing, oxyfuel and plasma cutting, and gas tungsten arc welding (GTAW).

U

ULTRAVIOLET RAYS

A type of invisible radiation produced by an electric arc. It will produce a sunburn on bare skin with an exposure of only a few minutes. Adequate precautions should be taken when near any type of electric welding, cutting, gouging, plasma welding, and plasma cutting to protect yourself from these rays.

UNDERCUT

A depression at the toe of the weld which is below the surface of the base metal.

V

VERTICAL POSITION

See the American Welding Society welding position diagrams in Chapter 1.

W

WELDING

The melting, flowing together and combining, and freezing of two materials under controlled conditions. Dissimilar metals can sometimes be welded together. Materials like plastic or glass are being welded.

WELDING ROD

The filler metal which is melted into the weld metal. Depending on the process, it may be bare or contain a coating.

WELDING SEQUENCE

Order in which the component parts of a structure are welded.

WELDMENT

Assembly of component parts joined together by welding.

WELD METAL

Fused portion of base metal or fused portion of the base metal and the filler metal.

Y

YIELD STRENGTH

The force in psi or kg/mm^2 at which a weld specimen starts to pull apart or assumes a specified limiting permanent set.

Internet Web Sites for Welding Resources

AGA Gas, Inc.
6225 Oak Tree Blvd.
Cleveland, OH 44131
(216) 642-6600
FAX: (216) 573-7870

Air Products & Chemicals
7201 Hamilton Boulevard
Allentown, PA 18195-1501
(616) 481-4911
Internet Web Site: www.airproducts.com

American Society for Nondestructive Testing, Inc. (ASNT), The
1711 Arlingate Lane
Columbus, OH 43228
(614) 274-6003
FAX: (614) 274-6899
E-Mail: embinder@asnt.org
Internet Web Site: www.asnt.org

American Welding Society, The
550 N.W. LeJeune Road
Miami, FL 33126
(305) 443-9353
FAX: (305) 443-7559
Internet Web Site: www.aws.org

ASM International/Advanced Materials & Processes
9639 Kinsman Road
Materials Park, OH 44073
(440) 338-5151
FAX: (440) 338-4634
Internet Web Site: www.asminternational.org

BOC Gases (Formerly AIRCO Gases)
575 Mountain Avenue
Murray Hill, NJ 07974
(908) 464-8100
FAX: (908) 771-1764
Internet Web Site: www.boc-gases.com

Canadian Welding Bureau
7250 West Credit Avenue
Mississauga, Ontario
L5N 5N1, Canada
(905) 542-1312
FAX: (905) 542-1318
E-Mail: info@cwbgroup.com
Internet Web Site: www.cwbgroup.com

CK Worldwide, Inc.
3501 "C" Street North East
Auburn, WA 98002
(800) 426-0877, (253) 854-5820
FAX: (253) 939-1746
Internet Web Site: www.ckworldwide.com

Cloos Robotic Welding, Inc.
911 Albion Avenue
Schaumburg, IL 60193
(847) 923-9988
FAX: (847) 923-9989
E-Mail: cloos@cloos-robot.com

Compressed Gas Association, Inc.
4221 Walney Road, 5th Floor
Chantilly, VA 20151
(703) 788-2700
FAX: (703) 961-1831
E-Mail: cga@cganet.com
Internet Web Site: www.cganet.com

Dataweld, Inc.
1909 Citizens Bank Drive
Bossier City, LA 71111
(318) 746-6111
FAX: (318) 746-0323
E-Mail: inform@dataweld.com
Internet Web Site: www.dataweld.com

DVS - German Welding Society
Aachener Strasse 172
Dusseldorf, 40223
Germany
49-0-2111591176
FAX: 49-0-2111591200
E-Mail: info@dvs-hg.de
Internet Web Site: www.die-verbindungs-spezialisten.de

ESAB Cutting Systems
411 S. Ebenezer Rd.
Florence, SC 29501
(843) 664-4394
FAX: (843) 664-4403
Internet Web Site: www.ESAB.com

Eureka Welding Alloys
2000 East Avis Drive
Madison Heights, MI 48071
(248) 588-0001
FAX: (248) 585-7111
E-Mail: eureka@eurekaweldingalloys.com
Internet Web Site: www.eurekaweldingalloys.com

EWF - European Welding Federation
Tagus Park
Apartado 012
P-2780-994 Porto Salvo
Portugal
E-Mail: mequinto@isq.pt

FANUC Robotics North America, Inc.
3900 W Hamlin Rd.
Rochester Hills, MI 48309
(800) 477-6268, (248) 377-7678
FAX: (248) 377-7365
E-Mail: ellen.sherwood@fanucrobotics.com
Internet Web Site: www.fanucrobotics.com

Genesis Systems Group
8900 Harrison Street
Davenport, IA 52806
(563) 445-5600
FAX: (563) 445-5699
Internet Web Site: www.genesis-systems.com

Hobart Institute of Welding Technology
400 Trade Square East
Troy, OH 45373
(800) 332-9448
FAX: (937) 332-5200
E-Mail: hiwt@welding.org
Internet Web Site: www.welding.org

Hypertherm, Incorporated
P.O. Box 5010
Etna Rd.
Hanover, NH 03755
(603) 643-3441
FAX: (603) 643-5352
E-Mail: customer.service@hypertherm.com
Internet Web Site: www.hypertherm.com

ITW Welding Products Group
1635 W. Spencer St.
Appleton, WI 54914
(920) 734-9821
FAX: (920) 735-4013
Internet Web Site: www.millerwelds.com

Lincoln Electric Company, The
22801 St. Clair Avenue
Cleveland, OH 44117-1199
(216) 481-8100
FAX: (216) 486-1751
Internet Web Site: www.lincolnelectric.com

Magnaflux
3624 West Lake Ave.
Glenview, IL 60026
(847) 657-5300
FAX: (847) 657-5388
Internet Web Site: www.magnaflux.com

Matheson Gas Products Inc.
166 Keystone Drive
Montgomeryville, PA 18936
(215) 641-2700
FAX: (215) 641-0656
E-Mail: info@mathesongas.com
Internet Web Site: www.mathesongas.com

MG Systems & Welding Inc.
W141 N9427 Fountain Blvd.
Menomonee Falls, WI 53051
(262) 255-5520
FAX: (262) 255-5170
E-Mail: sales@messer-mg.com
Internet Web Site: www.messer-mg.com

Miller Electric Manufacturing Company
1635 West Spencer Street
Appleton, WI 54914
(920) 734-9821
FAX: (920) 735-4013
Internet Web Site: www.millerwelds.com

MK Products, Inc.
16882 Armstrong Avenue
Irvine, CA 92606-4975
(949) 863-1234
FAX: (949) 474-1428
E-Mail: sales@mkprod.com
Internet Web Site: www.mkproducts.com

Motoman Inc.
805 Liberty Lane
West Carrollton, OH 45449
(937) 847-6200
FAX: (937) 847-6277
Internet Web Site: www.motoman.com

Nachi Robotic Systems, Inc.
22285 Roethel Drive
Novi, MI 48375
(248) 305-6545
FAX: (248) 305-6542
E-Mail: nrsrobot@ix.netcom.com
Internet Web Site: www.nachi.robotic.industry.net

Ohio State University, Welding Engineering
The Ohio State University
1248 A. E. Adams Drive
Columbus, OH 43211
(614) 292-2570
FAX: (614) 292-7852
E-Mail: albright.4@osu.edu
Internet Web Site: www-iwse.eng.ohio-state.edu/

Praxair, Inc.
39 Old Ridgebury Road
Danbury, CT 06810-5113
(800) 772-9247
FAX: (800) 772-9985
E-Mail: info@praxair.com
Internet Web Site: www.praxair.com

Preco, Incorporated (Formally Laser Machining, Inc.)
500 Laser Drive
Somerset, WI 54025
(715) 247-3285
FAX: (715) 247-5650
E-Mail: sales@preco.com
Internet Web Site: www.preco.com

Sellstrom Manufacturing Company
One Sellstrom Drive
Palatine, IL 60067
(847) 358-2000
FAX: (847) 358-8564
E-Mail: sellstrom@sellstrom.com
Internet Web Site: www.sellstrom.com

Stoody Company
16052 Swingley Ridge Road, Suite 300
St. Louis, MO 63017
(636) 728-3000
Internet Web Site: www.thermadyne.com/stoody

Thermadyne Industries
16052 Swingley Ridge Road, Suite 300
St. Louis, MO 63017
(636) 728-3000
Internet Web Site: www.thermadyne.com

Tweco, Products
16052 Swingley Ridge Road, Suite 300
St. Louis, MO 63017
(636) 728-3000
Internet Web Site: www.thermadyne.com/tweco/

Victor Equipment Co.
16052 Swingley Ridge Road, Suite 300
St. Louis, MO 63017
(636) 728-3000
Internet Web Site: www.thermadyne.com/victor/

Weld Mold Company
750 Rickett Road
Brighton, MI 48116
(800) 521-9755, (810) 229-9521
FAX: (810) 229-9580
E-Mail: sales@weldmold.com
Internet Web Site: www.weldmold.com

Wilson Industries, Inc.
123 Explorer Street
Pomona, CA 91768
(909) 468-3636
FAX: (909) 468-3640
E-Mail: sales@wilsonind.com
Internet Web Site: www.wilsonindustries.com

Answers to Test Yourself Questions

Chapter 1

1. Plasma arc cutting
   Oxyfuel cutting
2. Gas tungsten arc welding (GTAW)
3. Shielded metal arc welding (stick welding)
4. Semi-automatic
5. Oxygen
   Acetylene
6. Carbon steel
7. Fusion welding
8. Soldering 840°F/450°C or less
   Brazing 840°F/450°C or more
9. Capillary attraction
10. Thermal cutting processes
11. Heat
    Distance
    Angle
    Speed
    Manipulation
12. Increase tip size, open torch body valve slightly
13. Amperage
14. Face
    Legs
    Throat
    Root
    Toes
15. Reinforcement
16. Root opening
17. Butt joint
    "T" joint
    Lap joint
    Corner joint
    Edge joint
18. Length of legs
19. Penetration
20. Leg size; throat size
21. A — overhead
    F — fillet weld
22. Horizontal
23. Undercut
    Cold lap
    Concaved weld
    Excess convexity
    Off joint
24. Non-destructive testing
    Destructive testing
25. Bend test
    Tensile test
    Nick break test
26. American Society of Mechanical Engineers (ASME)
    American Welding Society (AWS)
    American Petroleum Institute (API)
    Military Code
27. Face bend
    Root bend
    Side bend

Chapter 2

1. Primary combustion (inner cone)
   Secondary Combustion (outer envelope)
2. Safety fuse plugs
3. Cylinders
   Pressure regulators
   Torch body welding tip
   Check valves and flashback arrestors
   Hoses and cylinder cart
4. Protect cylinder valve
5. 2200 psi at 70°F/15 200 KPa at 20°C
6. Double seated oxygen cylinder valve
7. Safety disk
8. 250 psi at 70°F/1724 kPa at 20°C
9. Slowly and all the way (valve should be seated open)
10. 1/7
11. 15 psi /100 kPa
12. Gas
    Liquid
13. Reduce cylinder pressure to safe working pressure
14. Single-stage
    Two-stage
15. Mix gases and control amount of gases going to the tip
16. Copper
17. Orifice size
18. Start with the smallest size tip cleaner
19. Rectangular
20. Hand tight
21. Tip too close to work piece
    Overheating tip
    Insufficient amount of gas flow to tip
22. Oxygen
23. Clogged orifice
    Incorrect oxygen and acetylene pressure (too low)
24. Stop the flashback from traveling past the point of installation
25. Groove in nut
    Left-hand threads

26. Green
27. 4 or 5 shade
28. Upright position
29. Brass
30. Valve stems
    Cylinder regulator connections
    Regulator hose connections
    Hose-to-torch body connections
    Torch body-to-tip connections
    Torch body valves (check at end of tip)
31. Carburizing flame
    Neutral flame
    Oxidizing flame
32. Heat [volume of heat]
    Distance [inner cone from welding puddle]
    Travel speed [movement along the weld joint]
    Angle [work and travel angle of torch and rod]
    Manipulation [movement of torch and rod].
33. Acetylene 5589°F or 3087°C
34. Mapp
    Propylene
    Propane
    Natural

Chapter 3

1. Stick
2. Type of base metal
3. Single- or three-phase
4. Ultraviolet rays
5. Form slag over weld bead to protect the weld from the atmospheric air.
   Forms a gaseous shield to protect the molten droplets of metal as they are transferred across the welding arc.
6. Provide a proper combination of high controllable amperage and low voltage for welding.
7. Constant current (CC)
8. Constant current (CC) adjusts amperage
   Constant voltage (CV) adjusts voltage
9. Shielded metal arc welding (SMAW)
   Gas tungsten arc welding (GTAW)
10. Approximately 175 amps
11. 4 min. out of 10 min.
    Approximately 160 amps
12. Input voltage
    Input amperage
    Duty cycle
    Max amperage output
13. Alternating current (AC only)
14. Alternating current (AC)
    Direct-current DCEP and DCEN (polarity)
15. Amperage-carrying output
16. 1/0
17. Loose connections, too-small cable, connection, or electrode holder
18. Tennis shoes
19. Safety glasses with side shields
20. 8 to 10
21. Turn off main power input
22. 1/4 inch
23. 30°
24. 1/16—1/8 inch
25. Base metal
    Current (polarity)
26. Manufactures
    American Welding Society
27. Minimum tensile strength (60 x 1000) or 60,000 psi
28. 8 (AC or DCEP)
29. E309 15 or 16
30. E6010 or E6011
31. Core wire diameter
32. Low hydrogen
33. Heat (amperage)
    Distance (arc length)
    Angles (work and travel angle of the electrode)
    Travel speed (movement along the metal)
    Manipulation (movement of the electrode to additional control of weld puddle)
34. 1. Base metal
    2. Position of welding
    3. Electrode size and type
35. 35—55°
36. 1-1/2 to 2 times the diameter of the core wire (coating not included)
37. Scratching method
    Tapping method
38. 1/8 inch
39. To obtain 100% penetration
40. Sketch illustration per Fig. 46.

Chapter 4

1. Dipmatic
   Microwire
   CO_2
   MIG
2. Electrode feeding and arc length
3. SMAW
   OAW
4. Gas metal arc welding short-circuit transfer
5. Short circuiting

6. Isles of silicone
7. 1/4—3/8
8. 3/8
9. GMAW-P
10. D. Pulses per second or PPS
11. Inverter
12. Spray
13. FCAW-G
FCAW-S
14. FCAW-S
15. Constant voltage
16. 100
17. Fan
18. Constant-speed feeders
Voltage-sensing feeders
19. Spool on gun
Push/pull
20. "U"
21. 100, CO_2
22. 30—50
23. CO_2
24. 70,000
25. Burns
Skin punctures
26. Point wire down away from personnel
27. Wire feed speed
28. Higher, lower
29. Electrode extension
Travel angle
Work angle
Travel speed
Joint alignment
30. 3/8
31. Decrease
32. Drag or pull
33. 0—20°
34. Arcing burns
Skin punctures
35. Clean base metal
Check gas flow or leak
Remove draft or wind
Get fresh flux-cored wire
36. Process variables
Consumables
Equipment
37. Contact tube
Gas nozzle or head tube
38. Flat; horizontal
39. Non-conductor; tension knob

Chapter 5

1. Heli arc
TIG
Argon arc
2. 11,000°F (6000°C)
6170°F (3410°C)
The electrode does not melt because of its high thermal conductivity. Plus, the torch is made of high-conductivity material.
3. Constant current
4. Sine wave
Square wave
Asymmetric wave
5. EN (electrode negative)
6. Cleaning; Heating
7. 200 amps
8. Stainless steel
9. Collet
10. EWCe-2, 1.8–2.2% Ceria
11. Pointed
12. 150 + 10% = 165
13. Inert
14. Helium
15. Argon
16. 2 or 3
17. Entrap gas; lead to weld defects
18. Puncture wounds (eye injury)
19. Oxyacetylene filler rod (or bailing wire or coat hangers).
20. Argon used in a confined space replaces oxygen and can cause a person to suffocate. Cylinders are under high pressure and can rupture when not handled properly.
21. Arc end; Primary power
22. 3
23. Gold to straw color
24. Semi-killed
25. 240
26. 150
27. 1 (equal the electrode diameter)
28. 10—20°
29. Clean
30. There is no metal transfer through the arc. Metal is deposited in the weld by direct contact.
31. Accidental arcing of the tungsten against the cylinder (which could cause the cylinder to rupture) if the torch touches the cylinder.

32. Uniform
    Crater filled
    Proper leg size
    Flat contour
    Root penetration
    Good fusion
33. It will melt and contaminate the base metal.

Chapter 6

1. Oxyfuel
   Plasma
   Laser beam.
2. SMAW is slow compared to the other processes and requires considerably more grinding, chipping and cleaning before a part is ready for welding.
3. Oxidation processes — heat from an oxyfuel flame takes carbon steel to its kindling temperature, and a stream of pure oxygen is concentrated on the hot spot which makes the metal burn.
   Thermal processes — a very intense heat melts an area and the force from the plasma stream or laser beam removes the molten metal.
4. The depth and amount of metal that is removed in cutting leaves a small kerf from the top to bottom of the metal, compared to gouging which leaves a much wider groove that does not go the entire depth of the metal.
5. An acetylene cutting tip will have 4 or 6 preheats and a fuel gas tip will have 8 preheats.
6. The center orifice is for the cutting oxygen and the other orifices are for the preheat flames.
7. The thickness of metal: thick metal requires larger preheat orifices and a larger oxygen orifice than thinner metal.
8. The hot area will explode, causing damage to the surface and burn injuries to anyone hit by the flying projectiles.
9. Carburizing — preheats contain acetylene feathers.
   Oxidizing — short narrow preheats.
   Neutral — preheats are the same length as those on a carburizing flame but do not contain acetylene feathers. This flame should be used for cutting.
10. The flame from a clean tip will have two straight bright lines in the center when the cutting oxygen lever is depressed.
11. The preheat flames should be held away from the surface 1/16—1/8 inch (1.6—3.2 mm).
12. Once a spot is hot, slowly depress the cutting oxygen lever partway (enough to make oxygen flow, but not a full stream) until the hole is pierced. Then depress the cutting lever all the way.
13. There will be a line that indicates the groove is through the weld and has reached the intersection of the two pieces of base metal.
14. The bottom of the groove should be smooth. Lines indicate unfused edges of metal, while dark spots indicate inclusions. The objective is to remove anything that could cause a weld defect before filling the groove with more weld.
15. A special direct-current power source, a torch, a gas supply, and interconnecting cables and hoses.
16. Plasma arc cutting is faster than oxyfuel cutting.
    Plasma arc cutting does not distort the metal as much as oxyfuel cutting.
17. Nitrides can cause weld porosity.
18. It takes 10 amps to cut 0.100 inch (2.5 mm)
19. Tungsten electrodes — use with nitrogen or argon/hydrogen.
    Hafnium electrodes — use with air or oxygen.
20. Torch positioning and travel speed.
21. Dross is material or "slag" which adheres to a piece of metal that has been cut with a plasma torch.

Chapter 7

1. Light amplification by stimulated emission of radiation
2. Wood
   Plastic
   Composites
3. Parallel
4. Monochromatic, one
5. Non-contact
6. 25
7. 0.020 inch (0.5 mm)
8. 1-1/4
9. Oxygen
10. Recast
11. 5
12. Some
13. 0.001
14. Loss of speed
    Direction sensitive
    Accumulation of oxide slag/wide kerf

Chapter 8

1. Situations that place a hardship on a human and in long-term repetitive jobs.
2. Electrically-powered robots and hydraulically-powered robots.
3. So welding can take place on one station while the fixture on the other station is being loaded.
4. Resistance welding and gas metal arc welding (GMAW).
5. The area in which the robot is capable of moving within.

6. Arc rays and persons entering the robot's work envelope while it is in operation.

Chapter 9

1. The metals to be joined are not melted in braze welding, while they are in fusion welding.
2. Because in braze welding there is less danger of destroying the properties of the base metal.
3. To eliminate oxides from the surface of the base metals (which could prevent a good bond).
4. Torch heating, furnace heating, induction heating.
5. Two carbon electrodes are used with AC units, while only one is used with DC units.
6. Braze welding uses a deposited bead at the joint instead of the thin bond in regular brazing.
7. Either:
   Soldering is not recommended for joints that must be very strong.
   Soldering should not be used where temperatures may approach the melting point of solder.
8. "Tin" and "lead."
9. It supplies a uniform heat.
10. Coating the copper with a thin coat of solder.
11. In sweat soldering, two surfaces are soldered without the solder being visible. In seam soldering, a larger layer of solder is run along the outside edge of a joint.

Chapter 10

1. Surfacing is a process of metal fusion (like welding) that surfaces a metal. Metallizing is the spray coating of a surface with metal particles.
2. Abrasion
   Impact
   Corrosion.
3. These metals can be annealed to reduce their hardness, or they can be treated by preheating and post-cooling gradually.
4. The arc process is much faster for depositing metal.
5. Amperage should be as low as possible.
6. The arc should be medium long.
7. The particles adhere because they are partly melted and are sprayed under pressure.
8. The surface must be cleaned and it must be roughened.

Chapter 11

1. Ferrous (The ferrous are the irons and iron alloys.)
   Nonferrous types
2. Mild steel
3. Iron is a refined iron ore, while steel is a refined iron plus an alloy of carbon and other substances.
4. a — 2
   b — 4
   c — 5
   d — 1
   e — 3
5. Mild steel
6. Softening
7. Malleable iron is cast iron which has been heat treated to make it more ductile (and more resistant to shock).
8. It loses its malleable qualities. (For this reason, malleable iron should never be welded — only brazed or braze welded.)
9. Three answers needed (choices are):
   Society of Automotive Engineers (SAE) 1018
   American Society for Testing and Materials (ASTM) A53 Grade B
   American Iron and Steel Institute (AISI) 1050
   Aluminum Association (AA) 6061 T-6
10. 35 points
11. 35%
12. Wrought aluminum, magnesium and silicon
13. S7
14. H13
15. 304
16. Extra low in carbon
17. Spark test
18. Members of the "white metals" can be separated from each other. The relative hardness of a metal and if the metal is heat treatable.
19. Zinc diecast
    Magnesium
    Aluminum
20. Zinc diecast

Chapter 12

1. (Any three of these.)
   Flexibility
   Stronger but lighter
   Stiffer
   Less cracking
   Less investment in equipment
   Freedom of design
   Resistance to shock.
2. For correct examples, see Fig. 2 through Fig. 4.
3. The V-type butt joint is best for joining heavier metals (because it allows more strength in the weld).
4. Welds which are transverse to lines of stress.
5. Welds should be spaced apart (for greater strength).
6. 3/4.

7. a — 4
   b — 2
   c — 3
   d — 1
8. Skip welding is the welding of short strips along the joint, then returning to fill in the openings. It is used to reduce the heat input in the base metal (preventing residual stress).

Chapter 13

1. Hard hat
   Goggles
   Helmet with dark lenses
   Ear protectors
   Ear plugs
   Face shield
   Safety glasses
   Respirator or self-contained breathing device
   Fire blanket and fire extinguisher
   Fire-resistant welding screen
2. Gloves
   Welding jacket or heat-resistant long sleeve shirt
   Vest
   Sleeves
   Long pants without cuffs
   Boots or high topped shoes.
3. Oxygen
4. American Welding Society's standard practice recommendations (A.W.S.749.1).
5. There may still be traces of oil, grease or other readily oxidized material under the seams.
6. Sparks may travel as far as 35 feet.
7. Sparks, glowing embers and flames
8. A spark can cause clothing in an oxygen atmosphere to burst into flames.
9. Oil, grease
10. Vertical
11. Burn with explosive violence if ignited.
12. Left, right, and left (in order)
13. Crack
14. One side
15. One and one-half turns, and preferably less than a single turn.
16. Completely
17. Damp, dry, electric shock (in order)
18. Fumes
19. Guarded
20. Compressed, cylinder (in order)

Introduction to Instructor

Use of Work Project Sheets

The project sheets which follow can be used by an instructor to customize assignments for welding students. For example, one set of options for Project #1 is circled. The instructions for Project #1 are listed as "Example" #1.

Basic SMAW Shop Project Options
Project #1—Beads

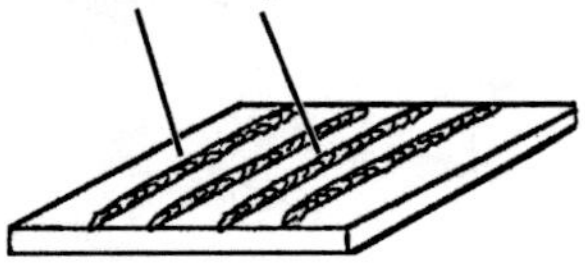

Options

Current

1. AC (circled)
2. DC

Filler Metal

--Type--

3. 6010/E6011 (E6011 circled)
4. 6013/E7014
5. 67018
6. Other --

--Diameter--

7. 3/32" (circled)
8. 1/8 inch

Material

Part--->		A	B
Thickness	(Min.)	3/16" (circled)	--
	(Max.)	3/8"	--
Width	(Min.)	3" (circled)	--
	(Max.)	4"	--
Length	(Min.)	4"	--
	(Max.)	6" (circled)	--

Suggestions for using the parameter records:

Transfer Information that is given in Project #1 to the parameter record. Assign values for the other items necessary to complete the record.

NOTE: *Amperage ranges from electrode charts should be narrowed within a range that will produce weld beads that are acceptable.*

Instructions for Project #1, Example #1

Process --->	SMAW	
	Material/Part	3/16"
	Type Weld Joint	Beads
	Electrode Type	6011
	Electrode Diameter	3/32"
	Equip. Manufacture & Model	Miller/Thunderbolt
	Type Current	AC
	Polarity	----
	Amperage Range	65 to 75

Other --

Arc Length 3/32", Travel Angle 60° to 80°

Direction — Push x Pull

NOTE: *Beads should be 1-1/2 to 2 times the width of the electrode core wire. Cool after making two beads.*

Suggestions for using the various options for each project:

Each project has many options for customizing an assignment. Circle the combinations which will best suit the needs of each situation with regard to the equipment, the availability of materials, and the length of class.

The following is an example of options for Project #1:

Example #1:

Project #1 would be completed with the following combination

1. AC

3. 6011

7. 3/32"

Material 3/16" x 3" x 6"

Welding Parameter Records

Process--> SMAW

Material/Part ______
Type Weld Joint ______
Electrode Type ______
Electrode Diameter ______
Equipment:
Manufacturer and Model ______
Type Current ______
Polarity ______
Amperage Range ______ to ______
Other ______
Arc Length ______ Travel Angle ______
Direction ______ Push ______ Pull
Notes ______

Process--> SMAW

Material/Part ______
Type Weld Joint ______
Electrode Type ______
Electrode Diameter ______
Equipment:
Manufacturer and Model ______
Type Current ______
Polarity ______
Amperage Range ______ to ______
Other ______
Arc Length ______ Travel Angle ______
Direction ______ Push ______ Pull
Notes ______

Process--> SMAW

Material/Part ______
Type Weld Joint ______
Electrode Type ______
Electrode Diameter ______
Equipment:
Manufacturer and Model ______
Type Current ______
Polarity ______
Amperage Range ______ to ______
Other ______
Arc Length ______ Travel Angle ______
Direction ______ Push ______ Pull
Notes ______

Process--> GMAW Short Circuit

Material/Part ______
Type Weld Joint ______
Wire Type ______
Wire Size ______
Equipment:
Manufacturer and Model ______
Type Gas ______ Flow ______ to ______
Wire Speed Range ______ to ______
Voltage Range ______ to ______
Amperage Range ______ to ______
Polarity ______
Slope ______ to ______
Inductance ______ to ______
Stick-Out ______ to ______
Other ______
Work Angle ______ Travel Angle ______
Direction ______ Push ______ Pull
Notes ______

Process--> GMAW Spray Transfer

Material/Part ______

Type Weld Joint ______

Wire Type ______

Wire Size ______

Equipment:

Manufacturer and Model ______

Type Gas ______ Flow ______ to ______

Wire Speed Range ______ to ______

Voltage Range ______ to ______

Amperage Range ______ to ______

Polarity ______

Slope ______ to ______

Inductance ______ to ______

Stick-Out ______ to ______

Other ______

Work Angle ______ Travel Angle ______

Direction ______ Push ______ Pull

Notes ______

Process--> FCAW

Material/Part ______

Type Weld Joint ______

Wire Type ______

Wire Size ______

Equipment:

Manufacturer and Model ______

Type Gas ______ Flow ______ to ______

Wire Speed Range ______ to ______

Voltage Range ______ to ______

Amperage Range ______ to ______

Polarity ______

Slope ______ to ______

Inductance ______ to ______

Stick-Out ______ to ______

Other ______

Work Angle ______ Travel Angle ______

Direction ______ Push ______ Pull

Notes ______

Process--> GMAW Pulse

Material/Part ______

Type Weld Joint ______

Wire Type ______

Wire Size ______

Equipment:

Manufacturer and Model ______

Program ______

Type Gas ______ Flow ______ to ______

Wire Speed Range ______ to ______

Voltage/Trim Range ______ to ______

Peak Amperage Range ______ to ______

Background Amperage Range ______ to ______

Stick-Out ______ to ______

Other ______

Work Angle ______ Travel Angle ______

Direction ______ Push ______ Pull

Notes ______

Process--> GTAW

Material/Part ______

Type Weld Joint ______

Type Gas ______ Flow ______ to ______

Filler Metal Type and Diameter ______

Tungsten Type and Diameter ______

Tungsten Shape ______ ______

Gas Cup Size ______

Equipment:

Manufacturer and Model ______

Type Current ______

Polarity ______

Amperage Range ______ to ______

Other ______

Work Angle ______ Travel Angle ______

Direction ______ Push ______ Pull

Notes ______

Process--> GTAW

Material/Part ______

Type Weld Joint ______

Type Gas ______ Flow ______ to ______

Filler Metal Type and Diameter ______

Tungsten Type and Diameter ______

Tungsten Shape ______ ______

Gas Cup Size ______

Equipment:

Manufacturer and Model ______

Type Current ______

Polarity ______

Amperage Range ______ to ______

Other ______

Work Angle ______ Travel Angle ______

Direction ______ Push ______ Pull

Notes ______

______

Process--> Oxyfuel Braze

Material/Part ______

Type Joint ______

Filler Metal Type ______

Filler Metal Diameter ______

Equipment:

Manufacturer and Model ______

Tip Size ______

Oxygen:

Cylinder On (Turns) ______

Pressure ______

Type Fuel Gas:

Cylinder On (Turns) ______

Pressure ______

Flame Distance ______

Direction ______ Push ______ Pull

Notes ______

______

Process--> Oxyacetylene Weld

Material/Part ______

Type Joint ______

Filler Metal Type ______

Filler Metal Diameter ______

Equipment:

Manufacturer and Model ______

Tip Size ______

Oxygen:

Cylinder On (Turns) ______

Pressure ______

Acetylene:

Cylinder On (Turns) ______

Pressure ______

Flame Distance ______

Direction ______ Push ______ Pull

Notes ______

______

Process--> Plasma Cut

Material/Part ______

Equipment:

Manufacturer and Model ______

Tip/Orifice Size ______

Pressures:

Plasma Gas or Air ______

Secondary Gas or Air ______

Amperage ______

Shutoff Distance ______

Torch Angles - Work Angles ______

Direction ______ Push ______ Pull

Notes ______

______

Process--> Plasma Gouge

Material/Part ______________________

Equipment:

Manufacturer and Model ______________

Tip/Orifice Size ________

Shield ________

Pressures:

Plasma Gas or Air ________

Secondary Gas or Air ________

Amperage ____________

Shutoff Distance ________

Torch Angles - Work Angles __________

Direction ______ Push ______ Pull

Notes ______________________________

____

Process--> Air Carbon Arc Gouge (Arc Air)

Material/Part ______________________

Equipment:

Manufacturer and Model ______________

Electrode Type ________

Electrode Diameter ________

Equipment:

Manufacturer and Model ______________

Duty Cycle ________

Type Current ________

Polarity ________

Amperage ________ __________

Other ____________

Work Angle ________ Travel Angle ________

Direction ______ Push ______ Pull

Notes ______________________________

____

Basic Oxyfuel Shop Project Options

Project #1—Joining Copper Tubing

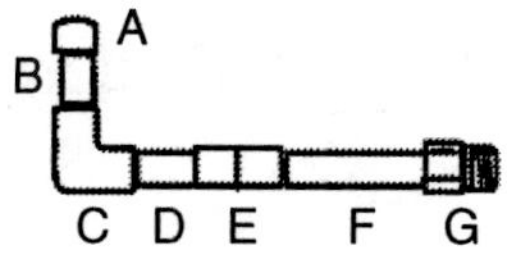

JDPX7281

--Options--

Process
1. Oxyacetylene
2. Other

Position
3. Moved to Flat
4. Fixed (Not Moved)

Filler Metal
5. Lead-Free Solder
6. Silver Solder

Material

Part-->	A	B	C	D	E	F	G
Thickness	--	--	--	--	--	--	--
Diameter	1/2"	1/2"	1/2"	1/2"	1/2"	1/2"	1/2"
Length	--	2"	--	3"	--	4"	--

Project #2—Heating

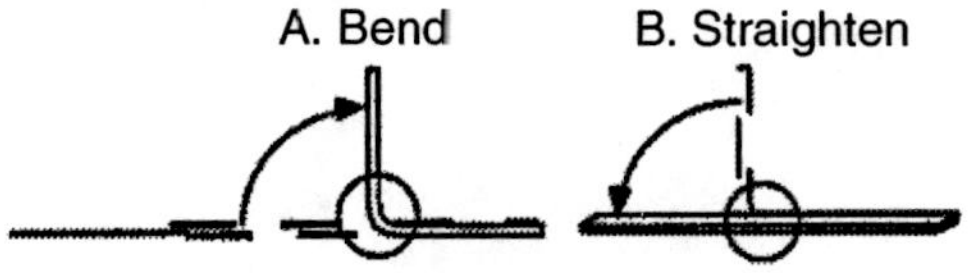

Tolerance = ±1/8" (3.2 mm)

JDPX7282

--Options--

Process
1. Oxyacetylene
2. Other
3. Bend to 90°
4. Straighten to 180°

Material

Part-->		A	B
Thickness	(Min.)	1/4"	1/4"
	(Max.)	3/8"	3/8"
Width	(Min.)	1"	1"
	(Max.)	1-1/2"	1-1/2"
Length	(Min.)	10"	10"
	(Max.)	12"	12"

Project #3—Melt Strips

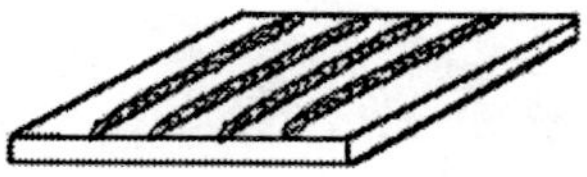

JDPX7283

--Options--

Process
1. Oxyacetylene
2. Other

Material

Part-->		
Thickness	(Min.)	1/16"
	(Max.)	1/8"
Width	(Min.)	4"
	(Max.)	6"
Length	(Min.)	4"
	(Max.)	6"

Project #4—Beads

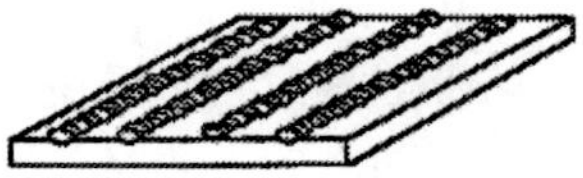

JDPX7284

--Options--

Process
1. Oxyacetylene
2. Other

Filler Metal
3. Steel Filler Rod
4. Brazing Rod

Material

Part-->		
Thickness	(Min.)	1/16"
	(Max.)	1/8"
Width	(Min.)	4"
	(Max.)	6"
Length	(Min.)	4"
	(Max.)	6"

Project #5—Butt Joints

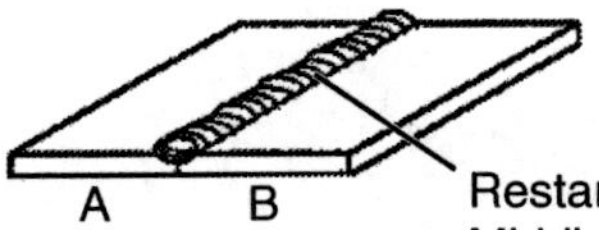

--Options--

Process
1. Oxyacetylene
2. Other

Filler Metal
3. Steel Filler Rod
4. Brazing Rod

Material

Part-->		A	B
Thickness	(Min.)	1/16"	1/16"
	(Max.)	1/8"	1/8"
Width	(Min.)	1-1/2"	1-1/2"
	(Max.)	2"	2"
Length	(Min.)	4"	4"
	(Max.)	6"	6"

Project #6—"T" Joint

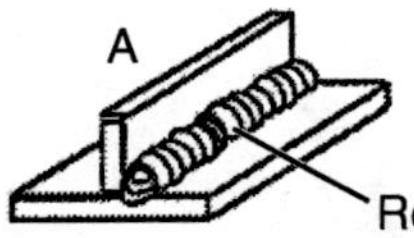

--Options--

Process
1. Oxyacetylene
2. Other

Position
3. Flat
4. Horizontal

Filler Metal
5. Steel Filler Rod
6. Brazing Rod

Material

Part-->		A	B
Thickness	(Min.)	1/16"	1/16"
	(Max.)	1/8"	1/8"
Width	(Min.)	1-1/2"	3"
	(Max.)	2"	4"
Length	(Min.)	4"	4"
	(Max.)	6"	6"

Project #7—Pipe "T"

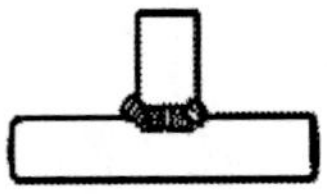

--Options--

Process
1. Oxyacetylene
2. Other

Position
3. Flat (Rolled)
4. Horizontal (Not Removed)

Filler Metal
5. Steel Filler Rod
6. Brazing Rod

Material

Part-->		A	B
Thickness	(Min.)	0.060"	0.060"
	(Max.)	0.145"	0.145"
Diameter	(Min.)	1"	1"
	(Max.)	1-1/2"	1-1/2"
Length	(Min.)	1-1/2"	4"
	(Max.)	3"	6"

Basic SMAW Shop Project Options

Project #1—Beads

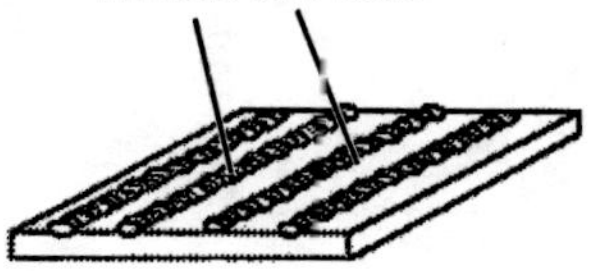

JDPX7288

--Options--

Current

1. AC
2. DC

Filler Metal

--Type--

3. 6010/E6011
4. 6013/E7014
5. 67018
6. Other --

--Diameter--

7. 3/32"
8. 1/8"

Material

Part--->		A	B
Thickness	(Min.)	3/16"	--
	(Max.)	3/8"	--
Width	(Min.)	3"	--
	(Max.)	4"	--
Length	(Min.)	4"	--
	(Max.)	6"	--

Project #2—"T" Joint

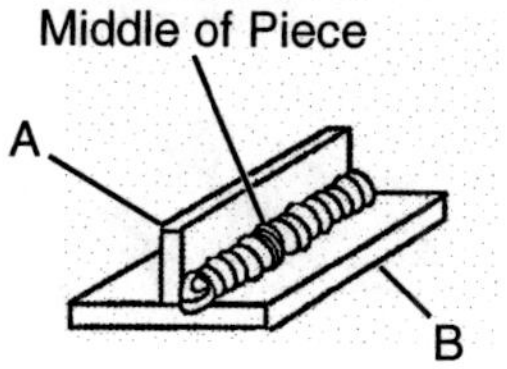

JDPX7289

--Options--

Current

1. AC
2. DC

Position

3. Flat
4. Horizontal

Filler Metal

--Type--

5. 6010/E6011
6. 6013/E7014
7. 67018
8. Other

--Diameter--

9. 3/32"
10. 1/8"

Material

Part--->		A	B
Thickness	(Min.)	1/8"	1/8"
	(Max.)	1/4"	1/4"
Width	(Min.)	1-1/2"	3"
	(Max.)	2"	4"
Length	(Min.)	4"	4"
	(Max.)	6"	6"

Project #3—Butt Joint

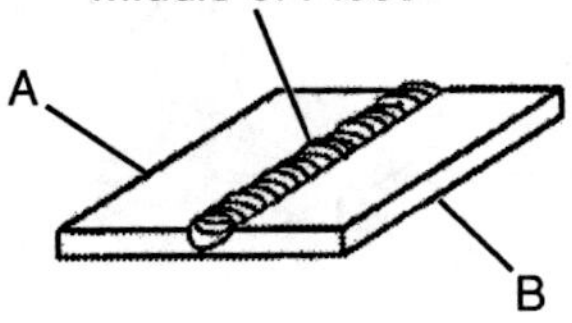

JDPX7290

--Options--

Current

1. AC
2. DC

Filler Metal

--Type--

3. 6010/E6011
4. 6013/E7014
5. 67018
6. Other

--Diameter--

7. 7/32"
8. 1/8"

Material

Part--->		A	B
Thickness	(Min.)	1/8"	1/8"
	(Max.)	1/4"	1/4"
Width	(Min.)	1-1/2"	3"
	(Max.)	2"	4"
Length	(Min.)	4"	4"
	(Max.)	6"	6"

Project #4—Single "V" Groove

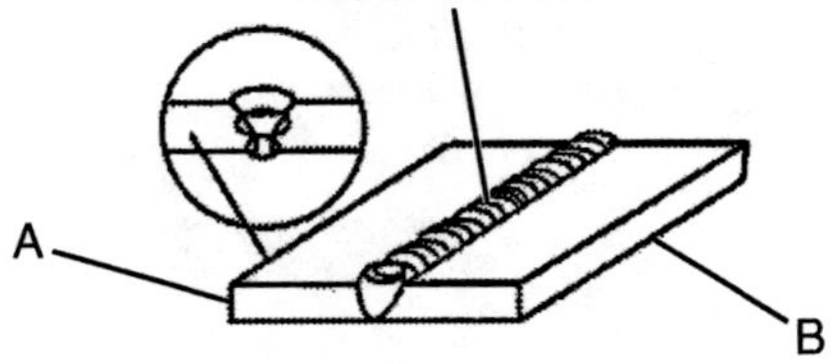

JDPX7291

--Options--

Current

1. AC
2. DC

Filler Metal

--Type--

3. 6010/E6011
4. 6013/E7014
5. 67018
6. Other

--Diameter--

7. 7/32"
8. 1/8"

Material

Part--->		A	B
Thickness	(Min.)	3/8"	3/8"
	(Max.)	1/2"	1/2"
Width	(Min.)	3"	3"
	(Max.)	4"	4"
Length	(Min.)	5"	5"
	(Max.)	7"	7"

Project #5—Pipe "T"

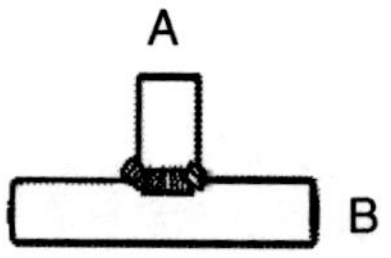

JDPX7292

--Options--

Current

1. AC
2. DC

Position

3. Flat (Rolled)
4. Horizontal (Not Moved)

Filler Metal

--Type--

5. 6010/E6011
6. 6013/E7014
7. 67018
8. Other

--Diameter--

9. 3/32"
10. 1/8"

Material

Part--->		A	B
Thickness	(Min.)	0.125"	0.125"
	(Max.)	0.250"	0.250"
Diameter	(Min.)	1"	1"
	(Max.)	1-1/2"	1-1/2"
Length	(Min.)	1-1/2"	4"
	(Max.)	3"	6"

Project #6—Pipe "V" Groove

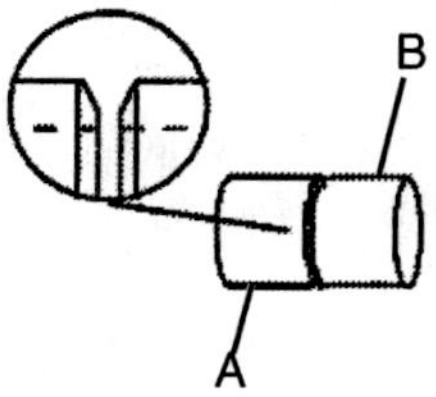

JDPX7293

--Options--

Current

1. AC
2. DC

Position

3. Flat (Rolled)

Filler Metal

--Type--

4. 6010/E6011
5. 6013/E7014
6. 67018
7. Other

--Diameter--

8. 3/32"
9. 1/8"

Material

Part--->		A	B
Thickness	(Min.)	0.125"	0.125"
	(Max.)	0.250"	0.250"
Diameter	(Min.)	2"	1"
	(Max.)	4"	1-1/2"
Length	(Min.)	3"	3"
	(Max.)	6"	6"

Basic GMAW and FCAW Shop Project Options

Project #1—Beads

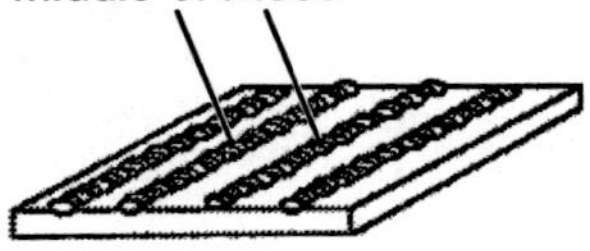

JDPX7294

--Options--

Current

1. DC

Filler Metal

--Type--

2. ER70S-3/ER70S-6
3. E70T-1/E71-T1
4. E70T-4/E71T-11
5. Other --

--Diameter--

6. 0.035"
7. 0.045"
8. Other

Material

Part--->		A	B
Thickness	(Min.)	1/8"	--
	(Max.)	1/4"	--
Width	(Min.)	3"	--
	(Max.)	4"	--
Length	(Min.)	4"	--
	(Max.)	6"	--

Project #2—"T" Joint

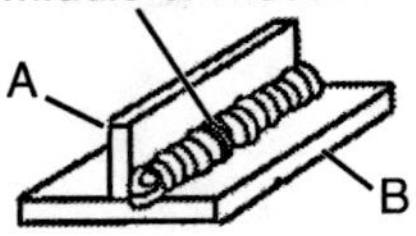

JDPX7295

--Options--

Current

1. DC

Position

2. Flat
3. Horizontal

Filler Metal

--Type--

4. ER70S-3/ER70S-6
4. E70T-1/E71-T1
6. E70T-4/E71T-11
7. Other

--Diameter--

8. 0.03511"
9. 0.0945"
10. Other

Material

Part--->		A	B
Thickness	(Min.)	1/8"	1/8"
	(Max.)	1/4"	1/4"
Width	(Min.)	1-1/2"	3"
	(Max.)	2"	4"
Length	(Min.)	4"	4"
	(Max.)	6"	6"

Project #3—Butt Joint

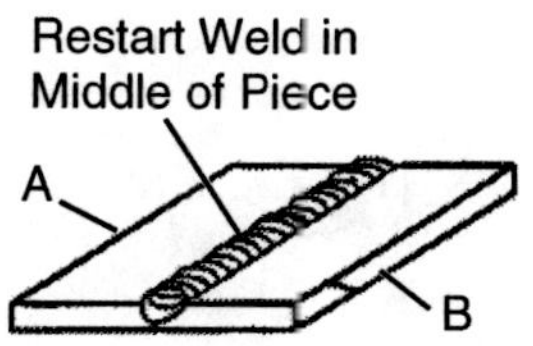

JDPX7296

--Options--

Current

1. DC

Filler Metal

--Type--

2. ER70S-3/ER70S-6
3. E70T-1/E71-T1
4. E70T-4/E71T-11
5. Other --

--Diameter--

6. 0.035"
7. 0.045"
8. Other

Material

Part--->		A	B
Thickness	(Min.)	1/8"	1/8"
	(Max.)	1/4"	1/4"
Width	(Min.)	1-1/2"	3"
	(Max.)	2"	4"
Length	(Min.)	4"	4"
	(Max.)	6"	6"

Project #4—Single "V" Groove

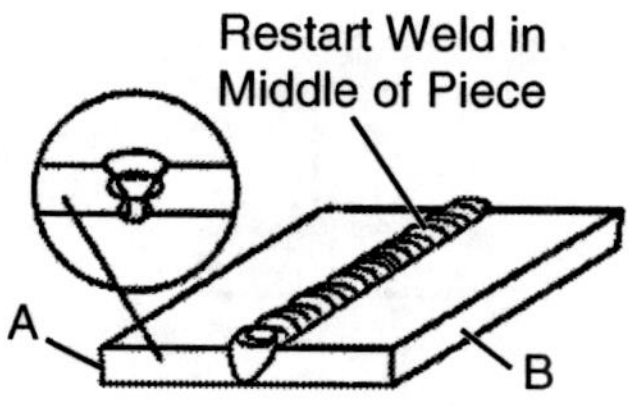

JDPX7297

--Options--

Current

1. DC

Filler Metal

--Type--

2. ER70S-3/ER70S-6
3. E70T-1/E71-T1
4. E70T-4/E71T-11
5. Other --

--Diameter--

6. 0.035"
7. 0.045"
8. Other

Material

Part--->		A	B
Thickness	(Min.)	3/8"	3/8"
	(Max.)	1/2"	1/2"
Width	(Min.)	3"	3"
	(Max.)	4"	4"
Length	(Min.)	5"	5"
	(Max.)	7"	7"

Project #5—Pipe "T"

JDPX7298

--Options--

Current

1. DC

Position

2. Flat

3. Horizontal

Filler Metal

--Type--

4. ER70S-3/ER70S-6
4. E70T-1/E71-T1
6. E70T-4/E71T-11
7. Other

--Diameter--

8. 0.035"
9. 0.045"
10. Other

Material

Part--->		A	B
Thickness	(Min.)	0.125"	0.125"
	(Max.)	0.250"	0.250"
Diameter	(Min.)	1"	1"
	(Max.)	1-1/2"	1-1/2"
Length	(Min.)	1-1/2"	4"
	(Max.)	3"	6"

Project #6—Pipe "V" Groove

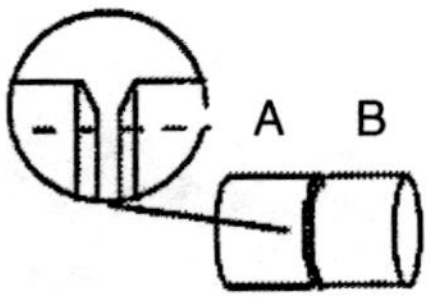

JDPX7299

--Options--

Current

1. DC

Position

2. Flat

3. Horizontal

Filler Metal

--Type--

4. ER70S-3/ER70S-6
4. E70T-1/E71-T1
6. E70T-4/E71T-11
7. Other

--Diameter--

8. 0.035"
9. 0.045"
10. Other

Material

Part--->		A	B
Thickness	(Min.)	0.125"	0.125"
	(Max.)	0.250"	0.250"
Diameter	(Min.)	2"	1"
	(Max.)	4"	1-1/2"
Length	(Min.)	3"	3"
	(Max.)	6"	6"

Basic GTAW Shop Project Options

Project #1—Metal Strips

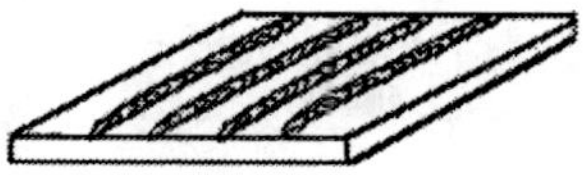

JDPX7300

--Options--

Current

1. ACHF[a]
2. DCEN

Base Metal

3. SS
4. CS
5. AL[a]
6. Other

Material

Part--->		A	B
Thickness	(Min.)	1/16"	--
	(Max.)	1/8"	--
Width	(Min.)	1-1/2"	--
	(Max.)	2"	--
Length	(Min.)	4"	--
	(Max.)	6"	--

SS = Stainless Steel
CS = Carbon Steel
AL = Aluminum

a. Proper Current/Base Metal Combinations

Project #2—Beads

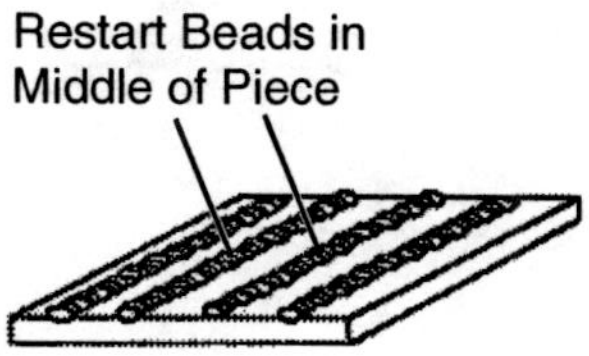

JDPX7301

--Options--

Current

1. ACHF[a]
2. DCEN

Base Metal	Filler Metal Type	Diameter	
3. SS	ER309L/ER316L	1/16"	3/32"
4. CS	ER70S-2	1/16"	3/32"
5. AL[a]	ER4043/ER5356	3/32"	1/8"
6. Other			

Material

Part--->		A	B
Thickness	(Min.)	1/16"	--
	(Max.)	1/8"	--
Width	(Min.)	3"	--
	(Max.)	4"	--
Length	(Min.)	4"	--
	(Max.)	6"	--

SS = Stainless Steel
CS = Carbon Steel
AL = Aluminum

a. Proper Current/Base Metal Combinations

Project #3—Butt Joint

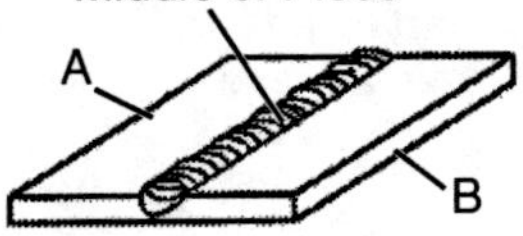

JDPX7302

--Options--

Current

1. ACHF[a]
2. DCEN

Base Metal	Filler Metal Type	Diameter	
3. SS	----------None-Fusion Weld-----------		
4. SS	ER309L/ER316L	1/16"	3/32"
5. CS	ER70S-2	1/16"	3/32"
6. AL[a]	ER4043/ER5356	3/32"	1/8"
7. Other			

Material

Part--->		A	B
Thickness	(Min.)	1/16"	1/16"
	(Max.)	1/8"	1/8"
Width	(Min.)	1-1/2"	1-1/2"
	(Max.)	2"	2"
Length	(Min.)	4"	4"
	(Max.)	6"	6"

SS = Stainless Steel
CS = Carbon Steel
AL = Aluminum

a. Proper Current/Base Metal Combinations

Project #4—"T" Joint

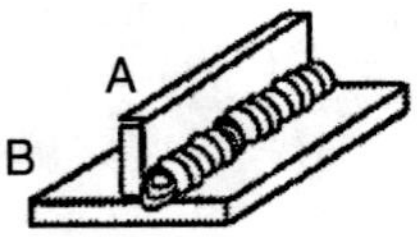

JDPX7303

--Options--

Current

1. ACHF[a]
2. DCEN

Position

3. Flat
4. Horizontal

Base Metal	Filler Metal Type	Diameter	
5. SS	ER309L/ER316L	1/16"	3/32"
6. CS	ER70S-2	1/16"	3/32"
7. AL[a]	ER4043/ER5356	3/32"	1/8"
8. Other			

Material

Part--->		A	B
Thickness	(Min.)	1/16"	1/16"
	(Max.)	1/8"	1/8"
Width	(Min.)	1-1/2"	3"
	(Max.)	2"	4"
Length	(Min.)	4"	4"
	(Max.)	6"	6"

SS = Stainless Steel
CS = Carbon Steel
AL = Aluminum

a. Proper Current/Base Metal Combinations

Project #5—Lap Joint

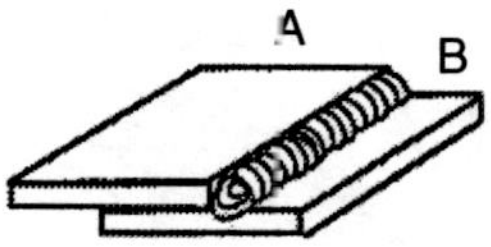

JDPX7304

--Options--

Current

1. ACHF[a]
2. DCEN

Position

3. Flat
4. Horizontal

Base Metal	Filler Metal Type	Diameter	
5. SS	----------None-Fusion Weld-----------		
6. SS	ER309L/ER316L	1/16"	3/32"
7. CS	ER70S-2	1/16"	3/32"
8. AL[a]	ER4043/ER5356	3/32"	1/8"
9. Other			

Material

Part--->		A	B
Thickness	(Min.)	1/16"	1/16"
	(Max.)	1/8"	1/8"
Width	(Min.)	2"	2"
	(Max.)	3"	3"
Length	(Min.)	4"	4"
	(Max.)	6"	6"

SS = Stainless Steel
CS = Carbon Steel
AL = Aluminum

a. Proper Current/Base Metal Combinations

Project #6—Pipe "T"

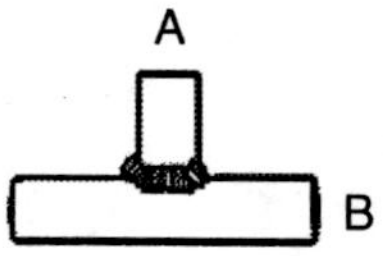

JDPX7305

--Option--s

Current

1. ACHF[a]
2. DCEN

Position

3. Flat (Rolled)
4. Fixed (Not Rolled)

Base Metal	Filler Metal Type	Diameter	
5. SS	ER309L/ER316L	1/16"	3/32"
6. CS	ER70S-2	1/16"	3/32"
7. AL[a]	ER4043/ER5356	3/32"	1/8"
8. Other			

Material

Part--->		A	B
Thickness	(Min.)	0.060"	0.060"
	(Max.)	0.250"	0.250"
Diameter	(Min.)	1"	1"
	(Max.)	1-1/2"	1-1/2"
Length	(Min.)	1-1/2"	4"
	(Max.)	3"	6"

SS = Stainless Steel
CS = Carbon Steel
AL = Aluminum

a. Proper Current/Base Metal Combinations

Project #7—Pipe Square Butt Joint

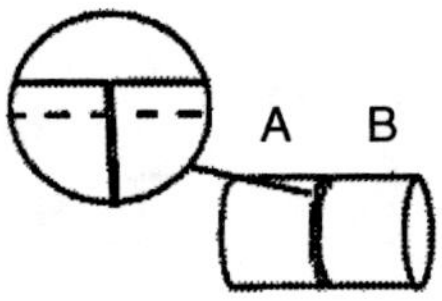

JDPX7306

--Options--

Current

1. ACHF[a]
2. DCEN

Position

3. Flat (Rolled)

Base Metal	Filler Metal Type	Diameter	
4. SS	----------None-Fusion Weld-----------		
5. SS	ER309L/ER316L	1/16"	3/32"
6. CS	ER70S-2	1/16"	3/32"
7. AL[a]	ER4043/ER5356	3/32"	1/8"
8. Other			

Material

Part--->		A	B
Thickness	(Min.)	0.060"	0.060"
	(Max.)	0.125"	0.125"
Diameter	(Min.)	1"	1"
	(Max.)	3"	3"
Length	(Min.)	2"	2"
	(Max.)	4"	4"

SS = Stainless Steel
CS = Carbon Steel
AL = Aluminum

a. Proper Current/Base Metal Combinations

Project #8—"V" Groove (2 Passes)

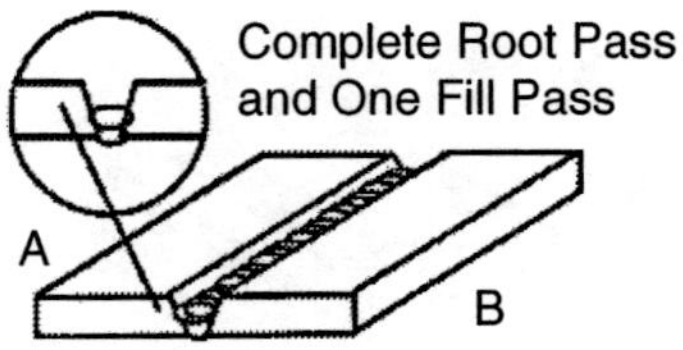

JDPX7307

--Options--

Current

1. DCEN

Position

2. Flat

Base Metal	Filler Metal Type	Diameter	
3. CS	ER70S-2	3/32"	1/8"
4. Other			

Material

Part--->		A	B
Thickness	(Min.)	0.250"	0.250"
	(Max.)	0.375"	0.375"
Width	(Min.)	2"	2"
	(Max.)	3"	3"
Length	(Min.)	4"	4"
	(Max.)	6"	6"
Bevel Angle	(Min.)	35°	35°
	(Max.)	37.5°	37.5°

Edge Preparation = Sharp Points (No Land)

Fit-Up

Root Opening = Rod Diameter
CS = Carbon Steel

Advanced SMAW, GTAW, and FCAW Shop Project Options

Project #1

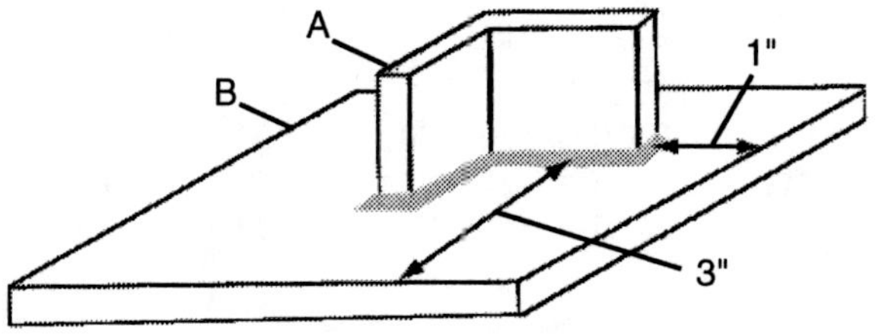

JDPX7308

--Options-- **Filler Metal**

Process	------Type-----		-----Diameter------	
1. SMAW	E6010	E6011	3/32"	1/8"
2. SMAW	E6013	E7014	3/32"	1/8"
3. SMAW	E7018		3/32"	1/8"
4. GMAW	ER70S-6	ER70S-6	0.035"	0.045"
5. FCAW	E70T-1	E71T-1	0.035"	0.045"
6. FCAW	E70T-4	E71T-11	0.035"	0.045"

Material

Part--->	A	B
Thickness	1/4"	1/4"
Width (Legs)	2" x 2"	5"
Length	2"	6"

Project #2

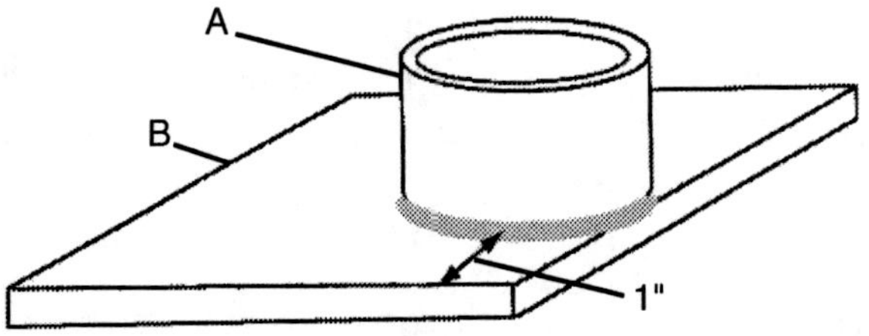

JDPX7309

--Options-- **Filler Metal**

Process	------Type-----		-----Diameter------	
1. SMAW	E6010	E6011	3/32"	1/8"
2. SMAW	E6013	E7014	3/32"	1/8"
3. SMAW	E7018		3/32"	1/8"
4. GMAW	ER70S-3	ER70S-6	0.035"	0.045"
5. FCAW	E70T-1	E71T-1	0.035"	0.045"
6. FCAW	E70T-4	E71T-11	0.035"	0.045"

Material

Part--->	A	B
Thickness	1/4"	1/4"
Diameter	3"	5"
Length	2"	6"

Project #3

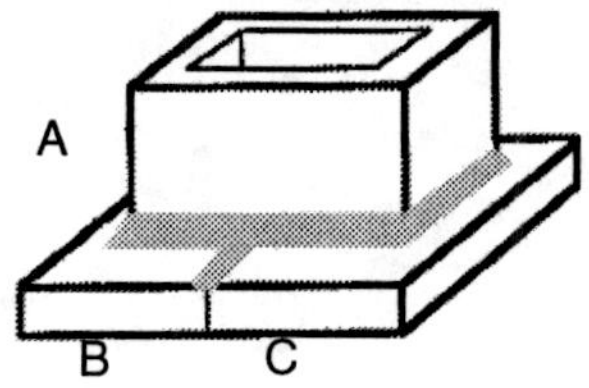

JDPX7310

--Options-- **Filler Metal**

Process	------Type-----		-----Diameter------	
1. SMAW	E6010	E6011	3/32"	1/8"
2. SMAW	E6013	E7014	3/32"	1/8"
3. SMAW	E7018		3/32"	1/8"
4. GMAW	ER70S-3	ER70S-6	0.035"	0.045"
5. FCAW	E70T-1	E71T-1	0.035"	0.045"
6. FCAW	E70T-4	E71T-11	0.035"	0.045"

Material

Part--->	A	B	C
Thickness	1/4"	1/4"	1/4"
Width	2" x 2"	2"	2"
Length	2"	4"	4"

Project #4

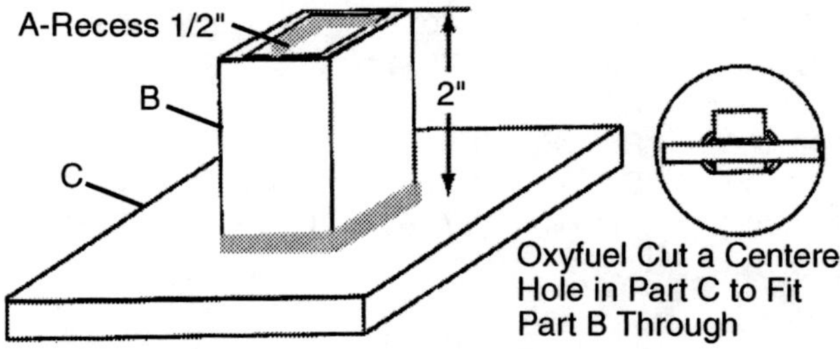

JDPX7311

--Options-- **Filler Metal**

Process	------Type-----		-----Diameter------	
1. SMAW	E6010	E6011	3/32"	1/8"
2. SMAW	E6013	E7014	3/32"	1/8"
3. SMAW	E7018		3/32"	1/8"
4. GMAW	ER70S-3	ER70S-6	0.035"	0.045"
5. FCAW	E70T-1	E71T-1	0.035"	0.045"
6. FCAW	E70T-4	E71T-11	0.035"	0.045"

Material

Part--->	A	B	C
Thickness	1/8"	1/4"	3/8"
Width	1-1/2"	2" x 4"	4"
Length	3-1/2"	4"	4"

Project #5

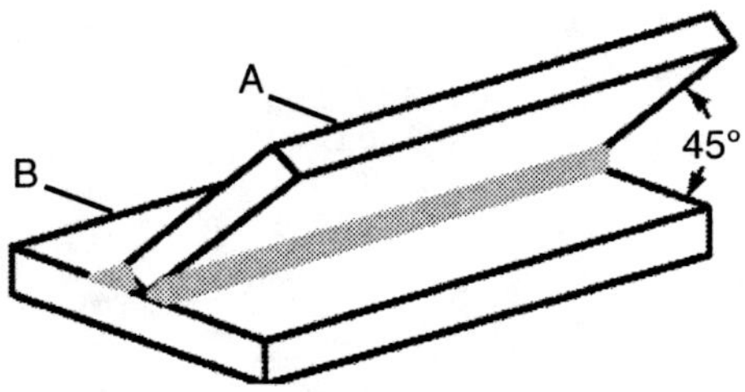

JDPX7312

--Options---

Process	Filler Metal Type		Diameter	
1. SMAW	E6010	E6011	1/8"	5/32"
2. SMAW	E6013	E7014	1/8"	5/32"
3. SMAW	E7018		1/8"	5/32"
4. GMAW	ER70S-3	ER70S-6	0.035"	0.045"
5. FCAW	E70T-1	E71T-1	0.035"	0.045"
6. FCAW	E70T-4	E71T-11	0.035"	0.045"

Material

Part--->	A	B
Thickness	3/8"	3/8"
Width	2"	4"
Length	6"	6"

Project #6

--Options--

Process	Filler Metal Type		Diameter	
1. SMAW	E6010	E6011	1/8"	5/32"
2. SMAW	E6013	E7014	1/8"	5/32"
3. SMAW	E7018		1/8"	5/32"
4. GMAW	ER70S-3	ER70S-6	0.035"	0.045"
5. FCAW	E70T-1	E71T-1	0.035"	0.045"
6. FCAW	E70T-4	E71T-11	0.035"	0.045"

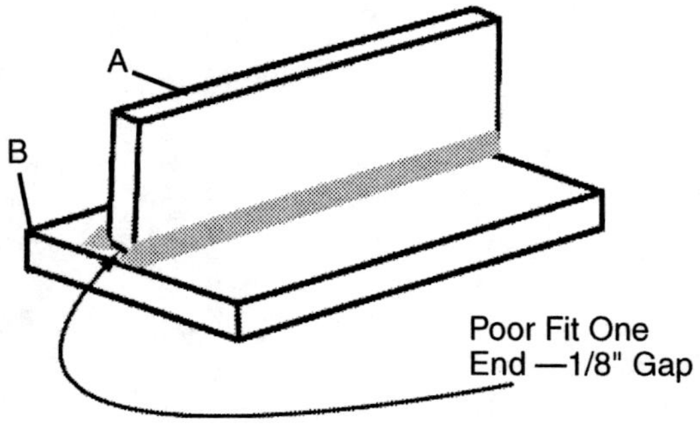

JDPX7313

Material

Part--->	A	B
Thickness	1/4"	1/4"
Width	2"	5"
Length	2"	6"

Project #7

--Options--

Process	Filler Metal Type		Diameter	
1. SMAW	E6010	E6011	1/8"	5/32"
2. SMAW	E6013	E7014	1/8"	5/32"
3. SMAW	E7018		1/8"	5/32"
4. GMAW	ER70S-3	ER70S-6	0.035"	0.045"
5. FCAW	E70T-1	E71T-1	0.035"	0.045"
6. FCAW	E70T-4	E71T-11	0.035"	0.045"

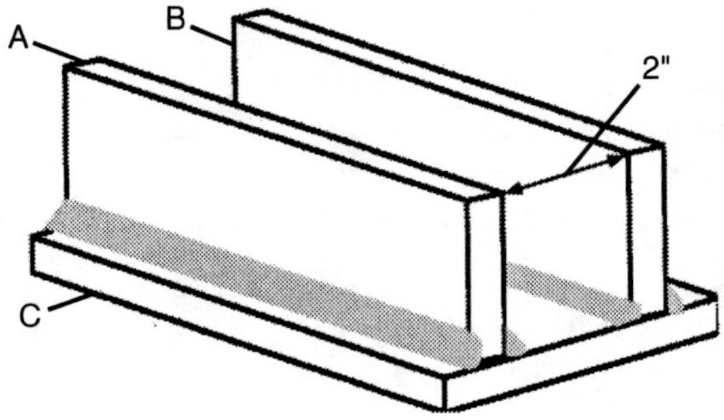

JDPX7314

Material

Part--->	A	B	C
Thickness (Min.)	1/4"	1/4"	3/8"
Width (Min.)	2"	2"	4"
Length (Min.)	6"	6"	6"

Project #8

--Options--

Process	Filler Metal Type		Diameter	
1. SMAW	E6010	E6011	3/32"	1/8"
2. SMAW	E6013	E7014	3/32"	1/8"
3. SMAW	E7018		3/32"	1/8"
4. GMAW	ER70S-3	ER70S-6	0.035"	0.045"
5. FCAW	E70T-1	E71T-1	0.035"	0.045"
6. FCAW	E70T-4	E71T-11	0.035"	0.045"

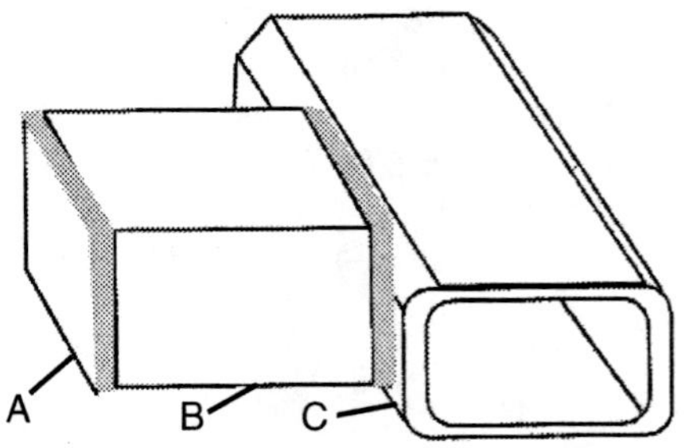

JDPX7315

Material

Part--->	A	B	C
Thickness (Min.)	1/8"	1/4"	1/4"
Width (Min.)	1-3/4"	2" x 4"	2" x 4"
Length (Min.)	3-3/4"	3"	6"

Project #9—Vertical Down "T"

--Options--

Filler Metal

Process	------Type-----		-----Diameter------	
1. SMAW	E6010	E6011	3/32"	1/8"
2. SMAW	E6013		3/32"	1/8"
3. GMAW[a]	ER70S-3	ER70S-6	0.035"	0.045"
5. FCAW	E71T-1		0.035"	0.045"
6. FCAW	E71T-11		0.035"	0.045"

Material

Part--->	A	B
Thickness (Min.)	1/8"	1/8"
Width (Min.)	3"	1-1/2"
Length (Min.)	6"	6"

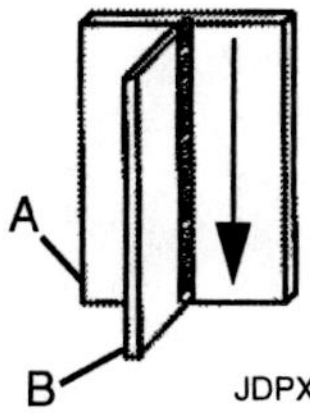

JDPX7316

a. Short Circuit

Project #10—Vertical Up "T"

--Options--

Filler Metal

Process	------Type-----		-----Diameter------	
1. SMAW	E6010	E6011	3/32"	1/8"
2. SMAW	E6013		3/32"	1/8"
3. SMAW	E7018		3/32"	1/8"
4. GMAW	ER70S-3	ER70S-6	0.035"	0.045"
5. FCAW	E70T-1	E71T-1	0.035"	0.045"
6. FCAW	E70T-4	E71T-11	0.035"	0.045"

Material

Part--->	A	B
Thickness (Min.)	1/4"	1/4"
Width (Min.)	3"	1-1/2"
Length (Min.)	6"	6"

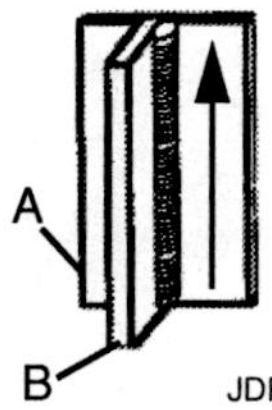

JDPX7317

Project #11—Overhead "T"

--Options--

Filler Metal

Process	------Type-----		-----Diameter------	
1. SMAW	E6010	E6011	3/32"	1/8"
2. SMAW	E6013		3/32"	1/8"
3. GMAW[a]	ER70S-3	ER70S-6	0.035"	0.045"
4. FCAW	E71T-1		0.035"	0.045"
5. FCAW	E71T-11		0.035"	0.045"

Material

Part--->	A	B
Thickness (Min.)	1/4"	1/4"
Width (Min.)	3"	1-1/2"
Length (Min.)	6"	6"

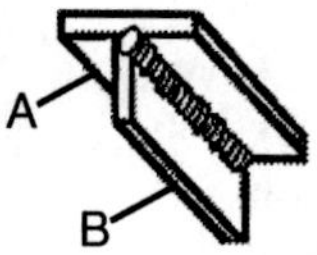

JDPX7318

a. Short Circuit

Project #12—Vertical Up Butt Joint

--Options--

Filler Metal

Process	------Type-----		-----Diameter------	
1. SMAW	E6010	E6011	3/32"	1/8"
2. SMAW	E6013		3/32"	1/8"
3. SMAW	E7018		3/32"	1/8"
4. GMAW[a]	ER70S-3	ER70S-6	0.035"	0.045"
5. FCAW	E70T-1	E71T-1	0.035"	0.045"
6. FCAW	E70T-4	E71T-11	0.035"	0.045"

Material

Part--->	A	B
Thickness (Min.)	1/4"	1/4"
Width (Min.)	2"	2"
Length (Min.)	6"	6"

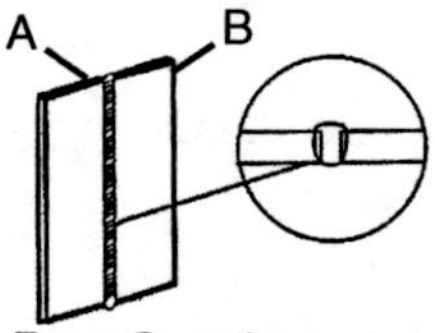

Root Opening (Space) 1/2 Metal Thickness

JDPX7319

a. Short Circuit

Project #13—Vertical "V" Groove

--Options-- **Filler Metal**

Process	------Type-----		-----Diameter------	
1. SMAW	E6010	E6011	3/32"	1/8"
2. SMAW	E7018		3/32"	1/8"
3. GMAW	ER70S-3	ER70S-6	0.035"	0.045"
4. FCAW	E71T-1		0.035"	0.045"
5. FCAW	E71T-11		0.035"	0.045"

Material

Part--->	A	B
Thickness (Min.)	3/8"	3/8"
Width (Min.)	3"	3"
Length (Min.)	6"	6"
Bevel Angle	30°—35°	30°—35°

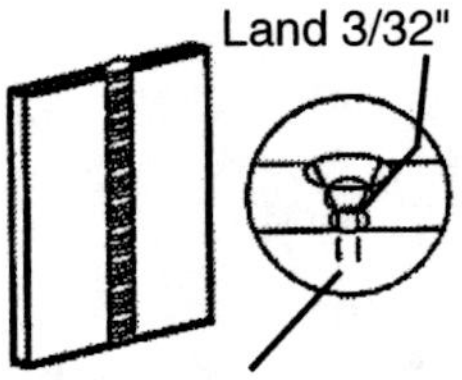

JDPX7320

Project #14—Overhead "V" Groove

--Options-- **Filler Metal**

Process	------Type-----		-----Diameter------	
1. SMAW	E6010	E6011	3/32"	1/8"
2. SMAW	E7018		3/32"	1/8"
3. GMAW	ER70S-3	ER70S-6	0.035"	0.045"
4. FCAW	E71T-1		0.035"	0.045"
5. FCAW	E71T-11		0.035"	0.045"

Material

Part--->	A	B
Thickness (Min.)	3/8"	3/8"
Width (Min.)	3"	3"
Length (Min.)	6"	6"
Bevel Angle	30°—35°	30°—35°

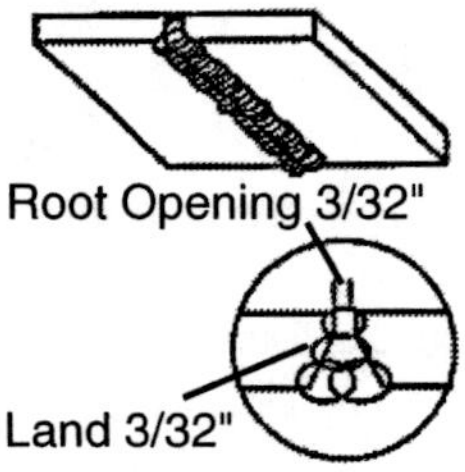

JDPX7321

Advanced GTAW Shop Project Options

Project #1—Thin to Thick

--Options--

Current
1. ACHF[a]
2. DCEN

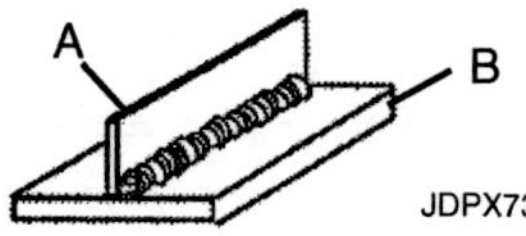

Filler Metal

Base Metal	------Type-----	-----Diameter------	
3. SS	ER309L/ER316L	1/16"	3/32"
4. CS	ER70S-2	1/16"	3/32"
5. AL[a]	ER4043/ER5356	3/32"	1/8"

Material

Part--->		A	B
Thickness	(Min.)	1/16"	1/8"
	(Max.)	1/8"	1/4"
Width	(Min.)	1-1/2"	3"
	(Max.)	2"	4"
Length	(Min.)	4"	4"
	(Max.)	6"	6"

SS = Stainless Steel
CS = Carbon Steel
AL = Aluminum

a. Proper Current/Base Metal Combinations

Project #2—Thin to Thick

--Options--

Current
1. ACHF[a]
2. DCEN

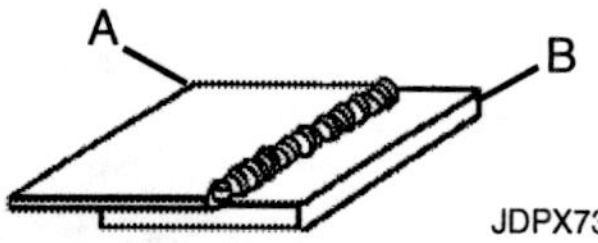

Filler Metal

Base Metal	------Type-----	-----Diameter------	
3. SS	ER309L/ER316L	1/16"	3/32"
4. CS	ER70S-2	1/16"	3/32"
5. AL[a]	ER4043/ER5356	3/32"	1/8"
6. Other			

Material

Part--->		A	B
Thickness	(Min.)	1/16"	1/8"
	(Max.)	1/8"	1/4"
Width	(Min.)	3"	3"
	(Max.)	4"	4"
Length	(Min.)	4"	4"
	(Max.)	6"	6"

SS = Stainless Steel
CS = Carbon Steel
AL = Aluminum

a. Proper Current/Base Metal Combinations

Project #3—Thick to Thin Butt Joint

--Options--

Current
1. ACHF[a]

No Root Opening (Space) Full (100%) Penetration

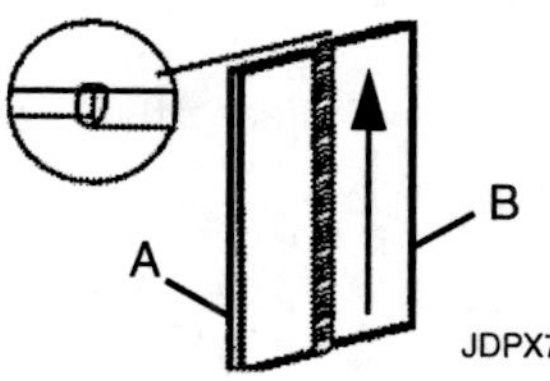

Position
2. Horizontal (2G)
3. Vertical Up (3G)

Filler Metal

Base Metal	------Type-----	-----Diameter------	
4. AL[a]	ER4043/ER5356	3/32"	1/8"

Material

Part--->		A	B
Thickness	(Min.)	1/16"	1/8"
	(Max.)	1/8"	1/4"
Width	(Min.)	1-1/2"	1-1/2"
	(Max.)	3"	3"
Length	(Min.)	4"	4"
	(Max.)	6"	6"

AL = Aluminum

a. Proper Current/Base Metal Combinations

Project #4—Weld Around

--Options--

Current
1. ACHF[a]
2. DCEN

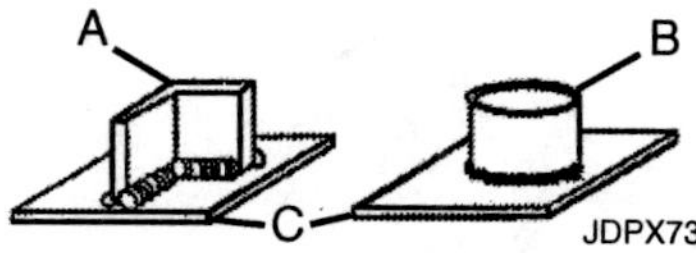

Filler Metal

Base Metal	------Type-----	-----Diameter------	
3. SS	----------None-Fusion Weld-----------		
4. SS	ER309L/ER316L	1/16"	3/32"
5. CS	ER70S-2	1/16"	3/32"
6. AL[a]	ER4043/ER5356	3/32"	1/8"

Material

Part--->		A	B	C
Thickness	(Min.)	1/16"	1/16"	1/16"
	(Max.)	3/16"	3/16"	3/16"
Size	(Min.)	1" x 1"	1-1/2"	3"
	(Max.)	2" x 2"	3"	3"
Length	(Min.)	2"	2"	4"
	(Max.)	4"	4"	6"

SS = Stainless Steel
CS = Carbon Steel
AL = Aluminum

a. Proper Current/Base Metal Combinations

Cutting and Gouging Shop Project Options

Project #1—Straight Line Cutting

JDPX7326

--Options--	--------Metals--------		
Process	Steel	Stainless Steel	Aluminum
1. Oxyfuel	A	--	--
2. Plasma	B	C	D
Operation			
3. Free Hand Cut	A, B	C	D
4. Use Straight Edge	A, B	C	D
Metal Thickness			
5. 1/16" or 1/8"	A, B	C	D
6. 1/4"	A, B	C	D
7. 3/8"	A, B	C	D
8. 1/2"	A, B	C	D
Length of Cut			
	Min. 5"		
	Max. 10"		

Project #2—Bevel Cutting

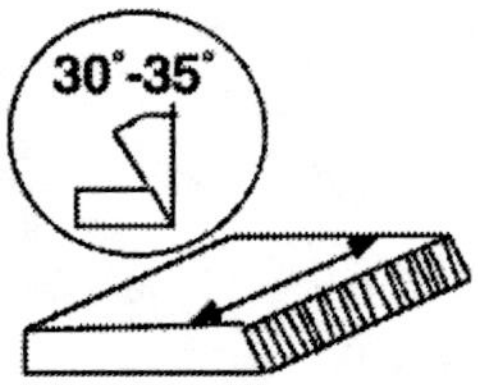

JDPX7327

--Options--	--------Metals--------		
Process	Steel	Stainless Steel	Aluminum
1. Oxyfuel	A	--	--
2. Plasma	B	C	D
Operation			
3. Free Hand Cut	A, B	C	D
4. Use Straight Edge	A, B	C	D
Metal Thickness			
5. 3/8"	A, B	C	D
6. 1/2"	A, B	C	D
Length of Cut	**Bevel**		
Min. 6"	30°		
Max. 10"	35°		

Project #3—Piercing and Cutting Holes

JDPX7328

--Options--	--------Metals--------		
Process	Steel	Stainless Steel	Aluminum
1. Oxyfuel	A	--	--
2. Plasma	B	C	D
Operation			
3. Free Hand Cut	A, B	C	D
4. Use Circle Cutter	A, B	C	D
5. Follow Template	A, B	C	D
Metal Thickness			
6. 1/16" or 1/8"	A, B	C	D
7. 1/4" or 1/8"	A, B	C	D
8. 1/2"	A, B	C	D
Hole Size			
9. 1" or 2"			
10. 3" or 4"			

Project #4—Cutting Slots

JDPX7329

--Options--	--------Metals--------		
Process	Steel	Stainless Steel	Aluminum
1. Oxyfuel	A	--	--
2. Plasma	B	C	D
Operation			
3. Free Hand Cut	A, B	C	D
4. Use Straight Edge	A, B	C	D
Metal Thickness			
5. 1/4"	A, B	C	D
6. 3/8"	A, B	C	D
7. 1/2"	A, B	C	D
Slot Size (Drill Holes)			
	Width	Length	Spacing
	3/8"	2" to 6"	3"
	1/2"	2" to 6"	3"

Project #5—Cut a Flange with Holes

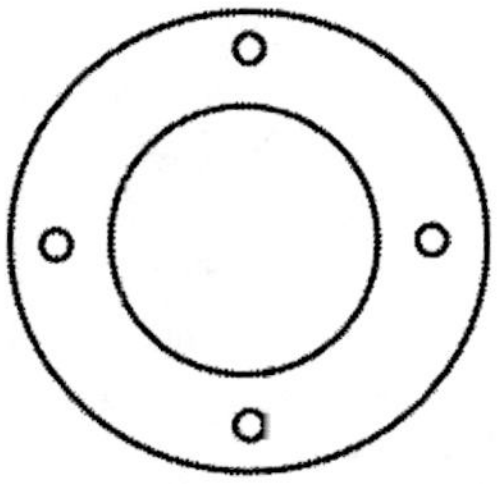

JDPX7330

--Options--	----------Metals----------		
Process	Steel	Stainless Steel	Aluminum
1. Oxyfuel	A	--	--
2. Plasma	B	C	D
Metal Thickness			
3. 3/16"	A, B	C	D
4. 3/8"	A, B	C	D
Diameter			
5. Inside	(Min.)	2"	
	(Max.)	3"	
6. Outside	(Min.)	5"	
	(Max.)	6"	
Hole Size		3/4"	

Project #6—Air Carbon Arc (Arc Air)

--Options--

Process Air Carbon Arc (Arc Air)

--Options--	----------Metals----------		
Operation	Steel	Stainless Steel	Aluminum
1. Cut Straight Line	B	C	D
2. Pierce Hole	B	C	D
3. Make "U" Groove	B	C	D
4. Remove Weld Defect	A, B	C	D
5. Remove Fillet Weld	A, B	C	D
6. Backgouge Butt Joint	A, B	C	D
Metal Thickness		C	D
7. 1/16" to 1/8"	A, B	C	D
8. 1/4" to 3/8"	A, B	C	D
Length (Min.)	4"	4"	4"

Shop Project Progress Record

Inspection	Acceptable =1	Meets Standards
Guidelines	Good = 2	Exceeds Standards
	Unacceptable = 0	Below Standards

Process	Process	Process
Completion Date	Completion Date	Completion Date
Project No.	Project No.	Project No.
Option No.	Option No.	Option No.
----Joint/Weld Number----	----Joint/Weld Number----	----Joint/Weld Number----

Appearance	1	2	3	4	1	2	3	4	1	2	3	4
Completed												
Uniform												
Crater Filled												
Size-Fillet Welds												
Leg Size (in.)												
Throat (in.)												
Size-Groove Welds												
Convexity												
Penetration												
Face Width												
Weld Test												
Root Bend												
Face Bend												
Side Bend												
Nick Bend												
Macro Etch												
Penetrant Die												
Other												
Leak Test												
A-Total Points---->												
Inspection	Acceptable = 0											
Guidelines	Unacceptable = 1											
Discontinuities												
Concave												
Undercut												
Overlap												
Porosity												
Arc Strikes												
Off Joint												
Oversize												
Spatter												
B-Total Points --->												
Total Score (A-B)												

AWS Acceptable and Unacceptable Weld Profiles

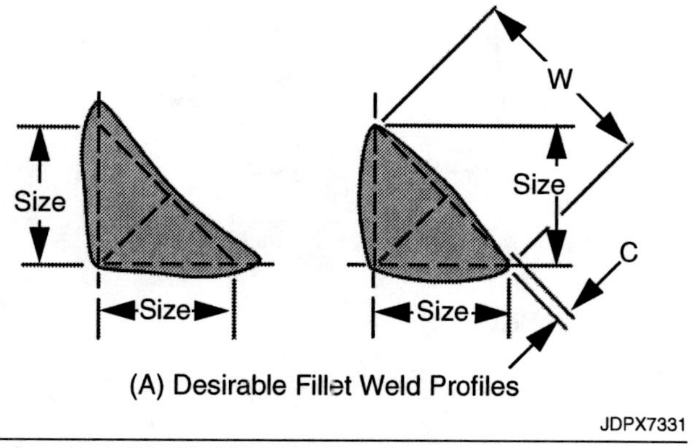

Fig. 2 — Desirable Fillet Weld Profiles (A)

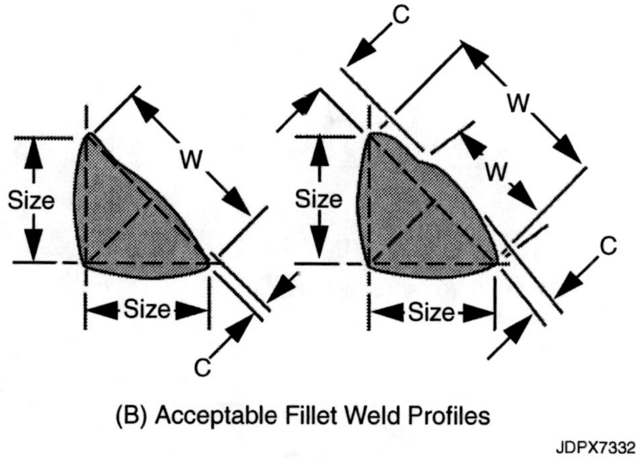

Fig. 3 — Acceptable Fillet Weld Profiles (B)

NOTE: Convexity (C), of a weld or individual surface bead with dimension (W) shall not exceed the value of the following table.

Width of Weld Face or Individual Surface Bead (W)	Maximum Convexity (C)
W < 5/16 in. (8 mm)	1/16 in. (1.6 mm)
W > 5/16 in. to W < 1 in. (25 mm)	1/8 in. (3 mm)
W > 1 in.	3/16 in. (5 mm)

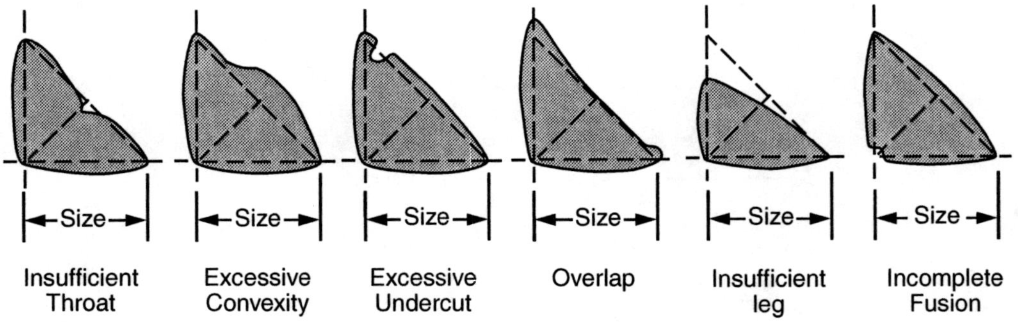

(C) Unacceptable Fillet Weld Profiles

JDPX7333

Fig. 4 — Unacceptable Fillet Weld Profiles (C)

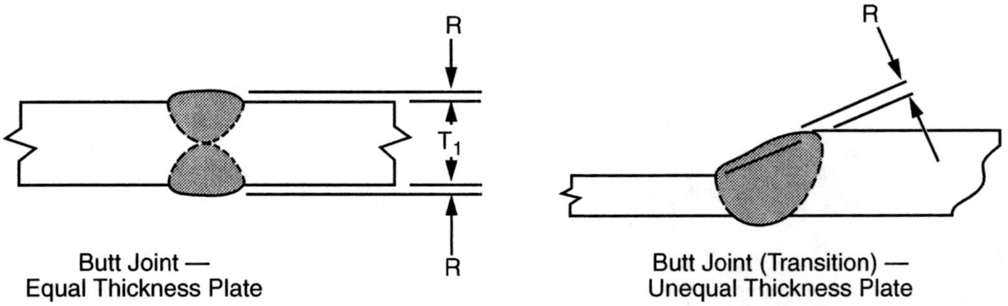

NOTE: Note: Reinforcement R Shall Not Exceed 1/8 in. (3 mm). See 5.24.4.

(D) Acceptable Groove Weld Profile in Butt Joint

JDPX7334

Fig. 5 — Acceptable Groove Weld Profile in Butt Joint (D)

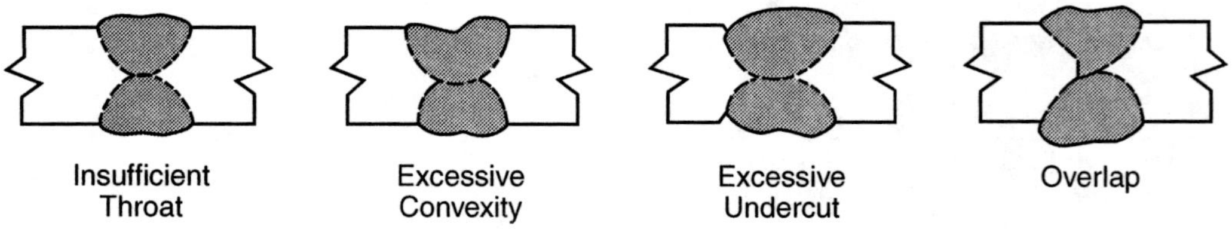

(E) Unacceptable Groove Weld Profiles in Butt Joint

JDPX7335

Fig. 6 — Unacceptable Groove Weld Profiles in Butt Joint (E)

Welder Qualification — Production Welding Positions Qualified by Plate, Pipe and Box Tube Tests[a]															
Qualification Test			Production Plate Welding Qualified			Production Plate Welding Qualified					Production Box Tube Welding Qualified				
	Weld Type	Positions 2	Groove CJP	Groove FJP	Fillet	Butt-Groove		T-, Y-, K-Groove		Fillet	Butt-Groove		T-, Y-, K-Groove		Fillet
						CJP	PJP	CJP	PJP		CJP	PJP	CJP	PJP	
PLATE	Groove[3]	1G 2G 3G 4G 3G + 4G	F F, H F, H, V F, OH All	F F, H F, H, V F. OH All	F, H F, H F, H, V F, H, OH All Note 9	F F, H F, H, V F, OH All Note 4	F F, H F, H, V F, OH All Note 4		F F, H F, H, V F, OH All Note 4, 6	F, H F, H F, H, V F, V, OH All Note 5	F F, H F, H, V F, OH All Note 5	F F, H F, H, V F, OH All		F F, H F, H, V F, OH All Note 6	F, H F, H F, H, V F, H, OH All Note 9
	Fillet	1F 2F 3F 4F 3F + 4F			F F, H F, H, V F, H, OH All Note 9					F F, H F, H, V F, H, OH All Note 9					F F, H F, H, V F, H, OH All Note 9
	Plug	Qualifies Plug and Slot Welding For Only The Positions Tested													
TUBULAR	Groove[3] (Pipe or Box)	1G Rotated 2G 5G 6G 2G + 5G Note 10	F F, H F, V, OH All All	F F, H F, V, OH All All	F, H F, H F, V, OH All All Note 9	F F, H F, V, OH All All Note 7	F F, H F, V, OH All All Note 7		F F, H F, V, OH All All Notes 6, 7	F, H F, H F, V, OH All All Note 9	F F, H F, V, OH All All	F F, H F, V, OH All All		F F, H F, V, OH All All Note 6	F, H F, H F, V, OH All All Note 9
		6GR (Fig, 4.27)	All	All	All Note 9	All Notes 5, 7	All Note 7	All Notes 6, 7	All Notes 6, 7	All Note 9	All Note 5	All		All Note 6	All Note 9
		6GR (Fig, 4.27 & 4.28)	All	All	All Note 9	All Notes 5, 7	All Note 7	All Notes 6, 7	All Notes 6, 7	All Note 9	All Note 5	All	All Notes 6, 8	All Note 6	All Note 9
	Pipe Fillet	1F Rotated 2F 2F Rotated 4F 5F			F F, H F, H F, H, OH All Note 9					F F, H F, H F, H, OH All Note 9					F F, H F, H F, H, OH All Note 9

CLP = Complete Joint Penetration
PJP = Partial Joint Penetration
R = Restriction

a. *All figure and table references in this table and the following forms refer to American Welding Society documents. For more information, see the American Welding Society, Inc., www.aws.org.*

Table 1 — Welder Qualification — Production Welding Positions Qualified by Plate, Pipe, and Box Tube Tests

NOTE: Notes shown at the bottom of a column box apply to all entries.

1. Not applicable for welding operator qualification (See Table 4.10)
2. See Figure 4.3 through 4.6.
3. Groove weld qualification also qualifies plug and slot welds for the test positions indicated.
4. Only qualified for pipe over 24 in. (610 mm) in diameter with backing, backgouging, or both.
5. Not qualified for joints welded from one side without backing, or welded from two sides without backgouging.
6. Not qualified for welds having groove angles less than 30° (see 4.12.4.2).
7. Qualification using box tubing (Figure 4.27) also qualifies welding pipe over 24 in. (610 mm) in diameter.
8. Pipe or box tubing is required for the 6GR qualification (Figure 4.27). If box tubing is used per Figure 4.27, the macroetch test may be performed on corners of the test specimen (similar to Figure 4.28).
9. See Figures 4.25 and 4.28 for dihedral angle restrictions for plate joints and tubular T-, Y-, and K-connections.
10. Qualification for welding production joints without backing or backgouging requires using the Figure 4.24 joint detail. For welding production joints with backing or backgouging, either the Figure 4.24 or Figure 4.25 joint detail can be used for qualification.

AWS Guided-Bend Test Jig

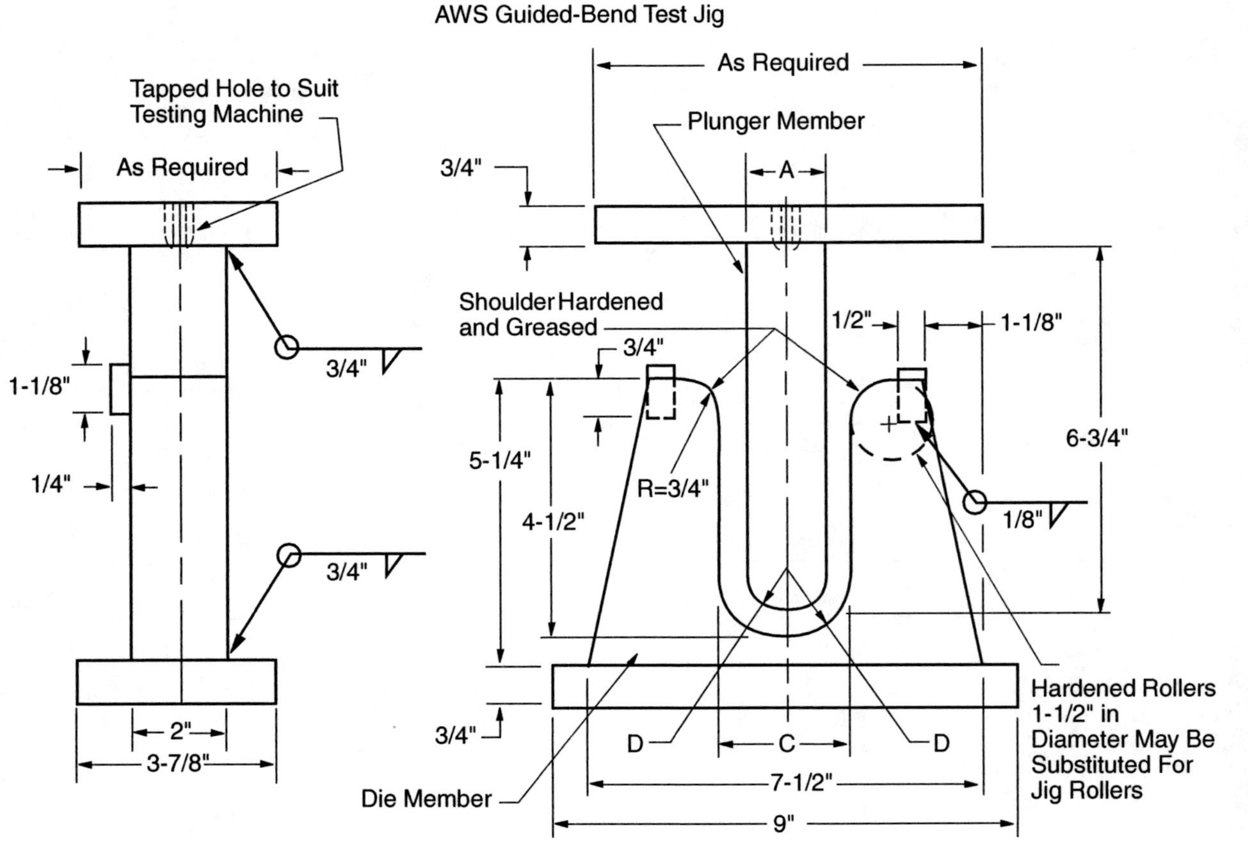

Fig. 7 — AWS Guided-Bend Test Jig

Specified or Actual Base Yield Strength (psi)	A	B	C	D
50,000 and Under	1-1/2 in.	3/4 in.	2-3/8 in.	1-3/16 in.
Over 50,000 to 90,000	2 in.	1 in.	2-7/8 in.	1-7/16 in.
90,000 and Over	2-1/2 in.	1-1/4 in.	3-3/8 in.	1-11/16 in.

NOTE: Plunger and interior die surfaces shall be machine-finished.

AWS Test Plates For Groove Welds

Test Plate of Limited Thickness — All positions — Welder Qualifications

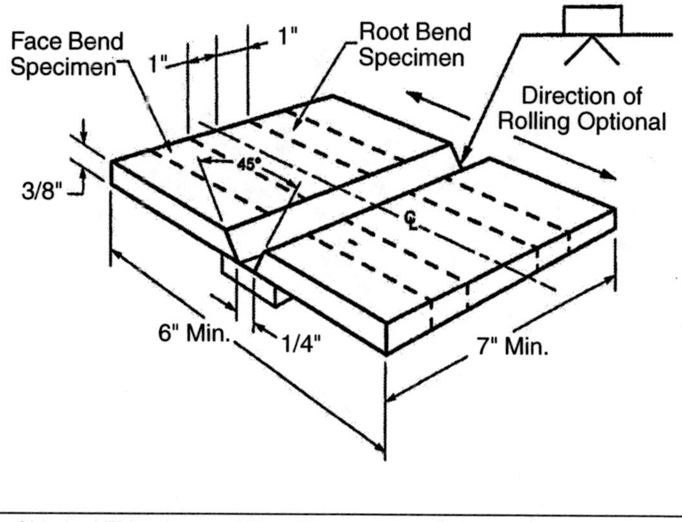

in.	mm
1/4	6
3/8	10
1	25
3	75
6	150
7	180

Fig. 8 — Test Plate of Limited Thickness — All positions — Welder Qualifications

1. When radiography is used for testing, no tack welds shall be in the test area.
2. The backing thickness shall be a minimum of 1/4 in. to a maximum of 3/8 in. Backing width shall be 3 in. minimum when not removed for radiography, otherwise 1 in. minimum.

Test Plate for Unlimited Thickness — Welder Qualifications

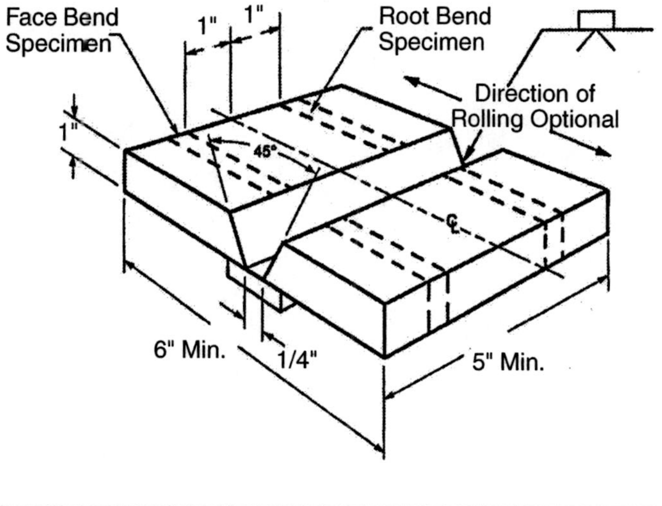

in.	mm
1/4	6
3/8	10
1	25
3	75
5	125
6	150

Fig. 9 — Test Plate for Unlimited Thickness — Welder Qualifications

1. When radiography is used for testing, no tack welds shall be in the test area.
2. The backing thickness shall be a minimum of 1/4 in. to a maximum of 3/8 in. Backing width shall be 3 in. minimum when not removed for radiography, otherwise 1 in. minimum.

AWS Test Plate For Fillet Welds

Fillet Weld Break and Macroetch Test Plate, Welder or Welding Operator Qualification — Option 1

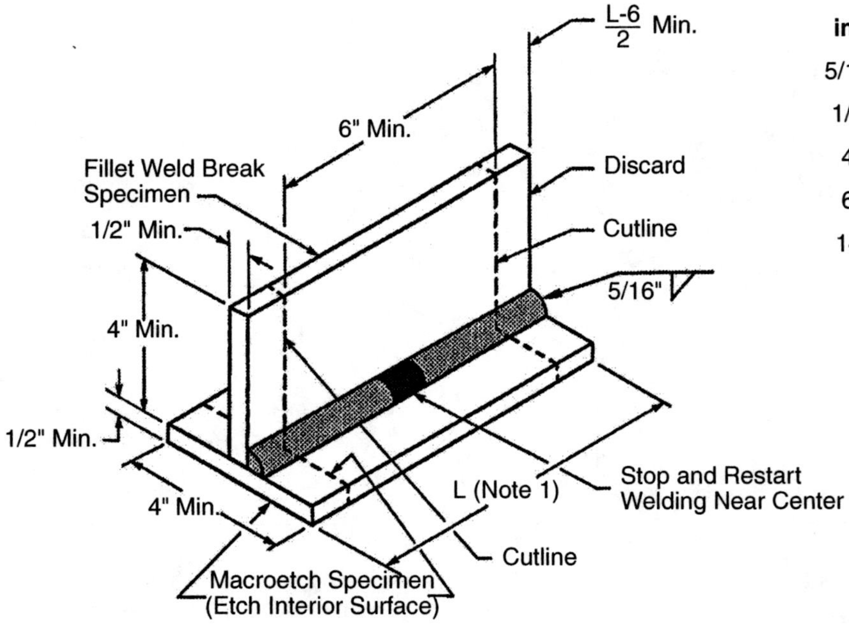

in.	mm
5/16	8
1/2	13
4	100
6	150
14	380

JDPX7351

Fig. 10 — Fillet Weld Break and Macroetch Test Plate, Welder or Welding Operator Qualification — Option 1

1. L = 8 inches minimum (welder), 15 inches minimum (welding operator).
2. Either end may be used for the required macroetch specimen. The other end may be discarded.

WELDING PROCEDURE SPECIFICATION (WPS) Yes ☐
PREQUALIFIED ________ QUALIFIED BY TESTING ________
or PROCEDURE QUALIFICATION RECORDS (PQR) Yes ☐

Identification # ________
Revison ______ Date ______ By ______
Authorized by ________ Date ______
Type—Manual ☐ Semi-Automatic ☐
Machine ☐ Automatic ☐

Company Name ________
Welding Process(es) ________
Supporting PQR No. (s) ________

JOINT DESIGN USED
Type: ________
Single ☐ Double Weld ☐
Backing: Yes ☐ No ☐
Backing Material: ________
Root Opening ______ Root Face Dimension ______
Groove Angle: ______ Radius (J-U) ______
Back Gouging: Yes ☐ No ☐ Method ______

POSITION
Position of Groove: ________ Fillet: ______
Vertical Progression: Up ☐ Down ☐

ELECTRICAL CHARACTERISTICS
____
Transfer Mode (GMAW) Short-Circuiting ☐
Globular ☐ Spray ☐
Current: AC ☐ DCEP ☐ DCEN ☐ Pulsed ☐
Other ________
Tungsten Electrode (GTAW)
Size: ________
Type: ________

BASE METALS
Material Spec. ________
Type or Grade ________
Thickness: Groove ______ Fillet ______
Diameter (Pipe) ________

FILLER METALS
AWS Specification ________
AWS Classification ________

TECHNIQUE
Stringer or Weave Bead: ________
Multi-pass or Single Pass (per side) ________
Number of Electrodes ________
Electrode Spacing Longitudinal ______
Lateral ______
Angle ______

SHIELDING
Flux ________ Gas ______
Compositon ______
Electrode-Flux (Class) ______ Flow Rate ______
________ Gas Cup Size ______

Contact Tube to Work Distance ________
Peening ________
Interpass Cleaning: ________

PREHEAT
Preheat Temp., Min ________
Interpass Temp., Min ________ Max ______

POSTWELD HEAT TREATMENT
Temp. ________
Time ________

WELDING PROCEDURE

Pass or Weld Layer(s)	Process	Filler Metals		Current		Volts	Travel Speed	Joint Details
		Class	Diam.	Type & Polarity	Amps or Wire Feed Speed			

Form E-1 (Front)

JDPX7338

WELDING PROCEDURE SPECIFICATION (WPS) Yes [X]
PREQUALIFIED ________ QUALIFIED BY TESTING X
or PROCEDURE QUALIFICATION RECORDS (PQR) Yes [ ]

Identification # PQR 231
Revison 1 Date 12-1-87 By W.Lye
Authorized by J. Jones Date 1-18-88
Type—Manual [ ] Semi-Automatic [X]
Machine [ ] Automatic [ ]

Company Name RED INC.
Welding Process(es) FCAW
Supporting PQR No. (s) PQR 231

JOINT DESIGN USED
Type: Butt
Single [X] Double Weld [ ]
Backing: Yes [X] No [ ]
Backing Material: ASTM A131A
Root Opening 1/4" Root Face Dimension –
Groove Angle: 52-1/2° Radius (J-U) –
Back Gouging: Yes [ ] No [X] Method –

BASE METALS
Material Spec. ASTM A131
Type or (Grade) A
Thickness: Groove 3/4–1 1/2" Fillet –
Diameter (Pipe) –

FILLER METALS
AWS Specification A5.20
AWS Classification E71T-1

SHIELDING
Flux –
Gas CO_2
Compositon
Electrode-Flux (Class)
Flow Rate 45 5C
Gas Cup Size #

PREHEAT
Preheat Temp., Min 60°
Interpass Temp., Min 60 Max 350°F

POSITION
Position of Groove: O.H. Fillet: –
Vertical Progression: Up [ ] Down [ ]

ELECTRICAL CHARACTERISTICS
Transfer Mode (GMAW) Short-Circuiting [ ] Globular [X] Spray [ ]
Current: AC [ ] DCEP [X] DCEN [ ] Pulsed [ ]
Other
Tungsten Electrode (GTAW)
Size:
Type:

TECHNIQUE
Stringer or Weave Bead: Stringer
Multi-pass or Single Pass (per side) Multipass
Number of Electrodes 1
Electrode Spacing Longitudinal –
Lateral –
Angle –

Contact Tube to Work Distance 1/2-1"
Peening None
Interpass Cleaning: Wire Brush

POSTWELD HEAT TREATMENT
Temp. N.A.
Time N.A.

WELDING PROCEDURE

Pass or Weld Layer(s)	Process	Filler Metals: Class	Filler Metals: Diam.	Current: Type & Polarity	Current: (Amps) or Wire Feed Speed	Volts	Travel Speed	Joint Details
All	FCAW	E71T-1	.045"	DC+	180-220A	25-26V	8-12 ipm	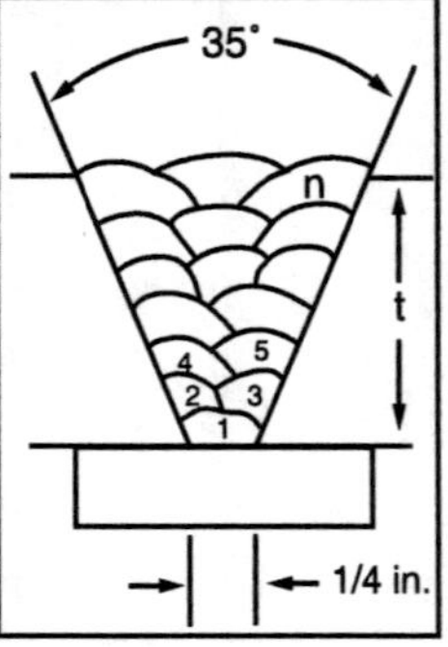

Form E-1 (Front)

JDPX7339

Procedure Qualification Record (PQR) # ________
Test Results

TENSILE TEST

Specimen No.	Width	Thickness	Area	Ultimate tensile load, lb	Ultimate unit stress, psi	Character of failure and location
231-1	.75"	1.00"	.75"	52 500	70 000	Ductile
231-3	.75"	1.00"	.75"	52 275	69 700	Ductile

GUIDED BEND TEST

Specimen No.	Type of bend	Result	Remarks
231-2	Side	Pass	
231-4	Side	Pass	Small (< 1/1[illegible] acceptable
231-6	Side	Pass	
231-5	Side	Pass	

VISUAL INSPECTION
Appearance ________ acceptable
Undercut ________ acceptable
Piping porosity ________ none
Convexity ________ none
Test date ________ 12-3-87
Witnessed by ________ D. Davis

Radiographic-ultrasonic examination
RT report no.: D231 Result passed
UT report no.: ________ Result ________

FILLET WELD TEST RESULTS

Minimum size multiple pass Macroetch		Maximum size single pass Macroetch	
1. ______	3. ______	1. ______	3. ______
2. ______		2. ______	

Other Tests

All-weld-metal tension test
Tensile strength, psi ________ 83,100
Yield point/strength, psi ________ 72,600
Elongation in 2 in., % ________ 28
Laboratory test no. ________ PW 231

Welder's name ________ W. Williams Clock no. 261 Stamp no. ________

Tests conducted by ________ RED Inc. & ABC Testing Laboratory

Test number PQR 231

Per D. Miller

We, the undersigned, certify that the statements in this record are correct and that the test welds were prepared, welded, and tested in accordance with the requirements of section 4 or ANSI/AWS D1.1, (________) Structural Welding Code—Steel.
(year)

Signed ________________________
Manufacturer or Contractor
By R. M. Boncrack
Title ________________________
Date ________________________

Form E-1 (Back)

JDPX7341

WELDER, WELDING OPERATOR, OR TACK WELDER QUALIFICATION TEST RECORD

Type of Welder ____________________

Name ____________________ Identification No. ____________________

Welding Procedure Specification No. __________ Rev __________ Date __________

Variables	Record Actual Values Used in Qualification	Qualification Range
Process/Type [Table 4.10, Item (2)]		
Electrode (single or multiple) [Table 4.9, Item (9)]		
Current/Polarity		
Position [Table 4.10, Item (5)]		
Weld Progression [Table 4.10, Item (7)]		
Backing (YES or NO) [Table 4.10, Item (8)]		
Material/Spec. [Table 4.10, Item (1)]	to	
Base Metal		
Thickness: (Plate)		
Groove		
Fillet		
Thickness: (Pipe/tube)		
Groove		
Fillet		
Diameter:(Pipe)		
Groove		
Fillet		
Filler Metal [Table 4.10, Item (3)]		
Spec. No.		
Class		
F-No.		
Gas/Flux Type [Table 4.10, Item (4)]		
Other		

VISUAL INSPECTION (4.8.1)

Acceptable YES or NO _____

Guided Bend Test Results (4.30.5)

Type	Result	Type	Result

Fillet Test Results (4.30.2.3 and 4.30.4.1)

Appearance ____________________ Fillet Size ____________________

Fracture Test Root Penetration ____________________ Macroetch ____________________

(Describe the location, nature, and size of any crack or tearing of the specimen.)

Inspected by ____________________ Test Number ____________________

Organization ____________________ Date ____________________

RADIOGRAPHIC TEST RESULTS (4.30.3.1)

Film Identification Number	Results	Remarks	Film Identification Number	Results	Remarks

Interpreted by ____________________ Test Number ____________________

Organization ____________________ Date ____________________

We, the undersigned, certify that the statements in this record are correct and that the test welds were prepared, welded, and tested in accordance with the requirements of section 4 or ANSI/AWS D1.1, (________ (year)) Structural Welding Code—Steel.

Manufacturer or Contractor ____________________ Authorized By ____________________

Form E-4 Date ____________________

JDPX7342

SAMPLE FORM FOR WELDING PROCEDURE SPECIFICATION (WPS)

Company Name ______________________ By ______________________
Welding Procedure Specification No. ______________________ Rev __________ Date __________
Supporting Procedure Qualification Test Record(s) Nol. ______________________
Welding Process(es) ______________________ Type ______________________
(Automatic, manual, etc.)
Mode of Transfer for GMAW ______________________
(short circuiting, spray, etc.)

JOINTS (Table 6.1)
Type of Welded Joint(s) ______________________
Backing (Yes) __________ (No) __________
Backing Material Type ______________________
Groove Welded From:
one side __________ both sides __________

COATING(S):
Type ______________________
Thickness ______________________

Sketch of Joint Details

BASE METAL (1.2)
Material specification type and grade:
Sheet steel __________ to __________
Support steel ______________________
Thickness Range:
Sheet Steel ______________________
Support Steel ______________________
Thickness ______________________
Base Metal Preparation ______________________

FILLER METAL (Table 5.1):
Specification ______________________
Classification ______________________

POSITIONS (Table 1.1)
Position of Groove ______________________
Position of Fillet ______________________
Progression ______________________

PREHEAT (1.1.1 and 4.1.3):
Preheat Temperature Min ______________________
Preheat Temperature Max ______________________

GAS (6.6)
Shielding Gas ______________________ Flow Rate ______________________
Percent Mixture ______________________

FLUX (6.6) ______________________

TECHNIQUE

Pass No.	Electrode Size	Welding Current		Travel Speed	Melting Rate	Wire Feed Speed
		Amperes	Volts			

This procedure may vary due to fabrication sequence, fit-up, pass size, etc, within the limitation of variables given in ANSI/AWS D1.3 (——), Structural Welding Code — Sheet Steel.
(Year)

Authorized by ______________________ Date __________

Form A-2

JDPX7344

SAMPLE FORM FOR WELDING PROCEDURE QUALIFICATION TEST RECORD (PQR)

Company Name ____________________
Procedure Qualification Test Record No. ____________ Date ________
Welding Procedure Specification No. ____________ Rev ________ Date ________
Welding Process(es) ____________ Type ____________
(Automatic, manual, etc.)
Mode of Transfer for GMAW ____________________
(short circuiting, spray, etc.)

JOINTS (Table 6.1)
Type of Welded Joint(s) ____________

Backing (YES) ________ (NO) ________
Backing Material Type ____________
Groove Welded From:
one side ________ both sides ________

BASE METAL (1.2)
Material specification type and grade:
Sheet steel ________ to ________
Thickness ____________
Support Steel
Thickness ____________
Base Metal Preparation ____________

POSITIONS (Table 1.1)
Position of Groove ____________
Position of Fillet ____________
Progression ____________

GAS (6.6)
Shielding Gas ________ Flow Rate ________
Percent Mixture ____________

FLUX (6.6)
Filler Metal (Table 5.1):
Specification ____________
Classification ____________

COATINGS(S)
Type ____________
Thickness ____________

VISUAL EXAMINATION RESULTS (6.7):
Specimen 1 ____________
Specimen 2 ____________
Diam of Arc Spot Nugget ____________
Test Conducted by ____________
per ____________
Lab Test No. ____________
Date of Test ____________

Sketch of Joint Details

TECHNIQUE

Pass No.	Electrode Size	Welding Current		Travel Speed	Melting Rate	Wire Feed Speed
		Amperes	Volts			

Welder or Welding Operator Name ____________________
Identification No. ____________ Date of Qualification ____________
Welder's Social Security No. ____________________

The undersigned certifies that the statements in this record are correct and that the test welds were prepared, welded, and tested in accordance with the requirements of 6.7 of ANSI/AWS D1.3 (____ (Year)), Structural Welding Code—Sheet Steel.

Authorized by ____________________ Date ____________

Form A-1

JDPX7343

SAMPLE FORM FOR WELDER AND WELDING OPERATOR QUALIFICATION TEST RECORD

Welder or Welding Operator's Name ______________________
Identification No. ______________________ Qualification Date ______________________
Welder's Social Security No. ______________________

In Accordance With WPS No. ______________________ Revision ______________________
Welding Process(es) ______________________ Type ______________________
(Automatic, Manual, etc)

Mode of Transfer for GMAW ______________________
(Short Circuiting, Spray, Globular)

VARIABLE	ACTUAL VARIABLE USED IN QUAL.	QUALIFICATION RANGE
JOINT:		
Joint Type	______	______
Backing Material Type	______	______
Groove Welded From: one side or both sides	______	______
BASE METAL:		
Material Specification		
Sheet Steel	______ to ______	______ to ______
Supporting Steel	______	______
Sheet Thickness:		
Groove	______	______
Fillet	______	______
Arc Spot	______	______
Arc Seam	______	______
COATING(S):		
Type	______	______
Thickness	______	______
POSITION:		
Groove	______	______
Fillet	______	______
Arc Spot	______	______
Arc Seam	______	______
Progression	______	______
GAS	______	______
ELECTRODE:		
Size	______	______
Group Designation	______	______

VISUAL EXAMINATION RESULTS (6.7)

Specimen 1 ______________________ Specimen 2 ______________________
Appearance ______________ Cracks ______________ Undercut ______________
Reinforcement ______________________ Diam of Arc Spot Nugget ______________________

Test Conducted By ______________________ Per ______________________
Laboratory Test No. ______________________ Date of Test ______________________

The undersigned certify that the statements in this record are correct and that the welds were prepared and tested in accordance with the requirements of 6.7 of ANSI/AWS D1.3 (_____), Structural Welding Code — Sheet Steel.
(Year)

Company ______________________ Authorized By ______________________

Form A-3

JDPX7345

INDEX

A

B

C

J

L

M

N

O

P

R

S

T